Probiotic Bacteria and Enteric Infections

Joshua J. Malago • Jos F. J. G. Koninkx
R. Marinsek-Logar
Editors

Probiotic Bacteria and Enteric Infections

Cytoprotection by Probiotic Bacteria

Springer

Editors:
Dr. Joshua J. Malago
Department of Veterinary Pathology
Faculty of Veterinary Medicine
Sokoine University of Agriculture
P.O. Box 3203, Chuo Kikuu, Morogoro
Tanzania
malagojj@yahoo.com

Dr. Jos F. J. G. Koninkx
Division Pathology
Department of Pathobiology
Faculty of Veterinary Medicine
Utrecht University
Yalelaan 1, 3508 TD, Utrecht
Netherlands
j.f.j.g koninkx@uu nl

Dr. R. Marinsek-Logar
Zootechnical Department
Biotechnical Faculty
University of Ljubljana
Domžale
Slovenia
romana marinsek@bf.uni-lj.si

ISBN 978-94-007-0385-8 e-ISBN 978-94-007-0386-5
DOI 10.1007/978-94-007-0386-5
Springer Dordrecht Heidelberg London New York

Cover design deblik, Berlin

Printed on acid-free paper

Springer is part of Springer Science+Business Media (www.springer.com)

Contents

Contributors

Hans Christian Beck The Proteomics Group, Danish Technological Institute, Holbergsvej 10, 6000 Kolding, Denmark
e-mail: hcb@teknologisk.dk

Giuseppe Caramia Hemeritus Head Physician of Paediatrics and Neonatology, Specialized Maternal-Infantil Hospital "G.Salesi", Via Toti 1, 60123 Ancona, Italy
e-mail: caramiagm@libero.it

Christophe Chassard Laboratory of Food Biotechnology, Institute of Food Science and Nutrition, ETH Zurich, Schmelzbergstrasse 7, LFV C20 CH-8092 Zurich, Switzerland
e-mail: christoph.chassard@ilw.agrl.ethz

J. Kevin Collins Department of Microbiology, University College Cork, Cork, Ireland
Cork Cancer Research Centre, University College Cork, Cork, Ireland
e-mail: microbiology@ucc.ie

Sebastiano Collino BioAnalytical Science, Metabolomics & Biomarkers, Nestlé Research Center, P.O. Box 44, 1000 Lausanne 26, Switzerland
e-mail: Sebastiano.Collino@rdls nestle.com

E. P. Culligan Alimentary Pharmabiotic Centre and Department of Microbiology, University College Cork, Cork, Ireland
e-mail: eamonnculligan@gmail.com

Rossana D'Arienzo Institute of Food Sciences, CNR, via Roma 64, 83100 Avellino, Italy
e-mail: rdarienzo@isa.cnr.it

Véronique Delcenserie Canadian Research Institute for Food Safety, University of Guelph, 43, McGilvray Street, Guelph, Ontario, Canada N1G 2W1
e-mail: vdelcens@uoguelph.ca

Anne Donnet-Hughes Nestlé Research Centre, P.O. Box 44, Vers-chez-les-Blanc, 1000 Lausanne, Switzerland
e-mail: anne.donnet@rdls.nestle.com

Joo-Ann Ewe School of Industrial Technology, Universiti Sains Malaysia, 11800 Penang, Malaysia
e-mail: jooann_ewe@hotmail.com

Søren Feddersen Odense University Hospital, Department of Clinical Chemistry and Pharmacology, Sdr. Boulevard 29, DK-5000 Odense C, Denmark
e-mail: soeren feddersen@ouh.regionsyddanmark.dk

Sofia D. Forssten Danisco Health & Nutrition, Sokeritehtaantie 20, 02460 Kantvik, Finland
e-mail: sofia.forssten@danisco.com

Melanie Gagnon Laboratory of Food Biotechnology, Institute of Food Science and Nutrition, ETH Zurich, Schmelzbergstrasse 7, 8092 Zurich, Switzerland
e-mail: melanie.gagnon@ilw.agrl.ethz.ch

Cormac G. M. Gahan Department of Microbiology, University College Cork, Cork, Ireland
e-mail: c.gahan@ucc.ie

C. Hill Alimentary Pharmabiotic Centre and Department of Microbiology, University College Cork, Cork, Ireland
e-mail: c.hill@ucc.ie

Lene Jespersen Department of Food Science, Food Microbiology, Faculty of Life Sciences, University of Copenhagen, Rolighedsvej 30, 1958 Frederiksberg, Denmark
e-mail: lj@life ku.dk

Ouafae Karimi Department of Pathology, VU University Medical Center, Laboratory of Immunogenetics, P.O. Box 7057, 1007 MB Amsterdam, The Netherlands
e-mail: ouafaekarimi@gmail.com

Sunil Kochhar BioAnalytical Science, Metabolomics & Biomarkers, Nestlé Research Center, P.O. Box 44, 1000 Lausanne 26, Switzerland
e-mail: sunil.kochhar@rdls.nestle.com

Jos F. J. G. Koninkx Division Pathology, Department of Pathobiology, Faculty of Veterinary Medicine, Utrecht University, Yalelaan 1, 3508 TD Utrecht, The Netherlands
e-mail: j.f.j.g.koninkx@uu nl

Christophe Lacroix Laboratory of Food Biotechnology, Institute of Food Science and Nutrition, ETH Zurich, Schmelzbergstrasse 7, 8092 Zurich, Switzerland
e-mail: christophe.lacroix@ilw.agrl.ethz.ch

Sampo J. Lahtinen Danisco Health & Nutrition, Sokeritehtaantie 20, 02460 Kantvik, Finland
e-mail: sampo.lahtinen@danisco.com

Vanessa Liévin-Le Moal Faculté de Pharmacie, Inserm Unité 756, 92296 Châtenay-Malabry, France
e-mail: vanessa.lievin-le-moal@u-psud.fr

Ting-Jin Lim School of Industrial Technology, Universiti Sains Malaysia, 11800 Penang, Malaysia
e-mail: ericlimtingjin@hotmail.com

Min-Tze Liong School of Industrial Technology, Universiti Sains Malaysia, 11800 Penang, Malaysia
e-mail: mintze.liong@usm my

Joshua J. Malago Department of Veterinary Pathology, Faculty of Veterinary Medicine, Sokoine University of Agriculture, 3203, Chuo Kikuu, Morogoro, Tanzania
e-mail: malagojj@yahoo.com

Huey-Shi Lye School of Industrial Technology, Universiti Sains Malaysia, 11800 Penang, Malaysia
e-mail: hueyshi_lye@yahoo.com

François-Pierre J. Martin BioAnalytical Science, Metabolomics & Biomarkers, Nestlé Research Center, P.O. Box 44, 1000 Lausanne 26, Switzerland
e-mail: Francois-Pierre.Martin@rdls.nestle.com

Bojana Bogovič Matijašić Department of Animal Science, Chair of Dairy Science, Biotechnical Faculty, University of Ljubljana, Groblje 3, 1230 Domžale, Slovenia
e-mail: bojana.bogovic@bf.uni-lj.si

Renato Morona Research Centre for Infectious Diseases, School of Molecular and Biomedical Science, University of Adelaide, Adelaide, SA 5005, Australia
e-mail: renato morona@adelaide.edu.au

Saloomeh Moslehi-Jenabian Department of Food Science, Food Microbiology, Faculty of Life Sciences, University of Copenhagen, Rolighedsvej 30, 1958 Frederiksberg, Denmark
e-mail: smje@life.ku.dk

Dennis Sandris Nielsen Department of Food Science, Food Microbiology, Faculty of Life Sciences, University of Copenhagen, Rolighedsvej 30, 1958 Frederiksberg, Denmark
e-mail: dn@life.ku.dk

Lay-Gaik Ooi School of Industrial Technology, Universiti Sains Malaysia, 11800 Penang, Malaysia
e-mail: ooilaygaik@yahoo.com

Gerald C. O'Sullivan Cork Cancer Research Centre, University College Cork, Cork, Ireland
Deptment of Surgery, Mercy University Hospital, Cork, Ireland
e-mail: gcosullivan@rcsi.ie

Arthur C. Ouwehand Danisco Health & Nutrition, Sokeritehtaantie 20, 02460 Kantvik, Finland
e-mail: arthur.ouwehand@danisco.com

Adrienne W. Paton Research Centre for Infectious Diseases, School of Molecular and Biomedical Science, University of Adelaide, Adelaide, SA 5005, Australia
e-mail: adrienne.paton@adelaide.edu.au

James C. Paton Research Centre for Infectious Diseases, School of Molecular and Biomedical Science, University of Adelaide, Adelaide, SA 5005, Australia
e-mail: james.paton@adelaide.edu.au

A. S. Peña Department of Pathology, VU University Medical Center, Laboratory of Immunogenetics, P.O. Box 7057, 1007 MB Amsterdam, The Netherlands
e-mail: pena.as@gmail.com

Jørgen Petersen The Proteomics Group, Danish Technological Institute, Holbergsvej 10, 6000 Kolding, Denmark
e-mail: jpn@teknologisk.dk

Serge Rezzi BioAnalytical Science, Metabolomics & Biomarkers, Nestlé Research Center, P.O. Box 44, 1000 Lausanne 26, Switzerland
e-mail: serge rezzi@rdls nestle.com

Irena Rogelj Department of Animal Science, Biotechnical Faculty, University of Ljubljana, Groblje 3, 1230 Domžale, Slovenia
e-mail: irena rogelj@bfro.uni-lj.si

Mauro Rossi Institute of Food Sciences, CNR, via Roma 64, 83100 Avellino, Italy
e-mail: mrossi@isa.cnr.it

Denis Roy Departement of Food science and Nutrition, Institute of Nutraceutical and Functional Foods, Université Laval, 2440 Boul. Hochelaga, G1 V 0A6 QC, Canada
e-mail: denis roy@inaf.ulaval.ca

Eduardo Jorge Schiffrin Nestlé Nutrition, R & D, Nestec Ltd, 22 Avenue Reller, 1800 Vevey, Switzerland
e-mail: eduardo.schiffrin@nestle.com

Kathryne B. Schwartz Institute of Food Sciences, CNR, via Roma 64, 83100 Avellino, Italy
e-mail: schwa21k@gmail.com

Alain L. Servin Faculté de Pharmacie, Inserm Unité 756, 92296 Châtenay-Malabry, France
e-mail: alain.servin@u-psud.fr

Stefania Silvi Department of "Scienze Morfologiche e Biochimiche Comparate", University of Camerino, Via Gentile III da Varano, 62032 Camerino, Italy
e-mail: stefania.silvi@unicam.it

R. D. Sleator Department of Biological Sciences, Cork Institute of Technology, Rossa Avenue, Bishopstown, Cork
e-mail: roy.sleator@cit.ie

Siok-Koon Yeo School of Industrial Technology, Universiti Sains Malaysia, 11800 Penang, Malaysia
e-mail: siokkoon85@yahoo.com

Annina Zihler Laboratory of Food Biotechnology, Institute of Food Science and Nutrition, ETH Zurich, Schmelzbergstrasse 7, 8092 Zurich, Switzerland
e-mail: zihlera@ilw.agrl.ethz.ch

Part I
Introduction and History of Probiotics

Chapter 1
Probiotics: From the Ancient Wisdom to the Actual Therapeutical and Nutraceutical Perspective

Giuseppe Caramia and Stefania Silvi

1.1 Probiotics

1.1.1 The Beginnings of Probiotics: The Fermented Milk

The recent history of probiotics began in the early 1900s. Thanks to Metchnikoff (1845–1916) (Fig. 1.1), professor of biology at the University of Odessa, who moved from Ukraine, his homeland, to Messina (Italy) for political reasons after the assassination of Czar Alexander II. In 1882 he discovered the mechanism of phagocytosis and cell-mediated immunity, for which he received the Nobel Prize in 1908, and in 1888, moved to Paris to work at the institute directed by Pasteur, pushed his research on the conditions and the organic alterations that promote aging. At Pasteur's death in 1895, he became the Director of the famous Pasteur Institute and continued his studies in various fields of knowledge and philosophy becoming famous among the general public for his books (*The Nature of Man*, 1904; *The Prolongation of Life*, 1906, etc.).

Starting from the studies of Pasteur on seething microorganisms, and of other researchers on the intestinal bacterial flora (Carre 1887; Tissier 1906), considering that the Caucasian shepherds had a longer average life than the inhabitants of Paris and, according to reports at the time, than the Americans (87 years against 48), he suggested that the shepherds' longevity depended on fermented milk, which they largely consumed, since it was a source of "good" and "anti-putrefactive" microorganisms. It was indeed known that the food wastes ferment in the colon due to some intestinal microorganisms and he was convinced that the putrefactive flora produces toxins, lethal in the long time.

G. Caramia (✉)
Hemeritus Head Physician of Paediatrics and Neonatology, Specialized Maternal-Infantil Hospital "G.Salesi", Via Toti 1, 60123 Ancona, Italy
Tel.: +39 071 36938, +39 335 6166470
e-mail: caramiagm@libero.it

J. J. Malago et al. (eds.), *Probiotic Bacteria and Enteric Infections*,
DOI 10.1007/978-94-007-0386-5_1, © Springer Science+Business Media B.V. 2011

Fig. 1.1 Elias Metchnikoff (1845–1916)

Really, the history of fermented milk and yogurt, with their excellent nutritional properties, was born together with man, in the earliest times of antiquity, most probably 500,000 years ago, when our ancient progenitors learned to light the fire defending themselves from the cold, keeping out animals, lighting the caverns, cooking the game and therefore many millennia before the beginning of the pasture and livestock. The use of fire, fermented milk and yogurt are thus part of human history and their role has been with humanity, to date, between legends and historical data (Flandrin and Montanari 1977; Perles 1977).

The need to keep such a precious food must have been felt since the beginning, and an ancient legend tells of a merchant who, having to cross the desert, brought some foods with him, including milk placed in a bag made with the dried stomach of a sheep. The enzymes remained on the wall of the sheep's stomach used as container, acidified milk and clotted its proteins in small lumps, giving rise to the curd and discovering cheese. The same phenomenon happened to the primordial yogurt derived from the acid fermentation of milk sugars. Thanks to the contamination with special milk enzymes, and a kind of liquid yogurt, used for many millennia by nomadic shepherds and people from the East. Certainly, it was used by the Indians and Sumerians in the fourth century BC, at the beginning of the Egyptian Civilization in the IV–III millennium BC, by the Phoenicians in the III–II millennium BC. The Bible, dated to the thirteenth century BC, reports that "Abraham offered to God, showed in an oakwood, fermented milk" (Genesis 18, 1–8) and Isaiah (VIII BC, 7:15) also says that "you will eat curdled milk and honey."

The Greek historian Herodotus (484–425 BC), Xenophon (430–355 BC), and Aristotle (384–322 BC) have spoken on the use of the yogurt (Bresciani 1977). At the time of the ancient Greeks and Romans, the consumption of fermented milk was recommended as a tonic, especially for children and convalescents, and the Greek physician Galen (129–216 AD), lived in the Imperial Rome, extensively spoke about the yogurt in one of his works, giving to it certain beneficial effects for both the liver and the stomach.

In the Middle Ages, fermented milk and cheese was mainly produced at the abbeys and convents, and they appear in the Crusaders' chronicles; later, we can find them in very distant populations such as Bulgarian shepherds, the Hindus, the Calmucchi, in France, at the court of Francis I (1494–1547), the Zulu, the Russians and other peoples of the Ottoman Empire that used yogurt, a term that derives from the Turkish *yogur* (kneading or mixing with a tool), as a panacea to purify the blood, to prevent tuberculosis, to solve some intestinal disorders and even to help sleeping.

It was known that fermentation is a very important aspect in the formation of yogurt, but the origin of such fermentation was still unclear.

The presence of invisible microorganisms (or micro-Dei), which can creep into our bodies causing diseases, is already present in trace in some Chinese legends and in ancient Egyptian medical texts. Afterwards, Marco Terenzio Varrone (116 BC–27 AD) before and Girolamo Fracastoro (1478–1553) later, talk about it openly. The existence of small organisms, called "animalculi", in the genesis of the diseases and of many other unclear phenomena, was firstly postulated by Lazzaro Spallanzani (1729–1799), who in 1780 coined and introduced into the medical literature the term "germ", so he is considered the founder of the experimental microbiology.

This was opposed to the "spontaneous generation" theory, for which the life is born in a "spontaneous" way from inert or inanimate matter by the effect of some "vital flows", a theory supported until then by the Aristotelian school disciples, by the Epicurean School, by famous philosophers of the Renaissance and in the eighteenth century by Georges-Louis Leclerc, Count of Buffon (1707–1788), and by John Turberville Needham (1713–1781). This dispute continued for many years and was finally permanently settled by Louis Pasteur (1822–1895) in 1864 which made light of that argument confirming the Spallanzani's thesis and thus winning the prize of the Science Academy of Paris for having clearly demonstrated the germs source. Pasteur arrived at such result, thanks to his studies on the fermentation of beer (1854), wine and vinegar (1861–1862) and on the deterioration of the wine by fungi or bacteria (1863–1864); findings confirmed in the following years by studies on silkworm disease (1865–1870), chickens cholera (1880), anthrax in bovines, sheep, horses (1881). In this route it was crucial, of course, the availability of the microscope, "small glasses to see minimal things nearly" that "multiplies things perhaps fifty thousand times" as his discoverer Galileo Galilei wrote (1564–1642) (Saggiatore: 1623), which significantly evolved over the past two centuries mainly thanks to Anton van Leeuwenhoek (1632–1723) and of his successors, thus triumphantly entering in the scientific research field (Caramia 2000).

1.1.2 From the Intuition to the Yogurt

Using bacterial strains selected from the milk of Caucasian and Bulgarian shepherds, through fermentation and acid coagulation of milk by the two microorganisms, *Streptococcus thermophilus* and *Lactobacillus delbruekii* subsp. *bulgaricus* (Fig. 1.2), is obtained a fermented milk, the "Lactobacilline", that in 1906 the

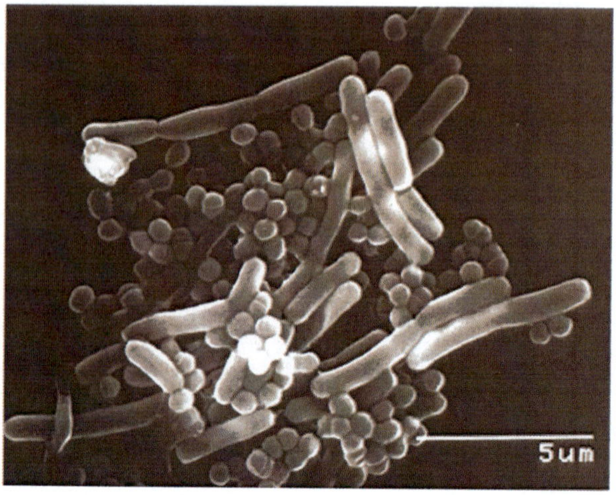

Fig. 1.2 *Streptococcus thermophilus* and *Lactobacillus bulgaricus* from yogurt matrix at scanning electron microscope. (By M. Benevelli—Dept. "Scienze degli alimenti", Bologna University, Italy)

French Society "Le Fermente" began to market and sold in pharmacies, according to the Metchnikoff's idea of helping children suffering from diarrhoea. The product obtained great success among the consumers: today French are the biggest consumers of yogurt compared with other European partners (including Italy), thanks also to the Greek entrepreneurs of Jewish origin, Isac and Daniel Carasso, who was born in Thessaloniki (in Spanish called Mr. Danone).

In 1907/1908 Metchnikoff in his book "The prolongation of life. Optimistic studies" confirms that not all microorganisms are harmful to human health and suggests that "The dependence of the intestinal microbes on the food makes it possible to adopt measures to modify the flora in our bodies and to replace the harmful microbes by useful microbes" (Metchnikoff 1907; Caramia 2008).

Some years later after his death, in 1925 it was sold a product called "yogurt" that rapidly spread in Europe and North America. However, there were also harsh critics since these microorganisms were not found in the faeces of "yogurt" consumers, than someone excluded any beneficial effect of the two seething bacteria. Metchnikoff's intuition, based on empiricism, scientific observations and ingenious intuition, was then mocked by the scientific community, but the beneficial properties of yogurt remained in the collective imaginary, so its use was increasingly widespread.

Always in the 20's, Minoru Shirota, a Japanese microbiologist at the University of Kyoto (Fig. 1.3), discovered that some bacteria of the intestinal flora contribute to bacterial pathogens defence. The following studies led to isolate and cultivate *Lactobacillus casei* (Lc) (Fig. 1.4), afterwards called Lc Shirota, and in 1935 in Japan began the production of a beverage containing this microorganism, called Yakult®, that over the years was spread throughout the world.

An important contribution to the Metchnikoff's theory came in 1936 from two veterinarians, Zobell and Andersen, who suggested the existence in the large intestine of a "microbial film" made by the population of intestinal microorganisms adhering to the intestinal mucosa, which represents a "complex ecosystem with intensive metabolic activities".

Fig. 1.3 Minoru Shirota (1899–1982)

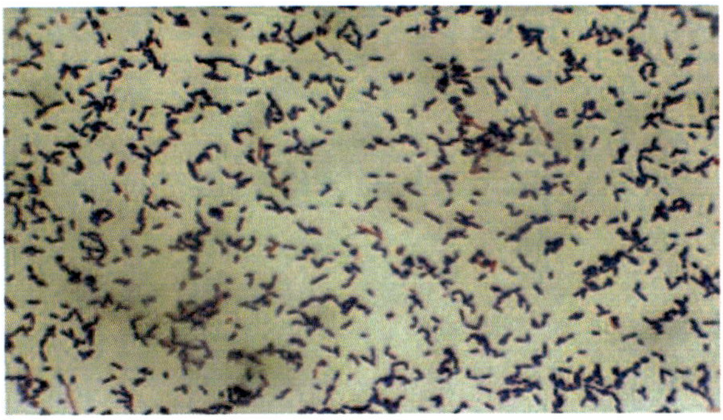

Fig. 1.4 Gram staining of *Lactobacillus casei Shirota*

1.1.3 The term "Probiotic" and its Technical-Scientific Evolution

Metchnikoff has the worth of having introduced the concept of probiotic microorganisms, from the Greek "pro-bios", for life, even if the origin of the term "probiotic" (to be distinguished from lactic ferments that are bacteria of not human origin and producing lactic acid) should be attributed for some to Kollath (1953) and for others to the German researcher Ferdinand Vergin, who in 1954 proposed to use the term "Probiotika" for the "active substances that are essential for a healthy development of life" (Vergin 1954).

In an article published in *Science* in 1962 two veterinarians, Lilly and Stillwell, very likely not knowing the Vergin's proposal, called "probiotics" the so-called "lactic ferments," that is "anaerobic bacteria able to produce lactic acid, starting from different dietary substrates, and to stimulate the growth of other microorganisms" (Lilly and Stillwell 1965).

The last term, also used in contrast to the antibiotic one (against life), which in 1960 was at its peak, thanks to the discovery and development of some important new drugs with antibacterial action that changed the history of the anti-infective therapy, comes in the current use, not only in medicine. With the advance of knowledge on the physiological and therapeutic role of probiotics, the probiotic definitions became increasingly elaborate and exhaustive. So Parker in 1974 was the first man to use that term to identify the microorganisms- based supplements used for zootechnical feeding, defining them as: "organisms and substances which contribute to intestinal microbial balance" (Parker 1974). This new concept has been successful, especially through the work of a British microbiologist, Roy Fuller, specialized in the study of lactic acid bacteria, who in 1989 deleted from the definition the "substances" giving probiotic capabilities to microorganisms only: "a live microbial feed supplement which beneficially affects the host animal by improving its intestinal microbial balance" (Fuller 1989).

Few years later, two Dutch researchers, Haven and Huis in't Veld, extended the definition including in the beneficial action of the probiotic microorganisms the microflora of both the uro-genital and the upper respiratory system. The probiotics become then: "mono-or mixed cultures of live microorganisms which when applied to animal or man, beneficially affects the host by improving the properties of the indigenous microflora" (Huis in't Veld et al. 1994).

It is currently accepted the probiotic definition formulated in 2001 by FAO/WHO "Live microorganisms which when administered in adequate amount confer a health benefit to the host" (FAO/WHO 2001). Respecting the "Guidelines on probiotics and prebiotics" their characteristics can be summarized as follows:

- Must not lose its properties during storage;
- Must be normally present in the human intestine;
- Must be able to survive, to overcome the gastric barrier, resisting to the action of digestive gastric juice, intestinal enzymes and bile salts and colonize the intestine: for this reason, the minimum effective dose, which is very indicative because it depends on the strain and preparation used, is 10^7 CFU/day;
- Must be able to adhere to and to colonize the intestinal cells: the bacterial membrane structure is involved in the mechanism of adhesion and direct switch with the mucosa, the surface proteins and possibly also the secreted ones. In this respect should be reported the possible apoptotic induction on neoplastic cell lines, recently highlighted, which opens possible therapeutic implications;
- Must exert metabolic functions at the enteric level, with beneficial effects for human health, and antagonism against pathogenic microorganisms by producing antimicrobial substances;
- Should not cause immune or otherwise harmful reactions and then be considered as safe (GRAS status: generally recognized as safe);
- Resistance to antibiotics must be intrinsic or due to genetic mutations, whereas if it is caused by a horizontal gene transfer (i.e. transposons, genomic DNA segments that breaks off to join another, conjugative plasmids carrying genes for resistance, virulent or temperate phages) his choice becomes more problematic;
- Must also be administered in adequate doses and have a favourable cost-efficacy ratio.

1.2 Prebiotics

Prebiotics are predominantly dietary fibers, particularly soluble, also called "colonic food", consisting of specific carbohydrates. Increasingly used by the food industry (beverages, sweets) since 1980 for modifying viscosity, emulsification capacity, gel formation, freezing point and colour of foods, prebiotics have been widely studied since the early 90's, while the spread of the probiotics use, to provide the optimal nutrients to encourage growth of beneficial intestinal microflora (symbionts).

In 1995 Gibson and Roberfroid defined prebiotics as "non-digestible substances that when consumed provide a beneficial physiological effect on the host by selectively stimulating the favourable growth or activity of a limited number of indigenous bacteria in the colon, and thus improves host health" (Gibson and Robertfroid 1995).

As beneficial effect of health by "selective stimulation of the growth" and "activity of a limited number of colonic bacteria" are difficult to verify, in recent years the authors revisited their concept and defined prebiotics as: "a selectively fermented ingredient that allows specific changes, both in the composition and/or activity in the gastrointestinal microflora that confers benefits upon host well-being and health" (Gibson et al. 2004; Roberfroid 2007; Kelly 2008).

Based on the last definition, prebiotics may have the following characteristics (Gibson and Robertfroid 1995; Gibson et al. 2004; Roberfroid 2007; Kelly 2008; de Vrese and Schrezenmeir 2008):

- must pass, almost undamaged and in adequate amount, the digestive processes occurring in the first section of the digestive tract (mouth, stomach and small intestine);
- must be a nutritional fermentable substrate for intestinal microflora, in order to selectively stimulate the growth and/or metabolism of one or a few bacterial species;
- should positively change the bacterial flora in favour of the acidophile protective one (bifidobacteria, lactobacilli); and finally they should induce systemic or luminal effects that are positive for the human health.

Prebiotics are present in many edible plants such as chicory, artichoke, onions, leeks, garlic, asparagus, wheat, bananas, oats, soybeans and other legumes. Many commercial prebiotics are obtained from vegetable raw materials, while others are produced by enzymatic way through the hydrolysis of complex polysaccharides or the trans-glycosylation of mono- or disaccharides, a beneficial system for mass production starting from simple sugars (sucrose and lactose). Anyway, the addition of prebiotics in foods must comply with the ESPGHAN (European Society for Paediatric Gatroenterology Hepatology and Nutrition) recommendations (Aggett et al. 2003; Roberfroid 2007) including:

- standard methods for the analysis of carbohydrates content in food;
- right labels with the indication of quality and quantity carbohydrates content;
- international databases;
- knowledge of the origin, specific effects and indications for the use of prebiotics.

The natural and commercial prebiotics consisting of oligo- and polysaccharides that are not, or only to a small extent, hydrolized by the digestive enzymes of the human upper intestinal tract and reach intact the colon where they are selectively fermented, particularly from indigenous and exogenous bifidobacteria and lactic acid bacteria, act as a fermentable carbon sources for the colonic microflora.

The most popular, most widely commercially available and the most researched prebiotic compounds are oligosaccharides oligofructose, fructooligosaccharides (FOS), metabolized by the β-fructofuranosidase (β-Fru) enzyme, the polysaccharide inulin, and partly the *trans*-galacto-oligo-saccharides (TOS) metabolized by the β-galactosidase (β-Gal) enzyme (Gibson and Robertfroid 1995; Bouhnik et al. 2006; Kolida and Gibson 2007; Roberfroid 2007; de Vrese and Schrezenmeir 2008; Kelly 2008). Oligofructose, fructooligosaccharides (FOS) (a mixture of oligosaccharides consisting of 3–10 carbohydrate monomers) and inulin (a mixture of fructooligo- and polysaccharides), are bifidogenic, but there is a great deal of intra-individual variability in bifidogenic and anaerobe responses to those inulin-type prebiotics (some experts consider oligofructose, FOS and inulin as synonymous terms for "inulin-type probiotics", oligo- or polysaccharide chains comprised primarily of linked fructose molecules, and inulin HP for the long-chain, high-molecular weight mixes of inulin-type fructans with a degree of polymerization (DP) > 10) (Roberfroid 2007; Kelly 2008, 2009). The effects on other gut microorganisms, as well as pathogenic organisms, are inconsistent but oligofructose and FOS show nutrition and health relevant properties like a low cariogenicity, a low calorimetric value and glycemic index, and a moderate sweetness (30–60% of the sucrose value = 1–2 kcal/g) (Kelly 2008). For this reason they are used as sweeteners in syrup, tablets or powder. Other candidates as prebiotics, for which there are already promising data, but for someone not yet sufficient, are the gluco-oligo-saccharides (GOS) which are oligo or polysaccharide chains comprised primarily of linked galactose units and which stimulate the growth of bifidobacteria and lactobacilli species, the soy-oligo-saccharides (SOS) raffinose and stachiose, metabolized by the α-galactosidase (α-Gal) enzyme, the iso-malt-oligo-saccharides and more (Roberfroid 2007; Kelly 2009; Bruzzese et al. 2009).

1.3 Synbiotics

An alternative chance to modulate or balance the intestinal microflora is the use of pro-and pre-biotic together making synbiotic compounds, that are alimentary or pharmaceutical preparations that containing either one or more probiotic strains and prebiotic ingredients, exploit the synergy between the microorganisms activity and their support for the benefit of the intestinal microflora and, consequently, of the whole body.

In 1995 Gibson and Roberfroid defined synbiotic as "a mixture of probiotics and prebiotics that beneficially affects the host by improving the survival and implantation of live microbial dietary supplements in the gastrointestinal tract, by selectively stimulating the growth and/or activating the metabolism of one or a limited number of health-promoting bacteria, and thus improving host welfare".

The simultaneous administration of both probiotics and a substrate that they can metabolize gives to the administered strains greater opportunities for the colonization and survival of probiotic organisms in the colon of the host by increasing or prolonging their beneficial effects: this is really the best strategy for their integration, because it improves the survival (increasing the product shelf life) and on the other hand it provides a specific substrate for the resident bacterial flora.

Theoretically, the synbiotics have better beneficial effect on intestinal flora than pro- and prebiotics by lowering the pH, promoting growth of potentially protective bifidobacteria and inhibiting of potentially pathogenic microorganisms, stabilizing the intestinal environment and releasing short-chain organic acids.

Inulin-type probiotics, FOS or GOS, as well as their synbiotic combination with probiotic bacteria, *L. plantarum, L. paracasei* or *B. bifidum* strains, increased bifidobacteria and lactobacilli and inhibited various human- and animal pathogenic bacterial strains (*Clostridium sp., E. coli, Campylobacter jejuni, Enterobacterium sp., Salmonella enteritidis* or *S. typhimurium*) (Kanamori et al. 2004).

The most used and already marketed synbiotics regard mixtures of oligofructose, FOS, GOS, with probiotic bacterial strains of *L. plantarum*, *L. paracasei*, *L. rhamnosus*, *B. bifidum* or *B. lactis*.

1.4 Various Genera of Probiotics

The majority of probiotic microorganisms belong to the genera *Lactobacillus* (Figs. 1.5 and 1.6) and *Bifidobacterium* (Fig. 1.7). There are also other genera of bacteria and some yeasts widely used and reported in Table 1.1 (Baffoni and Biavati 2008). Lactobacilli and bifidobacteria are Gram-positive lactic acid-producing bac-

Fig. 1.5 Morphology of *Lactobacillus rhamnosus* at scanning electron microscope. (By M. Benevelli—Dept. "Scienze degli alimenti", Bologna University, Italy)

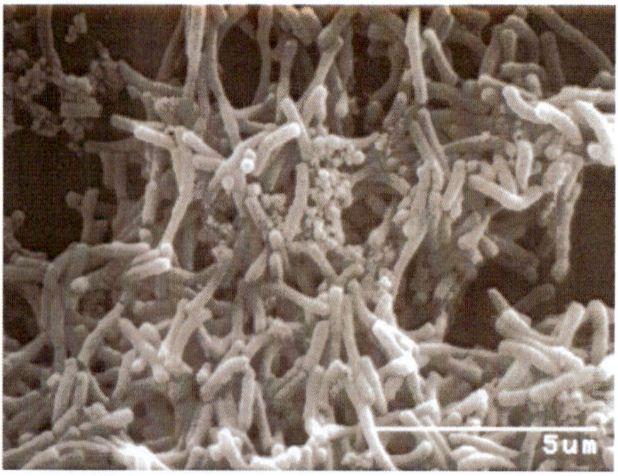

Fig. 1.6 Morphology of *Lactobacillus rhamnosus* from yogurt matrix at scanning electron microscope. (By M. Benevelli—Dept. "Scienze degli alimenti", Bologna University, Italy)

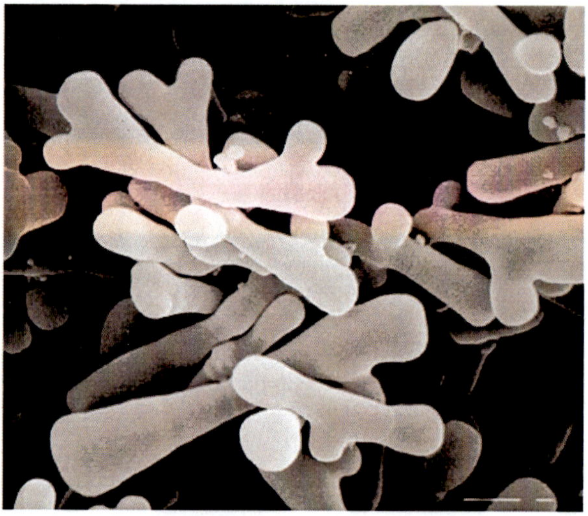

Fig. 1.7 Morphology of *Bifidobacterium* spp. at scanning electron microscope

teria that constitute a major part of the normal intestinal microflora in animals and humans. Lactobacilli are Gram-positive, non-spore forming rods or coccobacilli. They have complex nutritional requirements and are strictly fermentative, aerotolerant or anaerobic, aciduric or acidophilic. Lactobacilli are isolated from a variety of habitats where rich, carbohydrate-containing substrates are available, such as human and animal mucosal membranes, on plants or material of plant origin, sewage and fermented milk products, fermenting or spoiling food. Bifidobacteria constitute a major part of the normal intestinal microflora in humans throughout life. They appear in the faeces a few days after birth and increase in number thereafter. The number of Bifidobacteria in the colon of adults is 10^{10}–10^{11} CFU/g, but this number decreases with age. Bifidobacteria are non-motile, non-spore forming, Gram-positive rods with varying cell morphology. Most strains are strictly anaerobic.

Table 1.1 Microorganisms considered as probiotics. (By Baffoni and Biavati 2008, modified)

Lactobacillus	Bifidobacterium	Enterococcus	Streptococcus	Lactococcus
L. acidophilus	*B. adolescentis*	*E. faecalis*	*S. thermophilus*	*L. lactis subsp. cremoris*
L. brevis	*B. animalis*	*E. faecium*		*L. lactis subsp. lactis*
L. casei	*B. bifidum*			
L. curvatus	*B. breve*			
L. fermentum	*B. infantis*			
L. gasseri	*B. longum*			
L. johnsonii	*B. thermophilum*			
L. reuteri				
L. rhamnosus				
L. salivarius				
Propionibacterium		Yeast		Others
P. freudenreichii		*Kluyveromyces lactis*		*Leuconostoc mesenteroides*
P. freudenreichii subsp. shermanii		*Saccharomyces boulardii*		*Pediococcus acidilactici*
P. jensenii		*Saccharomyces cerevisiae*		

1.5 Probiotics as Therapy

The primordial milk enzymes at the beginning of last century, selected from the milk of the Caucasian and Bulgarian shepherds, have been sold according to the ideas of Metchnikoff and Tissier "to help children suffering from diarrhoea" and sold in pharmacies to bring "good and anti-putrefactive micro-organisms" because "not all microorganisms are harmful to human health".

In this light over the next few decades lactic acid bacteria with special features, now considered probiotics, kept the primary indication: the preventive-therapeutic use, particularly for some gastroenterological diseases, to try to restore and/or rebalance the functionality of microbiota, the intestinal mucosa and the immunological aspects, keeping in mind the indications listed in the guidelines about the evidence based medicine on the levels of scientific evidence and the strength of clinical recommendations.

1.5.1 Acute Infectious Diarrhoea

In most industrialized countries, acute infectious diarrhoea (AID) is now a minor disease because fatal cases are very exceptional. It is determined in about 70% by viral agents, such as rotavirus, which are responsible for 30–45% of all viral diarrhoea, calicivirus, including norwalk virus, enteric serotypes adenovirus 40 and 41, and Astrovirus; while among bacteria we should mention *Campylobacter jejuni* (main cause of diarrhoeal disease in adults in the US), *Salmonella, Shigella*, enteropathogens *Escherichia coli, and Yersinia enterocolitis.*

Firstly, it should be noted that not all probiotics are the same, because not all of them determine the same therapeutic effects, but, based on the levels of scientific evidence and the strength of clinical recommendations, it is believed appropriate to share in principle what was recently proposed by the ESPGHAN and by the European Society for Paediatric Infectious Disease (ESPID) and by many other scientists: "Probiotics may be an effective adjunct to the management of diarrhoea. However, because there is no evidence of efficacy for many preparations, we suggest the use of probiotic strains with proven efficacy and in appropriate doses for the management of children with acute gastroenteritis as an adjunct to rehydration therapy (levels of scientific evidence II and strength of clinical recommendations B). The following probiotics showed benefit in meta-analyses of RCTs: *Lactobacillus* GG (I, A), *L. reuteri* (I, A) and *Saccharomyces boulardii* (II, B)" (Floch et al. 2008; Guarino et al. 2008; Kligler and Cohrssen 2008).

In particular, *L. reuteri* has shown to shorten significantly the clinical course of rotavirus-induced gastroenteritis, as well as reducing incidence of acute diarrhoea (Figs. 1.8 and 1.9) (Shornikova et al. 1997a, b).

As for prevention of infectious diarrhoea, mostly of viral origin, which can be contracted at nursery schools, kindergartens or during hospitalization for other pathologies, it is not yet clear which probiotic or association of probiotics is more effective. Besides, the dose administered which must be equal to or greater than 5–10 billion CFU/day and the early initiation of therapy are important, so that the probiotic, with appropriate doses and immediately administered, may contrast the action of the pathogen (Floch et al. 2008; Guarino et al. 2008, 2009). More recently,

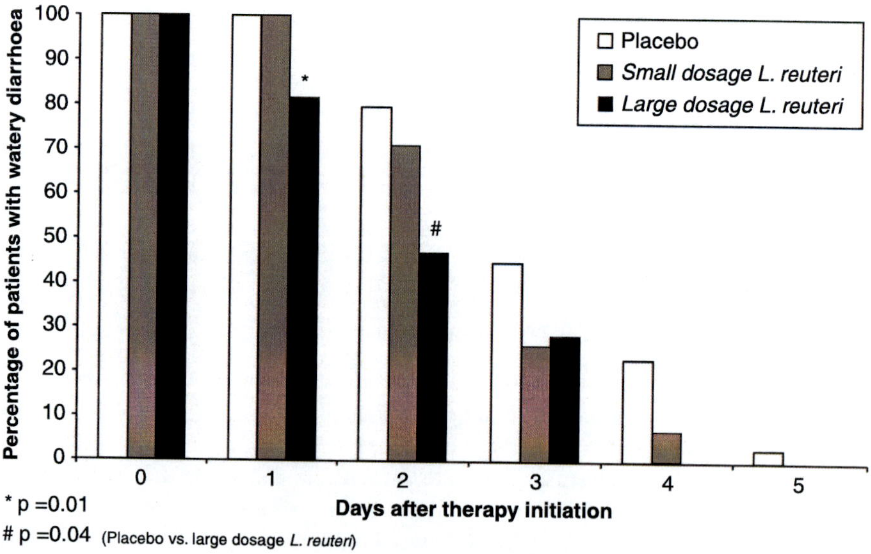

Fig. 1.8 Percentage of patients with persisting watery diarrhoea in the groups receiving placebo ($n = 25$) and small ($n = 20$) and large ($n = 21$) dosage of *L. reuteri* (Shornikova et al. 1997a)

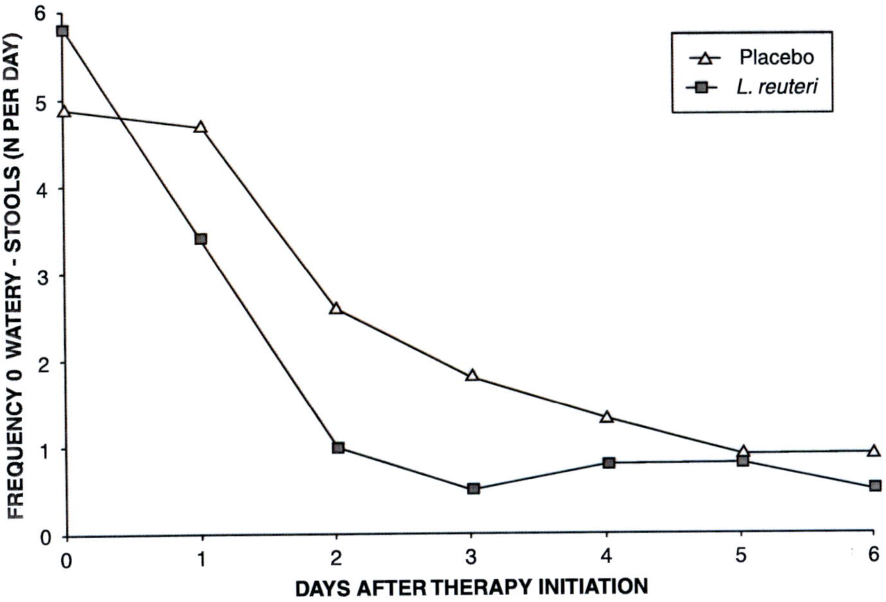

Fig. 1.9 Frequency of watery stools per 24-h period in patients receiving *L. reuteri* and placebo (Shornikova et al. 1997b)

Eom and colleagues showed the therapeutic effect of *L. reuteri*, administered at a dose of 2×10^8 CFU/die only, to significantly reduce acute diarrhoea in children (Eom et al. 2005).

1.5.2 Antibiotic Associated Diarrhoea

Antibiotics, aminopenicillins, cephalosporin, clindamycin etc., are much prescribed in all industrialized countries with several side effects especially in children: among which the most frequent is antibiotic associated diarrhoea (AAD). The resulting alteration of intestinal microflora reduces the development of anaerobic microflora, which leads to a reduced metabolism of carbohydrates and therefore to osmotic diarrhoea, favours the development of pathogens such as *Clostridium difficile, Salmonella, C. perfringens type A, Staphylococcus aureus* and *Candida albicans*. According to the recent studies, even in adults, there is level I of scientific evidence in favour of the use of probiotics in the treatment of AAD (Doron et al. 2008; Floch et al. 2008; Pham et al. 2008; Surawicz 2008). Therefore, there are grounds to recommend their use especially in risky cases, as in subjects where there is repeated use of antibiotics or in subjects with diarrhoeal episodes occurring after the administration of antibiotics. This in an attempt to prevent inflammatory processes of the intestinal mucosa in children that can often lead to chronic inflammatory disease of the large intestine (Crohn's disease, ulcerative colitis, pouchitis) in subsequent years (Caramia 2008; Floch et al. 2008; Guandalini 2008). In a randomized, double-blind, placebo-controlled pilot study, recently presented at the Clinical Nutrition Week 2009, patients receiving antibiotics were given *L. reuteri* (10^8 CFU b. i. d.) or an identical placebo for 4 weeks. Patients treated with *L. reuteri* had a significantly lower incidence of diarrhoea (only 7.7%) compared to patients receiving placebo (50%) (Cimperman et al. 2009).

1.5.3 *Clostridium difficile* Associated Diarrhoea

The *Clostridium difficile* is the main cause of diarrhoea caused by antibiotics (CDAD) and of nosocomial colitis. It has been indicated as responsible for between 10% and 20% of all cases of diarrhoea caused by antibiotics, 60% of antibiotic-associated colitis and nearly all cases of pseudo membrane colitis. The diarrhoeal disease caused by *C. difficile* is determined only by the *C. difficile* strains producing the toxin A, who plays a mild cytotoxic activity and causes damage to the mucous, inflammation and intestinal secretion, and by toxin B, one of the most powerful cytotoxin, which determines loss of intracellular potassium, inhibition of protein synthesis and nucleic acids.

Unfortunately, the diversity of probiotics, their doses and the heterogeneity of studies make it difficult to recommend a definitive therapy, and also to indicate which probiotics to use as an antibiotic treatment and for prevention of *C. difficile* associated diarrhoea and/or colitis. For this reason, despite there are many promising data, the level of scientific evidence in favour of the use of probiotics or a combination of antibiotic and probiotic in the treatment of CDAD is currently of type II only (Doron et al. 2008; Floch et al. 2008; Guandalini 2008; Hookman and Barkin 2009; Yangco et al. 2009).

1.5.4 Infection Caused by *Helicobacter pylori*

Helicobacter pylori (HP) infection affects over 50% of the world's population and covers 80% of the population in the developing countries. HP infection is the main cause of peptic ulcer disease (70–90% of cases), lymphoma and in 1% of infected persons, leads to the development of gastric cancer with remarkable increase in mortality (Kelly and LaMont 2008; Jarosz et al. 2009). In developed countries, the infection starts in childhood, where it seems to have an incidence of 10–15%, then rapidly increasing during evolution (Sabbi et al. 2008). The transmission is oro-faecal as the seed is located in the gingival bags and at the root of the tongue.

Several studies showed that patients treated with probiotics associated with the standard antibiotic therapy had higher rate of eradication with a minor number of side effects.

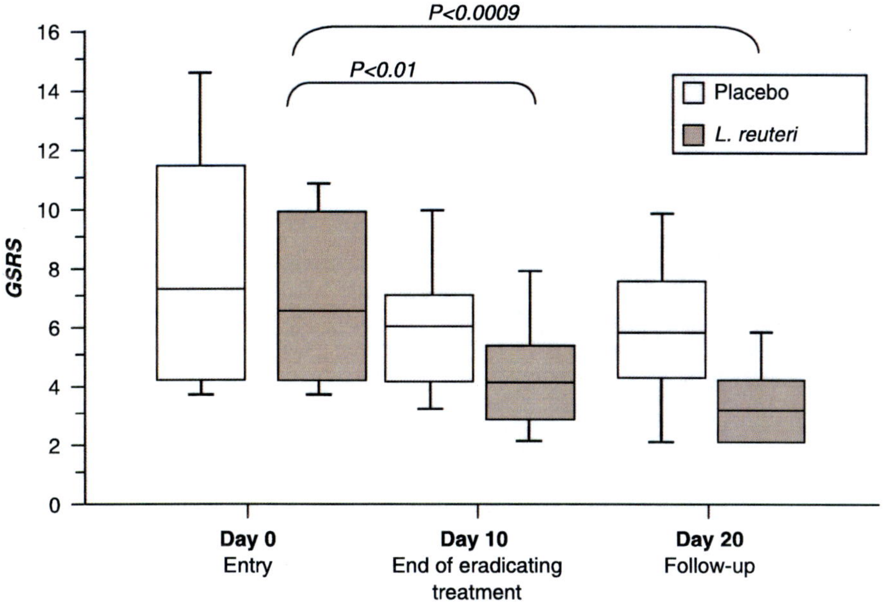

Fig. 1.10 Gastrointestinal symptom rating scale (GSRS) in children receiving *L. reuteri* or placebo. Comparison of continuous variables performed using Mann-Whitney. (Adapted from Lionetti et al. 2006)

A pilot study performed on 30 Hp positive adults treated with omeprazole + placebo or omeprazole + *L. reuteri* for 30 days, showed that 60% of patients in the *L. reuteri* group was eradicated, while none in the placebo group (Fig. 1.10) (Saggioro et al. 2005). Lionetti and colleagues in 2006 showed a reduction in gastro- intestinal symptoms by *L. reuteri* supplementation during and after the eradication therapy in a group of Hp infected children (Lionetti et al. 2006). Finally, Francavilla and colleagues in a recent pilot study conducted on 40 Hp positive adults, undergoing the standard eradication therapy, showed that the pre-administration of *L. reuteri* in the 4 weeks before treatment significantly reduces the bacterial load, and decreases the GI associated symptoms (Francavilla et al. 2008).

Excellent results are also reported by the administration of fermented milk enriched with probiotics (Tong et al. 2007; Sachdeva and Nagpal 2009). The HP infection in adults are related to the type and virulence of HP strain, the production of toxins A and B, the level of infection, the extent of inflammation, and the density of HP colonization. Therefore, there is a considerable interest in developing therapy to prevent HP infection. Probiotics intake, with suppression of HP, may have a favourable effect on HP infection and in decreasing the risk to develop related diseases; however, further, large, long-term and placebo-controlled studies are needed to confirm all those effects (Kelly and LaMont 2008; Selgrad and Malfertheiner 2008; McFarland 2009).

1.5.5 Traveller's Diarrhoea

The traveller's diarrhoea (TD) is the most common pathologic condition among travellers, and can occur in approximately 20–50% of the subjects during or immediately after a trip to a country with hot and humid weather conditions and inadequate sanitary conditions. More than one third of patients with TD had pathogen-negative illness but in the case of food contaminated with pathogens, the most frequent are enterotoxigenic *Escherichia coli, Campylobacter jejuni, Shighella, Salmonella*, sometimes virus, such as *rotavirus, calicivirus, enterovirus*, or parassites such as *Giardia lambdia, Entamoeba histolytica, Cryptosporidium parvum,* and *Cyclospora cayetanensis* are less common causes of TD (McFarland 2007; Johnson et al. 2009).

Sometimes serious but of transitory nature, it is characterized by non formed or liquid evacuations, often accompanied by nausea, vomiting, abdominal cramps, tenesmus, in 10% of cases stool blood, occasionally fever. The evolution is favourable in 1–5 days, but sometimes it needs up to 10 days. It has also been reported that 2–10% of travellers with TD develop persistent diarrhoea and that about 10% of them suffer from post infectious irritable bowel syndrome (Caramia 2008).

In a recent meta-analysis of probiotics for the prevention of TD on randomized, controlled, blinded efficacy trials in humans diarrhoea, several probiotics, *S. boulardii* and a combination of *L. acidophilus* and *B. bifidum*, had significant efficacy (Takahashi et al. 2007; Guslandi 2008).

There are well founded reasons to believe that probiotics may be a safe and effective strategy to prevent TD, to which one can associate racecadotril an inhibitor of intestinal encefalinasis and therefore antisecretive and antidiarrhoeic action, but continued research are needed (Tormo et al. 2008; Vandenplas et al. 2009).

1.5.6 Necrotizing Enterocolitis

The necrotizing enterocolitis (NEC), inflammatory intestine disease with an incidence of 1 to 3/1,000 live births, is relatively common in preterm in the first six weeks of life, and leads to death in 15–30% of those with lower weight to 1,500 g and the 20–40% of cases requiring surgery.

The causes of the disease appear to be multifactorial and are represented by prematurity, hypoxia-ischemia of the intestine, extensive use of antibiotics, reduced exposure to maternal microflora and excessive exposure to the typical sections of neonatal intensive care (NICU) (Staphylococci, Enterobacteriaceae, enterococci, *Candida* spp.), or to the use of sterile food as an alternative to mother's milk.

Based on several multicentre, randomized, double blind investigations there are reasonable grounds to believe that despite the differences between tested probiotics, the beginning of administration, dose and duration of the treatment and groups of participants, the enteral supplementation of some probiotics may reduce the risk of severe NEC and mortality in premature newborn with weight >1,000 g. There is therefore a level I of evidence, even if a deeper understanding of NEC pathogenesis and the mechanisms by which probiotics prevent this disease is needed (Dani et al. 2002; Bin-Nun et al. 2005; Deshpande et al. 2007; Alfaleh and Bassler 2008; Lin et al. 2008; Caplan 2009).

1.5.7 Bacterial Vaginosis

Bacterial vaginosis (BV) is the most prevalent vaginal infection worldwide and is characterized by depletion of the indigenous lactobacilli. The vaginal microflora is composed by several species of lactobacilli (*L. acidophilus, L. brevis, L. casei* spp. *paracasei, L. rhamnosus*) named "Doderlein" microflora. They create on the mucosa a bacterial biofilm able to inhibit the growth and the adhesion of pathogenic bacteria. This microflora presents variations related to the life style (diet, stress, sexual habits, etc.) which may cause quali-quantitative modifications of the normal environment and the introduction of several kinds of pathogens such as *Gardnerella vaginalis* and *Candida albicans*.

Antimicrobial therapy is often ineffective while the probiotic approach, either topic or combined with the oral administration, gave interesting results. This has been confirmed by *L. rhamnosus* strain (Lcr35) showing the ability to adhere to cervical and vaginal cells and to affect the viability of two main vaginosis-associated pathogens, *Prevotella bivia, Gardnerella vaginalis*, as well as *C. albicans* (Coudeyras et al. 2008).

In a trial eighty-four patients with bacterial vaginosis were randomized to receive either oral metronidazole 500 mg twice a day for seven days, or one vaginal tablet containing freeze-dried *L. rhamnosus* once a week at bedtime for two months starting one week after the last antibiotic administration. Chi-squared analysis showed a significant difference between the two treatment groups at day 90 ($p = 0.05$). Safe and effective long-term vaginal administration of *L. rhamnosus* appears to be a useful complementary approach in the management of bacterial vaginosis (Marcone et al. 2008).

Recently sixty-four women diagnosed with BV were randomly assigned to receive a single dose of tinidazole (2 g) supplemented with either 2 placebo capsules or 2 capsules containing *L. rhamnosus* GR-1 and *L. reuteri* RC-14 every morning for the following 4 weeks. After a treatment of 28 days, the probiotic group had a significantly higher cure rate of BV (87.5%) than the placebo group (50.0%) ($p = 0.001$). In addition, according to the Gram-stain Nugent score, more women were assessed with "normal" vaginal microbiota in the probiotic group (75.0% versus 34.4% in the placebo group; $p = 0.011$) (Martinez et al. 2009a). This study shows that probiotic lactobacilli can provide benefits to women with BV and that probiotic capsules containing *L. rhamnosus* GR-1 and *L. reuteri* RC-14 can increase the effectiveness of an antifungal pharmaceutical agent (fluconazole) (Martinez et al. 2009b).

1.5.8 Irritable Bowel Syndrome

The irritable bowel syndrome (IBS), a disorder of the function of the intestine, affects 3–25% of the general population and is characterized by pain at the abdomen, constipation alternating with periods of diarrhoea, presence of air and a sense of abdominal bloating. This is due to psychological factors, including sexual abuse, because the central nervous system interacts with intestinal neurotransmitters and hormones, to dietary factors resulting in alteration of intestinal motility, and abnormal dismicrobism, colic fermentation by bacteria and sometimes a secondary gastrointestinal infection (e.g. *Campylobacter* or *Salmonella*) (Belaise et al. 2002; Mättö et al. 2005).

Recent reviews of the literature provide interesting data also on gut motility and pain perception, however the results, sometimes contradictory, are not yet very clear and the effectiveness of probiotics in IBS show a level of scientific evidence of type I (Kunze et al. 2009). Very likely the strength of the recommendations of type B is due to the limited number of patients treated with different probiotics (Jiménez 2009). It is therefore considered useful to continue with more appropriate and thorough studies on those probiotics that have shown a more promising result.

1.5.9 Crohn's Disease

Crohn's disease, a gastrointestinal disorder characterized by chronic inflammation or ulceration involving all layers of the intestinal wall, is one of a group of diseases called inflammatory bowel disease (IBD) such as Crohn's disease (CD), ulcerative colitis (UC) and pouchitis. IBD are most common in the developed countries of Europe, the US, and Scandinavia, and less in Southern and Eastern Europe, in Asia or Africa, and the diseases are relatively uncommon in Cuba and Central and South America (Loftus 2004). This has given rise to a number of theories regarding IBD etiology and the significance of diet high in refined sugar, meat, milk, eggs and low in fiber, fruit, and vegetables.

CD can affect any area from the mouth to the anus but often affects the lower part of the small intestine, the ileum, while the colon is the second most common site of involvement. Prevalence of Crohn's disease is approx 0.18% and is more frequent in young adults between 15 and 30 years of age (Wallace 2009).

The intestinal microbiota play an important role in the pathogenesis and maintenance of disease. Although marked alterations occur in faecal and mucosal bacterial communities in CD and others IBD, it is unclear whether they are responsible for causing disease, or are due to changes in the gut environment that result from inflammatory reactions and extensive tissue destruction that later concur to maintain the pathological condition.

A study has been recently conducted on a limited number of patients treated for 13.0 ± 4.5 months with a mixture of probiotics composed of *B. breve* 30×10^9 CFU/

day plus *L. casei* 30 × 10^9 CFU/day plus *B. longum*, 15 × 10^9 CFU/day; 3.3 g of psyllium (*Plantago ovata*) were added to probiotics as prebiotics, diluted in 100 mL of water immediately before being given three times a day. The treatment led to satisfactory results with reduction of the CD activity index (CDAI), which takes into account eight parameters (the number of liquid discharges, abdominal pain, general condition, presence of complications of abdominal masses, low hematocrit, loss weight, treatment with opioids or similar), and of the International Organization for the Study of Inflammatory Bowel Disease (IOIBD) score. In particular there was a reduction in diarrhoea and abdominal pain in eight patients, complete remission in six cases and partial in one case, the suspension of cortico-therapy in two cases and dose reduction in four (Fujimori et al. 2007).

A Cochrane survey in the period 1966–2007 on several meta-analysis has found only one reliable study conducted on eleven patients for which it was not possible to draw any conclusion and a subsequent meta has found only two studies in which therapy with probiotics had positive results (Butterworth et al. 2008; Heilpern and Szilagyi 2008; Macfarlane et al. 2009b).

Other evidences suggest that the administration of select prebiotics, probiotics and synbiotics may improve the clinical outcome of patients with IBD. This suggests that there is potential for controlling these diseases through manipulation of the composition of the gut microbiota, and direct interactions with the gut immune system (Kanauchi et al. 2009; Macfarlane et al. 2009a). In addition, IBD patients are well known to carry a higher risk of developing colorectal cancer due to chronic inflammation (Kanauchi et al. 2009).

Therefore, probiotics and/or prebiotics may be appropriate treatments for prophylactic use due to their physiologic characteristics and lack of obvious toxicity.

1.5.10 Ulcerative Colitis

Ulcerative Colitis (UC) is a chronic intestinal inflammatory disease that affects the final part of the colon (sigma and rectum), but can be more extensive and involve the left side of the colon or the whole colon. Severe UC, a serious, potentially life threatening condition and hospitalization, should be considered in all patients who have more than 6–10 bloody stools per day, associated with fever, dehydration, tachycardia malaise and/or increased C-reactive protein.

The main therapeutic remedies are represented by optimal doses of oral 5-aminosalicylic acid (5-ASA), topical therapy with either 5-ASA or steroids, and systemic corticosteroids. Failure to induce remission will occur in only 40% of patients after a first course of oral systemic steroids and only 20% will have no improvement of symptoms whatsoever (Van Assche et al. 2008).

Among the newer agents, a probiotic preparation (VSL#3) may prove helpful in inducing remission also in paediatric patients with ulcerative colitis associated with 5-ASA (Huynh et al. 2009; Miele et al. 2009).

In a recent study three groups of 40 patients were treated with probiotic, prebiotic, or synbiotic therapy. The probiotic group ingested one daily capsule consisting of *B. longum* 2×10^9 CFU and the prebiotic group ingested daily 8.0-g doses of psyllium. The synbiotic group underwent both treatments. All patients completed Inflammatory Bowel Disease Questionnaires (IBDQs) at the onset of the trial, at the 2-week midpoint, and at the 4-week end of the trial.

Total IBDQ scores improved within groups by the end of the trial (probiotics 162–169, NS; prebiotics 174–182, NS; synbiotics 168–176, $p = 0.03$). C-reactive protein decreased significantly only with synbiotic therapy (from 0.59 to 0.14 mg/dL, $p = 0.04$). There were no adverse events and patients with UC on synbiotic therapy experienced greater quality of life changes than patients on probiotic or prebiotic treatment. These data suggest that synbiotic therapy may have a synergistic effect in the treatment of UC (Fujimori et al. 2009).

Probiotics, prebiotics, and synbiotics may improve the clinical outcome of patients with UC and may be appropriate treatments for prophylactic use due to their physiologic characteristics and lack of obvious toxicity (Kanauchi et al. 2009).

In patients with severe ulcerative colitis, the restorative proctocolectomy is often performed, it consists in removing the entire colon and rectum, with severe anatomo-pathological ulcerative lesions, and in making an ileal "reservoir or pouch", using the last ileal handle, which is anastomosed to the anus: it is thus created a pocket (pouch in Anglo-Saxon terminology) like a rectal ampoule, which acts as a reservoir for faecal content.

Pouchitis is the major long-term complication after ileal-pouch anal anastomosis for ulcerative colitis, and broad-spectrum antibiotics are the mainstay of treatment. Recently, it has been shown the efficacy of a highly concentrated probiotic preparation (VSL#3, 900 billions/sachet lyophilized viable bacteria containing four strains of *Lactobacillus*, three strains of *Bifidobacterium* and one strain of *Streptococcus thermophilus*) in preventing pouchitis onset, in preventing relapses of chronic pouchitis and in the treatment of mild pouchitis (Gionchetti et al. 2003, 2007). Patients receiving probiotics have shown a significant improvement in their quality of life and no any side effects or alteration of the laboratory parameters, however further controlled studies are warranted.

1.5.11 Atopic Dermatitis

The prevalence of atopic dermatitis (AD) has risen over the past decades, especially in western societies. This increase, according to the revised hygiene hypothesis, is caused by a changed intestinal colonization pattern during infancy, which has an impact on the immune system, so healthy subjects have an intact functional allergen-specific regulatory T-cell response, while allergic subjects have an impaired response. Manipulating the intestinal microflora with pro-, pre- or synbiotics is an innovative way to prevent or treat AD (von Hertzen et al. 2009).

In 2007 Abrahmsson and colleagues performed a double-blind, randomized, placebo-controlled trial, which comprised 232 families with allergic disease, of whom 188 completed the study. The mothers received *L. reuteri* (10^8 CFU/die) from the 36th gestational week until delivery (Abrahamsson et al. 2007). Their babies then continued with the same product from birth until 12 months of age and were followed up for another year. Primary outcome was allergic disease, with or without positive skin prick test or circulating IgE to food allergens. The cumulative incidence of eczema was similar in the two groups, however, the *L. reuteri* group had less IgE-associated eczema during the second year (8% versus 20%; $p = 0.02$). The skin prick test reactivity was also less common in the treated than in the placebo group, significantly so for infants with mothers with allergies, 14% versus 31% ($p = 0.02$). A recent prospective, double-blind, randomized, placebo-controlled clinical trial performed on 50 subjects aged 3–47 months with moderate/severe AD showed that 12 months of *L. reuteri* supplementation may be beneficial in the long-term control of eczema (Gromert and Axelsson 2009). Based on a recent review at this moment there is not enough evidence to support the use of pro-, pre- or synbiotics for the prevention or treatment of AD in children in clinical practice (van der Aa et al. 2009). On the contrary, some other published results suggest that the administration of selected strains of probiotics during the perinatal period may be helpful in the prevention of AD; however probiotic strain or the composition of different probiotics and/or prebiotics, the dose and the timing of supplementation still need to be determined (Ji 2009; Kopp and Salfeld 2009; von Hertzen et al. 2009). Results of these trials are still conflicting and more randomized controlled trials specifically designed for children with food allergy, with different probiotic strains, are necessary to draw definitive conclusions (van der Aa et al. 2009).

1.5.12 Dental Caries

Dental caries is a multifactorial disease that includes the participation of cariogenic and non-cariogenic bacteria, salivary components (proteins, enzymes, calcium, phosphate, fluoride) and dietary sources of fermentable carbohydrates (sucrose, glucose). Dental caries is one of the most prevalent diseases in humans, is second only to the common cold, affects 90% of late adolescents and young adults and more than 95% of adults experiencing caries on enamel and root surfaces (Islam et al. 2007; García-Godoy and Hicks 2008). Caries, gingivitis and periodontal disease occur because of the accumulation of pathogenic dental plaque. Prevalence and severity have increased during the past decade in young children, particularly in the primary dentition (García-Godoy and Hicks 2008). The changes in the homeostasis of the oral cavity with an overgrowth of *Streptococcus mutans* is recognized as the primary cause of the disease but cariogenic bacteria include *S. sobrinus, Lactobacillus* species and *Actinomyces* species, as well as to a lesser extent *S. mitis, S. oralis, S. gordonii* and *S. anginosus*. They strongly adhere and releases acids by the fermentation of carbohydrates, leading to the demineralization of the tooth. This attachment is mediated mostly by the interaction of surface proteins and bacterial polysaccharides. Early acquisition of *S. mutans* is a major risk factor for early childhood caries and predicts future caries experience.

Ambiguities in the basic treatment of dental caries, such as the use of fluoride and antibiotics, and increasing problems of resistance to synthetic antimicrobials, vitalize the deployment of probiotic therapies for elimination of the bacterium or suppression of its virulence (Kuramitsu et al. 2007; García-Godoy and Hicks 2008).

The probiotic approach to change the oral environment toward health by the deliberate implantation of specific oral streptococci that naturally lack the ability to produce lactic acid to encourage the shift from a pathogenic to a nonpathogenic plaque or biofilm, is one interesting strategy (Hillman et al. 2009; Zahradnik et al. 2009).

Daily applications of *S. rattus* strain JH145, a naturally occurring LDH-deficient variant of *S. rattus*, can compete with *S. mutans* for its habitat on the tooth surface and has been used to affect the numbers of an implanted *S. mutans* strain in a rat model. It has been then demonstrated that *S. rattus* JH145 competes with *S. mutans* for its habitat on the tooth surface and as a probiotic has a prevention role in dental caries (Hillman et al. 2009).

A pilot clinical trial has been performed in humans to assess the safety and to test the ability of a probiotic "mouthwash" ProBiora containing three specific strains of naturally occurring oral bacteria *S. oralis* strain KJ3sm, *S. uberis* strain KJ2sm, and the spontaneous lactic acid-deficient variant of *S. rattus*, strain JH145, to affect the levels of *S. mutans* and certain known periodontal pathogens in the mouth when administered twice daily over a period of 4 weeks. The results of this pilot human study suggest that the probiotic mouthwash product may be safe for daily use as an aid in maintaining both dental and periodontal health.

Two interesting studies have been also performed on the short-term effect of chewing gums containing the probiotic *L. reuteri* on both the levels of salivary mutans streptococci in young adults (Table 1.2) (Caglar et al. 2006) and more recently

Table 1.2 Distribution of salivary mutans streptococci counts before and after 3 weeks of daily ingestion of a probiotic strain (*Lactobacillus reuteri*) with straw and tablets. The values denote the number of subjects. (Caglar et al. 2006)

Administration/time	n	Mutans streptococci score (cfu)				
		0 No growth	1 $\leq 10^4$	2 10^5	3 $\geq 10^6$	p
Group A *Probiotic straw*						
Baseline	30	6	4	5	15	
End	30	9	6	14	1	<0.05
Group B *Placebo straw*						
Baseline	30	6	9	9	6	
End	30	8	7	8	7	NS
Group C *Probiotic tablet*						
Baseline	30	9	5	5	11	
End	30	17	2	8	3	<0.01
Group D *Placebo tablet*						
Baseline	30	13	5	5	7	
End	30	12	6	6	6	NS

Cfu colony forming units, *NS* not significant

Table 1.3 Concentration of inflammatory mediators (pg/ml) (SD) in gingival crevicular fluid at baseline and at designated time intervals during and after use of chewing gums containing *Lactobacillus reuteri*. (Twetman et al. 2009)

Variable/time	A/P (active + placebo) (n = 13, 26 sites)	A/A (active + active) (n = 13, 26 sites)	P/P (placebo + placebo) (n = 13, 24 sites)
IL-1β			
Baseline	10.5 (10.5)	9.0 (7.6)	10.4 (12.3)
1 week	10.1 (9.8)	4.7 (6.9)	13.8 (14.0)
2 week	9.1 (8.9)	4.4 (4.7)	9.3 (10.5)
4 week	10.2 (11.6)	9.8 (11.1)	10.9 (13.3)
TNF-α			
Baseline	0.23 (0.24)	0.41 (0.29)	0.45 (0.52)
1 week	0.32 (0.31)	0.14 (0.15[a])	0.43 (0.49)
2 week	0.41 (0.34)	0.24 (0.26)	0.39 (0.36)
4 week	0.22 (0.19)	0.33 (0.36)	0.40 (0.38)
IL-6			
Baseline	0.92 (1.67)	1.90 (3.28)	1.53 (2.48)
1 week	0.67 (0.68)	0.94 (1.70)	1.60 (2.49)
2 week	0.76 (1.23)	0.95 (1.69)	1.91 (2.50)
4 week	0.99 (1.20)	0.74 (1.56[a])	1.03 (1.87)
IL-8			
Baseline	116.2 (51.8)	103.3 (36.8)	107.3 (69.9)
1 week	102.2 (43.7)	86.2 (54.4)	82.9 (32.9)
2 week	95.2 (60.2)	77.4 (31.8[a])	85.8 (42.0)
4 week	97.7 (69.0)	105.8 (63.8)	106.7 (60.6)
IL-10			
Baseline	0.28 (0.27)	0.24 (0.38)	0.32 (0.44)
1 week	0.32 (0.38)	0.19 (0.24)	0.28 (0.31)
2 week	0.32 (0.38)	0.22 (0.37)	0.24 (0.26)
4 week	0.20 (0.23)	0.19 (0.20)	0.20 (0.20)

[a] Statistically significant difference compared to baseline ($p < 0.05$, Wilcoxon signed rank test)

on the levels of inflammatory mediators in gingival crevicular fluid (Table 1.3) (Twetman et al. 2009).

Further research is required involving double-blind randomized controlled trials in order to bring further benefits of more effective caries control (Zahradnik et al. 2009).

1.6 Advances Made from the Traditional Use to the Modern Application in Medical Field

In the recent years, probiotics, together with prebiotics, met a growing scientific interest in the context of the so-called functional foods and/or nutriceuticals, attractive and imaging names that reveal their important implications for the human health.

The main function of probiotics, evoked by Metnikoff, has been to reduce the putrefactive intestinal flora responsible of, according to the popular theory of the end of nineteenth century, the auto-intoxication and the occurrence of disease due to the release of substances harmful for the body (LaMont 2000). Today, following the recent scientific knowledge, probiotics are considered as useful to promote a mutual beneficial balance between man and his intestinal bacterial microflora, selectively stimulating the growth and/or the metabolic activity of microbial groups, which are important for digestive processes, intestinal wall permeability, resistance to pathogenic invasion, lipid metabolism, intestinal and general mucosal immunity, etc. The improvement of clinical and biochemical intestinal function was also obtained in some pathologic conditions such as cystic fibrosis patients (Infante et al. 2008).

Prebiotics have received a great scientific interest as well as selectively stimulate the growth and/or metabolism of the microbial groups relevant to the host health (bifidobacteria, lactobacilli) and inhibit the growth and activity of pathogenic microorganisms of the proteolytic putrefactive flora, and lead to an increase in survival and colonization of probiotic microorganisms, administered as commercial products. These effects are important because they help to enhance the beneficial action by preventing the mucosal inflammation and the permeability alterations. It should be noted, however, that by their fermentation probiotics prevent both the formation of amines, which are harmful because alter blood circulation and muscle activity and are mutagenic and carcinogenic, especially in cases where there is a liver failure which compromises the detoxification; and the production of phenols, mainly scatole and indole highly carcinogenic; and the secondary bile acids that may promote colon cancer (Fotiadis et al. 2008).

Even the short-chain fatty acids (SCFA) that are formed (acetic, propionic and butyric acid) play an important role because they are excellent nutrients for a better trophism and function of colonic mucosal cells, they play a protective action against inflammatory bowel diseases and a preventive action on the development of colon cancer as well as reducing the proliferation of pathogens and having antiputrefactive properties. All results in a better absorption of nutrient substances versus the toxic ones playing a preventive role with obvious implications for the whole body health (Parracho et al. 2007).

To confirm these data also in preterm and term infants, a formula was supplemented with a prebiotic mixtures of galactose-based prebiotics, long-chain (GOS) and high-molecular weight mixture of inulin-type fructans containing 0.4–1 g/100 ml in a 9:1 ratio, and it was administered for 3–12 weeks, leading to a significant increase in bifidobacteria amount in the infant faecal flora from 20% to approx 60% (breast-fed babies ~80%), and giving some metabolic and clinical changes similar to those caused by breast milk (Moro et al. 2002; Knol et al. 2005; Boehm and Moro 2008; Kelly 2008).

These and many other metabolic and clinical effects on the immune system, on inflammatory processes, on lipid or mineral metabolism, etc., have induced many researchers to administer in various clinical conditions the associations of probiotics and prebiotics, the so-called synbiotics.

Several studies of the recent literature reveals evidence that support the positive impact of synbiotics on intestinal microflora of preterm and term infants and in adult, on immunonutritional parameters and on prevention of eczema, especially the atopic eczema (Rastall and Maitin 2002; Bartosch et al. 2005; Casiraghi et al. 2007; Kukkonen et al. 2007; Uchida et al. 2007; Panigrahi et al. 2008; Underwood et al. 2009). It has been also reported that synbiotics administered to newborn infants seem to increase resistance to respiratory infections during the first 2 years of life, reduce the incidence and severity of respiratory diseases during the cold season, decrease the incidence of septic complications in patients with severe systemic inflammatory response syndrome and are safe (Kukkonen et al. 2008; Pregliasco et al. 2008; Shimizu et al. 2009). Leyer and colleagues reported that also daily dietary probiotic supplementation for 6 months was a safe effective way to reduce fever, rhinorrea, and cough incidence and duration and antibiotic prescription incidence, as well as the number of missed school days attributable to illness, for children 3–5 years of age (Leyer et al. 2009). An interesting reported activity, that must however be confirmed, is the CD remission (Fujimori et al. 2007). Moreover some studies suggest that synbiotic therapy could prove more effective in the treatment of UC than therapies limited to probiotics or prebiotics, C-reactive protein decreased significantly only with synbiotic therapy and patients with UC on synbiotic therapy experienced greater quality-of life-changes than patients on probiotic or prebiotic treatment suggesting that synbiotic therapy may have a synergistic effect in the treatment of UC (Fujimori et al. 2007; Kanauchi et al. 2009; Macfarlane et al. 2009a).

Recent experimental and clinical studies support the fact that, in critically ill patients, early enteral nutrition enriched with synbiotics should restore the balance of microbial communities in a beneficial way with positive effects on intestinal permeability and bacterial translocation and may reduce systemic inflammation, improve the immunological status of the intestinal mucosa and help prevent infections (Manzanares and Hardy 2008). A positive effect of synbiotics has been noted also in multiple trauma patients and in patients with high-risk operations (Rayes et al. 2009).

So it seems clear that while probiotics have to compete with already established bacterial communities, prebiotics have the advantage that they target bacteria already present in the bowel. This makes prebiotics a potentially more efficient and practical way of manipulating the gut microbiota. However, if for any reason the target bacteria are absent from the gut, either due to disease, aging, or antibiotic therapy, then prebiotics alone are not likely to be effective, at least in the short term.

For this reason synbiotics, a combinations of prebiotic and probiotic, may be useful in several conditions and have also a synergistic role. The rationale for their use is that the prebiotic enhances growth of the probiotic component in the gut, giving it a competitive advantage while also stimulating the growth and metabolism of autochthonous microorganisms.

There are reasonable grounds for believing that, besides seeking new and more effective probiotics that it is hoped will be available in the near future but not so easy to obtain, we must not neglect the development of optimal synbiotics that may prove useful in various clinical-pathological conditions.

1.7 Advances Made from the Traditional Use to the Modern Application in Food Industry

In the industrial field a number of food bio-products have been employed or are in progress of being developed to enhance their usage as delivery vehicles of probiotic cell fed humans. Most of these products are of dairy origin and include fresh milk, fermented milk, powdered milk and cheese. Since few years, the food companies involved in the probiotic field following the new demands of consumers promote the researches on novel probiotic foods such as beverages, cookies, ice-cream, dairy dessert, sausages and others.

The possibility of influencing the composition of intestinal microflora by consuming probiotic bacteria partially depends on the dose level. It is generally recognized that 10^8–10^9 bacteria are necessary at the time of consumption. Therefore the probiotic culture must remain viable in the food carrier up to consumption. Shelf-life of a product is defined as the time that the product can be stored, during which the defined quality of a specified proportion of the goods (in case of probiotic, viability of culture) remains acceptable under expected or specified conditions of distribution, storage and display. In some cases, modification of traditional processing protocol is necessary to enhance the viability of probiotic bacteria. In addition to maintaining the viability of probiotic bacteria in the cultured products through the time of the consumption, it is imperative that the incorporation of the probiotic bacteria does not adversely affect the flavour, the texture and other quality attributes of food product.

In recent years some yogurt have been reformulated to include live strains of *L. acidophilus* and species of *Bifidobacterium* (known as AB-culture) in addition to conventional yogurt microorganism, *S. thermophilus* and *L. bulgaricus*. For the production of AB-yogurt, similar processing procedures to traditional yogurt are applied with the exception of incorporation of live probiotic starter cultures. *L. acidophilus* and *B. bifidum* have to retain viability and activity in the food carrier to meet the "therapeutic minimum" at the time of the consumption. Yogurt acidity, strains selection, co-culture and species interaction, inoculation practice, dissolved oxygen and storage conditions are all factors affecting the viability of probiotic species in yogurt (Lourens and Viljoen 2001).

A study from Estonia (Songisepp et al. 2004) tried to develop an original probiotic cheese based on Estonian open-texture, smear-ripened, semisoft cheese "Pikantne". Cheese was produced by two methods using cheese starter cultures in combination with 0.04% of probiotic *L. fermentum* strain ME-3 with high antimicrobial activity and antioxidative properties. The probiotic *Lactobacillus* was added into milk simultaneously with start cultures (cheese A) and into drained curd (cheese B). Cheese A, B, and the control, the original cheese "Pikantne" where no probiotics were added, were described to be of commercial grade with respect to sensory criteria after one month of ripening. Both cheese variants with probiotic additive were found to have flavour and texture comparable to the control cheese. The probiotic strain was found to withstand the technological processing of cheese, surviving and sustaining moderate antimicrobial and high antioxidative activity throughout ripening and storage. The ripened cheese contained approximately 5×10^7 CFU/g viable ME-3 cells but the viability of ME-3 strain incorporated into cheese showed a slight decrease between day 24 and 54 after preparation. Hence semisoft cheese "Pikantne" serves as suitable carrier of *L. fermentum* ME-3.

New researches from Brazil (Aragon-Alegro et al. 2007) suggest that, among chilled foods, chocolate mousse may be a suitable vehicle for probiotics, despite the delicate processes involved in the manufacture. For this study, three mousses have been prepared; one containing no live cultures or prebiotic fibres (control mousse), one containing *L. paracasei*, and the third containing *L. paracasei* plus inulin, a prebiotic fiber.

The product has been monitored for 28 days to assess the population of the probiotic *L. paracasei* as well as contaminants, during storage at +5°C. Data reported that probiotic was still viable after 28 days, maintaining population level about seven log CFU per gram. Moreover in this study detrimental effects, due to probiotic organism presence, were not observed on flavour of final product. Forty-two consumers were recruited to taste and evaluate the three chocolate mousses and no significant differences were reported although the probiotic mousse was considered the most preferable. For this reason the authors concluded that chocolate mousse could be an excellent vehicle for the incorporation of *L. paracasei* strain tested and the prebiotic ingredient inulin not interfere in this viability.

To use as probiotic vehicles, great potential has the ice-cream, it is a relatively innovative matrix for the application of probiotics, with the added advantage of being appreciated by people belonging to all age groups and social levels. However, the development of ice-creams containing probiotic bacteria requires the overcoming of certain technological intrinsic requirements related to their processing stages. Also, it is important to confirm if, after long storage periods, the probiotic cultures are still able to confer the same health benefits already observed in other foods with shorter shelf-lives and higher storage temperatures, such as yogurts and fermented milks (Başyiğit et al. 2006; Cruz et al. 2009).

The commercial application of probiotic microorganisms in fermented sausages is not common yet. There are both advantages and disadvantages connected to fermented meat matrices. They are adequate for the carriage of probiotic bacteria since they are usually not or only mildly heated and may promote the survival of probiotic bacteria in the gastrointestinal tract. In contrast, bacterial viability may be reduced due to the high content in curing salt and the low water activity and pH. Therefore, results are expected to be strain-dependent. Up till now, several approaches have been followed but most results are too preliminary to be able to evaluate the effect of probiotic fermented meats on human health. De Vuyst and colleagues obtained through screening for technological requirements among bacteria that are naturally present in the meat or that originate from meat starter cultures, possible probiotic candidates (De Vuyst et al. 2008). Alternatively, existing probiotic bacteria have been applied in meat products. The probiotic strains have been used not only as starters for fermented meat preparation but also for preservation of salami (Pidcock et al. 2002) from the contamination of *Listeria monocytogenes* and the enterohemorrhagic *E. coli* (EHEC). As one possible mode of action for probiotics is the production of antimicrobial compounds, lactic acid bacteria may act as both probiotic and bioprotective culture as well as fermenting agent in meat product, such as dry sausage (Työppönen et al. 2003). Also the vegetable foodstuff have been used as carriers for probiotic such as artichokes (Valerio et al. 2006) used in an artichoke human feeding study involving four volunteers, and giving good recovering from stool. Even fibers were used as carriers for *L. rhamnosus* during freeze-drying and storage in apple juice and chocolate-coated breakfast cereals (Saarela et al. 2006) giving good opportunity to solve the problems, often present, of preservation of probiotic into particular kinds of foods and during special producing processes. At last there are authors such as Verdenelli et al. (2009), who use a series of probiotic enriched-foods, from typical kinds of Italian cheese to classical dark chocolate bar, as probiotic intake in a volunteers study on colonisation properties of *L. rhamnosus* IMC 501 and *L. paracasei* IMC 502.

However, considering whatever food as carrier for probiotic it is important to guarantee the viability of the probiotic strains during the whole shelf-life of the product, the maintenance of the beneficial effects and, the preservation of nutritional and quality food characteristics. Moreover in order to respect the new dispositions of the European Community, whatever the health claims made directly or indirectly, companies will need to do the human studies to verify them.

The use of probiotics is also widely diffused in feeds for animals, replacing antibiotic treatments for long time widely used, with the aim of both increasing animal development and improving its health status. While numerous studies were conducted on a large variety of animals (Strompfová et al. 2004; Carnevali et al. 2006; Silvi et al. 2008; Taheri 2009), the major consumption happen on chickens, pigs and ruminants. Different compositions of the probiotic feed may be directed to different kinds of animals and the supplied probiotic may be related to the age of farming animal (Holzapfel and Naughton 2005). In the “animal production”, the use of safe and high quality fodders is a concept characterized by close links with the “protection of human life and health.” These fodders are used to promote the nutritional efficiency of the food principles of portion, to regulate the processes of digestion and assimilation and to stimulate and increase specific functional and/or productive performance of the animals. These auxinic substances with production purposes, also known as “growth or performance promoters”, have been represented for many years by antibiotics used at low dosages, from minerals and other substances whose residues or metabolites, however, could find themselves in foodstuffs or in the manure and then into the environment, representing a risk for humans and encouraging the antibiotic resistance phenomenon in pathogens.

For this reason, since the 1st of January 2006 was banned its use throughout the European Community and has gone to an increasing use of “additives without residues”, probiotics and associations between probiotics and prebiotics with the known benefits on the limitation and elimination of pathogens (decreased incidence of diarrhoea and other illnesses and decreased pathogen contamination of livestock products such as meat, eggs, and milk), on the improvement of the intestinal and general health status, on the stimulation of the animal growth and finally on the decreased of the environmental impact of livestock farming improving security since "auxinics without residues". A further aspect particularly interesting and in harmony with very recent researches on microbiology applied in space field, is the use of probiotic bacteria as dietary supplement during space missions, with the aim to improve the astronaut health and to protect them against gastro-intestinal disorders (Canganella 2008).

1.8 Final Considerations

During the last decades, the role of the intestinal microflora in health and disease has become increasingly recognized. Much interest exists in modulating the composition of the gut towards a potentially more beneficial community as probiotics and prebiotics that share a unique role in human nutrition, largely centering on manipulation of populations or activities of the bacteria that colonize our body. In fact, our digestive system is a microbiological-chemical and immunological laboratory in a continuous activity from the earliest periods of fetal life. During the physiological birth, the intestinal tract of foetus, sterile until then, undergoes a protective contamination, initially dominated by facultative anaerobic strains such as *E. coli*. Thereafter, differences exist in the species composition that develops, which is largely governed by the type of diet. In a very short time, however, more than 400–500 bacterial species proliferate and, after the first-second childhood, the microbial complex consists of 1–2 kg of regulatory microorganisms in a state of vital balance for the health. In this balance take part chemical substances such as the digestive secretions, which in adults amount to about 10 l per day and are formed from saliva (about 0.5 l), gastric juices (2.5 l), bile (0.5 l), pancreatic juices (1.5 l), small intestinal/colonic juices (1–5 l). Such secretions, together with the microbiota, obviously influenced by all assumed substances, food or not, are essential for the food digestion and both for the immune system activity, located for about 70% in the gastro-intestinal tract, and for stimulating the immune response, represented by immunoglobulins, transferrin, lysozyme, fibronectin, etc.

The better outcome may be achieved by using targeted dietary supplementation with functional foods, dietary ingredients that have a cellular or physiological effect above basic nutritional value. Recognition of the health-promoting properties of specific commensal microorganisms has encouraged modulation of the human intestinal microflora towards a more beneficial composition and metabolism, by using probiotics, prebiotics and synbiotics.

References

Abrahamsson TR, Jakobsson T, Böttcher MF et al (2007) Probiotics in prevention of IgE-associated eczema: a double-blind, randomised, placebo-controlled trial. J Allergy Clin Immunol 119(5):1174–1180

Aggett PJ, Agostoni C, Axelsson I et al (2003) Non-digestible carbohydrates in the diets of infants and young children: a commentary by the ESPGHAN committee on nutrition. J Ped Gastr Nutr 36(3):329–337

Alfaleh K, Bassler D (2008) Probiotics for prevention of necrotizing enterocolitis in preterm infants. Cochrane Database Syst Rev (1). doi:10.1002/14651858.CD005496.pub2 (CD005496)

Aragon-Alegro LC, Alegro JH, Cardarelli HR et al (2007) Potentially probiotic and synbiotic chocolate mousse. Food Sci Technol 40:669–675

Baffoni L, Biavati B (2008) Ecologia microbica dell'apparato digerente. In: Biavati B, Sorlini C (eds) Microbiologia agroambientale. Casa Editrice Ambrosiana, Milan, pp 147–162

Bartosch S, Woodmansey EJ, Paterson JC et al (2005) Microbiological effects of consuming a synbiotic containing *Bifidobacterium bifidum, Bifidobacterium lactis*, and oligofructose in elderly persons, determined by real-time polymerase chain reaction and counting of viable bacteria. Clin Infect Dis 40(1):28–37

Başyiğit G, Kuleaşan H, Karahan AG (2006) Viability of human-derived probiotic lactobacilli in ice cream produced with sucrose and aspartame. J Ind Microbiol Biotechnol 33:796–800

Belaise C, Romans S, Martin J et al (2002) Sindrome dell'intestino irritabile ed esperienze di abuso. Uno studio epidemiologico. Neurogastroenterologia 8:47–53

Bin-Nun A, Bromiker R, Wilschanski M et al (2005) Oral probiotics prevent necrotizing enterocolitis in very low birth weight neonates. J Pediatr 147:192–196

Boehm G, Moro G (2008) Structural and functional aspects of prebiotics used in infant nutrition. J Nutr 138:1818S–1828S

Bouhnik Y, Raskine L, Simoneau G et al (2006) The capacity of short-chain fructo-oligosaccharides to stimulate faecal bifidobacteria: a dose-response relationship study in healthy humans. Nutr J 28:5–8

Bresciani E (1977) La cultura alimentare degli egiziani antichi. In: Flandrin J-L, Montanari M (eds) Storia dell'alimentazione. Ed. Laterza, Roma-Bari, pp 37–45

Bruzzese E, Volpicelli M, Squeglia V et al (2009) A formula-containing galacto- and fructo-oligosaccharides prevents intestinal an extra-intestinal infections: an observational study. Clin Nutr 28(2):156–161

Butterworth AD, Thomas AG, Akobeng AK (2008) Probiotics for induction of remission in Crohn's disease. Cochrane Database Syst Rev (3). doi:10.1002/14651858.CD006634.pub2 (CD006634)

Caglar E, Cildir SK, Ergeneli S, Sandalli N, Twetman S (2006) Salivary mutans streptococci and lactobacilli levels after ingestion of the probiotic bacterium *Lactobacillus reuteri* ATCC 55730 by straws or tablets. Acta Odontol Scand 64(5):314–318

Canganella F (2008) Microbiologia degli ambienti estremi. In: Biavati B, Sorlini C (eds) Microbiologia agroambientale. Casa Editrice Ambrosiana, Milan, pp 51–87

Caplan MS (2009) Probiotic and prebiotic supplementation for the prevention of neonatal necrotizing enterocolitis. J Perinatol 29(2):S2–S6

Caramia G (2000) L'antibiotico terapia. Un'avventura nella storia della medicina. Ospedale & Territorio 1(1):87–98

Caramia G (2008) Metchnikoff e il centenario dei probiotici: dall'intuizione alla scienza. Ped Med Chir 30:215–219

Carnevali O, de Vivo L, Sulpizio R et al (2006) Growth improvement by probiotic in European sea bass juveniles (*Dicentrarchus labrax*, L.) with particular attention to IGF-1, myostatin and cortisol gene expression. Aquaculture 258:430–438

Carre C (1887) Ueber Antagonisten unter den Bacterien. Corresp-Blatt Schweiz Aerzte 17:385–392

Casiraghi MC, Canzi E, Zanchi R et al (2007) Effects of a synbiotic milk product on human intestinal ecosystem. J Appl Microbiol 103:499–506

Cimperman L, Best K, Oster M et al (2009) A randomized, double-blind, placebo-controlled pilot study of *Lactobacillus reuteri* for the prevention of antibiotic-associated diarrhoea in hospitalised adults. J Parenter Enteral Nutr 33:229 (abstr SP-31)

Coudeyras S, Jugie G, Vermerie M et al (2008) Adhesion of human probiotic *Lactobacillus rhamnosus* to cervical and vaginal cells and interaction with vaginosis-associated pathogens. Infect Dis Obstet Gynecol 2008;549–640

Cruz AG, Antunes AEC, Sousa ALOP et al (2009) Ice-cream as a probiotic carrier. Food Res Int 42(9):1233–1239

Dani C, Biadaioli R, Bertini G et al (2002) Probiotics feeding in prevention of urinary tract infection, bacterial sepsis and necrotizing enterocolitis in preterm infants. Biol Neonate 82:103–108

Deshpande G, Rao S, Patole S (2007) Probiotics for prevention of necrotising enterocolitis in preterm neonates with very low birth weight: a systematic review of randomised controlled trials. Lancet 369:1614–1620

de Vrese M, Schrezenmeir J (2008) Probiotics, prebiotics, and synbiotics. Adv Biochem Eng Biotechnol 111:1–66

De Vuyst I, Falony G, Leroy F (2008) Probiotics in fermented sausages. Meat Sci 80:75–78

Doron SI, Hibberd PL, Gorbach SL (2008) Probiotics for prevention of antibiotic-associated diarrhoea. J Clin Gastroenterol 42(2):S58–S63

Eom TH, Oh EY, Kim YH et al (2005) The therapeutic effect of *Lactobacillus reuteri* in acute diarrhoea in infants and toddlers. Korean J Pediatr 48:986–990

FAO/WHO (2001) Health and nutritional properties of probiotics in food including powder milk with live lactic acid bacteria. Report of a Joint FAO/WHO expert consultation on evaluation of health and nutritional properties of probiotics in food including powder milk with live lactic acid bacteria

Flandrin JL, Montanari M (1977) Storia dell'alimentazione. Edizioni Laterza, Roma-Bari

Floch MH, Walker WA, Guandalini S et al (2008) Recommendations for probiotic use. J Clin Gastroenterol 42(2):S104–S108

Fotiadis CI, Stoidis CN, Spyropoulos BG et al (2008) Role of probiotics, prebiotics and synbiotics in chemoprevention for colorectal cancer. World J Gastroenterol 14(42):6453–6457

Francavilla R, Lionetti E, Castellaneta SP et al (2008) Inhibition of *Helicobacter pylori* infection in humans by *Lactobacillus reuteri* ATCC 55730 and effect on eradication therapy: a pilot study. Helicobacter 13(2):127–134

Fujimori S, Gudis K, Mitsui K et al (2009) A randomised controlled trial on the efficacy of synbiotic versus probiotic or prebiotic treatment to improve the quality of life in patients with ulcerative colitis. Nutrition 25(5):520–525

Fujimori S, Tatsuguchi A, Gudis K et al (2007) High dose probiotic and prebiotic cotherapy for remission induction of active Crohn's disease. J Gastroenterol Hepatol 22:1199–1204

Fuller R (1989) Probiotics in man and animals. J Appl Bacteriol 66:365–378

García-Godoy F, Hicks MJ (2008) Maintaining the integrity of the enamel surface: the role of dental biofilm, saliva and preventive agents in enamel demineralisation and remineralisation. J Am Dent Assoc 139:25S–34S

Gibson GR, Probert HM, Van Loo JAE et al (2004) Dietary modulation of the human colonic microbiota: updating the concept of prebiotics. Nutr Res Rev 17:257–259

Gibson GR, Roberfroid MB (1995) Dietary modulation of the human colonic microbiota: introducing the concept of prebiotics. J Nutr 125(6):1401–1412

Gionchetti P, Rizzello F, Helwig U et al (2003) Prophylaxis of pouchitis onset with probiotic therapy: a double-blind, placebo-controlled trial. Gastroenterology 124:1202–1209

Gionchetti P, Rizzello F, Morselli C et al (2007) High-dose probiotics for the treatment of active pouchitis. Dis Colon Rectum 50(12):2075–2082

Gromert N, Axelsson I (2009) Dietary supplementation with *Lactobacillus reuteri* ATCC 55730 and its effect on atopic eczema in childhood. *Espghan.* 42nd European Society for Pediatric Gastroenterology Hepatology and Nutrition

Guandalini S (2008) Probiotics for children with diarrhoea. An Update. J Clin Gastroenterol 42:S53–S57

Guarino A, Albano F, Ashkenazi S et al (2008) ESPGHAN/ESPID evidence-based guidelines for the management of acute gastroenteritis in children in Europe expert working group. European society for paediatric gastroenterology, hepatology, and nutrition/European society for paedi-

atric infectious diseases evidence-based guidelines for the management of acute gastroenteritis in children in Europe: executive summary. J Pediatr Gastroenterol Nutr 46:619–621
Guarino A, Vecchio AL, Canani RB (2009) Probiotics as prevention and treatment for diarrhoea. Curr Opin Gastroenterol 25:18–23
Guslandi M (2008) Prevention of traveler's diarrhoea with probiotics. J Clin Gastroenterol 42(9):1066
Heilpern D, Szilagyi A (2008) Manipulation of intestinal microbial flora for therapeutic benefit in inflammatory bowel diseases: review of clinical trials of probiotics, prebiotics and synbiotics. Rev Recent Clin Trials 3:167–184
Hillman JD, McDonell E, Cramm T et al (2009) A spontaneous lactate dehydrogenase deficient mutant of Streptococcus rattus for use as a probiotic in the prevention of dental caries. J Appl Microbiol Apr 24. [Epub ahead of print]
Holzapfel WH, Naughton PJ (2005) Microbial ecology in growing animals. Elsevier, China
Hookman P, Barkin JS (2009) *Clostridium difficile* associated infection, diarrhoea and colitis. World J Gastroenterol 15(13):1554–1580
Huis in't Veld JH, Havenaar R, Marteau P (1994) Establishing a scientific basis for probiotic R&D. Trends Biotechnol 12:6–8
Huynh HQ, deBruyn J, Guan L et al (2009) Probiotic preparation VSL#3 induces remission in children with mild to moderate acute ulcerative colitis: a pilot study. Inflamm Bowel Dis 15(5):760–768
Infante PD, Redecillas FS, Torrent VA, Segarra CO et al (2008) Improvement of intestinal function in cystic fibrosis patients using probiotics. An Pediatr (Barc) 69(6):501–505
Islam B, Khan SN, Khan AU (2007) Dental caries: from infection to prevention. Med Sci Monit 13(11):RA196–RA203
Jarosz M, Rychlik E, Siuba M et al (2009) Dietary and socio-economic factors in relation to *Helicobacter pylori* re-infection. World J Gastroenterol 15(9):1119–1125
Ji GE (2009) Probiotics in primary prevention of atopic dermatitis. Forum Nutr 61:117–128
Jiménez MB (2009) Treatment of irritable bowel syndrome with probiotics. An etiopathogenic approach at last? Rev Esp Enferm Dig 101(8):553–564
Johnson AM, Kaushik RS, Rotella NJ et al (2009) Enterotoxigenic *Escherichia coli* modulates host intestinal cell membrane asymmetry and metabolic activity. Infect Immun 77:341–347
Kanamori Y, Sugiyama M, Hashizume K et al (2004) Experience of long-term synbiotic therapy in seven short bowel patients with refractory enterocolitis. J Pediatr Surg 39(11):1686–1692
Kanauchi O, Mitsuyama K, Andoh A (2009) The therapeutic impact of manipulating microbiota in inflammatory bowel disease. Curr Pharm Des 15(18):2074–2086
Kelly CP, LaMont JT (2008) *Clostridium difficile*—more difficult than ever. N Engl J Med 359:1932–1940
Kelly G (2008) Inulin-type prebiotics—a review: Part I. Altern Med Rev 13(4):315–329
Kelly G (2009) Inulin-type prebiotics—a review: Part II. Altern Med Rev 14(1):36–55
Kligler B, Cohrssen A (2008) Probiotics. Am Fam Physician 78:1073–1078
Knol J, Scholtens P, Kafka C et al (2005) Colon microflora in infants fed formula with galacto- and fructo-oligosaccharides: more like breast-fed infants. J Pediatr Gastroenterol Nutr 40(1):36–42
Kolida S, Gibson GR (2007) Prebiotic capacity of inulin-type fructans. J Nutr 137 (11 Suppl):2503S–2506S
Kollath W (1953) The increase of the diseases of civilization and their prevention. Munch Med Wochenschr 95:1260–1262
Kopp MV, Salfeld P (2009) Probiotics and prevention of allergic disease. Curr Opin Clin Nutr Metab Care 12(3):298–303
Kukkonen K, Savilahti E, Haahtela T et al (2007) Probiotics and prebiotic galacto-oligosaccharides in the prevention of allergic diseases: a randomised, double-blind, placebo-controlled trial. J Allergy Clin Immunol 119(1):192–198
Kukkonen K, Savilahti E, Haahtela T et al (2008) Long-term safety and impact on infection rates of postnatal probiotic and prebiotic (synbiotic) treatment: randomised, double-blind, placebo-controlled trial. Pediatrics 122(1):8–12

Kunze WA, Mao YK, Wang B et al (2009) *Lactobacillus reuteri* enhances excitability of colonic AH neurons by inhibiting calcium dependent potassium channel opening. J Cell Mol Med 13(8B):2261–2270. doi:10.1111/j.1582-4934.2009.00686 x

Kuramitsu HK, He X, Lux R et al (2007) Interspecies interactions within oral microbial communities. Microbiol Mol Biol Rev 71(4):653–670

LaMont JT (2000) The renaissance of probiotics and prebiotics. Gastroenterology 19:291

Leyer GJ, Li S, Mubasher ME et al (2009) Probiotic effects on cold and influenza-like symptom incidence and duration in children. Pediatrics 124(2):172–179

Lilly DM, Stillwell RH (1965) Probiotics: growth-promoting factors produced by microorganisms. Science 147:747–748

Lin HC, Hsu CH, Chen HL et al (2008) Oral probiotics prevent necrotizing enterocolitis in very low birth weight preterm infants: a multicenter, randomised, controlled trial. Pediatrics 122:693–700

Lionetti E, Miniello VL, Castellaneta SP et al (2006) *Lactobacillus reuteri* therapy to reduce side-effects during anti-*Helicobacter pylori* treatment in children: a randomised placebo controller trial. Alim Pharmacol Ther 24:1461–1468

Loftus EJ (2004) Clinical epidemiology of inflammatory bowel disease: incidence, prevalence and environmental influences. Gastroenterology 126:1504–1517

Lourens HA, Viljoen BC (2001) Yogurt as probiotic carrier food. Int Dairy J 11:1–17

Macfarlane GT, Blackett KL, Nakayama T et al (2009a) The gut microbiota in inflammatory bowel disease. Curr Pharm Des 15(13):1528–1536

Macfarlane S, Steed H, Macfarlane GT (2009b) Intestinal bacteria and inflammatory bowel disease. Crit Rev Clin Lab Sci 46(1):25–54

Manzanares W, Hardy G (2008) The role of prebiotics and synbiotics in critically ill patients. Curr Opin Clin Nutr Metab Care 11(6):782–789

Marcone V, Calzolari E, Bertini M (2008) Effectiveness of vaginal administration of *Lactobacillus rhamnosus* following conventional metronidazole therapy: how to lower the rate of bacterial vaginosis recurrences. New Microbiol 31(3):429–433

Martinez RC, Franceschini SA, Patta MC et al (2009a) Improved cure of bacterial vaginosis with single dose of tinidazole (2 g), *Lactobacillus rhamnosus* GR-1, and *Lactobacillus reuteri* RC-14: a randomized, double-blind, placebo-controlled trial. Can J Microbiol 55(2):133–138

Martinez RC, Franceschini SA, Patta MC et al (2009b) Improved treatment of vulvovaginal candidiasis with fluconazole plus probiotic *Lactobacillus rhamnosus* GR-1 and *Lactobacillus reuteri* RC-14. Lett Appl Microbiol 48(3):269–274

Mättö J, Maunuksela L, Kajander K et al (2005) Composition and temporal stability of gastrointestinal microbiota in irritable bowel syndrome—a longitudinal study in IBS and control subjects. FEMS Immunol Med Microbiol 43:213–222

McFarland LV (2007) Meta-analysis of probiotics for the prevention of traveller's diarrhoea. Travel Med Infect Dis 5:97–105

McFarland LV (2009) Renewed interest in a difficult disease: *Clostridium difficile* infections—epidemiology and current treatment strategies. Curr Opin Gastroenterol 25:24–35

Metchnikoff E (1907) The prolongation of life. Optimistic studies. William Heinemann, London, pp 1–38

Miele E, Pascarella F, Giannetti E et al (2009) Effect of a probiotic preparation (VSL#3) on induction and maintenance of remission in children with ulcerative colitis. Am J Gastroenterol 104(2):437–443

Moro G, Minoli I, Mosca M et al (2002) Dosage related bifidogenic effects of galacto- and fructo-oligosaccharides in formula-fed term infants. J Pediatr Gastroenterol Nutr 34:291–295

Panigrahi P, Parida S, Pradhan L et al (2008) Long-term colonization of a *Lactobacillus plantarum* synbiotic preparation in the neonatal gut. J Pediatr Gastroenterol Nutr 47(1):45–53

Parker RB (1974) The other half of the antibiotic story. Anim Nut Health 29:4–8

Parracho H, McCartney AL, Gibson GR (2007) Probiotics and prebiotics in infant nutrition. Proc Nutr Soc 66(3):405–411

Perles C (1977) Le strategie alimentari nella preistoria In: Flandrin J-L, Montanari M (eds) Storia dell'alimentazione. Ed. Laterza, Roma-Bari, pp 12–25

Pham M, Lemberg DA, Day AS (2008) Probiotics: sorting the evidence from the myths. Med J Aust 188:304–308

Pidcock K, Heard GM, Henriksson A (2002) Application of non-traditional meat starter cultures in production of Hungarian salami. Int J Food Microbol 76:75–81

Pregliasco F, Anselmi G, Fonte L et al (2008) A new chance of preventing winter diseases by the administration of symbiotic formulations. J Clin Gastroenterol 42(3):S224–S233

Rastall RA, Maitin V (2002) Prebiotics and synbiotics: towards the next generation. Curr Opin Biotechnol 13:490–496

Rayes N, Seehofer D, Neuhaus P (2009) Prebiotics, probiotics, synbiotics in surgery—are they only trendy, truly effective or even dangerous? Langenbecks Arch Surg 394(3):547–555

Roberfroid M (2007) Prebiotics: the concept revisited. J. Nutr 137:830S–837S

Saarela M, Virkajärvi I, Nohynek L et al (2006) Fibres as carriers for *Lactobacillus rhamnosus* during freeze-drying and storage in apple juice and chocolate-coated breakfast cereals. Int J Food Microbiol 112:171–178

Sabbi T, De Angelis P, Dall'Oglio L (2008) *Helicobacter pylori* infection in children: management and pharmacotherapy. Expert Opin Pharmacother 9:577–585

Sachdeva A, Nagpal J (2009) Effect of fermented milk-based probiotic preparations on *Helicobacter pylori* eradication: a systematic review and meta-analysis of randomised-controlled trials. Eur J Gastroenterol Hepatol 21:45–53

Saggioro A, Caroli M, Girardi L et al (2005) *Helicobacter pylori* eradication with *Lactobacillus reuteri*. A double-blind placebo-controlled study. Abstract presented at SIGE Congress, 12–16 March 2005

Selgrad M, Malfertheiner P (2008) New strategies for *Helicobacter pylori* eradication. Curr Opin Pharmacol 8:593–597

Shimizu K, Ogura H, Goto M et al (2009) Synbiotics decrease the incidence of septic complications in patients with severe SIRS: a preliminary report. Dig Dis Sci 54(5):1071–1078

Shornikova AV, Casas IA, Isolauri E et al (1997a) *Lactobacillus reuteri* as a therapeutic agent in acute diarrhoea in young children. J Pediatr Gastroenterol Nutr 24(4):399–404

Shornikova AV, Casas IA, Mykkänen H et al (1997b) Bacteriotherapy with *Lactobacillus reuteri* in rotavirus gastroenteritis. Pediatr Infect Dis J 16(12):1103–1107

Silvi S, Nardi M, Sulpizio R et al (2008) Effect of the addition of *Lactobacillus delbrueckii* subsp. *delbrueckii* on the gut microbiota composition and contribution to the well-being of European sea bass (*Dicentrarchus labrax*, L.). Microb Ecol Health Dis 20:53–59

Songisepp E, Kullisaar T, Hutt P et al (2004) A new probiotic cheese with antioxidative and antimicrobial activity. J Dairy Sci 87:2017–2023

Strompfová V, Lauková A, Ouwehand AC (2004) Selection of enterococci for potential canine probiotic additives. Vet Microbiol 100(1–2):107–114

Surawicz CM (2008) Role of probiotics in antibiotic associated diarrhoea, *Clostridium difficile* associated diarrhoea and recurrent *Clostridium difficile* diarrhoea. J Clin Gastroenterol 42:S64–S70

Taheri HR, Moravej H, Tabandeh F et al (2009) Screening of lactic acid probiotic bacteria toward their selection as a source of selective chicken probiotics. Poult Sci 88(8):1586–1593

Takahashi O, Noguchi Y, Omata F et al (2007) Probiotics in the prevention of traveller's diarrhoea: meta-analysis. J Clin Gastroenterol 41:336–337

Tissier H (1906) Tratement des infections intestinales par la méthode de la flore bactérienne de l'intestin. Crit Rev Soc Biol 60:359–361

Tong JL, Ran ZH, Shen J et al (2007) Meta-analysis: the effect of supplementation with probiotics on eradication rates and adverse events during *Helicobacter pylori* eradication therapy. Aliment Pharmacol Ther 25:155–168

Tormo R, Polanco I, Salazar-Lindo E et al (2008) Acute infectious diarrhoea in children: new insights in antisecretory treatment with racecadotril. Acta Paediatr 97:1008–1015

Twetman S, Derawi B, Keller M et al (2009) Short-term effect of chewing gums containing probiotic *Lactobacillus reuteri* on the levels of inflammatory mediators in gingival crevicular fluid. Acta Odontol Scand 67(1):19–24

Työppönen S, Petäjä E, Mattila-Sandholm T (2003) Bioprotectives and probiotics for dry sausages. Int J of Food Microbiol 83:233–244

Uchida K, Takahashi T, Inoue M et al (2007) Immunonutritional effects during synbiotics therapy in paediatric patients with short bowel syndrome. Pediatr Surg Int 23(3):243–248

Underwood MA, Salzman NH, Bennett SH et al (2009) A randomised placebo-controlled comparison of 2 prebiotic/probiotic combinations in preterm infants: impact on weight gain, intestinal microbiota, and faecal short-chain fatty acids. J Pediatr Gastroenterol Nutr 48(2):216–225

Valerio F, De Bellis P, Lonigro SL et al (2006) *In vitro* and *in vivo* survival and transit tolerance of potentially probiotic strains carried by artichoke in the gastrointestinal tract. Appl Environ Microbiol 72(4):3042–3045

Van Assche G, Vermeire S, Rutgeerts P (2008) Treatment of severe steroid refractory ulcerative colitis. World J Gastroenterol 14(36):5508–5511

Vandenplas Y, Brunser O, Szajewska H (2009) *Saccharomyces boulardii* in childhood. Eur J Pediatr 168(3):253–265

van der Aa LB, Heymans HS, van Aalderen WM et al (2009) Probiotics and prebiotics in atopic dermatitis: review of the theoretical background and clinical evidence. Pediatr Allergy Immunol. July 2. [Epub ahead of print]

Verdenelli MC, Ghelfi F, Silvi S et al (2009) Probiotic properties of *Lactobacillus rhamnosus* and *Lactobacillus paracasei* isolated from human faeces. Eur J Nutr 48:355–363

Vergin F (1954) Antibiotics and probiotics. Hippokrates 25:116–119

von Hertzen LC, Savolainen J, Hannuksela M et al (2009) Scientific rationale for the finnish allergy programme 2008–2018: emphasis on prevention and endorsing tolerance. Allergy 64(5):678–701

Wallace B. (2009) Clinical use of probiotics in the paediatric population. Nutr Clin Pract 24(1):50–59

Yangco BG, Sher G, Bardin MC (2009) Nitrazoxanide and probiotics for the treatment of recurrent *Clostridium difficile* infection in a peritoneal dialysis patient. South Med J 102(7):746–747

Zahradnik RT, Magnusson I, Walker C et al (2009) Preliminary assessment of safety and effectiveness in humans of ProBiora(3), a probiotic mouthwash. J Appl Microbiol 107(2):682–690

Part II
The Gut Microorganisms and Probiotics

Chapter 2
The Intestinal Microbiota and Probiotics

Sofia D. Forssten, Sampo J. Lahtinen and Arthur C. Ouwehand

2.1 Introduction

The human gastrointestinal (GI) tract consists of the upper and lower GI tracts. The stomach and the small and large intestines are the main organs of the GI tract, which is about 7 m long in adults. Approximately a total of 10^{14} bacteria (10^{11} cells per gram of faeces) are estimated to be present in the gut microbiota of an adult individual (Savage 1977). The microbial richness is maximal in the colon, where hundreds of different bacterial species present contain 100-fold more genes than the human genome. It is estimated that the microbial population within a human outnumbers the human cells by a factor of ten (O'Hara and Shanahan 2006). Most of the bacterial taxa present in the GI tract have not yet been successfully cultured, identified, or otherwise characterized (Hayashi et al. 2002). Normally, the majority of bacteria in the GI tract are beneficial or harmless to the host health, while only a minority represents potentially harmful bacteria. The GI microbiota is a complex system that has a major influence on human health. It is known to contribute to e.g. maturation of the gut, nutrition of the host, resistance to pathogens and the maintenance of host health (Stecher and Hardt 2008).

2.2 Intestinal Microbiota in Newborns, Children, Adults and Elderly

The microbiota in the GI tract undergoes changes throughout life. The foetus is sterile, but during delivery it is colonized by bacteria from the birth canal, and subsequently exposed to bacteria from other humans as well as the surroundings. The upper GI tract consisting of the stomach, duodenum, jejunum, and upper ileum

S. D. Forssten (✉)
Danisco Health & Nutrition, Sokeritehtaantie 20, 02460 Kantvik, Finland
Fax: +358 9 298 2203
e-mail: sofia forssten@danisco.com

J. J. Malago et al. (eds.), *Probiotic Bacteria and Enteric Infections*,
DOI 10.1007/978-94-007-0386-5_2, © Springer Science+Business Media B.V. 2011

contains a sparse microbiota with bacterial concentrations less than 10^4 organisms per ml of digesta. Most of the bacteria in the upper GI tract are transient species that pass through the gut via food. The microbial concentrations increase towards the end of the small intestine, reaching levels ranging between 10^7 and 10^8 cell per gram, while concentrations of 10^{11} bacteria per gram of stool are present in the large intestine. Humans have to some extent a stable microbiota, and the host's individuality has the largest effect on the structure of the microbial community. The stable part of the intestinal microbiota is referred to as the individual core microbiota (Zoetendal et al. 2008). It is also known that age-related changes occur in the abundance of GI bacteria in healthy humans.

Directly after a vaginal delivery, the sterile upper GI tract of the newborn begins to be colonized by vaginal and faecal microbiota of the mother. Infants that are born by caesarean section are exposed to the microbiota of the mother. However, the primary microbial exposure of infants delivered by caesarean section probably originates from the surrounding environment, for example from the air and from the nursing staff (Schwiertz et al. 2003). For vaginally born infants it takes about one month before their primary intestinal microbiota is established, while for infants born by cesarean delivery the GI microbiota may be disturbed for up to 6 months post-birth. After birth, oral and cutaneous bacteria are transferred from the mother to the infant but also bacteria from the nosocomial environment. The initial bacterial count is rather unstable during the first week after birth but is thereafter in most infants in the range of 10^9–10^{10} per gram of faeces (Mackie et al. 1999; Palmer et al. 2007; Schwiertz et al. 2003). The first colonizers are aerobes such as enterobacteria and streptococci, while strict anaerobes such as eubacteria and clostridia represent the late colonizers (Favier et al. 2002; Orrhage and Nord 1999; Palmer et al. 2007; Park et al. 2005). The bacterial colonization seems to be host-specific (Mackie et al. 1999) and has been shown to fluctuate; *Bacteroides* might dominate the microbiota in some babies, and be nearly absent in others. Other taxa like *Prevotella, Acinetobacter, Desulfovibrio* and *Veillonella* tend to appear transiently within the first year of life of a baby. Lactobacilli have been shown to reach a peak at the age of 6 (Adlerberth et al. 2010) months and *Lactobacillus rhamnosus* and *Lactobacillus gasseri* being among the most common species (Ahrne et al. 2005; Mitsou et al. 2008).

Differences in the species composition of the GI microbiota of formula-fed infants and breast-fed infants have been detected. The GI microbiota of breast-fed babies is dominated by bifidobacteria and lactobacilli (Chen et al. 2007; Orrhage and Nord 1999; Satokari et al. 2002), while the microbiota of formula-fed infants is more diverse with high numbers of *Enterobacteriaceae*, enterococci, bifidobacteria, *Bacteroides*, and clostridia (Fanaro et al. 2003; Harmsen et al. 2000). Staphylococci on the other hand are recognized as characteristic first-colonizers of the newborn gut, regardless of the mode of delivery (Adlerberth et al. 2010). Differences between formula-fed infants and breast-fed infants may also exist at species-level, for example among *Bifidobacterium* spp. The microbiota of breast-fed infants becomes similar to that of formula-fed infants after introduction of solid food and weaning.

When a child is about two years old the faecal microbiota is similar to the microbiota of adults. The majority of the microbiota is believed to be obligate anaerobes, and dominated by species from the phylum *Firmicutes* (mostly *Clostridium*; 50–70% of total bacterial count), *Bacteroidetes* (10–30%), *Proteobacteria* (≤10%), and *Actinobacterium* (≤5%) (Eckburg et al. 2005; Lay et al. 2005). Regarding the microbial concentrations in children, it was reported by Enck and co-workers that the total microbial count and individual microbial species were highest during the first year of life. In infants, the total microbial count rose during the first five months and then declined with weaning, and peaked at the age of 9–10 months. The microbial count then decreased within the first two years, while remaining stabilized for the rest of childhood. *Bacteroides* sp. and *Lactobacilli* increased with age, while *Enterococci* and *Escherichia coli* decreased, and bifidobacteria remained stable (Enck et al. 2009). Geographical variations in the composition of the human gastrointestinal microbiota can be seen. For example, considerable differences in the composition of species in Estonian and Swedish one-year-old infants have been reported (Sepp et al. 1997). While the Estonian children were intensively colonized with lactobacilli and eubacteria, the Swedish infants had increased numbers of clostridia, particularly *Clostridium difficile*, as well as bacteroides and other anaerobes.

In adults, aerotolerant bacteria such as lactobacilli and streptococci dominate the small intestinal microbiota, while the microbiota is more complex in the junction (the ileocaecal valve) of the small intestine (ileum) and the large intestine. In the large intestine, anaerobes like bifidobacteria, *Bacteroides, Clostridium* (clusters XIVa and IV), *Eubacterium,* and *Ruminococcus* are dominant. Facultatives which principally include the *Enterobacteriaceae* family (e.g. *E. coli, Klebsiella* and *Enterobacter*) and enterococci represent less than 1% of the total microbiota (Finegold et al. 1983; Gavini et al. 2001). Low numbers (10^3–10^6 cells per gram) of transient staphylococci can be found (Finegold et al. 1983). Fungi and yeasts such as *Candida albicans* can also be isolated from healthy individuals, but in low numbers.

In a microarray study of the faecal microbiota, clostridia and *Bacteroidetes* were the major classes identified (Paliy et al. 2009). In addition, the gut microbiota of healthy adults contained more clostridia and less *Bacteroidetes* and *Proteobacteria* compared to children.

The composition of the intestinal microbiota in elderly is changed by factors related to ageing, like mobility, nutrition and reduced intestinal functionality. Changes of the small bowel motor pattern, together with a decreased degree of mobility may influence the motility of the bowel and have a negative effect on digestion, causing constipation, and may hence be associated with changes in the gut microbiota (Holdeman et al. 1976). Furthermore, reduced secretion of saliva, gastric and pancreatic juice may lead to an increased influx of exogenous bacteria and nutrients through the digestive tract. A decrease in bifidobacteria counts is one of the most commonly mentioned changes in the composition of the intestinal microbiota of elderly, together with the shifts observed in the genus *Bacteroides* (Gavini et al. 2001; Hopkins et al. 2001; Woodmansey et al. 2004). In particular in studies in which culture based techniques have been used, the faecal *Bifidobacterium* levels seem to be reduced compared to younger adults. However, studies that have used culture based techniques do often not observe this reduction. This discrepancy can be explained by a difference in composition of the bifidobacterial microbiota rather than a difference in numbers (Lahtinen et al. 2009; Tiihonen et al. 2010). Whilst the genus *Bifidobacterium* may contain up to four or five different species continuously in adult humans, the diversity of bifidobacterial species in the elderly is reduced to one or two dominant organisms. *Bifidobacterium adolescentis* or the phenotypically similar *Bifidobacterium angulatum* and *Bifidobacterium longum* have been reported to be the dominant species in elderly (Gavini et al. 2001; Hopkins et al. 2001; Hopkins and Macfarlane 2002). A high individual variation of *Bacteroides* species in the elderly has been reported, although *Bacteroides thetaiotaomicron* can be found in all individuals (Layton et al. 2006). A decrease in the diversity of *Bacteroides* species has been reported with an increasing age (Bartosch et al. 2004; Hopkins and Macfarlane 2002). Higher *Bacteroides* levels have been detected in males than females (Mueller et al. 2006). *Clostridium* counts have been found to be significantly higher in elderly compared with younger subjects. Conversely, levels of *Clostridium* cluster XIVa seem to decrease in elderly compared to younger persons (Hayashi et al. 2003). Furthermore, *C. difficile* is more frequently isolated from elderly than younger adults, partly as a result of hospitalization (Ljungberg et al. 1990). An increased diversity of *Eubacteria* (Hopkins and Macfarlane 2002) as well as an increase of some eubacteria that are phylogenetically related to clostridia (Woodmansey et al. 2004) have also been observed in elderly.

Lower levels of *Lactobacillus* have also been observed in elderly subjects compared to healthy adults (Hopkins et al. 2001). In a study by Mitsuoka et al. (1996), the levels of *Bacteroides*, enterococci, *Enterobacteriaceae* and clostridia in elderly were not different from adults, while an increase of *Lactobacillus* and *Enterococci* was detected. Another study reported higher levels of *Ruminococcus* and *Enterobacteriaceae*, as well as lower levels of *Eubacterium* and *Bacteroides* in elderly compared to younger adults (He et al. 2001). A higher level of aerobes in the microbiota of elderly than among the microbiota of adults has also been observed (Guigoz et al. 2008; Tiihonen et al. 2008). Similarly, a higher frequency of gammaproteobacteria, such as *Klebsiella* spp. was found in a study by Hayashi et al. (2003). In a study of elderly aged 70–100 years (with a median of 86 years) by van Tongeren et al. (2005) the prevalence of *Bacteroides/Prevotella, Eubacterium rectale/Clostridium coccoides*, and *Ruminococcus* were rather similar to the levels in healthy adults aged between 20 and 55 years in a study by Harmsen et al. (2002). However differences in the composition of the faecal microbiota can be observed between e.g. low and high frailty elderly (van Tongeren et al. 2005). Of the most predominant bacterial groups, the percentages of *Bacteroides/Prevotella* (from 11.0% to 4.5%), *E. rectale/C. coccoides* (from 10.6% to 6.7%) and *Faecalibacterium prausnitzii* (from 1.2% to 0.3%) decreased in the high-frailty elderly, while *Ruminococcus* (from 6.3% to 12.7%) and *Atopobium* (from 2.1% to 4.3%) increased in the high-frailty volunteers.

Through the use of molecular analysis such as rDNA gene based analysis in addition to conventional culture-based methods, more detailed information on both cultured and uncultured microorganisms can be obtained. Depending on the method, 10–50% of the microbes in the human GI tract have been reported to be cultivable (Lay et al. 2005; Zoetendal et al. 2004). With the current large research projects on human microbiome, in which include metagenomic and genomic DNA sequencing studies of human-associated microbial populations, it is possible to investigate whether most individuals share similar core human microbiome. In addition, changes in the human microbiome and correlations with changes in human health can be recognized and understood.

Thus far, it has been shown that the microbiota of the GI differs substantially between individuals at species level. Recent studies suggest that at the level of microorganisms, it appears that there is no core microbiome that is shared between most individuals; Turnbaugh and co-workers reported that no single bacterial phylotype was present at abundant frequency in all of the 154 sampled individuals within the study (Turnbaugh et al. 2009). Instead, a core gut microbiome may exist at the level of shared genes. By analyzing 16S rRNA the microbiota has been reported to be quite stable over time of an individual with only minor changes occurring over time (Zoetendal et al. 1998). However, since the stability has been examined from samples collected within weeks, months or at most few years, the variability over a longer timescale is still unknown (Franks et al. 1998; Lofmark et al. 2006; Matto et al. 2005; Nyberg et al. 2007; Vanhoutte et al. 2004; Zoetendal et al. 1998). The microbiota responds to physiological, dietary, and environmental factors (Bartosch et al. 2004), and the response has an effect on the diversity and the individuality of the intestinal microbiota.

2.3 Environment Factors

In addition to intrinsic factors, such as aging and non-infectious diseases, also external events can influence the composition and activity of the intestinal microbiota. As discussed above, a well known example of this is the difference in microbiota composition between breast fed and formula fed infants, although continuous product development of infant formulas makes the difference smaller (Boehm et al. 2005).

Major environmental factors affecting the composition and activity of the intestinal microbiota are diet; e.g. consumption of pre- and/or probiotics, stress, infectious diseases and medication.

2.3.1 The Influence of Prebiotics on the Intestinal Microbiota

Prebiotics are defined as 'non-digestible food ingredients that, when consumed in sufficient amounts, selectively stimulate the growth and/or activity of one or a limited number of microbes in the colon resulting in documented health benefits' (Ouwehand et al. 2006). Updates of the definition have been suggested; but the essence of prebiotics remains the same. Although prebiotics have much in com-

Table 2.1 Examples of prebiotic and proposed prebiotic dietary components

	Monomer	Degree of polymerisation	Linkage
Galacto-oligosaccharides (GOS)	Glucose-Galactose$_n$, Galactose$_n$	1–4	β(1→6); β(1→3); β(1→4)
Fructo-oligosaccharides (FOS)	Glucose-Fructose$_n$, Fructose$_n$	1–9	β(2→1)
Inulin	Glucose-Fructose$_n$	10–60	β(2→1)
Xylo-oligosaccharides (XOS)	Xylose	1–8	β(1→4)
Lactitol	Galactose-Glucitol	2	β(1→4)
Lactulose	Galactose-Fructose	2	β(1→4)
Polydextrose	(Sorbitol)-Glucose	2–30	β(1→6); β(1→4); β(1→3); β(1→2)
Partially hydrolysed guar gum	Mannose-Galactose	2–7	β(1→4), α(1→6)

mon with dietary fiber, an important difference is the selective fermentability by the colonic microbiota. Most prebiotics would classify as dietary fiber, but not all fiber has prebiotic properties. Examples of prebiotics are presented in Table 2.1. Different production processes and sources may yield prebiotics with a different degree of polymerization, different distribution of linkages etc. It is currently not known whether these variants of the same prebiotic class have similar functional properties.

For the functionality of prebiotics, in particular the ability of the component to increase the level of faecal bifidobacteria has received much attention. In particular inulin and fructo-oligosaccharides (FOS) (Kelly 2008) and to a lesser extend galacto-oligosaccharides (GOS) (Playne and Crittenden 2008) have been documented to increase the level of faecal bifidobacteria. But also other oligosaccharides have been reported to increase faecal bifidobacteria (de Vrese and Schrezenmeir 2008), Table 2.1. The increase in *Bifidobacterium* levels is very much dependent on initial levels and an inverse correlation exists between initial *Bifidobacterium* levels and the increase due to prebiotic consumption (Tuohy et al. 2001). However, bifidogenic activity is not a health benefit *per se*; instead, improvement levels of bifidobacteria may serve as an indicator for intestinal health. A reduction in the levels of pathogenic and opportunistic microbes, and/or improved resistance against pathogenic and opportunistic microbes maybe more important for the host health. In this respect, it is important to realize that many pathogens have their target in the upper gastrointestinal tract; such as *Helicobacter pylori* (stomach), enterotoxigenic *Escherichia coli* (small intestine) etc. Prebiotics are, in general not expected to be active at these sites. However, colonic pathogens may provide a potential target for prebiotics and protection from antibiotic associated diarrhoea with FOS has been reported (Lewis et al. 2005a) although failures have been reported as well (Lewis et al. 2005b). GOS has been reported to reduce the incidence and duration of traveler's diarrhoea, although the drop-out rate was high in this study (Drakoularakou et al. 2010). Under certain circumstances, prebiotics have been observed to enhance the colonization of pathogens. FOS has been reported to increase translocation of *Salmonella* in rats (Bovee-Oudenhoven et al. 2003), and FOS and xylo-oligosaccharides have been observed to increase *Salmonella* translocation in mice, while apple pectin was found to increase *Salmonella* colonization in mice (Petersen et al. 2009).

Prebiotics in general also change the metabolism of the intestinal microbiota; this may have more relevance from a health perspective than a change in its composition. Increases in acetic, propionic and butyric acid have been frequently observed and result in a lowering of the faecal pH. This is thought to suppress the growth and activity of potential pathogenic members of the microbiota and may improve the solubility of minerals such as calcium; enabling an increased absorption of these minerals (de Vrese and Schrezenmeir 2008). The organic acids serve further more as energy sources; acetic acid is metabolised by the liver and muscle tissue, propionic acid is metabolized by the liver as part of gluconeogenesis and butyric acid is the main energy source for colonocytes (Scheppach and Weiler 2004).

By serving as a source of fermentable energy, prebiotics and also other fermentable fibers, direct the metabolism of the colonic microbiota towards so-called saccharolytic direction, instead of proteolytic one. The latter, also referred to as putrefaction, leads to the formation of biogenic amines, phenols and indoles, with potentially detrimental effects on the host. This is particularly true for long chain and complex prebiotics such as inulin and polydextrose (Mäkeläinen et al. 2007).

Health benefits related to the consumption of selected prebiotics include improved mineral absorption and improved bowel function. Other, as yet, less well established health benefits are: modulation of the immune system, reduction of certain markers for colorectal cancer, and positive influences on lipid metabolism. The amounts of prebiotics that need to be consumed in order to obtain a physiological effect are in general at least 4 g (de Vrese and Schrezenmeir 2008). Prebiotics are very safe; they are intrinsic components of the normal diet. Overconsumption of purified prebiotics may lead to diarrhoea which subsides as soon as consumption is stopped.

2.3.2 The Influence of Probiotics on the Intestinal Microbiota

Probiotics have been defined in many ways over the years. The most widely accepted definition is currently: 'Live microorganisms which when administered in adequate amounts confer a health benefit on the host' (FAO/WHO 2002). Probiotics are thus not restricted to food or feed applications. It is, however important to realise that health benefits cannot be extrapolated from one probiotic strain to another, not even when they are strains of the same species. Examples of probiotics are given in Table 2.2. The selection of probiotics will be described in Sect. 2.4.

Considering the amount of microbes present in the intestine, see above, it is almost surprising that consumption of probiotics (normally at a dose of 10^9–10^{10} CFU/day) affects the composition and activity. Nevertheless, a number of strains have been shown to increase the faecal levels of endogenous bifidobacteria and/or lac-

Table 2.2 Examples of probiotics

Genus	Species	Strain
Lactobacillus	*acidophilus*	NCFM, La-5, R-52, La-14
	casei[a]	Lc-11, DN-114 001, Shirota
	rhamnosus	GG, Lr-32, GR-1, Lc-705, R-11, HN001
	johnsonii	La-1
	paracasei	Lpc-37, F-19
	plantarum	299v, Lp-115
	salivarius	Ls-33, UCC-118
	reuteri	RC-14
Bifidobacterium	*animalis* subsp. *lactis*	Bb-12, 420, Bl-04, Bi-07, HN019, DN-173 010
	longum	BB536, 2C, 46, Bl-05
Bacillus	*coagulans*[b]	BC30, or usually not referred to by strain
Propionibacterium	*freudenreichii* subsp. *shermanii*	JS
Saccharomyces	*cerevisiae* (*boulardii*)[c]	Usually not referred to by strain
Escherichia	*coli*	Nissle 1917

[a] Many stains that are referred to as *L. casei* are actually *L. paracasei*
[b] Often, incorrectly, referred to as *Lactobacillus sporogenes*
[c] *S. boulardii* is taxonomically considered to belong to the species *S. cerevisiae*

tobacilli (Gopal et al. 2003; Tannock et al. 2000). As with prebiotics, this does not necessarily imply a clinical health benefit, but may rather serve as a marker for improved gut environment.

In contrast to prebiotics, however, substantial indications and proof exists that specific strains of prebiotics are able to protect against or reduce the level and/or activity of pathogens and opportunistic pathogens. Several probiotic strains; e.g. *L. rhamnosus* GG, *B. lactis* Bb-12, *L. reuteri*, etc. have been documented to shorten the duration of rotavirus diarrhoea (Szajewska and Mrukowicz 2001). A number of strains have been shown to reduce the risk and duration of antibiotic associated diarrhoea. In particular *Saccharomyces cerevisiae* (*boulardii*) has been shown to shorten *C. difficile* diarrhoea (McFarland 2006). A potential mechanism of the observed effects is the stabilisation of the endogenous microbiota during antibiotic treatment (Engelbrektson et al. 2005). *Helicobacter pylori* is a pathogen of the stomach and thought to be the aetiological agent of gastric ulcer and carcinoma. In general, probiotics have been unsuccessful in the eradication of *H. pylori*. But, some strains have been shown to reduce the activity of the pathogen. Whether that is sufficient to control disease risk is not known. Better substantiated, however, is the positive effect several strains of probiotics can have on reducing the side effects of anti-*Helicobacter* treatment (Zou et al. 2009). For traveller's diarrhoea, the effects of probiotics are less clear. The reason for this may be the different aetiologies of this type of diarrhoea and the, some times high drop out rate (Sazawal et al. 2006).

In addition to their effects on the microbiota and pathogens of the gastrointestinal tract, selected probiotic strains have been suggested to affect also pathogens of the oral cavity and reduce the colonisation level and activity of cariogenic streptococci and reduce levels of *Candida* (Meurman 2005). There is also ample evidence that selected probiotics may reduce the incidence and/or duration of upper respiratory tract infections (Leyer et al. 2009). It is not clear through what mechanism this effect is obtained. Probiotics have been shown to modulate the microbiota of the upper respiratory tract (Glück and Gebbers 2003), but it cannot be excluded that a modulation of the immune system is a contributing factor. In particular two strains; *L. rhamnosus* GR-1 and *L. reuteri* RC-14 have been investigated for their effects on vaginal infections (Reid 2008).

2.3.3 The Influence of Medication on the Intestinal Microbiota

Medication may have a substantial effect on the intestinal microbiota and *vice versa*. The clearest example of this is of course alterations in the microbiota caused by antibiotics. Antibiotics, especially those that are poorly absorbed from the gastrointestinal tract can cause substantial disturbance of the microbiota. This disturbance may lead to an increase of antibiotic resistant strains that arise among the bacteria of normal microbiota. This may in turn lead to diarrhoea and infections of other organs (Sullivan et al. 2001).

Also other medication may influence the composition of the intestinal microbiota. Proton pump inhibitors (PPI) disable the antimicrobial capacity of normal gastric secretions. Not surprisingly, their use leads to substantial changes in the gastro-intestinal microbiota composition. Organisms originating from the oral cavity, e.g. α-haemolytic streptococci and corynebacteria, are normally killed by the action of gastric juice; but with the use of PPI they survive and colonise the stomach (Karmeli et al. 1995). This has been hypothesised to lead to increased risk for upper respiratory tract infections and gastroenteritis (Canani et al. 2006). Non-steroidal anti-inflammatory drug (NSAID), have been observed to influence the composition of the intestinal microbiota in elderly subjects. Moreover, use of NSAID by elderly subjects has been found to reduce the level of iso-butyric, iso-valeric and L-lactic acid (Tiihonen et al. 2008).

2.3.4 The Intestinal Microbiota and Stress

Psychological and physical stress is accompanied with a number of substantial changes in the gastrointestinal tract physiology; inhibition of gastric acid release, alterations in gastrointestinal motility (and thus in transit time), changes in duodenal bicarbonate release, hormonal changes and reduction in mucosal immune function. These changes have mainly been observed to reduce faecal *Lactobacillus* levels and increases in *Bacteroides* (Hawrelak and Myers 2004). It is, however, likely that more changes in the microbiota take place during stress, but in the past the focus has been limited mainly to these genera.

2.4 Selection of Probiotics for Gut Health

The rapid growth of probiotic market, the growing understanding of the importance of gut microbes on human health, and the new discoveries of the health benefits of probiotics have inspired many researchers to search for new potential probiotic strains. In addition, existing strains of probiotics are finding new areas of application, with new targeted health benefits and novel probiotic product applications. Not all probiotics are equal and therefore the suitability of probiotic strains needs to be assessed separately for each new application. In the field of gut health, several widely used and established selection criteria are available. These tools are also applicable for the selection of probiotics against enteric infections. Here, commonly used selection criteria for probiotics are reviewed.

2.4.1 Viability During Storage in Probiotic Products

According to the current definition, probiotics need to be viable (FAO/WHO 2002). Therefore, viability of probiotics is of technologic, economic, regulatory and biologic importance. In products, probiotics are often stored in unfavorable conditions, for example in the presence of acids in fermented products. Good resistance to environmental stresses is therefore required to ensure that probiotics remain viable in the products through-out the shelf-life. Unfortunately this is not always the case, as there are reports suggesting that in some commercial products, the level of viable probiotics does not meet the regulatory criteria in the end of the storage (Gueimonde et al. 2004; Hamilton-Miller et al. 1999).

Stability tests in different product matrices are carried out routinely during the screening of new probiotics and when testing the suitability for established probiotics for new product applications (Laine et al. 2003; Madueira et al. 2005). The tests applied include normal stability tests as well as accelerated stability tests e.g. in elevated storage temperatures. Normally, the viability of probiotics during these tests is monitored using the traditional culture methods, usually plate counting. Recent studies have shown however that plate counting may not always be the optimal choice for determining probiotic viability in food products. Several studies have shown that during storage, probiotics may enter a state in which they cease to grow on plates yet remain viable (Amor et al. 2002; Lahtinen et al. 2006). New culture-independent methods have been developed which are very useful in the viability assays of such probiotics. However, currently the traditional plating techniques dominate the field. Despite that, according to the current definition, probiotics need to be viable to even be considered as probiotics. Interestingly it appears that also non-viable probiotics may have certain health benefits (Ouwehand and Salminen 1998). However, as long as the definition of probiotics includes the requirement for viability, the viability of probiotics during storage will remain an important screening step for the selection of new probiotics.

2.4.2 *Resistance to Acid and Bile and Survival in the GI Tract*

For probiotics intended for gut health applications and against enteric infections, the behavior and the fate of the probiotics during the GI passage is particularly important. The mechanisms of the probiotic action against enteric pathogens may involve production of antimicrobial agents locally in the gut—a task which can only be carried out by viable and active probiotics. *In vitro* tests assessing the effect of exposure to different pH values and to the presence of bile, typically mimicking the conditions during the GI transit, are routinely used as a screening tool for potential probiotics (Jacobsen et al. 1999). The results of these *in vitro* screenings usually correlate relatively well with *in vivo* survival of probiotics (Dunne et al. 2001). In addition to simple survival tests, the fate of probiotics in the GI tract can also be assessed with more elaborate *in vitro* assays. Several *in vitro* simulators of the GI tract (Kontula et al. 1998) or the colon (Mäkeläinen et al. 2009) have been developed, and these have been utilized in the analysis of the fate of the probiotic bacteria during simulated GI passage. In addition to *in vitro* testing, probiotic viability during GI transit can also be assessed *in vivo*. In these assessments, probiotics are administered orally and later re-isolated from faecal samples in order to demonstrate that the cells, or at least a proportion of them, have survived during the GI transit. To this regard, several animal models have been utilized (Kaplan et al. 2001; Pavan et al. 2003; Wagner et al. 1997). The gold standard for demonstrating the survival during GI passage is a clinical intervention trial, where faecal colonization of probiotics can be used to assess the survival of the probiotic in the gut, as well as the compliance of the volunteers participating in the study (Fujiwara et al. 2001; Tannock et al. 2000). In some cases, survival of probiotics in the gut has also been assessed with intestinal biopsies (Alander et al. 1997; Valeur et al. 2004).

2.4.3 *Adhesion to Host Tissues*

Adhesion of probiotics to host tissues, in particular to intestinal mucus and epithelial cells, is an important selection criterion for probiotics. It is believed that adhesion is important to ensure efficient host-microbial interactions, for example direct interactions between the bacteria and the cells of the host immune system. It has also been suggested that adhesion to host tissues may prolong the persistence of the probiotics in the gut, although it is generally accepted that orally administered probiotics do not colonize the host permanently, and continuous intake of probiotics is required to ensure prolonged presence of the cells in the gut. While it may be postulated that good adhesion properties are needed to ensure direct interactions with the bacteria and the host immune cells, the adhesion to host mucus may not be a strict requirement for probiotic efficacy, as demonstrated by the probiotic strain *L. casei* Shirota. This strain has been linked to different gut health benefits although it does not seem to be a particularly adherent strain (Juntunen et al. 2001). Certain health benefits of probiotic bacteria may be mediated through mechanisms independent of mucus adhesion. Such mechanisms may involve direct cell-to-cell interactions between the gut bacteria, see below. In addition, the health effects may be mediated by the metabolites produced by the probiotic bacteria. In healthy colon, a direct interaction with host epithelial cells and gut microbes, including ingested probiotics, seems unlikely, as it is currently believed that in a healthy colon, a tight inner mucus layer covering the epithelial cells, impassable to microbes due to its tight structure, isolates the host epithelial cells and the bacteria attached to the loose outer mucus layer of the colon (Johansson et al. 2008). In the colon, the probiotic interactions may therefore be mediated mainly by the metabolites of the bacteria. Similar tight inner mucus layer is not present in the small intestine, where direct associations between the microbes and the epithelial cells are plausible.

2.4.4 Interactions with the Host Immune System

Probiotic bacteria, like other bacteria in the gut, may interact with the cells of the host immune system. For example, dendritic cells have been described to sample the contents of the gut lumen by extending the dendrites past the epithelial cells into the lumen, while M-cells of the Peyer's patches are specialized in antigen presentation from the gut (Mowat 2003). Moreover, bacteria may elicit immune signaling by interacting with the epithelial cells. By interacting with the host cells in the gut, probiotic bacteria may modulate the host immune responses for example by enhancing host innate immune functions (Gill et al. 2001) or by modulating the inflammatory status of the host (Foligne et al. 2007). Immunomodulatory properties are therefore an important selection criterion for probiotic bacteria. Several models for screening of the immunomodulatory properties of probiotics are available. Cell culture studies with epithelial cells (Wong and Ustunol 2006) or immune cells (Braat et al. 2004; Haller et al. 2000; Konstantinov et al. 2008) can be used to give first indication of the immunomodulatory properties of probiotics, as well as to investigate the potential mechanisms behind the host-bacterial interactions. *In vivo* animal models have also proved to be useful in this regard (Daniel et al. 2006; Gill et al. 2000; Kirjavainen et al. 1999; Madsen et al. 2001; Wagner et al. 1997). In addition to preclinical trials, clinical interventions have also been used as a selection for probiotics aimed at immune modulation (Paineau et al. 2008; Schiffrin et al. 1997). Immune cells and the epithelial cells are not the only host cells which can be directly affected by gut microbes and probiotics; probiotic strain *L. acidophilus* NCFM has recently been shown to be able to interact directly also with host pain receptors (Rousseaux et al. 2006).

2.4.5 *Interactions Among Gut Bacteria*

Production of anti-microbial compounds has been a popular screening tool for probiotics for gut health (Bevilacqua et al. 2003). In addition to the production of metabolites such as organic acids, probiotic bacteria may also produce specific antibacterial agents, such as bacteriocins (Dobson et al. 2007; Kwak et al. 2001; Tahara et al. 1996). While specific anti-pathogenic action has been demonstrated for many strains, the antimicrobial agent responsible for the activity has not always been characterized in detail, especially in the case of bacteriocins produced by bifidobacteria (Yildirim and Johnson 1998). In addition to production of bacteriocins and bacteriocin-like compounds, and by reducing intestinal pH, probiotic bacteria may also elicit anti-pathogenic effects by blocking the adhesion of the pathogenic strains (prevention or displacement) and by co-aggregating with the pathogens (Collado et al. 2007). Screening of anti-pathogenic actions of probiotics can also be carried out in animal models of pathogen challenge (Shu et al. 2001; Wagner et al. 2000). Pre-clinical trials predominate the screening of anti-pathogenic probiotics. Although clinical intervention trials are the gold standard for documenting probiotic efficacy, human trials are not usually used as screening step for identification of anti-pathogenic strains. While effects of probiotics on the resident beneficial or commensal bacteria have been studied in clinical studies as well as in laboratory fermentations and simulations, such effects have not been used as commonly as a screening tool for probiotics. In the case of prebiotics, the effects on beneficial microbes are used routinely as a screening tool.

2.4.6 *Anti-carcinogenic and Anti-toxic Effects*

Effects of probiotics on colon enzyme activity and faecal water genotoxicity have been among the first health effects of probiotics investigated (Goldin and Gorbach 1984). It is postulated that by improving colonic microbiota and metabolic activity, probiotics may reduce the level of genotoxic and carcinogenic compounds in the gut. *In vitro* assessments of anti-carcinogenic effects relying on cell cultures have been utilized as a screening tool for probiotics (Burns and Rowland 2004; Choi et al. 2006). *In vitro* binding tests of mutagenic compounds such as heterocyclic amines (Orrhage et al. 1994) and mycotoxins (El Nezami et al. 1998; Haskard et al. 2001) by probiotic bacteria have also been used in the screening. Moreover, screening of anti-genotoxic and anti-carcinogenic properties of probiotics with animal models has shown that the effects observed *in vitro* may also be seen *in vivo* (Pool-Zobel et al. 1996). Human trials in this area have mainly focused on biomarker analyses rather than clinical end-points. In addition to mycotoxins, the potential of probiotics to remove other toxins such as cyanotoxins (Nybom et al. 2008) or heavy metals (Halttunen et al. 2007) has been used as a screening tool for probiotics.

2.4.7 *Origin of the Probiotic Strains*

An old dogma of probiotic selection has been that the probiotic strains should be of "human origin". One may argue that from evolutionary point of view, describing bacteria to be of human origin does not make much sense at all. The requirement for probiotics to be of human origin relates actually to the isolation of the strain rather than the "origin" itself. Usually, the strains claimed to be "of human origin" have been isolated from faecal samples of healthy human subjects, and have therefore been considered to be "part of normal healthy human gut microbiota". In reality the recovery of a strain from a faecal sample does not necessarily mean that this strain is part of the normal microbiota of this individual, since microbes passing the GI tract transiently can also be recovered from the faecal samples. This can be demonstrated also by the established probiotic strains: if an individual consumes a yoghurt with a certain probiotic strain, and this strain is later isolated from a faecal sample from the same individual, this does not mark a discovery of a "new strain of human origin", but rather a re-isolation of the original probiotic strain transiently passing through the GI tract. In practice it is impossible to know the actual origin of the probiotic strains, regardless of whether they have been isolated from faecal samples, fermented dairy products or any other source for that matter. Isolation of a strain from faeces of a healthy individual is also not a guarantee of the safety of the strain—such a sample will also always contain commensal microbes which can act as opportunistic pathogens, or even low levels of true pathogens, which are present in the individual at sub-clinical levels. The authors therefore recommend that instead of concentrating on the first point of isolation, the selection processes for new potential probiotic strains should mainly focus on the functional properties of the probiotic strains rather than the "origin" (Ouwehand and Lahtinen 2008).

2.4.8 *Safety*

Safety is a critically important feature of probiotics. Most commonly used probiotics belong to *Lactobacillus* spp. or *Bifidobacterium* spp., which have a long history of safe human consumption and are generally considered as safe (EFSA 2007). Despite the excellent safety record of probiotics, safety remains to be an important selection criterion for probiotics. Especially in the case of probiotics which do not belong to *Lactobacillus* spp. or *Bifidobacterium* spp., demonstration of safety is of particular importance. For example, in some cases bacteria belonging to *Enterococcus* spp. are used as probiotics, which has raised concerns of potential spread of antibiotic resistance genes and presence of virulence determinants (Eaton and Gasson 2001; Vankerckhoven et al. 2008). Also strains belonging to *Bacillus* spp. and strains of *E. coli* have been used as probiotics. *In vitro* safety screenings of probiotics may include, among others, antibiotic resistance assays, screenings for virulence factors, resistance to host defence mechanisms and induction of haemolysis (Asahara et al. 2003; Eaton and Gasson 2001; Tompkins et al. 2008; Vankerckhoven et al. 2008; Vesterlund et al. 2007). Several different animal models have been utilized in the safety assessment of probiotics. These include models of immunodeficiency (Wagner et al. 1997), endocarditis (Asahara et al. 2003), colitis (Daniel et al. 2006) and liver injury (Osman et al. 2005). In some cases even acute toxicity of probiotics has been assessed (Zhou et al. 2000). Last but not least, also clinical intervention trials have yielded evidence on the safety of probiotics for human consumption (Dekker et al. 2009; Laitinen et al. 2005; Mäkeläinen et al. 2003; Saarela et al. 2007; Wolf et al. 1995).

2.4.9 Suitability for Commercial Production

A critically important and sometimes overlooked aspect of probiotic selection is the suitability of the strain for large-scale industrial production. While a new probiotic candidate strain may have interesting properties in laboratory analyses and in pilot trials, it can never become a commercial probiotic consumed daily by customers, if it cannot be produced cost-efficiently in an industrial scale. Discovery of such strain will be of academic interest only. Strains which have too stringent requirements for rapid growth cannot be produced efficiently in large scale and their cost-in-use may therefore be too high. In addition to growth requirements, technological properties such as stability during processing and in finished formulations are of utmost importance.

2.5 Conclusions

The intestinal microbiota has a major role in regulation of GI function and host health. The sterile GI tract in a foetus is rapidly colonized after birth by opportunistic bacteria originating from the mother and the surrounding environment. The microbiota develops in a predictable fashion into a complex system with increasing age, determined by internal and external factors. This complex system is quite stable with a majority of obligate anaerobes, and dominated by species from the phylum *Firmicutes* and *Bacteroidetes*. However, the composition and activity of the intestinal microbiota can be influenced by external or environmental factors such as, medication or food. The probiotic market has increased rapidly during the last years, and a large variety of probiotic foods and supplements are available. With probiotics it is possible to alter the intestinal microbiota in order to improve the health of the host or the gut microbiota. For example, the endogenous microbiota can be stabilised during an antibiotic treatment with probiotics, and the incidence and/or duration of infections may be reduced. The growing understanding of the importance of gut microbes on human health leads to new application areas for probiotics, with new targeted health benefits and novel probiotic product applications. However, it is still important to ensure the safety and the suitability for commercial production when selecting probiotics for gut health.

References

Adlerberth I, Lindberg E, Aberg N, Hesselmar B, Saalman R, Strannegård I, Wold A (2010) Reduced enterobacterial and increased staphylococcal colonization of the infantile bowel: an effect of hygienic lifestyle? Pediatr Res 59(1):96–101

Ahrne S, Lonnermark E, Wold AE, Aberg N, Hesselmar B, Saalman R, Strannegard IL, Molin G, Adlerberth I (2005) Lactobacilli in the intestinal microbiota of Swedish infants. Microb Infect 7(11–12):1256–1262

Alander M, Korpela R, Saxelin M, Vilpponen-Salmela T, Mattila-Sandholm T, von Wright A (1997) Recovery of *Lactobacillus rhamnosus* GG from human colonic biopsies. Lett Appl Microbiol 24(5):361–364

Amor KB, Breeuwer P, Verbaarschot P, Rombouts FM, Akkermans AD, de Vos WM, Abee T (2002) Multiparametric flow cytometry and cell sorting for the assessment of viable, injured, and dead *Bifidobacterium* cells during bile salt stress. Appl Environ Microbiol 68(11):5209–5216

Asahara T, Takahashi M, Nomoto K, Takayama H, Onoue M, Morotomi M, Tanaka R, Yokokura T, Yamashita N (2003) Assessment of safety of *Lactobacillus* strains based on resistance to host innate defence mechanisms. Clin Diagn Lab Immunol 10(1):169–173

Bartosch S, Fite A, Macfarlane GT, McMurdo MET (2004) Characterization of bacterial communities in faeces from healthy elderly volunteers and hospitalized elderly patients by using real-time PCR and effects of antibiotic treatment on the faecal microbiota. Appl Environ Microbiol 70(6):3575–3581

Bevilacqua L, Ovidi M, Di Mattia E, Trovatelli LD, Canganella F (2003) Screening of *Bifidobacterium* strains isolated from human faeces for antagonistic activities against potentially bacterial pathogens. Microbiol Res 158(2):179–185

Boehm G, Stahl B, Jelinek J, Knol J, Miniello V, Moro GE (2005) Prebiotic carbohydrates in human milk and formulas. Acta Paediatr Suppl 94(449):18–21

Bovee-Oudenhoven IMJ, ten Bruggencate SJM, Lettink-Wissink MLG, van der Meer R (2003) Dietary fructo-oligosaccharides and lactulose inhibit intestinal colonisation but stimulate translocation of *Salmonella* in rats. Gut 52:1572–1578

Braat H, van den Brande J, van Tol E, Hommes D, Peppelenbosch M, van Deventer S (2004) *Lactobacillus rhamnosus* induces peripheral hyporesponsiveness in stimulated CD4+ T cells via modulation of dendritic cell function. Am J Clin Nutr 80(6):1618–1625

Burns AJ, Rowland IR (2004) Antigenotoxicity of probiotics and prebiotics on faecal water-induced DNA damage in human colon adenocarcinoma cells. Mutat Res 551(1–2):233–243

Canani RB, Cirillo P, Roggero P, Romano C, Malamisura B, Terrin G, Passariello A, Manguso F, Morelli L, Guarino A (2006) Therapy with gastric acidity inhibitors increases the risk of acute gastroenteritis and community-acquired pneumonia in children. Pediatrics 117(5):e817–e820

Chen JJ, Cai W, Feng Y (2007) Development of intestinal bifidobacteria and lactobacilli in breast-fed neonates. Clin Nutr 26(5):559–566

Choi SS, Kim Y, Han KS, You S, Oh S, Kim SH (2006) Effects of *Lactobacillus* strains on cancer cell proliferation and oxidative stress in vitro. Lett Appl Microbiol 42(5):452–458

Collado MC, Meriluoto J, Salminen S (2007) Role of commercial probiotic strains against human pathogen adhesion to intestinal mucus. Lett Appl Microbiol 45(4):454–460

Daniel C, Poiret S, Goudercourt D, Dennin V, Leyer G, Pot B (2006) Selecting lactic acid bacteria for their safety and functionality by use of a mouse colitis model. Appl Environ Microbiol 72(9):5799–5805

de Vrese M, Schrezenmeir J (2008) Probiotics, prebiotics, and synbiotics. Adv Biochem Eng/Biotechnol 111:1–66

Dekker JW, Wickens KW, Black PN, Stanley TV, Mitchell EA, Fitzharris P, Tannock GW, Purdie G, Crane J (2009) Safety aspects of probiotic bacterial strains *Lactobacillus rhamnosus* HN001 and *Bifidobacterium animalis* subsp. *lactis* HN019 in human infants aged 0–2 years. Int Dairy J 19:149–154

Dobson AE, Sanozky-Dawes RB, Klaenhammer TR (2007) Identification of an operon and inducing peptide involved in the production of lactacin B by *Lactobacillus acidophilus*. J Appl Microbiol 103(5):1766–1778

Drakoularakou A, Tzortzis G, Rastall RA, Gibson GR (2010) A double-blind, placebo-controlled, randomized human study assessing the capacity of a novel galacto-oligosaccharide mixture in reducing travellers' diarrhoea. Eur J Clin Nutr 64:146–152

Dunne C, O'Mahony L, Murphy L, Thornton G, Morrissey D, O'Halloran S, Feeney M, Flynn S, Fitzgerald G, Daly C, Kiely B, O'Sullivan GC, Shanahan F, Collins JK (2001) In vitro selection criteria for probiotic bacteria of human origin: correlation with in vivo findings. Am J Clin Nutr 73(2 Suppl):386S–392S

Eaton TJ, Gasson MJ (2001) Molecular screening of *Enterococcus* virulence determinants and potential for genetic exchange between food and medical isolates. Appl Environ Microbiol 67(4):1628–1635

Eckburg PB, Bik EM, Bernstein CN, Purdom E, Dethlefsen L, Sargent M, Gill SR, Nelson KE, Relman DA (2005) Diversity of the Human Intestinal Microbial Flora. Science 308(5728):1635–1638

EFSA (2007) European Food Safety Authority Scientific Committee (EFSA) public consultation on the Qualified Presumption of Safety (QPS) approach for the safety assessment of microorganisms deliberately added to food and feed. Annex 3: Assessment of gram positive nonsporulating bacteria with respect to a qualified presumption of safety. http://www.efsa.europa.eu/en/science/sc_commitee/sc_consultations/sc_consultation_qps html

El Nezami H, Kankaanpää P, Salminen S, Ahokas J (1998) Ability of dairy strains of lactic acid bacteria to bind a common food carcinogen, aflatoxin B1. Food Chem Toxicol 36(4):321–326

Enck P, Zimmermann K, Rusch K, Schwiertz A, Klosterhalfen S, Frick J (2009) The effects of maturation on the colonic microflora in infancy and childhood. Gastroenterol Res Pract. Published online 16 Sept 2009, doi:10.1155/2009/752401

Engelbrektson AL, Nguyen HD, Lung DY, Korzenik JR, Sanders ME, Leyer G, Klaenhammer TR, Kitts CL (2005) The effect of probiotics on the species distribution of *Bifidobacterium* and *Lactobacillus* in healthy subjects undergoing antibiotic therapy. Conference Proceeding

Fanaro S, Chierici R, Guerrini P, Vigi V (2003) Intestinal microflora in early infancy: composition and development. Acta Paediatr 92:48–55

FAO/WHO (2002) Guidelines for the evaluation of probiotics in food. http://www.who.int/foodsafety/publications/fs_management/probiotics2/en/. pp 1–11

Favier CF, Vaughan EE, de Vos WM, Akkermans ADL (2002) Molecular monitoring of succession of bacterial communities in human neonates. Appl Environ Microbiol 68(1):219–226

Finegold SM, Sutter VL, Mathisen GE (1983) Normal indigenous intestinal flora. In: Hentges DJ (ed) Human intestinal microflora in health and disease. Academic Press, New York, pp 3–31

Foligne B, Nutten S, Grangette C, Dennin V, Goudercourt D, Poiret S, Dewulf J, Brassart D, Mercenier A, Pot B (2007) Correlation between in vitro and in vivo immunomodulatory properties of lactic acid bacteria. World J Gastroenterol 13(2):236–243

Franks AH, Harmsen HJ, Raangs GC, Jansen GJ, Schut F, Welling GW (1998) Variations of bacterial populations in human faeces measured by fluorescent in situ hybridization with group-specific 16S rRNA-targeted oligonucleotide probes. Appl Environ Microbiol 64(9):3336–3345

Fujiwara S, Seto Y, Kimura A, Hashiba H (2001) Intestinal transit of an orally administered streptomycin-rifampicin-resistant variant of *Bifidobacterium longum* SBT2928: its long-term survival and effect on the intestinal microflora and metabolism. J Appl Microbiol 90(1):43–52

Gavini F, Cayuela C, Antoine J-M, Lecoq C, Lefebvre B, Membreacute J-M, Neut C (2001) Differences in the distribution of bifidobacterial and enterobacterial species in human faecal microflora of three different (children, adults, elderly) age groups. Microb Ecol Health Dis 13(1):40–45

Gill HS, Rutherfurd KJ, Prasad J, Gopal PK (2000) Enhancement of natural and acquired immunity by *Lactobacillus rhamnosus* (HN001), *Lactobacillus acidophilus* (HN017) and *Bifidobacterium lactis* (HN019). Br J Nutr 83(2):167–176

Gill HS, Rutherfurd KJ, Cross ML, Gopal PK (2001) Enhancement of immunity in the elderly by dietary supplementation with the probiotic *Bifidobacterium lactis* HN019. Am J Clin Nut 74(6):833–839

Glück U, Gebbers J-O (2003) Ingested probiotics reduce nasal colonization with pathogenic bacteria (Staphylococcus aureus, Streptococcus pneumoniae, and β-hemolytic streptococci). Am J Clin Nutr 77:517–520

Goldin BR, Gorbach SL (1984) The effect of milk and lactobacillus feeding on human intestinal bacterial enzyme activity. Am J Clin Nutr 39(5):756–761

Gopal P, Prasad J, Gill HS (2003) Effects of the consumption of *Bifidobacterium lactis* HN019 (DR10™) and galacto-oligosaccharides on the microflora of the gastrointestinal tract in human subjects. Nutr Res 23:1313–1328

Gueimonde M, Delgado S, Baltasar M, Ruas-Madiedo P, Margolles A, los Reyes-Gavilan CG (2004) Viability and diversity of probiotic *Lactobacillus* and *Bifidobacterium* populations included in commercial fermented milks. Food Res Int 37:839–850

Guigoz Y, Dore J, Schiffrin EJ (2008) The inflammatory status of old age can be nurtured from the intestinal environment. Curr Opin Clin Nutr Metab Care 11(1):13–20

Haller D, Blum S, Bode C, Hammes WP, Schiffrin EJ (2000) Activation of human peripheral blood mononuclear cells by nonpathogenic bacteria in vitro: evidence of NK cells as primary targets. Infect Immun 68(2):752–759

Halttunen T, Salminen S, Tahvonen R (2007) Rapid removal of lead and cadmium from water by specific lactic acid bacteria. Int J Food Microbiol 114(1):30–35

Hamilton-Miller JM, Shah S, Winkler JT (1999) Public health issues arising from microbiological and labelling quality of foods and supplements containing probiotic microorganisms. Public Health Nutr 2(2):223–229

Harmsen HJM, Wildeboer-Veloo ACM, Raangs GC, Wagendorp AA, Klijn N, Bindels JG, Welling GW (2000) Analysis of intestinal flora development in breast-fed and formula-fed infants by using molecular identification and detection methods. J Pediat Gastroenterol Nutr 30(1):61–67

Harmsen HJM, Raangs GC, He T, Degener JE, Welling GW (2002) Extensive set of 16S rRNA-based probes for detection of bacteria in human faeces. Appl Environ Microbiol 68(6):2982–2990

Haskard CA, El Nezami HS, Kankaanpää PE, Salminen S, Ahokas JT (2001) Surface binding of aflatoxin B1 by lactic acid bacteria. Appl Environ Microbiol 67(7):3086–3091

Hawrelak JA, Myers SP (2004) The causes of intestinal dysbiosis: a review. Altern Med Rev 9(2):180–197

Hayashi H, Sakamoto M, Benno Y (2002) Phylogenetic analysis of the human gut microbiota using 16S rDNA clone libraries and strictly anaerobic culture-based methods. Microbiol Immunol 46(8):535–548

Hayashi H, Sakamoto M, Kitahara M, Benno Y (2003) Molecular analysis of faecal microbiota in elderly individuals using 16S rDNA library and T-RFLP. Microbiol Immunol 47(8):557–570

He F, Ouwehand AC, Isolauri E, Hosoda M, Benno Y, Salminen S (2001) Differences in composition and mucosal adhesion of bifidobacteria isolated from healthy adults and healthy seniors. Curr Microbiol 43(5):351–354

Holdeman LV, Good IJ, Moore WE (1976) Human faecal flora: variation in bacterial composition within individuals and a possible effect of emotional stress. Appl Environ Microbiol 31(3):359–375

Hopkins MJ, Macfarlane GT (2002) Changes in predominant bacterial populations in human faeces with age and with *Clostridium difficile* infection. J Med Microbiol 51(5):448–454

Hopkins MJ, Sharp R, Macfarlane GT (2001) Age and disease related changes in intestinal bacterial populations assessed by cell culture, 16S rRNA abundance, and community cellular fatty acid profiles. Gut 48(2):198–205

Jacobsen CN, Rosenfeldt Nielsen V, Hayford AE, Moller PL, Michaelsen KF, Paerregaard A, Sandstrom B, Tvede M, Jakobsen M (1999) Screening of probiotic activities of forty-seven strains of *Lactobacillus* spp. by in vitro techniques and evaluation of the colonization ability of five selected strains in humans. Appl Environ Microbiol 65(11):4949–4956

Johansson ME, Phillipson M, Petersson J, Velcich A, Holm L, Hansson GC (2008) The inner of the two Muc2 mucin-dependent mucus layers in colon is devoid of bacteria. Proc Natl Acad Sci U S A 105(39):15064–15069

Juntunen M, Kirjavainen PV, Ouwehand AC, Salminen SJ, Isolauri E (2001) Adherence of probiotic bacteria to human intestinal mucus in healthy infants and during rotavirus infection. Clin Diagn Lab Immunol 8(2):293–296

Kaplan CW, Astaire JC, Sanders ME, Reddy BS, Kitts CL (2001) 16S ribosomal DNA terminal restriction fragment pattern analysis of bacterial communities in faeces of rats fed *Lactobacillus acidophilus* NCFM. Appl Environ Microbiol 67(4):1935–1939

Karmeli Y, Stalnikowitz R, Eliakim R, Rahav G (1995) Conventional dose of omeprazole alters gastric flora. Dig Dis Sci 40(9):2070–2073

Kelly G (2008) Inulin-type prebiotics—a review: Part I. Alt Med Rev 13(4):315–329

Kirjavainen PV, El Nezami HS, Salminen SJ, Ahokas JT, Wright PF (1999) The effect of orally administered viable probiotic and dairy lactobacilli on mouse lymphocyte proliferation. FEMS Immunol Med Microbiol 26(2):131–135

Konstantinov SR, Smidt H, de Vos WM, Bruijns SC, Singh SK, Valence F, Molle D, Lortal S, Altermann E, Klaenhammer TR, van Kooyk Y (2008) S layer protein A of *Lactobacillus acidophilus* NCFM regulates immature dendritic cell and T cell functions. Proc Natl Acad Sci U S A 105(49):19474–19479

Kontula P, Jaskari J, Nollet L, De Smet I, von Wright A, Poutanen K, Mattila-Sandholm T (1998) The colonization of a simulator of the human intestinal microbial ecosystem by a probiotic strain fed on a fermented oat bran product: effects on the gastrointestinal microbiota. Appl Microbiol Biotechnol 50(2):246–252

Kwak G-S, Kim S-K, Jun H-K (2001) Purification and characterization of bacteriocin J105 produced by *Lactococcus lactis* subsp. *lactis* J105 isolated from Kimchi. J Microbiol Biotechnol 11(2):275–280

Lahtinen SJ, Ouwehand AC, Reinikainen JP, Korpela JM, Sandholm J, Salminen SJ (2006) Intrinsic properties of so-called dormant probiotic bacteria, determined by flow cytometric viability assays. Appl Environ Microbiol 72(7):5132–5134

Lahtinen SJ, Tammela L, Korpela J, Parhiala R, Ahokoski H, Mykkanen H, Salminen SJ (2009) Probiotics modulate the *Bifidobacterium* microbiota of elderly nursing home residents. Age (Dordr.) 31(1):59–66

Laine R, Salminen S, Benno Y, Ouwehand AC (2003) Performance of bifidobacteria in oat-based media. Int J Food Microbiol 83(1):105–109

Laitinen K, Kalliomaki M, Poussa T, Lagstrom H, Isolauri E (2005) Evaluation of diet and growth in children with and without atopic eczema: follow-up study from birth to 4 years. Br J Nutr 94(4):565–574

Lay C, Rigottier-Gois L, Holmstrom K, Rajilic M, Vaughan EE, de Vos WM, Collins MD, Thiel R, Namsolleck P, Blaut M, Dore J (2005) Colonic microbiota signatures across five northern European countries. Appl Environ Microbiol 71(7):4153–4155

Layton A, McKay L, Williams D, Garrett V, Gentry R, Sayler G (2006) Development of Bacteroides 16S rRNA gene TaqMan-Based Real-Time PCR assays for estimation of total, human, and bovine faecal pollution in water. Appl Environ Microbiol 72(6):4214–4224

Lewis S, Burmeister S, Brazier J (2005a) Effect of the prebiotic oligofructose on relapse of *Clostridium difficile*-associated diarrhoea: a randomized, controlled study. Clin Gastroenterol Hepatol 3(5):442–448

Lewis S, Burmeister S, Cohen S, Brazier J, Awasthi A (2005b) Failure of dietary oligofructose to prevent antibiotic-associated diarrhoea. Aliment Pharmacol Ther 21(4):469–477

Leyer GJ, Li S, Mubasher ME, Reifer C, Ouwehand AC (2009) Probiotic effects on cold and influenza-like symptom incidence and duration in children. Pediatrics 124(2):e172–e179

Ljungberg B, Nilsson-Ehle I, Edlund C, Nord CE (1990) Influence of ciprofloxacin on the colonic microflora in young and elderly volunteers: no impact of the altered drug absorption. Scand J Infect Dis 22(2):205–208

Lofmark S, Jernberg C, Jansson JK, Edlund C (2006) Clindamycin-induced enrichment and long-term persistence of resistant Bacteroides spp. and resistance genes. J Antimicrob Chemother 58(6):1160–1167

Mackie RI, Sghir A, Gaskins HR (1999) Developmental microbial ecology of the neonatal gastrointestinal tract. Am J Clin Nutr 69(5):1035S–1045S

Madsen K, Cornish A, Soper P, McKaigney C, Jijon H, Yachimec C, Doyle J, Jewell L, De Simone C (2001) Probiotic bacteria enhance murine and human intestinal epithelial barrier function. Gastroenterology 121(3):580–591

Madueira AR, Giao MS, Pintado ME, Gomes AM, Freitas C, Malcata FX (2005) Incorporation and survival of probiotic bacteria in whey cheese matrices. J Food Sci 70(3):M160–M165

Mäkeläinen H, Tahvonen R, Salminen S, Ouwehand AC (2003) In vivo safety assessment of two *Bifidobacterium longum* strains. Microbiol Immunol 47(12):911–914

Mäkeläinen H, Forssten S, Olli K, Granlund L, Rautonen N, Ouwehand AC (2009) Probiotic lactobacilli in a semi-soft cheese survive in the simulated human gastrointestinal tract. Int Dairy J 19:675–683

Mäkeläinen HS, Mäkivuokko HA, Salminen SJ, Rautonen NE, Ouwehand AC (2007) The effects of polydextrose and xylitol on microbial community and activity in a 4-stage colon simulator. J Food Sci 72(5):M153–M159

Matto J, Maunuksela L, Kajander K, Palva A, Korpela R, Kassinen A, Saarela M (2005) Composition and temporal stability of gastrointestinal microbiota in irritable bowel syndrome—a longitudinal study in IBS and control subjects. FEMS Immunol Med Microbiol 43(2):213–222

McFarland LV (2006) Meta-analysis of probiotics for the prevention of antibiotic associated diarrhoea and the treatment of *Clostridium difficile* disease. Am J Gastroenterol 101(4):812–822

Meurman JH (2005) Probiotics: do they have a role in oral medicine and dentistry? Eur J Oral Sci 113(3):188–196

Mitsou EK, Kirtzalidou E, Oikonomou I, Liosis G, Kyriacou A (2008) Faecal microflora of Greek healthy neonates. Anaerobe 14(2):94–101

Mitsuoka T (1996) Intestinal flora and human health. Asia Pacific J Clin Nutr 5(1):2–9

Mowat AM (2003) Anatomical basis of tolerance and immunity to intestinal antigens. Nat Rev Immunol 3(4):331–341

Mueller S, Saunier K, Hanisch C, Norin E, Alm L, Midtvedt T, Cresci A, Silvi S, Orpianesi C, Verdenelli MC, Clavel T, Koebnick C, Zunft HJ, Dore J, Blaut M (2006) Differences in faecal microbiota in different European study populations in relation to age, gender, and country: a cross-sectional study. Appl Environ Microbiol 72(2):1027–1033

Nyberg SD, Osterblad M, Hakanen AJ, Lofmark S, Edlund C, Huovinen P, Jalava J (2007) Long-term antimicrobial resistance in *Escherichia coli* from human intestinal microbiota after administration of clindamycin. Scand J Infect Dis 39(6–7):514–520

Nybom SM, Salminen SJ, Meriluoto JA (2008) Specific strains of probiotic bacteria are efficient in removal of several different cyanobacterial toxins from solution. Toxicon 52(2):214–220

O'Hara AM, Shanahan F (2006) The gut flora as a forgotten organ. EMBO Rep 7(7):688–693

Orrhage K, Nord CE (1999) Factors controlling the bacterial colonization of the intestine in breast-fed infants. Acta Paediatr 88:47–57

Orrhage K, Sillerström E, Gustafsson JA, Nord CE, Rafter J (1994) Binding of mutagenic heterocyclic amines by intestinal and lactic acid bacteria. Mutat Res 311(2):239–248

Osman N, Adawi D, Ahrne S, Jeppsson B, Molin G (2005) Probiotic strains of *Lactobacillus* and *Bifidobacterium* affect the translocation and intestinal load of *Enterobacteriaceae* differently after D-galactosamine-induced liver injury in rats. Microb Ecol Health Dis 17(1):40–46

Ouwehand AC, Lahtinen SJ (2008) The (non-) sense of human origin of probiotics. NutraCos 7:8–10

Ouwehand AC, Salminen S (1998) The health effects of cultured milk products with viable and non-viable bacteria. Int Dairy J 8:749–758

Ouwehand AC, Mäkeläinen H, Tiihonen K, Rautonen N (2006) Digestive health. In: Mitchell H (ed) Sweeteners and sugar alternatives in food technology, 1st edn. Blackwell Publishing, Oxford, pp 44–53

Paineau D, Carcano D, Leyer G, Darquy S, Alyanakian MA, Simoneau G, Bergmann JF, Brassart D, Bornet F, Ouwehand AC (2008) Effects of seven potential probiotic strains on specific immune responses in healthy adults: a double-blind, randomized, controlled trial. FEMS Immunol Med Microbiol 53(1):107–113

Paliy O, Kenche H, Abernathy F, Michail S (2009) High-throughput quantitative analysis of the human intestinal microbiota with a phylogenetic microarray. Appl Environ Microbiol 75(11):3572–3579

Palmer C, Bik EM, DiGiulio DB, Relman DA, Brown PO (2007) Development of the human infant intestinal microbiota. PLoS Biol 5(7):e177

Park HK, Shim SS, Kim SY, Park JH, Park SE, Kim HJ, Kang BC, Kim CM (2005) Molecular analysis of colonized bacteria in a human newborn infant gut. J Microbiol 43(4):345–353

Pavan S, Desreumaux P, Mercenier A (2003) Use of mouse models to evaluate the persistence, safety, and immune modulation capacities of lactic acid bacteria. Clin Diagn Lab Immunol 10(4):696–701

Petersen A, Heegaard PM, Pedersen AL, Andersen JB, Sorensen RB, Frokiaer H, Lahtinen SJ, Ouwehand AC, Poulsen M, Licht TR (2009) Some putative prebiotics increase the severity of *Salmonella enterica* serovar *Typhimurium* infection in mice. BMC Microbiol 9:245

Playne MJ, Crittenden RG (2008) Galacto-oligosaccharides. Production, properties and applications in foods. NutraFoods 7(1):9–18

Pool-Zobel BL, Neudecker C, Domizlaff I, Ji S, Schillinger U, Rumney C, Moretti M, Vilarini I, Scassellati-Sforzolini R, Rowland I (1996) *Lactobacillus*- and *Bifidobacterium*-mediated antigenotoxicity in the colon of rats. Nutr Cancer 26(3):365–380

Reid G (2008) Probiotic Lactobacilli for urogenital health in women. J Clin Gastroenterol (42 Suppl, 3 Pt) 2:S234–S236

Rousseaux C, Thuru X, Gelot A, Barnich N, Neut C, Dubuquoy L, Dubuquoy C, Merour E, Geboes K, Chamaillard M, Ouwehand A, Leyer G, Carcano D, Colombel JF, Ardid D, Desreumaux P (2006) *Lactobacillus acidophilus* modulates intestinal pain and induces opioid and cannabinoid receptors. Nat Med 13(1):35–37

Saarela M, Maukonen J, von Wright A, Vilpponen-Salmela T, Patterson AJ, Scott KP, Hämynen H, Mätto J (2007) Tetracycline susceptibility of the ingested *Lactobacillus acidophilus* LaCH-5 and *Bifidobacterium animalis* subsp. *lactis* Bb-12 strains during antibiotic/probiotic intervention. Int J Antimicrob Agent 29(3):271–280

Satokari RM, Vaughan EE, Favier CF, Doré J, Edwards C, Vos WM (2002) Diversity of *Bifidobacterium* and *Lactobacillus* spp. in breast-fed and formula-fed infants as assessed by 16S rDNA sequence differences. Microb Ecol Health Dis 14(2):97–105

Savage DC (1977) Microbial ecology of the gastrointestinal tract. Ann Rev Microbiol 31(1):107–133

Sazawal S, Hiremath G, Dhingra U, Malik P, Deb S, Black RE (2006) Efficacy of probiotics in prevention of acute diarrhoea: a meta-analysis of masked, randomised, placebo-controlled trials. Lancet Infect Dis 6(6):374–382

Scheppach W, Weiler F (2004) The butyrate story: old wine in new bottles. Curr Opin Clin Nutr Metabol Care 7:563–567

Schiffrin EJ, Brassart D, Servin AL, Rochat F, Donnet-Hughes A (1997) Immune modulation of blood leukocytes in humans by lactic acid bacteria: criteria for strain selection. Am J Clin Nutr 66(2):515S–520S

Schwiertz A, Gruhl B, Lobnitz M, Michel P, Radke M, Blaut M (2003) Development of the intestinal bacterial composition in hospitalized preterm infants in comparison with breast-fed, full-term infants. Pediatr Res 54(3):393–399

Sepp E, Julge K, Vasar M, Naaber P, Bjorksten B, Mikelsaar M (1997) Intestinal microflora of Estonian and Swedish infants. Acta Paediatr 86(9):956–961

Shu Q, Qu F, Gill HS (2001) Probiotic treatment using *Bifidobacterium lactis* HN019 reduces weanling diarrhoea associated with rotavirus and *Escherichia coli* infection in a piglet model. J Pediatr Gastroenterol Nutr 33(2):171–177

Stecher B, Hardt WD (2008) The role of microbiota in infectious disease. Trends Microbiol 16(3):107–114

Sullivan Å, Edlund C, Nord CE (2001) Effect of antimicrobial agents on the ecological balance of human microflora. Lancet Infect Dis 1(2):101–114

Szajewska H, Mrukowicz JZ (2001) Probiotics in the treatment and prevention of acute infectious diarrhoea in infants and children: a systematic review of published randomized, double blind, placebo-controlled trials. J Pediatr Gastroenterol Nutr 33:S17–S25

Tahara T, Oshimura M, Umezawa C, Kanatani K (1996) Isolation, partial characterization, and mode of action of acidocin J1132, a two component bacteriocin produced by *Lactobacillus acidophilus* JCM 1132. Appl Environ Microbiol 62(3):892–897

Tannock GW, Munro K, Harmsen HJM, Welling GW, Smart J, Gopal PK (2000) Analysis of the faecal microflora of human subjects consuming a probiotic product containing *Lactobacillus rhamnosus* DR20. Appl Environ Microbiol 66(6):2578–2588

Tiihonen K, Tynkkynen S, Ouwehand A, Ahlroos T, Rautonen, N. (2008) The effect of ageing with and without non-steroidal anti-inflammatory drugs on gastrointestinal microbiology and immunology. Br J Nutr 100(1):130–137

Tiihonen K, Ouwehand AC, Rautonen N (2010) Human intestinal microbiota and healthy ageing. Ageing Res Rev 9(2):107–116

Tompkins TA, Hagen KE, Wallace TD, Fillion-Forte V (2008) Safety evaluation of two bacterial strains used in Asian probiotic products. Can J Microbiol 54(5):391–400

Tuohy KM, Kolida S, Lustenberger AM, Gibson GR (2001) The prebiotic effects of biscuits containing partially hydrolysed guar gum and fructo-oligosaccharides—a human volunteer study. Br J Nutr 86:341–348

Turnbaugh PJ, Hamady M, Yatsunenko T, Cantarel BL, Duncan A, Ley RE, Sogin ML, Jones WJ, Roe BA, Affourtit JP, Egholm M, Henrissat B, Heath AC, Knight R, Gordon JI (2009) A core gut microbiome in obese and lean twins. Nature 457(7228):480–484

Valeur N, Engel P, Carbajal N, Connolly E, Ladefoged K (2004) Colonization and immunomodulation by *Lactobacillus reuteri* ATCC 55730 in the human gastrointestinal tract. Appl Environ Microbiol 70(2):1176–1181

van Tongeren SP, Slaets JPJ, Harmsen HJM, Welling GW (2005) Faecal microbiota composition and frailty. Appl Environ Microbiol 71(10):6438–6442

Vanhoutte T, Huys G, De Brandt E, Swings J (2004) Temporal stability analysis of the microbiota in human faeces by denaturing gradient gel electrophoresis using universal and group-specific 16S rRNA gene primers. FEMS Microbiol Ecol 48(3):437–446

Vankerckhoven V, Huys G, Vancanneyt M, Snauwaert C, Swings J, Klare I, Witte W, Van Autgaerden T, Chapelle S, Lammens C, Goossens H (2008) Genotypic diversity, antimicrobial resistance, and virulence factors of human isolates and probiotic cultures constituting two intraspecific groups of *Enterococcus faecium* isolates. Appl Environ Microbiol 74(14):4247–4255

Vesterlund S, Vankerckhoven V, Saxelin M, Goossens H, Salminen S, Ouwehand AC (2007) Safety assessment of *Lactobacillus* strains: presence of putative risk factors in faecal, blood and probiotic isolates. Int J Food Microbiol 116(3):325–331

Wagner RD, Warner T, Roberts L, Farmer J, Balish E (1997) Colonization of congenitally immunodeficient mice with probiotic bacteria. Infect Immun 65(8):3345–3351

Wagner RD, Pierson C, Warner T, Dohnalek M, Hilty M, Balish E (2000) Probiotic effects of feeding heat-killed *Lactobacillus acidophilus* and *Lactobacillus casei* to *Candida albicans*-colonized immunodeficient mice. J Food Protect 63(5):638–644

Wolf BW, Garleb KA, Ataya DG, Casas IA (1995) Safety and tolerance of *Lactobacillus reuteri* in healthy adult male subjects. Microb Ecol Health Dis 8(2):41–50

Wong C, Ustunol Z (2006) Mode of inactivation of probiotic bacteria affects interleukin 6 and interleukin 8 production in human intestinal epithelial-like Caco-2 cells. J Food Protect 69(9):2285–2288

Woodmansey EJ, McMurdo MET, Macfarlane GT, Macfarlane S (2004) Comparison of compositions and metabolic activities of faecal microbiotas in young adults and in antibiotic-treated and non-antibiotic-treated elderly subjects. Appl Environ Microbiol 70(10):6113–6122

Yildirim Z, Johnson MG (1998) Characterization and antimicrobial spectrum of bifidocin B, a bacteriocin produced by *Bifidobacterium bifidum* NCFB 1454. J Food Protect 61(1):47–51

Zhou JS, Shu Q, Rutherfurd KJ, Prasad J, Gopal PK, Gill HS (2000) Acute oral toxicity and bacterial translocation studies on potentially probiotic strains of lactic acid bacteria. Food Chem Toxicol 38(2–3):153–161

Zoetendal EG, Akkermans AD, de Vos WM (1998) Temperature gradient gel electrophoresis analysis of 16S rRNA from human faecal samples reveals stable and host-specific communities of active bacteria. Appl Environ Microbiol 64(10):3854–3859

Zoetendal EG, Collier CT, Koike S, Mackie RI, Gaskins HR (2004) Molecular ecological analysis of the gastrointestinal microbiota: a review. J Nutr 134(2):465–472

Zoetendal EG, Rajilic-Stojanovic M, de Vos WM (2008) High-throughput diversity and functionality analysis of the gastrointestinal tract microbiota. Gut 57(11):1605–1615

Zou J, Dong J, Yu X (2009) Meta-analysis: *Lactobacillus* containing quadruple therapy versus standard triple first-line therapy for *Helicobacter pylori* eradication. Helicobacter 14(5):97–107

Chapter 3
Ecology of Probiotics and Enteric Protection

Melanie Gagnon, Annina Zihler, Christophe Chassard and Christophe Lacroix

Abbreviations

BLIS	Bacteriocin-like inhibitory substance
CFU	Colony forming unit
FOS	Fructooligosaccharides
GI	Gastrointestinal
GRAS	Generally recognized as safe
IBS	Irritable bowel syndrome
IFN	Interferon
IgA	Immunoglobulin A
IL	Interleukin
LAB	Lactic acid bacteria
LGG	*Lactobacillus rhamnosus* GG
MIC	Minimal inhibitory concentration
NF-κB	Nuclear factor-κB
SHIME	Simulator of the human intestinal microbial ecosystem
SCFA	Short chain fatty acid
TEER	Transepithelial electrical resistance
TNF	Tumour necrosis factor

3.1 Introduction

The human gastrointestinal (GI) tract harbours a complex community of microorganisms that confers a wide range of health benefits on the host. During the past few years there has been increased interest in influencing this microbial ecosystem

C. Lacroix (✉)
Laboratory of Food Biotechnology, Institute of Food Science and Nutrition,
ETH Zurich, Schmelzbergstrasse 7, 8092 Zurich, Switzerland
e-mail: christophe.lacroix@ilw.agrl.ethz.ch

J. J. Malago et al. (eds.), *Probiotic Bacteria and Enteric Infections*,
DOI 10.1007/978-94-007-0386-5_3, © Springer Science+Business Media B.V. 2011

to support and enhance enteric protection. Enteric infectious diseases are a major cause of global human morbidity and mortality, resulting in more than 2 million deaths each year. Effective vaccines are not yet available for many important diarrhoeal diseases and, conventional antimicrobial therapy is increasingly complicated by escalating rates of drug resistance (Paton et al. 2006). There is therefore a major need for alternative approaches to standard antibiotic therapies that can prevent and/or treat infectious diseases in the future.

Probiotic bacteria exhibit a high potential in this respect because they possess an array of strategies to compete within the intestinal microbiota and exogenous invaders for the colonization of specific ecological niches. Some *in vivo* studies confirm the potential of probiotics for prevention of antibiotic-related diarrhoea in infant or prevention of traveller's diarrhoea (Johnston et al. 2007; McFarland 2007). While these studies have been important in demonstrating probiotic efficacy against enteric pathogens, few have specifically identified a mechanistic basis supporting the observed benefits, and many rely on *in vitro* testing. Possible mechanisms include direct antagonism by inhibitory substances, competitive exclusion and potentially host-mediated effects such as improved barrier function and altered immune response (O'Toole and Cooney 2008; Corr et al. 2009). It should be emphasized that properties of probiotics, including the inhibition of pathogens, are strain-specific, and further understanding of underlying mechanisms is required to specifically target GI pathogens through the use of appropriate probiotic strains.

This chapter discusses the ecology and taxonomy of probiotics in relation to their physiology and their potential for enteric protection from selection to probiosis. These will be studied in advanced models and in humans.

3.2 Probiotics from Taxonomy to Functionality

Probiotics are defined as live microorganisms that, when administered in adequate amounts, confer a health benefit on the host (FAO/WHO 2002). They are becoming increasingly important in human nutrition and health, as scientific evidence is gathered on functional properties and benefits of probiotics for health promotion and well-being. Targeted design and conception of probiosis have however evolved, especially during the two last decades. The main target for probiotics is healthy people and, concerning infections and others diseases, consumption is obviously related to the preventive potential of probiotics incorporated in regular food. Probiotics are also good candidates for pharmaceutical applications and clinical use, providing efficient alternative therapies for some of the most prevalent human diseases. Current criteria for selection of probiotics with targeted efficacy include safety, technological and functional aspects (Fig. 3.1). As the number of potential probiotic microorganisms increase the safety of each must be carefully tested.

Fig. 3.1 Current criteria for selection of probiotics with targeted efficacy

Safety criteria

- Human origin
- Strain typing
- Virulence factors
- Antibiotic resistances
- Genome sequencing

Technological criteria

- Easy propagation
- High viability
- Stability in products
- Good physiology

Functional criteria

- Survival in the GIT
- Adhesion to epithelial cells
- Immuno-stimulatory with appropriate cytokines stimulation
- Inhibit GI pathogens
- Desirable metabolic activities e.g. carbohydrate metabolism
- Anti-mutagenic and anti-carcinogenic properties

3.2.1 *Origin and Taxonomy of Probiotics*

The origin of the microorganisms used as probiotics is very important. A major criterion for selection of probiotics for human use is their human origin. Only few probiotic strains have another origin such as animal digestive ecosystems or fermented food stuffs. The main strains recognized as probiotics for humans are bacteria belonging to the genera *Lactobacillus, Bifidobacterium, Enteroccocus, Streptoccocus* and the yeasts of *Saccharomyces*. Among these microorganisms, bacteria belonging to the genera *Lactobacillus* and *Bifidobacterium* are most frequently used as human probiotics (Table 3.1).

Lactobacilli are rod-shaped bacteria phylogenetically related to the Firmicutes phylum. They are mainly found in plants and they play an important role, contributing to acidification and preservation, aroma and texture development in fermented foods but probiotic strains have mainly a human origin. Lactobacilli have complex nutritional requirements and are aerotolerant or anaerobic, aciduric or acidophilic. Bifidobacteria are rods of variable morphology, with characteristic Y shape, belonging to the Actinobacteria phylum. They have an almost ubiquitous origin from digestive ecosystems. A comprehensive review of taxonomy and physiology of lactobacilli and bifidobacteria is given by Biavati and Mattarelli (2006), Felis and Dellaglio (2007) and Walter (2008). Bifidobacteria diversity in the human gut was extensively investigated using culture and culture-independent methods. A culture-independent study analyzing the diversity of several 16S rRNA gene-based libraries recently revealed the presence of novel bifidobacterial phylotypes, which had not been found earlier and, represent novel taxa within the genus *Bifidobacterium* (Turroni et al. 2009). This study demonstrates that bifidobacterial diversity is higher than expected and suggests that new species could be isolated from the human intestine. Improvement of growth conditions including media and abiotic parameters could provide new tools and strategies to cultivate new *Bifidobacterium* strains from the human gut (Chassard et al. 2011).

Since *Lactobacillus* and *Bifidobacterium* are a major part of the normal intestinal microflora in humans, these probiotics should be able to compete with autochtho-

Table 3.1 A list of potential probiotic strains and their observed beneficial effects. (Adapted from Corr et al. 2009)

Organisms	Effects	Mechanisms	References
Lactobacilli			
L. rhamnosus R0011	Improves epithelial barrier function	Prevents the pathogen-induced drop in transepithelial resistance	Sherman et al. (2005)
L. rhamnosus GG	Inhibition of virulence factor expression Immunomodulation	Reduces expression of genes encoding shiga toxin Activates NF-κB and regulates inflammatory response in macrophages	Carey et al. (2008) Miettinen et al. (2000)
L. plantarum 299v	Improves epithelial barrier function	Increases extracellular secretion of mucin, MUC3	Mack et al. (2003)
L. plantarum ITM21B	Inhibition of virulence factor expression	Inhibition of urease activity in Y. enterocolitica	Lavermicocca et al. (2008)
L. johnsonii NCC533	Secretion of inhibitory substances	Production of hydrogen peroxide	Pridmore et al. (2008)
L. salivarius UCC118	Secretion of inhibitory substances	Production of bacteriocin	Corr et al. (2007b)
L. casei NCDO1205	Immunomodulation	Decrease IL-8 and increase IL-10 response	Corr et al. (2007a)
Bifidobacteria			
B. breve Yakult	Secretion of inhibitory substances	Production of acetic acid and thus lowering of pH	Asahara et al. (2004)
B. lactis Bb-12	Immunomodulation	Stimulates sIgA	Fukushima et al. (1998)
Others			
E. coli strain Nissle 1917	Improves epithelial barrier function	Increases tight junction protein expression, ZO-2	Zyrek et al. (2007)
VSL#3 probiotic mixture	Improves epithelial barrier function	Induces human β-defensin, hBD-2 gene expression	Schlee et al. (2008)

nous lactic acid bacteria (LAB) and bifidobacteria strains for survival and exerting their probiotic properties. Bifidobacteria and lactobacilli population levels are however very different in humans where the lactobacilli population is estimated around 10^6 colony forming unit (CFU) per gram using culture methods (Finegold et al. 1974; Walter 2008). Data obtained using culture independent methods confirmed these results, with lactobacilli accounting for about 0.01% of the total population (Sghir et al. 2000; Harmsen et al. 2002; Furet et al. 2009). Probes used for molecular lactobacilli enumeration are however not specific for lactobacilli enumeration and also detect *Enteroccoccus* (Harmsen et al. 2002), *Streptoccoccus* (Sghir et al. 2000) or *Leuconostoc-Pediococcus*. Bifidobacteria counts are estimated around 10^8 cells in healthy adults. Population is however strongly influenced by diseases, age or diet, and as example, bifidobacteria population is found higher in breast fed babies than in formula fed babies (Franks et al. 1998; Rigottier-Gois et al. 2003; Chassard et al. 2008; Le Blay et al. 2010).

Bifidobacteria and lactobacilli are commonly added in functional foods especially in fermented dairy products although for many strains found in products, beneficial health effects are poorly demonstrated or even absent (Lacroix and Yildirim 2007). It is recognized that the properties, functionality, and benefits of each probiotic are unique. Benefits of a particular strain therefore cannot be extrapolated to other strains, even within the same species (Pineiro and Stanton 2007). *Lactobacillus* and *Bifidobacterium* strains are also used for pharmaceutical and clinical applications. The probiotic potential of many strains is currently investigated while few characterized probiotic strains are already present on the pharmaceutical market.

3.2.2 Safety Assessment of Probiotics

Microorganisms used in food fermentation have a long history of safe use and are therefore considered to be non-pathogenic with "generally recognized as safe" (GRAS) status. The emergence of new strains however, increases the panel of potential probiotics. Proof of absence of pathogenicity and infectivity is required for probiotic safety. The safety of these new candidates needs therefore to be carefully evaluated, especially for microorganisms not commonly used in food and those most recently isolated from gut ecosystems. Taxonomic identification is a critical step in the assessment of safety of potential novel probiotic strains (Holzapfel et al. 2001). It is best done with a polyphasic approach combining phenotypic and physiological methods with molecular studies since physiological characterization alone, for example by fermentation profiling is insufficient to achieve reliable identification (von Ah et al. 2007). Antibiotic resistances must also be carefully checked in probiotics although in the context of probiotics used to treat enteric infection, this characteristic may be beneficial enabling the development of combined antibiotic-probiotic treatments. Antibiotic resistance gene should be very stable and not transferable and the potential for transmission of genetic elements to other intestinal/food microorganisms needs to be carefully considered in the future (Pineiro and Stanton 2007). Even if the safety of the current products is excellent, probiotics could, as living microorganisms, theoretically be responsible for side effects in susceptible individuals and must be investigated (Marteau and Shanahan 2003). Factors to be addressed in probiotic safety evaluation include pathogenicity, infectivity, and the virulence factors of the microbes (Ishibashi and Yamazaki 2001). Methods for safety assessment range from *in vitro* studies to *in vivo*, animal and human clinical studies (Donohue and Salminen 1996; Abe et al. 2009).

3.2.3 Viability and Functionality of Probiotics

High survival through the upper GI tract and high viability at its site of action are also required, together with high activity in the gut environment. Indeed, the viability of probiotics is a key parameter for developing probiotic foods or supplements. Probiotic cells must be viable in most of the reported beneficial health effects even if non-viable forms or cell-wall components of probiotics can also exhibit relevant functional health properties (Salminen et al. 1999). The amount of cells required to produce therapeutic benefits is not known and might largely vary as a function of the strain, product matrix composition, and the health effect desired. However in general a minimum level of more than 10^6 viable probiotic bacteria per milliliter or gram of food product is accepted (Lacroix and Yildirim 2007).

Probiotics must overcome the two main biological barriers of stomach acidity and bile secretion in the small intestine to reach the lower part of the GI tract in a viable state. Probiotic bacteria vary considerably in their tolerance to low pH and bile salts; this can be evaluated *in vitro*, by controlled incubation in gastric juice followed by bile treatment to reflect physiological conditions found in the upper part of the GI tract (Dunne et al. 2001). Tolerance to GI stress is generally low and it is important to find new strategies to improve probiotic strain resistance (Lacroix and Yildirim 2007). Bacterial mechanisms have been recently investigated (Whitehead et al. 2008), demonstrating the role of multidrug resistance transporters conferring bile resistance to probiotics and subsequently, a strong ecological advantage (Gueimonde et al. 2009).

Once probiotics have reached the colon they should compete efficiently with gut microbiota to survive and temporarily colonize it. It is therefore important to study physiological parameters like substrate utilization and to use such markers to help predict colonization success. The ability of probiotics to use a broad range of substrates provides more opportunities to colonize a specific ecological niche and limit competition with predominant gut bacteria. Lactobacilli have very limited abilities to utilize complex carbohydrates (Walter 2008), in contrast to bifidobacteria that plays an important role in the breakdown and fermentation of complex carbohydrates (Amaretti et al. 2006). Bifidobacteria utilize a variety of plant dietary fibers due to their ability to produce several cell-bound and extracellular depolymerizing enzymes (Rossi et al. 2005). Many glycosyl hydrolases of the genus *Bifidobacterium* have been characterized, such as β-galactosidase (Moller et al. 2001), β-fructofuranosidase (Janer et al. 2004) or α-amylase (Lee et al. 1997; Wang et al. 1999). Alternatively, lactobacilli can utilize simple sugars liberated from the degradation of complex carbohydrates, by other microbes opening specific niches in the GI tract since enteric pathogens such as enterobacteria occupy similar niches. High competition between probiotic lactobacilli and enteric pathogens is expected. However, some species such *Lactobacillus acidophilus* or *Lactobacillus plantarum* are also able to metabolize complex prebiotic carbohohydrates such as fructans that remain untouched by human enzymes that can serve as nutrients in the colon (Saulnier et al. 2007; Walter 2008).

Probiotics growth could therefore be stimulated by selected prebiotics providing a specific substrate. This combination of probiotics and prebiotics is referred to as synbiotics. The synbiotic approach can potentiate the effect of probiotics with prebiotics. Promising results have been shown *in vitro* (Saulnier et al. 2008) and *in vivo* in animals (Frece et al. 2009), and also confirmed *in vivo* in humans. Most studies demonstrate that probiotic/prebiotic combinations were more effective than placebo, and probiotics or prebiotics provided alone (De Vrese and Schrezenmeir 2008). More work is however needed to understand the effects of selected synbiotics on microbial communities in the gut, and their interactions with the host's immune system. Until then, the impact of fructans and other prebiotics on enterobacteria population remains unclear and controversial. Several studies reported that enterobacteria could also use prebiotics as a substrate for their growth (Hartemink et al. 1997; Martín-Peláez et al. 2008), suggesting that a synbiotic strategy may be not suitable or safe to treat or prevent enteric infection since pathogens could be stimulated generating an opposite effect. Future investigators should use suitable *in vitro* and *in vivo* models prior to starting *in vivo* studies including prebiotics or synbiotics in those studies to treat and/or prevent enteric infections.

Furthermore, antimicrobial properties primarily targeted against specific pathogens could promote colonization and enhance effectiveness of probiotics (Cleusix et al. 2007; Le Blay et al. 2007). Organic acids produced by probiotics especially acetate and lactate, can inhibit bacterial growth but specific microbial products such as bacteriocins (antimicrobial peptides) are however also naturally produced by many strains of LAB and bifidobacteria to counter bacterial growth of closely related strains and provide a competitive advantage to producer strains in its natural environment. The most recent developments in this potent field of research will be discussed later in the chapter.

3.2.4 Probiotics and Technological Aspects

Probiotics should have good technological properties to be produced on a large scale and incorporated into food products without losing viability, ecological properties and especially functionality or creating unpleasant flavour or texture (Lacroix and Yildirim 2007). A probiotic must exhibit high cell yields during fermentation and high survival rates in downstream processes (such as centrifugation and drying) and in food products during storage.

Commercial strains are currently largely selected for their technological properties since they must be developed economically on a large scale. This rules out some strains with promising health properties but with limited technological potential. Understanding mechanisms that enhance cell survival could lead to more efficient applications of probiotics in products with high functionality (Lacroix and Yildirim 2007). Consequently, industry demand for new technologies that promote high viable cell yields at large scale and to ensure probiotic stability in food remains strong, because many strains of intestinal origin are difficult to propagate and, high survival is important for both commercial production and health. More efficient technologies could also lead to greater product efficacy and strain diversification with the development of technologically unsuitable strains into products. Sublethal stress application for stress adaptation and, new technologies such as continuous cultures and cell immobilization are promising and may be used to profile cell physiology to optimize survival and functionality in the gut (Lacroix and Yildirim 2007). Physical protection of probiotic cells from their external environment using encapsulation techniques, also referred as microencapsulation referring to the size of produced capsules, are also increasingly used in industry to preserve cell viability under adverse conditions, such as in the presence of oxygen, acidic environment, simulated GI juices, etc. (Doleyres and Lacroix 2005). Current studies focus on optimizing technology and selecting the best matrix for the controlled delivery of microencapsulated cells in the GI tract and improving ecological features of probiotics in relation to their functionality.

3.3 Interaction of Probiotics with Pathogens and Gut Microbiota

In response to the increasing incidence of antibiotic resistant pathogens, probiotics are of great interest to influence gut microbiota composition and function in a manner that supports and enhances enteric protection. It is therefore essential to understand how probiotics interact with gut microbiota. The assessment of new probiotics with specific anti-pathogenic activity is normally based on scientific evidence established step-by-step; first *in vitro*, confirmed *in vivo* through animal models and finally validated in clinical trials (Chassard et al. 2011). All these methods have advantages and limits but are generally combined for more effectively studying the impact of probiotics on the gut microbiota and the mechanisms involved in enteric protection. The different models used for probiosis assessment are discussed with emphasis on the interaction of probiotics with pathogens and the gut microbiota.

3.3.1 Dose-dependent Effects of Probiotics on Pathogens and Gut Microbiota

The ability to confer benefits on the host through growth and/or activity in the human body presupposes that probiotics reach the specific site of action in viable, adequate numbers for optimal functionality. Determination of the required probiotic dose for mediating a desired efficacy is indeed a challenging task, incorporating both commercial (cost) and scientific interests.

Generally, a minimal number of 10^8–10^9 bacteria is recommended to be consumed daily for achieving a desired effect (Gill and Prasad 2008). The optimal concentration depends on several factors however, including the kind of probiotic organism, claimed health effect, duration of application, delivery matrix (food or pharmaceutical), and target population group (De Vrese and Schrezenmeir 2008). These interactive factors have been studied within the scope of different human dietary intervention studies yielding different results. A two-week therapy with *Bifidobacterium lactis* HNO19, for example, positively influenced the intestinal microbiota of elderly subjects (>60 years old), significantly elevating bifidobacteria and lowering enterobacteria counts, independently of the probiotic concentration consumed (5×10^9, 1×10^9 or 6.5×10^7 CFU/day) (Ahmed et al. 2007). In contrast, the recovery of *Bifidobacterium animalis* ssp. *lactis* (Bb-12) from faeces of healthy young adults was shown to be highly dose-dependent. In fact, after a three-week consumption of Bb-12 together with *Lactobacillus paracasei* spp. *paracasei* (CRL-431) at concentrations of 10^8, 10^9, 10^{10} and 10^{11} CFU/day, the faecal recovery of Bb-12 increased significantly with increasing dose, whereas CRL-431 was not recovered from any of the faecal samples (Larsen et al. 2006). Recently, a dose-efficacy study revealed that a one-week *Lactobacillus rhamnosus* GG (LGG) treatment at two dosages (10^{10} and 10^{12} CFU twice per day) was equally effective in decreasing the frequency and duration of rotavirus-induced acute watery diarrhoea in Indian children (Basu et al. 2009), but data of the effects of LGG on the gut microbiota are lacking.

There are no clinical trials investigating the dose-dependent effects of probiotics on both enteric infection and gut microbiota composition. This is mainly due to the fact that most of the current probiotics were not selected for specific purposes, e.g. for their ability to inhibit pathogens, but a broad-range activity was expected and tested empirically. There is however a general trend in developing a new generation of "tailored probiotics" with enhanced target-specific efficacy, e.g. for improved enteric protection (Chassard et al. 2011). Advanced *in vitro* and *in vivo* models combined with basic tests are therefore required and represent important tools to assess the functional activity of probiotics.

3.3.2 *In vitro* Assessment of Probiotic Functional Activity

Various probiotics produce a number of antimicrobial substances inhibiting bacteria that compete for similar intestinal niches or bacterial invaders such as *Listeria* or *Salmonella*. Antimicrobial properties, in fact, are an important characteristic for assessing probiotics and are often evaluated *in vitro* using simple inhibition tests. Such testing may include both pathogenic bacteria and strains representative of the predominant human intestinal microbiota. Agar-well diffusion or critical dilution assays are for example performed to determine the activity spectra of a bacteriocin (purified or present in culture supernatants) by comparing the diameters of the inhibition zone around the spot where the active substance was applied or, by optically comparing the number of wells with inhibited growth, respectively (Turcotte et al. 2004). For the latter, determination of the minimal inhibitory concentration (MIC) may be possible by applying serial dilutions of the antimicrobial molecule at known concentrations. For example, *Ruminococcus gnavus* E1, a potential probiotic bacteria isolated from human faeces, was shown to produce a bacteriocin (Ruminococcin A) highly active against *Clostridium perfringens* and various other enteric pathogens as well as phylogenetically-related *R. gnavus* strains but not against other prominent members of the gut microbiota (Dabard et al. 2001). Similar results were obtained for Pediocin PA-1, a bacteriocin produced by *Pediococcus acidilactici* and active against *Listeria* spp. and other food pathogens. In contrast to Nisin A, which is used as natural food preservative, Pediocin PA-1 did not inhibit any of the tested species of the human intestine (Le Blay et al. 2007).

Such "static" tests, in fact, are inexpensive to perform and may help to screen for new probiotic bacteria with specific anti-pathogen activity, and assess the activity spectrum on gut bacteria in pure culture. Gastrointestinal environments are however much more complex and dynamic interactions between microorganisms are largely involved in the balance and barrier effects of the gut microbiota *in vivo*. In this respect, more elaborated tests working with complex microbiota and environmental conditions akin to the GI tract are required to get a more accurate view of antimicrobial properties of probiotics. Continuous fermentation models mimicking the digestive tract may be useful in this respect and have been applied to investigate specific impact of probiotics on gut microbiota, including metabolic and antimicrobial properties. *In vitro* models have no ethical guidelines and are well-suited for mechanistic studies since the use of radioactive or toxic components is facilitated (Macfarlane and Macfarlane 2007). For example, increased bifidobacteria counts and butyrate concentrations were observed after addition of a synbiotic mixture of *Lactobacillus acidophilus* 74-2 and fructooligosaccharides (FOS), which is used in probiotic products, in a complex multistage fermentation system called "Simulator of the Human Intestinal Microbial Ecosystem (SHIME)" (Gmeiner et al. 2000). More recently, an *in vitro* model of colonic fermentation with immobilized human faeces was developed for the controlled cultivation of gut microbiota (Cinquin et al. 2004, 2006) and successfully used to demonstrate *in-situ* production and activity of reuterin by *Lactobacillus reuteri* after glycerol addition to the reactor (Cleusix et al. 2008). Presence of reuterin (3-hydroxypropionaldehyde), a broad-range potent antimicrobial compound significantly and selectively decreased *Escherichia coli* without affecting other bacterial populations. A similar model with immobilized human faeces was recently developed to simulate colonic *Salmonella* infection in children (Le Blay et al. 2009). Addition of *Salmonella* induced an increase of *Enterobacteriaceae, Clostridium coccoides, Eubacterium rectale* and *Atopobium* populations and a decrease of bifidobacteria. The same model also revealed the probiotic potential of *Bifidobacterium thermophilum* RBL67, producing a bacteriocin-like inhibitory substance (BLIS) that inhibits *Salmonella* without changing the gut microbiota composition as is seen with antibiotics. Although *in vitro* fermentation models may facilitate testing functional mechanisms of probiotics responsible for enteric protection within a complex microbial ecosystem, they have several limitations. In fact, *in vitro* intestinal fermentation models do not include host immune and neuroendocrine responses, making the use of *in vitro* epithelial cell models indispensable (as discussed later). Furthermore, they require fresh and high quality faecal inoculum and other biotic factors that are difficult to reproduce. *In vivo* studies are therefore needed for probiosis assessment.

3.3.3 *In vivo Assessment of Probiotic Functional Activity*

Animal models are essential to establish safety and functionality of probiotics and offer direct access to intestinal contents as well as tissues and organs at autopsy. They are of great value performing procedures that involve a level of harm that would not be ethical to inflict on a human. They constitute optimal tools for studying the effects of probiotics not only on the gut microbiota but also in enteric protection while investigating the corresponding mechanisms involved. For example, *Lactobacillus salivarius* UCC118, a probiotic strain of human origin, was shown to produce a bacteriocin *in vivo* that significantly protect mice against infection with *Listeria monocytogenes* (Corr et al. 2007a). In another study, *Bifidobacterium longum* Bb46 administered as prophylaxis to mice before being intragastrically challenged with *Salmonella typhimurium* resulted in higher survival rates of mice. This protective effect was related to a reduced inflammatory response mediated by the probiotic treatment rather than to a population antagonism (Silva et al. 2004). Both studies however, did not evaluate possible probiotic effects on the gut microbiota. Recently, the anti-listerial effect of treatments with purified Pediocin PA-1 and its producing strain, *Pediococcus acidilactici* UL5, were investigated *in vivo* in mice. Whereas repeated doses of Pediocin PA-1 significantly reduced faecal *Listeria monocytogenes* counts and slowed pathogen translocation into liver and spleen, administration of a single dose (4×10^{10} CFU) of the producing strain had no positive effects on listerial counts even though probiotic bacteria isolated from mice faeces were still producing the bacteriocin. More importantly, both treatment strategies did not affect the composition of the mouse intestinal flora (Dabour et al. 2009). Data from animal models for probiotics designated for human use have to be interpreted with caution however, since host anatomy and physiology may influence bacterial interactions. Human studies are therefore of great value.

3.3.4 *Human Studies—Ultimate Demonstration*

Very few human studies investigating the effects of probiotic intervention on both the microbiota and enteric protection have been done, and they often focus on one target (O'Toole and Cooney 2008). For example, lactobacilli and enterococci were detected more frequently and at higher numbers among ten subjects consuming a milk product containing *Lactobacillus rhamnosus* DR20 (Tannock et al. 2000). Whereas shedding of DR20 in faeces stopped soon after consumption in most subjects, it persisted in one person for two months after the test period, suggesting that ecological competitiveness of probiotics depends on specific host factors, e.g. microbiota composition before probiotic intake. In another study, consumption of yoghurt containing *Lactobacillus casei, Lactobacillus bulgaricus* and *Streptococcus thermophilus* significantly reduced the incidence of antibiotic- and *Clostridium difficile*-associated diarrhoea in elderly subjects (mean age 74) (Hickson et al. 2007). Direct effects on the microbiota were however not studied. Administration of a multispecies probiotic formula consisting of *Lactobacillus rhamnosus* GG, *L. rhamnosus* Lc705, *Propionibacterium freudenreichii* ssp. *shermanii* JS and *Bifidobacterium breve* Bb99 was shown to alleviate symptoms in patients suffering from irritable bowel syndrome (IBS) while having negligible effects upon healthy gut microbiota composition and metabolic activity (Kajander et al. 2007, 2008). To our knowledge, this is the only human study explicitly relating probiotic administration and functionality to enteric protection and gut microbiota integrity.

There are few *in vitro* and *in vivo* studies that have rigorously characterized the effects of probiotic administration for enteric protection on gut microbiota. Further investigations could contribute significantly to our understanding of the complex triangular relationship of probiotics, pathogens and gut microbiota within the human GI tract. Additional studies are also needed to characterize the mechanisms involved for enteric protection.

3.4 Intestinal Cytoprotection of Probiotics in Prophylactic Uses

Most data available on mechanistic effects of probiotics come from *in vitro* studies using epithelial cell monolayers (Mennigen and Bruewer 2009). These models are useful to study bacterial interactions with intestinal epithelium in particular states such as enteric infection. Different human intestinal cell lines such as Caco-2, HT-29 and HT29-MTX have been used, each one with specific characteristics (absorptive or globlet cellular type, secretion of mucus, production of immune molecules). Care must therefore be taken in the choice of the cell model for testing properties such as bacterial adhesion or immune modulation (Laparra and Sanz 2009). For example, the parental cell line HT-29 secretes a lower amount of mucus compared with the HT29-MTX and normal epithelial cells and, this characteristic could influence the output of the test and suitability of the model to reflect *in vivo* physiological conditions. Although comparison of different cell models are lacking, cell assays can provide important information on human enteric pathogen's interaction with the host, as in Fig. 3.2.

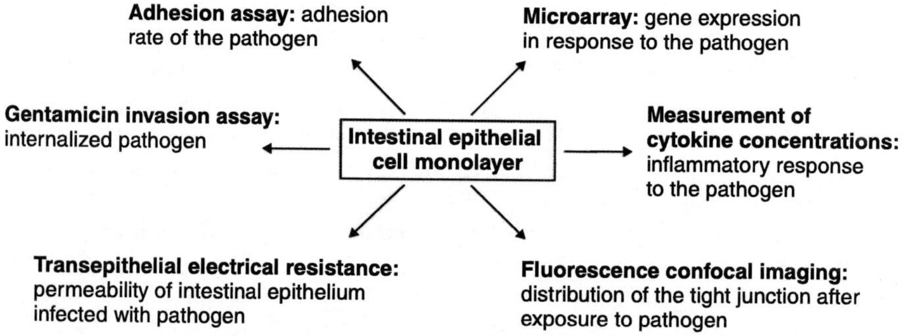

Fig. 3.2 Assays performed with human intestinal cells for characterizing human enteric pathogens

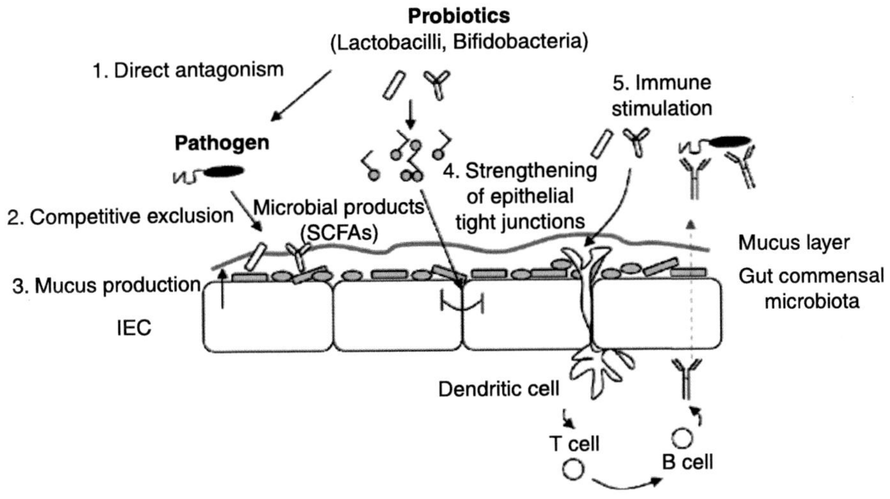

Fig. 3.3 Schematic representation of the interaction of microbiota with rationally selected probiotics to inhibit pathogens to the intestinal epithelial cells (IEC) through (*1*) direct antagonism (*2*, *3*, *4*) competitive exclusion and improvement of epithelial barrier function and (*5*) immunomodulation via immune cells activation. (Adapted from O'Toole and Cooney 2008)

Probiotic bacteria have multiple and diverse effects on the host in addition to the substantial role of the intestinal microbiota for the cytoprotection of the intestinal epithelial cells (Koninkx and Malago 2008). Figure 3.3 summarizes the principal mechanisms by which probiotics could exert their anti-pathogen activity. These modes of action will be discussed with some examples in relation with human enteric pathogens.

3.4.1 *Protection via Secretion of Antimicrobial Substances and Acids*

Probiotic bacteria produce a variety of substances that inhibit pathogenic bacteria by direct antagonism. These inhibitory compounds include organic acids and bacteriocins. The ability of bacteriocins to fight pathogens is well documented. They are therefore prime candidates as anti-pathogenics. The active antimicrobial molecules produced by some probiotics can kill pathogens before they could adhere to the intestinal epithelium. The probiotic *E. coli* strain Nissle 1917 was shown to interfere with *Salmonella typhimurium* invasion of human embryonic intestinal epithelial INT407 cells via secretion of inhibitory substances even when the probiotic was physically separated from the bacteria by a nonpermeable membrane (Altenhoefer et al. 2004). In a previous study Corr et al. (2007a) used a similar transwell chamber system to demonstrate that *Lactobacillus* and *Bifidobacterium* strains (*L. casei,*

L. acidophilus, L. salivarius, B. breve, B. infantis, B. longum) are capable of inhibiting *Listeria monocytogenes* invasion of C2Bbe1 epithelial cells through secretion of proteinaceous molecule(s), active at low pH in lactobacilli strains tested. However, the nature of the proteinaceous substances studied needs to be identified. *Lactobacillus* and *Bifidobacterium* spp. have also been shown to prevent infection of human intestinal cells by enterohemorrhagic *Escherichia coli* O157:H7 by the combined action of lactic acid and proteinaceous substances (Gopal et al. 2001). An *in vitro* study of the ability of *L. rhamnosus* DR20 and *B. lactis* DR10 to impede infection of differentiated human intestinal cell lines by *E. coli* O157:H7 found that pretreatment of *E. coli* with concentrated cell-free culture supernatants from these probiotic bacteria significantly reduced numbers of culturable *E. coli* and the invasiveness of this strain (Gopal et al. 2001).

The secretion of antimicrobial molecules by the probiotics needs to be confirmed in *in vivo* studies in the presence of complex intestinal microbiota. Furthermore, it is possible that the inhibition of enteric pathogens could be due to a combination of mechanisms that act through highly localized effects on intestinal epithelial cells.

3.4.2 Competitive Exclusion and Improvement of Epithelial Barrier Function

Competitive exclusion for bacterial adhesion sites on intestinal epithelial surfaces occurs mostly in preventive studies when the probiotic is added before the pathogen challenge. The adhesion ability of some probiotic strains allows them to compete with pathogenic bacteria for receptors expressed on epithelial cells, thus blocking contact between epithelial cells and pathogenic bacteria (Corr et al. 2009). Different strains of probiotic bacteria are not equally effective in blocking the adhesion sites for pathogens. Only strongly adhering probiotic strains could interfere with the pathogen-receptor interactions. For example, the penetration of an enteroinvasive *E. coli* strain in cell culture monolayers was significantly reduced when the monolayers were challenged with specific strains of probiotic lactobacilli (*L. rhamnosus, L. acidophilus* or *L. plantarum*) before pathogen infection (Resta-Lenert and Barrett 2003).

Probiotics can also induce other physiological changes for the host intestinal cells by induction or overexpression of mucin. The intestinal mucins are large, carbohydrate-rich, high molecular-weight glycoproteins which are the major components of the mucous layer overlying the intestinal epithelium. Mucin forms a physicochemical barrier that protects epithelial cells and limits microbial adhesion and subsequent invasion. At least 12 mucin genes have been identified and of these, MUC2 and MUC3 are the predominant ileocolonic mucins (Mack et al. 2003). In a previous study, two strains of *Lactobacillus* (*L. plantarum* 299v and *L. rhamnosus* GG) were found to inhibit an *E. coli* diarrheogenic strain by inducing MUC-2 gene expression using a HT-29 cell model (Mack et al. 1999).

Some fermentation metabolites like the short chain fatty acids (SCFA) (mainly acetic, butyric, and propionic acids) appear to maintain the health of the epithelial mucosa. Since butyrate is a fermentation product of the intestinal flora, it is tempting to speculate that modification of gut flora to produce more butyric acid (e.g. by probiotics) could reduce invasive infections by *Salmonella* (Gantois et al. 2006). A recent study suggests that the effect of butyrate on the epithelial barrier may be mediated by regulation of the tight junction proteins (Peng et al. 2009). Indeed, it was concluded that butyrate enhances the intestinal barrier by regulating the assembly of tight junctions.

Intestinal bacteria and transient flora including probiotics also contribute to the strengthening of the gut mucosal barrier. Harmful bacteria may increase the permeability facilitating the passage of bacteria through the mucosal wall. Several probiotic bacteria have been shown to prevent and repair such mucosal damage (Corr et al. 2009). One study showed that pretreatment of intestinal cell monolayers with probiotic *E. coli* strain Nissle 1917 prevents the decline in transepithelial electrical resistance (TEER) and the increase of epithelial permeability to small molecules and macromolecules induced by infection with enteropathogenic *E. coli* (Zyrek et al. 2007). DNA-microarray analysis identified more than 300 genes exhibiting altered expression following incubation of the epithelial cells with the *E. coli* strain Nissle 1917, including expression and distribution of zonula occludens-2 (ZO-2), a tight-junction protein. Another recent study showed that pretreatment with *L. rhamnosus* GG prevent *E. coli* O157:H7-induced morphological redistribution of intercellular tight junction proteins as well as decreases in ZO-1 expression (Johnson-Henry et al. 2008).

Together these observations suggest that specific probiotic strains exhibit a potential for pathogen exclusion via indirect mechanisms.

3.4.3 Stimulation of Immunity

Administration of probiotic strains causes a range of non-specific and host-specific immune responses in diseased and healthy subjects (Rastall et al. 2005). These immunological responses include, the enhancement of phagocytic activity of peripheral blood leukocytes, natural killer cell activity, regulatory T cells, interleukin-10 (Gill and Prasad 2008; Sherman et al. 2009). Additionally, stimulation of both non-specific secretory IgA and specific antibody responses, especially mucosal IgA, to *Salmonella typhi* and enteric pathogens such as rotavirus has been seen (Wold 2001). Increased cytokine production *in vivo* (IFN-γ, IFN-α, IL-2) and by peripheral blood mononuclear cells *ex vivo* (IL-1β, TNF-α, IL-6, IL-10, IFN-α, IFN-γ) have been reported following appropriate probiotic stimulation (Wold 2001). The question is if the reactions can be predicted for a given subject and if they can be effectively directed? Few studies have examined the anti-infective properties and host immune responses in the same subjects following consumption of probiotics. Further studies in humans are needed to elucidate the role of probiotics-stimulated immunological mechanisms in protection against enteric pathogens.

3.5 Intervention Designed for Specific Populations

An important factor in probiotic therapies is the target group to which the probiotic is given (De Vrese and Schrezenmeir 2008). For example, the ecological interaction of the probiotic with the intestinal microbiota can be different if the strain is given to infants, adults or elderly due to the modification of the gut microbiota composition with age. Interventions designed for newborns and infants are different than for other age groups. Breast-fed infants have a colonic microbial community more dominated by bifidobacteria than do their formula-fed counterparts, who harbour a more complex and adult-like microbial community co-dominated by bifidobacteria, bacteroides and to a lesser extent clostridia (Rastall et al. 2005). Introduction of probiotic bifidobacteria in a microbiota that already contain much of these bacteria may be more difficult because of the pre-occupied ecological niche. With advancing years the gut microbial community changes to a more diverse composition (Hebuterne 2003). Bifidobacterial levels markedly decrease after 55–60 years of age, for reasons unlikely to be explained by changes in diet or hormones (as the microbial community of men and women alter), but may be associated with immunological, physiological and/or lifestyle factors (Rastall et al. 2005). Such microbiota changes could render subjects more susceptible to GI problems. Functional foods containing probiotics may have a particular application in this high-risk group as well as infant where the gut microbiota is less diverse, the barrier effect lower and the immune system still immature (Penna et al. 2008).

3.6 Conclusion

There is increasing evidence supporting a prophylactic role of probiotics and possible therapeutic uses in combination with antibiotics in enteric infections. Efficient probiotic dosing in this case requires elucidation of the precise mechanism for each probiotic strain with anti-pathogen effect. Recently there has been significant progress towards understanding how probiotics exert their beneficial effect using *in vitro* models. In many cases the mode of action is not entirely known because these models yield only single interactions between pathogen and probiotic in contrast with the *in vivo* gut system which is a much more complex environment. Modern molecular tools can contribute to a more complete view of how the human body functions with its intestinal microbial community and how such interactions can be modulated during enteric infection. Probiotic strains could then be developed on established mechanisms to target specifically pathogenic microorganisms and a minimal effective contribution could be defined. The potential certainly exists for targeting of these probiotics for specific population groups, but this can only be realized by the generation of clinical or consumer documentation, adherence to strict guidelines and attainment of high quality clinical trials. The selection of probiotics for specific enteric pathogens also needs to be explored. There is an urgent demand for rigorous experimentation to evaluate the safety and efficacy of new probiotics before recommendations can be made.

References

Abe F, Muto M, Yaeshima T, Iwatsuki K, Aihara H, Ohashi Y, Fujisawa T (2009) Safety evaluation of probiotic bifidobacteria by analysis of mucin degradation activity and translocation ability. Anaerobe 16(2):131–136

Ahmed M, Prasad J, Gill H, Stevenson L, Gopal P (2007) Impact of consumption of different levels of *Bifidobacterium lactis* HN019 on the intestinal microflora of elderly human subjects. J Nutr Health Aging 11:26–31

Altenhoefer A, Oswald S, Sonnenborn U, Enders C, Schulze J, Hacker J, Oelschlaeger TA (2004) The probiotic *Escherichia coli* strain Nissle 1917 interferes with invasion of human intestinal epithelial cells by different enteroinvasive bacterial pathogens. FEMS Immunol Med Microbiol 40:223–229

Amaretti A, Tamburini E, Bernardi T, Pompei A, Zanoni S, Vaccari G, Matteuzzi D, Rossi M (2006) Substrate preference of *Bifidobacterium adolescentis* MB 239: compared growth on single and mixed carbohydrates. Appl Microbiol Biotechnol 73:654–662

Asahara T, Shimizu K, Nomoto K, Hamabata T, Ozawa A, Takeda Y (2004) Probiotic bifidobacteria protect mice from lethal infection with Shiga toxin-producing *Escherichia coli* O157:H7. Infect Immun 72:2240–2247

Basu S, Paul DK, Ganguly S, Chatterjee M, Chandra PK (2009) Efficacy of high-dose *Lactobacillus rhamnosus* GG in controlling acute watery diarrhoea in Indian children: a randomized controlled trial. J Clin Gastroenterol 43:208–213

Biavati B, Mattarelli P (2006) The family Bifidobacteriaceae. In: Dworkin M, Falkow S, Rosenberg E, Schleifer K-H, Stackebrandt E (eds) The Prokaryotes, vol 3, Chap. 1.1.2. Springer, Singapore, pp 322–382

Carey CM, Kostrzynska M, Ojha S, Thompson S (2008) The effect of probiotics and organic acids on Shiga-toxin 2 gene expression in enterohemorrhagic *Escherichia coli* O157:H7. J Microbiol Methods 73:125–132

Chassard C, Scott KP, Marquet P, Martin JC, Del'homme C, Dapoigny M, Flint HJ, Bernalier-Donadille A (2008) Assessment of metabolic diversity within the intestinal microbiota from healthy humans using combined molecular and cultural approaches. FEMS Microbiol Ecol 66:496–504

Chassard C, Grattepanche F, Lacroix C (2011) Probiotics and health claims-challenges for tailoring their efficacy. In: Salminen S, Kneifel W (eds) Probiotics and health claims. Wiley Blackwell, UK

Cinquin C, Le Blay G, Fliss I, Lacroix C (2004) Immobilization of infant faecal microbiota and utilization in an *in vitro* colonic fermentation model. Microb Ecol 48:128–138

Cinquin C, Le Blay G, Fliss I, Lacroix C (2006) New three-stage *in vitro* model for infant colonic fermentation with immobilized faecal microbiota. FEMS Microbiol Ecol 57:324–336

Cleusix V, Lacroix C, Vollenweider S, Duboux M, Le Blay G (2007) Inhibitory activity spectrum of reuterin produced by *Lactobacillus reuteri* against intestinal bacteria. BMC Microbiol 7:101 (Nov 12)

Cleusix V, Lacroix C, Vollenweider S, Le Blay G (2008) Glycerol induces reuterin production and decreases *Escherichia coli* population in an *in vitro* model of colonic fermentation with immobilized human faeces. FEMS Microbiol Ecol 63:56–64

Corr SC, Gahan CG, Hill C (2007a) Impact of selected *Lactobacillus* and *Bifidobacterium* species on *Listeria monocytogenes* infection and the mucosal immune response. FEMS Immunol Med Microbiol 50:380–388

Corr SC, Li Y, Riedel CU, O'Toole PW, Hill C, Gahan CG (2007b) Bacteriocin production as a mechanism for the antiinfective activity of *Lactobacillus salivarius* UCC118. Proc Natl Acad Sci U S A 104:7617–7621

Corr SC, Hill C, Gahan CG (2009) Understanding the mechanisms by which probiotics inhibit gastrointestinal pathogens, Chap. 1. Adv Food Nutr Res 56:1–15

Dabard J, Bridonneau C, Phillipe C, Anglade P, Molle D, Nardi M et al. (2001) Ruminococcin A, a new lantibiotic produced by a *Ruminococcus gnavus* strain isolated from human faeces. Appl Environ Microbiol 67:4111–4118

Dabour N, Zihler A, Kheadr E, Lacroix C, Fliss I (2009) *In vivo* study on the effectiveness of pediocin PA-1 and *Pediococcus acidilactici* UL5 at inhibiting Listeria monocytogenes. Int J Food Microbiol 133:225–233

De Vrese M, Schrezenmeir J (2008) Probiotics, prebiotics, and synbiotics. Adv Biochem Eng/Biotechnol 111:1–66

Doleyres Y, Lacroix C (2005) Technologies with free and immobilised cells for probiotic bifidobacteria production and protection. Int Dairy J 15:973–988

Donohue DC, Salminen S (1996) Safety of probiotic bacteria. Asia Pac J Clin Nutr 5:25–28

Dunne C, O'Mahony L, Murphy L, Thornton G, Morrissey D, O'Halloran S, Feeney M, Flynn S, Fitzgerald G, Daly C, Kiely B, O'Sullivan GC, Shanahan F, Collins JK (2001) *In vitro* selection criteria for probiotic bacteria of human origin: correlation with *in vivo* findings. Am J Clin Nutr 73:386S–392S

FAO/WHO (2002) Joint working group report on guidelines for the evaluation of probiotics in food. London, Ontario, Canada, 30 April and 1 May. Available at: http:www.who.int/foodsafety/publications/fs_management/probiotics2/en/index html

Felis GE, Dellaglio F (2007) Taxonomy of *Lactobacilli* and *Bifidobacteria*. Curr Issues Intest Microbiol 8:44–61

Finegold SM, Attebery HR, Sutter VL (1974) Effect of diet on human faecal flora: comparison of Japanese and American diets. Am J Clin Nutr 27:1456–1469

Franks AH, Harmsen HJ, Raangs GC, Jansen GJ, Schut F, Welling GW (1998) Variations of bacterial populations in human faeces measured by fluorescent *in situ* hybridization with group-specific 16S rRNA-targeted oligonucleotide probes. Appl Environ Microbiol 64:3336–3345

Frece J, Kos B, Svetec IK, Zgaga Z, Beganović J, Lebos A, Susković J (2009) Synbiotic effect of *Lactobacillus helveticus* M92 and prebiotics on the intestinal microflora and immune system of mice. J Dairy Res 76:98–104

Fukushima Y, Kawata Y, Hara H, Terada A, Mitsuoka T (1998) Effect of a probiotic formula on intestinal immunoglobulin A production in healthy children. Int J Food Microbiol 42:39–44

Furet JP, Firmesse O, Gourmelon M, Bridonneau C, Tap J, Mondot S, Doré J, Corthier G (2009) Comparative assessment of human and farm animal faecal microbiota using real-time quantitative PCR. FEMS Microbiol Ecol 68:351–362

Gantois I, Ducatelle R, Pasmans F, Haesebrouck F, Hautefort I, Thompson A, Hinton JC, Van Immerseel F (2006) Butyrate specifically down-regulates *Salmonella* pathogenicity island 1 gene expression. Appl Environ Microbiol 72:946–949

Gill H, Prasad J (2008) Probiotics, immunomodulation, and health benefits. Adv Exp Med Biol 606:423–454

Gmeiner M, Kneifel W, Kulbe KD, Wouters R, De Boever P, Nollet L, Verstraete W (2000) Influence of a synbiotic mixture consisting of *Lactobacillus acidophilus* 74-2 and a fructooligosaccharide preparation on the microbial ecology sustained in a simulation of the human intestinal microbial ecosystem (SHIME reactor). Appl Microbiol Biotechnol 53:219–223

Gopal PK, Prasad J, Smart J, Gill HS (2001) *In vitro* properties of *Lactobacillus rhamnosus* DR20 and *Bifidobacterium lactis* DR10 strains and their antagonistic activity against an enterotoxigenic *Escherichia coli*. Int J Food Microbiol 67:207–216

Gueimonde M, Garrigues C, van Sinderen D, de los Reyes-Gavilan, CG, Margolles A (2009) Bile-inducible efflux transporter from *Bifidobacterium longum* NCC2705, conferring bile resistance. Appl Environ Microbiol 75:3153–3160

Harmsen HJ, Raangs GC, He T, Degener JE, Welling GW (2002) Extensive set of 16S rRNA-based probes for detection of bacteria in human faeces. Appl Environ Microbiol 68:2982–2990

Hartemink R, Van Laere KM, Rombouts FM (1997) Growth of enterobacteria on fructo-oligosaccharides. J Appl Microbiol 83:367–374

Hebuterne X (2003) Gut changes attributed to ageing: effects on intestinal microflora. Curr Opin Nutr Metab Care 6:49–54

Hickson M, D'Souza AL, Muthu N, Rogers TR, Want S, Rajkumar C, Bulpitt CJ (2007) Use of probiotic Lactobacillus preparation to prevent diarrhoea associated with antibiotics: randomised double blind placebo controlled trial. BMJ 335:80

Holzapfel WH, Haberer P, Geisen R, Björkroth J, Schillinger U (2001). Taxonomy and important features of probiotic microorganisms in food and nutrition. Am J Clin Nutr 73:365S–373S

Ishibashi N, Yamazaki S (2001) Probiotics and safety. Am J Clin Nutr 73:465S–470S

Janer C, Rohr LM, Pelaez C, Laloi M, Cleusix V, Requena T, Meile L (2004) Hydrolysis of oligofructoses by the recombinant β-fructofuranosidase from *Bifidobacterium lactis*. Sys Appl Microbiol 27:279–285

Johnson-Henry KC, Donato KA, Shen-Tu G, Gordanpour A, Sherman PM (2008) Lactobacillus rhamnosus strain GG prevents enterohemorrhagic *Escherichia coli* O157:H7-induced changes in epithelial barrier function. Infect Immun 76:1340–1348

Johnston BC, Supina AL, Ospina M, Vohra S (2007) Probiotics for the prevention of pediatric antibiotic-associated diarrhoea. Cochrane Database Syst Rev (2). doi:10.1002/14651858.CD004827.pub2 (CD004827)

Kajander K, Krogius-Kurikka L, Rinttila T, Karjalainen H, Palva A, Korpela R (2007) Effects of multispecies probiotic supplementation on intestinal microbiota in irritable bowel syndrome. Aliment Pharmacol Ther 26:463–473

Kajander K, Myllyluoma E, Rajilic-Stojanovic M, Kyronpalo S, Rasmussen M, Jarvenpaa S et al. (2008) Clinical trial: multispecies probiotic supplementation alleviates the symptoms of irritable bowel syndrome and stabilizes intestinal microbiota. Aliment Pharmacol Ther 27:48–57

Koninkx JFJG, Malago JJ (2008) The protective potency of probiotic bacteria and their microbial products against enteric infections. Folia Microbiol 53:189–194

Lacroix C, Yildirim S (2007) Fermentation technologies for the production of probiotics with high viability and functionality. Curr Opin Biotechnol 18:176–183

Laparra JM, Sanz Y (2009) Comparison of *in vitro* models to study bacterial adhesion to the intestinal epithelium. Lett Appl Microbiol 49(6):695–701

Larsen CN, Nielsen S, Kaestel P, Brockmann E, Bennedsen M, Christensen HR et al. (2006) Dose-response study of probiotic bacteria *Bifidobacterium animalis* subsp. *lactis* BB-12 and *Lactobacillus paracasei* subsp *paracasei* CRL-341 in healthy young adults. Eur J Clin Nutr 60:1284–1293

Lavermicocca P, Valerio F, Lonigro SL, Di Leo A, Visconti A (2008) Antagonistic activity of potential probiotic Lactobacilli against the ureolytic pathogen *Yersinia enterocolitica*. Curr Microbiol 56:175–181

Le Blay G, Lacroix C, Zihler A, Fliss I (2007) *In vitro* inhibition activity of nisin A, nisin Z, pediocin PA-1 and antibiotics against common intestinal bacteria. Lett Appl Microbiol 45:252–257

Le Blay G, Rytka J, Zihler A, Lacroix C (2009) New *in vitro* colonic fermentation model for *Salmonella* infection in the child gut. FEMS Microbiol Ecol 67:198–207

Le Blay G, Chassard C, Baltzer S, Lacroix C (2010) Set up of a new *in vitro* model to study dietary fructans fermentation in formula-fed babies. Br J Nutr 103(3):403–411

Lee SK, Kim YB, Ji GE (1997) Note: purification of amylase secreted from *Bifidobacterium adolescentis*. J Appl Microbiol 83:267–272

Macfarlane GT, Macfarlane S (2007) Models for intestinal fermentation: association between food components, delivery systems, bioavailability and functional interactions in the gut. Curr Opin Biotechnol 18:156–162

Mack DR, Michail S, Wel S, McDougall L, Hollingsworth MA (1999) Probiotics inhibit enteropathogenic *E. coli* adherence *in vitro* by inducing intestinal mucin gene expression. Am J Physiol 276:G941–G950

Mack DR, Ahrne S, Hyde L, Wei S, Hollingsworth MA (2003) Extracellular MUC3 mucin secretion follows adherence of *Lactobacillus* strains to intestinal epithelial cells *in vitro*. Gut 52:827–833

Marteau P, Shanahan F (2003) Basic aspects and pharmacology of probiotics: an overview of pharmacokinetics, mechanisms of action and side-effects. Best Pract Res Clin Gastroenterol 17:725–740

Martín-Peláez S, Gibson GR, Martín-Orúe SM, Klinder A, Rastall RA, La Ragione RM, Woodward MJ, Costabile A (2008) *In vitro* fermentation of carbohydrates by porcine faecal inocula and their influence on *Salmonella typhimurium* growth in batch culture systems. FEMS Microbiol Ecol 66:608–619

McFarland LV (2007) Meta-analysis of probiotics for the prevention of traveler's diarrhoea. Travel Med Infect Dis 5:97–105

Mennigen R, Bruewer M (2009) Effect of probiotics on intestinal barrier function. Ann N Y Acad Sci 1165:183–189

Miettinen M, Lehtonen A, Julkunen I, Matikainen S (2000) Lactobacilli and Streptococci activate NF-kappa B and STAT signaling pathways in human macrophages. J Immunol 164:3733–3740

Moller PL, Jorgensen F, Hansen OC, Madsen SM, Stougaard P (2001) Intra- and extracellular beta-galactosidases from *Bifidobacterium bifidum* and *B. infantis*: molecular cloning, heterologous expression, and comparative characterization. Appl Environ Microbiol 67:2276–2283

O'Toole PW, Cooney JC (2008) Probiotic bacteria influence the composition and function of the intestinal microbiota. Interdiscip Perspect Infect Dis 2008:175285

Paton AW, Morona R, Paton JC (2006) Designer probiotics for prevention of enteric infections. Nat Rev Microbiol 4:193–200

Peng L, Li Z R, Green RS, Holzman IR, Lin J (2009) Butyrate enhances the intestinal barrier by facilitating tight junction assembly via activation of AMP-activated protein kinase in Caco-2 cell monolayers. J Nutr 139:1619–1625

Penna FJ, Péret LA, Vieira LQ, Nicoli JR (2008) Probiotics and mucosal barrier in children. Curr Opin Clin Nutr Metab Care 11:640–644

Pineiro M, Stanton C (2007) Probiotic bacteria: legislative framework-requirements to evidence basis. J Nutr 137:850S–853S

Pridmore RD, Pittet AC, Praplan F, Cavadini C (2008) Hydrogen peroxide production by *Lactobacillus johnsonii* NCC 533 and its role in anti-*Salmonella* activity. FEMS Microbiol Lett 283:210–215

Rastall RA, Gibson GR, Gill HS, Guarner F, Klaenhammer TR, Pot B, Reid G, Rowland IR, Sanders ME (2005) Modulation of the microbial ecology of the human colon by probiotics, prebiotics and synbiotics to enhance human health: an overview of enabling science and potential applications. FEMS Microbiol Ecol 52:145–152

Resta-Lenert S, Barrett KE (2003) Live probiotics protect intestinal epithelial cells from the effects of infection with enteroinvasive *Escherichia coli* (EIEC). Gut 52:988–997

Rigottier-Gois L, Bourhis AG, Gramet G, Rochet V, Doré J (2003) Fluorescent hybridisation combined with flow cytometry and hybridisation of total RNA to analyse the composition of microbial communities in human faeces using 16S rRNA probes. FEMS Microbiol Ecol 43:237–245

Rossi M, Corradini C, Amaretti A, Nicolini M, Pompei A, Zanoni S, Matteuzzi D (2005) Fermentation of fructooligosaccharides and inulin by bifidobacteria: a comparative study in pure and faecal cultures. Appl Environ Microbiol 71:6150–6158

Salminen S, Ouwehand A, Benno Y, Lee YK (1999) Probiotics: how should they be defined? Trends Food Sci Technol 10:107–110

Saulnier DM, Molenaar D, de Vos WM, Gibson GR, Kolida S (2007) Identification of prebiotic fructooligosaccharide metabolism in *Lactobacillus plantarum* WCFS1 through microarrays. Appl Environ Microbiol 73:1753–1765

Saulnier DM, Gibson GR, Kolida S (2008) *In vitro* effects of selected synbiotics on the human faecal microbiota composition. FEMS Microbiol Ecol 66:516–527

Schlee M, Harder J, Koten B, Stange EF, Wehkamp J, Fellermann K (2008) Probiotic lactobacilli and VSL#3 induce enterocyte beta-defensin 2. Clin Exp Immunol 151:528–535

Sghir A, Gramet G, Suau A, Rochet V, Pochart P, Dore J (2000) Quantification of bacterial groups within human faecal flora by oligonucleotide probe hybridization. Appl Environ Microbiol 66:2263–2266

Sherman PM, Johnson-Henry KC, Yeung HP, Ngo PS, Goulet J, Tompkins TA (2005) Probiotics reduce enterohemorrhagic *Escherichia coli* O157:H7- and enteropathogenic *E. coli* O127:H6-induced changes in polarized T84 epithelial cell monolayers by reducing bacterial adhesion and cytoskeletal rearrangements. Infect Immun 73:5183–5188

Sherman PM, Ossa JC, Johnson-Henry KC (2009) Unraveling mechanisms of action of probiotics. Nutr Clin Pract 24:10–14

Silva AM, Barbosa FH, Duarte R, Vieira LQ, Arantes RM, Nicoli JR (2004) Effect of *Bifidobacterium longum* ingestion on experimental salmonellosis in mice. J Appl Microbiol 97:29–37

Tannock GW, Munro K, Harmsen HJ, Welling GW, Smart J, Gopal PK (2000) Analysis of the faecal microflora of human subjects consuming a probiotic product containing *Lactobacillus rhamnosus* DR20. Appl Environ Microbiol 66:2578–2588

Turcotte C, Lacroix C, Kheadr E, Grignon L, Fliss I (2004) A rapid turbidometric microplate bioassay for accurate quantification of lactic acid bacteria bacteriocins. Int J Food Microbiol 90:283–293

Turroni F, Marchesi JR, Foroni E, Gueimonde M, Shanahan F, Margolles A, van Sinderen D, Ventura M (2009) Microbiomic analysis of the bifidobacterial population in the human distal gut. ISME J 3:745–751

von Ah U, Mozzetti V, Lacroix C, Kheadr EE, Fliss I, Meile L (2007) Classification of a moderately oxygen-tolerant isolate from baby faeces as *Bifidobacterium thermophilum*. BMC Microbiol 7:79

Walter J (2008) Ecological role of lactobacilli in the gastrointestinal tract: implications for fundamental and biomedical research. Appl Environ Microbiol 74:4985–4996

Wang X, Conway PL, Brown IL, Evans AJ (1999) *In vitro* utilization of amylopectin and high-amylose maize (Amylomaize) starch granules by human colonic bacteria. Appl Environ Microbiol 65:4848–4854

Whitehead K, Versalovic J, Roos S, Britton RA (2008) Genomic and genetic characterization of the bile stress response of probiotic *Lactobacillus reuteri* ATCC 55730. Appl Environ Microbiol 74:1812–1819

Wold AE (2001) Immune effects of probiotics. Scand J Nutr 45:76–85

Zyrek AA, Cichon C, Helms S, Enders C, Sonnenborn U, Schmidt MA (2007) Molecular mechanisms underlying the probiotic effects of *Escherichia coli* Nissle 1917 involve ZO-2 and PKCzeta redistribution resulting in tight junction and epithelial barrier repair. Cell Microbiol 9:804–816

Part III
Pathophysiology of Enteric Disorders Due to Disturbed Microbiota

Chapter 4
Factors Causing Disturbances of the Gut Microbiota

Joshua J. Malago and Jos F. J. G. Koninkx

4.1 Introduction

The gastrointestinal (GI) tract represents the largest contact area between the body and the external environment (Mackie et al. 1999). Several studies have estimated the total number of adult human body to be 10^{14} cells, of which 90% are microbiota cells, mostly inhabiting the large intestine, and only 10% constitute the body proper. There are approximately 10–20 genera with at least 500 species of microbiota in the gut. The gene number of this microbiota is about 50–100 times that of the human genome (Hooper et al. 1998). Although these facts are often overlooked, they allude to a significant host-microbiota interaction that is vital to the human health, especially that of the gut.

The most common commensal bacteria found in the intestines are *Bifidobacterium, Clostridium, Bacteroides, Eubacterium, Escherichia, Enterococcus, Streptococcus, Klebsiella,* and *Lactobacillus*. *Bifidobacterium* and *Lactobacillus* are considered the most important health-beneficial bacteria for the human host (Rastall 2004). Major functions of the beneficial bacteria include; production of short chain fatty acids (SCFA) which supply a valuable proportion of human energy requirements especially to the colonocytes, production of B vitamins and vitamin K, participation in the metabolism of drugs, hormones and carcinogens, protection of the host from infection by pathogenic bacteria (e.g. through competing for space and production of anti-bacterial substances like bacteriocins), maintenance of a healthy intestinal acidic pH, and enhancing the immune function. Some members of the intestinal microbiota are however, not beneficial or do so only when their numbers are kept low by competing beneficial bacteria. Most of them are potentially pathogenic. They include coliform bacteria, staphylococci, clostridia, and yeasts or fungi (Rastall 2004).

J. J. Malago (✉)
Department of Veterinary Pathology, Faculty of Veterinary Medicine,
Sokoine University of Agriculture, 3203, Chuo Kikuu, Morogoro, Tanzania
e-mail: malagojj@yahoo.com

J. J. Malago et al. (eds.), *Probiotic Bacteria and Enteric Infections*,
DOI 10.1007/978-94-007-0386-5_4,

The gut microbiota is considered to be a functionally and metabolically adaptable and rapidly renewable organ of the body (Thompson-Chagoyán et al. 2005). Because of this adaptability, intestinal infection or translocation of bacterial agents is an exception, not the rule, and mostly limited to highly pathogenic bacteria or predisposing disease states even though the GI tract is constantly colonized and challenged by a multitude of different microorganisms. This is achieved by multiple factors that contribute to intestinal barrier integrity and limit invasion and adherence of pathogenic as well as commensal bacteria. Alterations in the GI adaptability and its barrier integrity allow entry of the microorganisms and trigger inflammatory response leading to illness. The illness can emanate from potentially harmful microbiota members that produce toxins, invade the epithelial mucosa, or activate neoplastic and inflammatory responses. Additionally, perturbations of the non-pathogenic microbiota members disrupt their beneficial attributes to the host, e.g. nutrient supply, physiology, barrier integrity, and immunity, and thus predispose the host to illness. These perturbations occur when the amount of beneficial bacteria is reduced and the potentially pathogenic organisms are able to increase in number and become dominant.

Several factors have been implicated in perturbations of the intestinal microbiota. They include the use of antimicrobials (mainly antibiotics), abnormal intestinal motility, intestinal inflammation, alteration in GI physiology, allergies, host genetic factors, increased exposure to enteric pathogens, reduced gastric acidity, immunodeficiency, diet, psychological stress and behavioural changes, use of birth control pills, use of hormones (especially immunosuppressants like steroids), and alcohol. The disruption of the normal intestinal microbiota balance by these factors, also called dysbiosis, has been categorized into four types; putrefaction, fermentation, deficiency, and sensitization.

4.2 Types of Microbiotal Disruption

4.2.1 Putrefaction

Disturbances of microbiota by putrefaction develop following consumption of diets high in fat and animal flesh but low in insoluble fibre. Such diets increase transit time of ingested materials allowing them to putrefy in the colon. This results in an increased concentration of *Bacteroides* species in the colon which are known for their nutritional versatility and capability to utilize a wide variety of different carbon sources (Woodmansey 2007). In contrast, the concentrations of beneficial bacteria like Bifidobacteria species decrease. This imbalance leads to an increase in bacterial enzymes and other products that are responsible for disorders such as increased cancer causing substances and the associated colorectal and breast cancers, development of inflammatory bowel disease (IBD), diarrhoea, allergies, and hormonal imbalances (Malinen et al. 2005; Woodmansey 2007; Kassinen et al. 2007).

4.2.2 Fermentation

Fermentation disturbances primarily involve overgrowth of bacteria in the small intestine rather than the colon. Affected individuals develop intolerance to carbohydrate. Any carbohydrate ingested is fermented by the bacteria and produces toxic waste products like organic acids (e.g. acetic and lactic acid) and hydrogen sulphide. These compounds can lead to acidosis. In addition, the overgrowing bacteria compete with the patient for nutrients, potentially leading to malnutrition, and may also damage the enterocytes (Singh and Toskes 2004).

4.2.3 Deficiency

Deficiency of intestinal microbiota occurs as a result of antibiotic usage or intake of a diet low in soluble fibre. Absolute deficiency of *Bifidobacterium, Lactobacillus*, and *E. coli* has been observed (Bartosch et al. 2004; Tiihonen et al. 2008). The outcome of this deficiency manifests as malfunction of the microbiota such as deprivation of the host nutrients usually supplied by the microbiota, weakening of the immune system and the subsequent reduction in resistance to infection, and food intolerance.

4.2.4 Sensitization

This refers to increased host immune response against the intestinal microbiota. Conditions like IBD, spondyloarthropathies, and some skin disorders like psoriasis or acne may be associated with microbiota sensitization (Macpherson et al. 1996). During sensitization, the host may lose its tolerance to microbiota and thus the immune system overreacts to the bacteria themselves, or substances produced by them. Intestinal bacteria may play a part in autoimmune diseases as the immune system first reacts to bacterial antigens and then cross reacts with the body's own cells with a similar protein structures.

4.3 Factors Disturbing the Intestinal Microbiota

4.3.1 Nutrition

Daily changes in a normal diet are associated with slight modifications in the microbiota that have no effect on the health of an individual. Enteric disorders may occur following more significant modifications of the microbiota caused by extreme changes in the diet, such as fasting or switching to an elemental or artificial diet without fibre (Schneider et al. 2000).

4.3.1.1 Disturbances Leading to Colorectal Cancers

Probably the most common disorder due to dietary-induced disturbances of microbiota is the colorectal cancer in humans. The cancer may develop following disturbances of microbiota composition and their metabolic activities caused by dietary changes. Diets rich in fat and high consumption of red meat, especially processed meat, are associated with high risk of the cancer, while a high intake of fruits and vegetables, whole grain cereals, fish, and calcium reduces the risk (Rafter and Glinghammar 1998; Bingham 1999). Consumption of high amounts of dietary fibre, especially grain fibre and fibre from fruits and veggies is positively associated with numbers of lactic acid bacteria, fewer Clostridium and low colorectal cancer (Burkitt 1971; Mai et al. 2009). On the contrary, fat-rich diet and meat disrupt the microbiota and elevate the levels of N-nitroso compounds in the intestine. These compounds are genotoxic and capable of initiating and promoting colon cancers. The disrupted microbiota act on the N-nitroso compounds to produce carcinogens, cocarcinogens, or procarcinogens and thereby playing a part in initiation of colon cancer (Hughes et al. 2001). In a study to assess association between dietary habits and gut microbiota composition in African and Caucasian American volunteers, Mai et al. (2009) observed that the former volunteers, which suffer from an increased incidence and mortality of colorectal cancer, had high intake of 2-amino-1-methyl-6-phenyl-imidazo[4,5-b]pyridine (PhIP), and 2-amino-3,8-dimethylimidazo[4,5-f]quinoxaline (MeIQx). The PhIP and MeIQx are heterocyclic aromatic amines (HCA) formed during meat preparation with longer cooking times, high temperatures of 150–200°C, and greater external char, typically achieved in preparations of barbeque. These and other HCA are potential human carcinogens of dietary origin and may mediate their effect via disruption of the microbiota (Butler et al. 2003). The disruption could favour proliferation of microbiota strains that are known to strongly increase damage to DNA in colon cells induced by the HCA while suppressing other bacterial strains that can uptake and detoxify the compounds (Wollowski et al. 2001). Thus a community predominated with strains favouring DNA damage will increase the chances of an individual to develop colorectal cancers. Such organisms could belong to the genera *Bacteroides* and *Clostridium* which are known to increase the incidence and growth rate of colonic tumours, whereas the suppressed bacteria could include the genera *Lactobacillus* and *Bifidobacteria* that prevent tumourigenesis (Guarner and Malagelada 2003). In fact, colonic presence of microbiota *Bacteroides vulgatus* and *Bacteroides stercoris* has been shown to be associated with high risk of colon cancer in humans while *Lactobacillus acidophilus, Lactobacillus S06* and *Eubacterium aerofaciens* are associated with low risk (Moore and Moore 1995).

4.3.1.2 Disturbances Due to Artificial Nutrition

Nutritional disturbances to microbiota have also been observed in patients receiving exclusive artificial nutrition in clinical practice such as those under total enteral (TEN) or parenteral (TPN) nutrition. In these patients, bacterial translocation can occur in addition to diarrhoea. The diarrhoea is thought to be related to a decrease in the production of SCFA that are known to enhance colonic absorption of water. And this is a direct effect of reduced number of fermenters. In a study by Schneider et al. (2000), subjects in TEN group fed a commercial polymeric diet, fibre-, lactose-, and gluten-free given through gastrostomy, jejunostomy or nasogastric tubes, had an increase in the aerobic bacteria, mainly *E. coli, Enterococcus faecalis,* and *Lactobacillus* spp and a decrease in anaerobes, *Eubacterium* spp., *Bifidobacterium* spp., *Clostridium* spp., *Bacteroides* spp., *Prevotella* spp., and *Porphyromonas levii,* compared to controls. The TPN patients received a ternary solution containing non-protein energy as lipids (long-chain triglycerides), glucose and amino acid that was infused for an average of 6 ± 1 weeks. In these patients, both aerobic and anaerobic bacteria were dramatically decreased. The study also showed that Clostridia were the predominant anaerobic bacteria in both TEN and TPN patients. Further, the total SCFA concentration in faeces was significantly lower in TPN patients than in TEN patients, indicating a drop in the number of anaerobic fermenters. It was concluded that the TPN- and TEN-induced imbalances of the intestinal microbiota may play a role in the pathophysiology of diarrhoea.

Besides effects in patients, significant changes in microbiotal composition also occur in healthy individuals. Studies conducted on healthy human volunteers fed a chemically defined diet showed a reduction in faecal volume as well as a decrease in the number of enterococci and an increase in the number of enterobacteria (Attebery et al. 1972; Crowther et al. 1973).

4.3.1.3 Nutritional Disturbances in Elderly

Another common nutritional-induced microbiotal disturbance is malnutrition occurring mainly in elderly individuals. It can occur in two major ways; firstly, it follows consumption of a narrow, nutritionally imbalanced diet caused by increased thresholds for taste and smell, resulting in food tasting bland and uninteresting; masticatory dysfunction due to loss of teeth and muscle bulk; and swallowing difficulties (Woodmansey 2007). Secondly, it can develop due to atrophic gastritis and its subsequent hypochlorhydria associated with decreased absorption of calcium, ferric iron, and vitamin B_{12} and possibly, other micronutrients (Russell 1992). The malnutrition could contribute to the reduction in the numbers and diversity of many protective commensal anaerobes like *Bacteroides* and *Bifidobacteria* and high levels of potentially pathogenic bacteria like *E. coli* and some *Bacteroidetes* species observed in elderly subjects (Hopkins et al. 2001; Woodmansey 2007).

4.3.2 Antimicrobials

The amount of change caused by an antimicrobial agent is dependent on various factors, including the antimicrobial activity spectrum, pharmokinetics, dose, route of administration and concentration reached in the intestine. Disruption to the normal microbiota is usually through incomplete absorption of orally administered antibiotics, as well as secretion of the antimicrobial by the salivary glands, in the bile or from the intestine.

4.3.2.1 Antibiotics

Broad-spectrum antibiotics, particularly those with activity against anaerobic bacteria, alter the intestinal microbiota, leading to dysbiosis, impaired microbiota activity, and disruption of GI tract barrier effect. They kill off intestinal bacteria causing exclusion of some members of the microbiota. This leads to a decreased number of particular species. The decrease is usually followed by an overgrowth of other species less sensitive to the antibiotics. Thus, in most cases, the total density of the microbiota following antibiotic use does not vary much from the normal density. However, in severe and continual use of antibiotics the exclusion may be severe that many species are affected and the overgrowth could not retain the normal density. In such cases a more serious illness could develop. Antibiotic therapy therefore represents a strong perturbation of the gut microbiota that shifts the relative proportion of community members, allowing opportunists to establish. The exclusion could either be by direct eradication of the microbiota members or indirectly via breaking mutualistic interactions (Donskey et al. 2003).

A significant decrease in the intestinal level of anaerobic bacteria can occur within 3 days of initiation of antibiotic therapies. At this time, the bifidobacteria populations can decrease between 56% and 100% and lactobacilli can drop to 57% of baseline levels (Seki et al. 2003; Imase et al. 2008). Since anaerobes play critical role in prevention of pathogenic colonization of the GI tract, their decrease leads to suppression of the microbiota-mediated intestinal protection and may result into continual survival and excretion of pathogenic bacteria like *Salmonella* that would otherwise be eliminated by the antibiotics. Microbiota disruption partly accounts for some failures of antibiotics to reduce clinical illness in various intestinal disorders.

The commonly used amoxicillin-clavulanic acid combination to treat various bacterial infections often affects the intestinal microbiota. The combined drugs cause increased numbers of aerobic enterococci, *Enterobacteriaciae* (particularly *E. coli*) and resistant enterobacteria strains; can decrease as well as shift members of the *Bacteroides* group from exclusively *B. fragilis* to almost all *B. distasonis*; and can decrease to 0% all members of the Clostridial cluster XIVa, Bifidobacteria groups and *Prevotella* spp (Young and Schmidt 2004; Engelbrektson et al. 2009).

Examples of Antibiotic-Mediated Disorders

Examples of enteric disorders caused by antibiotic-mediated disruption of microbiota include experimental colitis of mice. In this model, clinical levels of metronidazole, that are effective against anaerobic bacteria, induce a shift in microbiota composition but maintain total numbers of intestinal bacteria. This perturbation causes alterations in the homeostatic state of the mucosal immune system and results in increased severity of *Citrobacter rodentium*-induced colitis (Wlodarska and Finlay 2010). Another example is the common antibiotic-associated diarrhoea (AAD) developing following oral antibiotics like cephalosporins, clindamycin and broad-spectrum penicillins (Wiström et al. 2001). About 5–25% of patients on antibiotic therapy develop AAD ranging from a mild, self-limiting illness to a serious and progressive pseudomembranous colitis. AAD occurs due to altered function of the disrupted microbiota, direct effect on host tissue, and colonization by opportunistic pathogens, including *Salmonella* spp., *Clostridium perfringens, Klebsiella oxytoca, S. aureus, Candida albicans*, and *C. difficile* that invade the altered microbial community (Beaugerie and Petit 2004). Sometimes disruption of the microbiota without pathogenic invasion can be enough by itself to cause AAD. For instance, elimination of fermenters in the colon may result in accumulation of carbohydrates that lead to an osmotic diarrhoea.

To the present, the most severe antibiotic-induced illness mediated through disturbances of microbiota is *Clostridium difficile*-pseudomembranous colitis. The condition develops due to alteration of microbiota following uptake of broad-spectrum antibiotics such as clindamycin, cephalosporins and chinolonics. The shift in the microbiota in this case is the predominance *C. difficile*. It starts by the antibiotics killing off competing microbiota in the intestine leaving behind bacteria with less competition for space and nutrients. This permits much more extensive growth than normal of *C. difficile* that dominates the microbiota (Hookman and Barkin 2007). The highly proliferating *C. difficile* produces UDP-glucose hydrolases and glucosyltransferases toxins A and B, which cause intense inflammation of the colonic mucosa with fluid and electrolyte secretion (Warny et al. 2005).

4.3.2.2 Other Drugs

Several other drugs directed against various conditions affect the microbiota and hamper their benefit to the human host. Examples include the anticancers daunorubicine and etoposide that impart a negative effect on the growth of both anaerobic and aerobic bacteria. A study by van Vliet et al. (2009) done in paediatric patients with acute myeloid leukemia indicates that chemotherapy with these anticancer drugs reduces the total density of microbiota by 100-fold lower than in healthy children. The decrease is due to a reduction of up to 10,000-fold in anaerobic bacteria (*Bacteroides* species, *Clostridium* cluster XIVa, *Faecalibacterium prausnitzii*, and *Bifidobacterium* species), partly compensated for by a 100-fold increase in potentially pathogenic enterococci (e.g *Enterococcus faecium* and *Enterococcus faecalis*). On the contrary, the number of streptococci is decreased by 100–1,000-fold, compared with healthy controls. Several other drugs, including antimycotics (e.g. miconazole, nystatin), anti-inflammatories (e.g. non-steroidal anti-inflammatory drugs) and many others are implicated in microbiotal disturbances. Most of them have been linked with enteric disorders (Penders et al. 2006; Mäkivuokko et al. 2010).

4.3.3 *Alterations of Intestinal Motility*

Both decreased and increased intestinal motility lead to changes in the composition of microbiota. Decreased motility results in faecal impaction, constipation, low faecal weight and reduced excretion of bacterial matter while diarrhoea and increased excretion of bacterial matter result from increased motility (Stephen et al. 1987; Woodmansey et al. 2004).

4.3.3.1 Decreased Intestinal Motility

Experimental acute pancreatitis provides an interesting model to understand the intestinal microbiota disturbances induced by alterations of the intestinal motility. The pancreatitis can be induced in animals through bile duct ligation, leading to impaired intestinal motility caused by inhibition of the intestinal migrating myoelectric complexes that characterise interdigestive motility (Li et al. 1993). Normally the interdigestive motilities are lumen-obliterating sweeping contractions that propagate from the stomach or duodenum to the terminal ileum. In so doing, they propel the microbiota to the distal gut. When they are inhibited, the gut microbiota is no longer contained in the distal gut, resulting in small intestinal bacterial overgrowth, SIBO (Lee and Pimentel 2006). According to Moody et al. (1995) the impaired intestinal motility and SIBO result in bacterial translocation. Other subsequent models of acute pancreatitis not requiring bile duct ligation have consistently substantiated the intestinal motility disruption and SIBO (van Felius et al. 2003).

Another study by Gorbach and Goldin (1992) found that intestinal motility is decreased in individuals presenting moderate to severe protein-energy malnutrition. The reduced motility contributes to the excessive growth of anaerobic microorganisms in the GI tract. The authors also observed reduced gastric, biliary, pancreatic and intestinal secretions in these individuals that together with reduced motility and microbiota, accounted for impeded absorption of carbohydrates, lipids, vitamin B_{12} and proteins. This resulting malabsorption, in addition to the slow intestinal transit and its subsequent increase in faecal retention time, leads to increased bacterial protein fermentation (putrefaction) producing ammonia and phenols. These compounds alter the acidic pH of the gut leading to disturbances of the microbiota.

4.3.3.2 Increased Intestinal Motility

Microbiotal disturbance due to increased intestinal motility is best depicted in diarrhoea. Diarrhoea, defined as increased frequency of loose or watery bowel movements, can be due to many causes, both infectious and non-infectious. All the aetiological agents of diarrhoea can disrupt the anatomo-physiology of the digestive system, causing, for example, atrophy of villi, hyperplasia of crypts, and loss of enzyme activity on the brush border. Diarrhoea increases the intestinal motility leading to an increase in the frequency of flatus. It also decreases the time for microbiota bacterial colonization. Because of increased loss of fluid and electrolytes, diarrhoea results in increased pH of the faeces, which is unfavourable to lactic acid bacteria. All these effects of diarrhoea culminate in reduction in the number of anaerobes and the subsequent production of SCFA in the intestine, paving way to increased opportunistic pathogens (Salminen et al. 1995), which may affect the normal intestinal bacterial flora. The increased stool frequency and volume also increases the rate of elimination of commensal bacteria leading to decreased total counts of normal intestinal microbiota.

4.3.3.3 Miscellaneous Causes of Intestinal Motility

Other causes of intestinal motility disorders that may alter the microbiota composition include hypothyroidism and some drugs, chronic intestinal pseudo-obstruction occurring in patients with severe comorbid clinical conditions or after traumas, surgery, or in patients with other underlying medical diseases; irritable bowel syndrome (IBS) in which case the patients with IBS are more sensitive and reactive to mild stimuli than usual or could be due to immature status of muscles and nerves in the intestinal wall of these patients; faecal incontinence caused by aging, dementia, strokes, parkinsonism, spinal cord injuries, rectal tears during birthing, diabetes, surgical complications, and neuromuscular disorders like myasthenia gravis, consumption of lactose by lactase deficient individuals; and constipation due to diet very poor in fibre and high in animal fats and refined sugars, and pregnancy.

4.3.4 Age

Establishment of the intestinal microbiota has been shown to be a progressive process. And age-related bacteriological changes have been shown by several studies. These changes highlight the ongoing microbial succession throughout the life of an individual. Studies on infant microbiota have revealed that the GI tract is first colonized by facultative anaerobes like *E. coli* (Salminen and Isolauri 2006; Mariat et al. 2009). After the completion of weaning at about 2 years of age, a pattern of gut microbiota that resembles the adult pattern becomes established. The microbiota colonization then reaches climax near the end of adolescence. At this time, the microbiota ecosystem displays a high stability in healthy adults (Frank et al. 1998) and remains relatively stable throughout adult life before it is modified in elderly individuals. In the elderly, the microbiota changes include reduction in numbers and species diversity of many beneficial or protective anaerobes, such as bacteroides and bifidobacteria, and shifts in the dominant bacterial species, favouring colonization of potential pathogens. Mariat et al. (2009) observed a different *E. coli* microbiota profile in elderly subjects compared to younger adults. The drop in number and diversity of protective bacteria and increase in potentially pathogenic organisms can help to understand the decreased functionality of the microflora in some elderly people. Several factors affect the composition of intestinal microbiota in different age groups (infant, adult and elderly).

4.3.4.1 Factors Affecting the Microbiota of the Infant

Immediately before birth, the GI tract of a baby is thought to be sterile. During the normal process of birth, it encounters bacteria from the digestive tract, skin and environment of the mother and starts to become colonised. Thus, the optimum human infant microbiota is considered to be the one seen in a healthy vaginally-delivered and breast-fed infant. It is dominated by Bifidobacteria, accounting for 60–90% of total bacteria in the infant from 2 to 4 weeks of age (Mariat et al. 2009). Other species include *Lactobacillus* and lesser amounts of *Bacteroides* such as *Bacteriodes fragilis*. The potential pathogens like *Clostridia* are almost non-detectable. Deviations from natural birth and feeding infants with diets other than breast-feeding disturb the colonization of infants and may have tremendous effect in later life of an individual. In fact, more and more evidence is emerging which suggests that the establishment of an appropriate intestinal microbiota early in life may be a significant in subsequent healthy development. Thus, it has been shown that the composition of the infant microbiota can be affected by various factors including the mode of delivery, type of infant feeding, infant hospitalization, use of antibiotics by the infant, the environment during birth, prematurity/gestational age, and hygiene measures. Most of these factors are, however, clustered. For instance, infants born through caesarean section need to stay in the hospital more often and receive antibiotics more frequently than do infants born vaginally.

Mode of Delivery

In early life, one of the first major factors determining the gut microbiota is the mode of delivery. Vaginally born infants are colonized at first by faecal and vaginal bacteria of the mother. On the contrary, infants born through cesarean section are exposed initially to bacteria originating from the hospital environment and health care workers (Bezirtzoglou 1997; Gronlund et al. 1999). Since the composition of faecal and vaginal bacteria of the mother differs from that of the hospital environment, significant differences in the microbiota of vaginally versus caesarean section born infants have been observed. Penders et al. (2006) enumerated faecal microbiota by quantitative real-time polymerase chain reaction assays from 1,032 infants of 1 month of age and found that infants born through caesarean section had lower colonization rates and counts of bifidobacteria and *Bacteroides fragilis*-group species, whereas prevalence and counts of *C. difficile* and *E. coli* were higher, compared with vaginally born infants. The most-pronounced differences in colonization were seen for the *B. fragilis* group and *C. difficile*; in which case the median counts of *B. fragilis*-group bacteria and *C. difficile* were about 100-fold lower and about

100-fold higher, respectively, for infants born through cesarean section, as compared with infants born vaginally at home. According to these researchers, the colonization rate and counts of the *B. fragilis* group differed most markedly between vaginally delivered infants and infants born through caesarean section, supporting several earlier studies that also found greatly reduced levels of members of the *B. fragilis* group as a result of caesarean section (Bennet and Nord 1987; Gronlund et al. 1999).

Type of Infant Feeding

Differences in intestinal colonization by infant microbiota have been observed in supplemented or unsupplemented formula-fed or breast fed infants. Infants exclusively fed on unsupplemented formula diet are more often colonized with *E. coli, C. difficile, Bacteroides*, and lactobacilli, compared with breast-fed infants. When the formula diet is supplemented with a mixture of galactooligosaccharides and fructooligosaccharides, higher counts of bifidobacteria and lactobacilli are observed in the stools of these infants, compared with infants fed an unsupplemented formula (Penders et al. 2006). Other studies have shown that supplementing diets with oligosaccharides increases the numbers of lactic acid-producing bacteria (Moro et al. 2003; Rinne et al. 2005). Further, it has been demonstrated that human milk contains not only oligosaccharides but also bifidobacteria. At the same time, genomic studies have convincingly shown that bifidobacteria present in the gut of breast-fed infants, such as *Bifidobacterium longum*, is specially equipped to utilize breast-milk oligosaccharides as nutrients. This bacterium is also adapted to the conditions in the large intestine where energy harvest from slowly absorbable carbohydrates takes place. These facts account for optimum microbiota balance in breast-fed infants as opposes to supplemented or unsupplemented formula-fed infants. Indeed, the microbiota of breast-fed infants is dominated by bifidobacteria where as its numbers of *Clostridium difficile* and *E. coli* are significantly lower than those of formula-fed infants (Penders et al. 2006). This microbiota is enhanced by breast feeding (Harmsen et al. 2000).

Infant Hospitalization

Hospitalization and prematurity are associated with higher prevalence and counts of *C. difficile* because of the hospital environment itself. It is now evident that infants are being colonized with *C. difficile* through the hospital environment since vaginal swabs collected just before delivery are uniformly negative for this organism (Al Jumaili et al. 1984) whereas the hands and stools of healthy hospital personnel and hospital neonatal intensive care unit, NICU, are positive (Kim et al. 1981). Premature infants have the highest carriage rate of *C. difficile* that is strongly associated with hospitalization (Penders et al. 2006). *C. difficile* colonization rate during hospitalization after birth increases at about 13% per day of hospitalization.

Use of Antibiotics

Antibiotic use by the infant is associated with decreased numbers of beneficial and protective *Bifidobacterium* and *Bacteroides*. Although the use of antibiotics may have a major effect on the composition of the gut microbiota, the effect differs between antibiotics (Edlund and Nord 2000). Generally the effect of antibiotics rest on decreasing the anaerobic microbiota, with counts of *Bifidobacterium* and *Bacteroides* being highly affected, e.g. oral use of amoxicillin by the infant during the first 1 month of life results in significant decrease in the numbers of bifidobacteria and *B. fragilis*-group species (Penders et al. 2006).

Maternal Diet

Maternal diet tends to influence her infant's gut microbiotic composition. It has been shown in univariate analysis that faecal *E. coli* counts were lower and the prevalence and counts of *B. fragilis*-group species also tended to be lower for infants of mothers consuming organic diets (>50% of meat, eggs, vegetables, fruit, and milk of organic or biodynamic origin) than did infants whose mothers consumed regular diet. In addition, mothers consuming organic diet breast-fed their infants more often than their counterparts. Although these results should be interpreted with caution because the type of infant feeding underlies the cause of the association, it is still convincing that there could be an association between maternal diet and the infant microbiotic composition. Extra studies are needed to elucidate this. The effect of organic diet to the microbiota has been explained by others. According to Alm et al. (2002), an organic diet (foods that are produced without the use of synthetic inputs like synthetic fertilizers and pesticides, veterinary drugs, genetically modified seeds and breeds, preservatives, additives, and irradiation) influences the gut microbiota, because it contains spontaneously fermented vegetables containing lactobacilli. Additionally, Peltonen et al. (1992) showed that an extreme vegan diet changes the bacterial fatty acid profiles in the faeces. This is a result of changes in the microbial profile in the gut.

Home Environment

Home environment focusing on sibling has been found to be associated with variations in the composition of infant microbiota and occurrence of diseases. This sibling effect has been hypothesised to be a marker for infections early in life. According to Penders et al. (2006), infants with older siblings tend to have lower total bacterial counts per gram of faeces and also have a greater proportion of bifidobacteria than infants without siblings. Studies on various infant diseases have clearly linked sibling with development of diseases, mainly allergic conditions that are associated with disturbed infant microbiota. Adlerberth et al. (2007) observed that infants without siblings (i.e. first born or only infants) that developed atopic eczema in Sweden, Italy and the United Kingdom had colonisation pattern resembling that of infants born by caesarean section. A sibling effect has also been noted in atopic dermatitis, asthma and wheezing. In these cases, older siblings are inversely related to the disorders (Crane et al. 1994; Koppelman et al. 2003). The sibling effect was earlier on thought to be linked to lower incidence of infection in first born children. However, a study by Gibbs et al. (2004) showed that none of the measures of infection reduced the odds of atopic dermatitis. This means the reduced incidence of these disorders in younger siblings cannot be unequivocally explained by a higher incidence of infections. It could be interesting to elucidate the relationship between disrupted microbiota and sibling and the mechanisms of the various disorders, which at the moment, remain enigmatic.

4.3.4.2 Intestinal Microbiota in Adult Individuals

Mariat et al. (2009) assessed by quantitative polymerase chain reaction the microbiota of human in different age groups and found that the microbiota composition of major bacterial groups; *Clostridium leptum, Clostridium coccoides, Bacteroidetes, Bifidobacterium, Lactobacillus* and *Escherichia coli* changes throughout life, from early childhood to old age. In the adult the microbiota is represented by four dominant groups; *C. leptum, C. coccoides, Bacteroides* and *Bifidobacterium*. It also has subdominant groups of *Lactobacilli, Enterobacteriaceae, Desulfovibrio, Sporomusa, Atopobium* and groups of *Clostridium* clusters XI, XIVb, and XVIII (Hayashi et al. 2002; Rigottier-Gois et al. 2003) colonizing the gut. Other studies have indicated that in healthy adults, 80% of the faecal microbiota belongs to three dominant phyla; *Bacteroidetes, Firmicutes* and *Actinobacteria* and that the *Firmicutes* to *Bacteroidetes* ratio is generally of significant relevance in composition and signalling human gut microbiota status (Lay et al. 2005; Ley et al. 2006). According to Ley et al. (2006) the ratio is lowest (0.4) in infants, peaks (10.9) in adults and declines to 0.6 in elderly individuals. The adult ration of this type of microbiota is thought to be the most stable and adaptive to the host. It is in optimum balance that provides the best microbial benefits to the host.

4.3.4.3 Intestinal Microbiota in Elderly Individuals

Changes in specific bacterial genera and species in elderly subjects occur with extensive interindividual variations and relatively stable mean total counts of anaerobic bacteria (Woodmansey et al. 2004). Shifts in composition of different genera are often observed. A significant decline in both total number and species diversity of beneficial bacteroides and bifidobacteria has been observed (Woodmansey et al. 2004; Woodmansey 2007). On the contrary, facultative anaerobes, eubacteria, and proteolytic bacteria such as fusobacteria, clostridia, and propionibacteria tend to increase. The increased proteolytic bacteria ferment amino acids that result in production of harmful products like ammonia and indoles which have ill-health effect (Woodmansey et al. 2004).

Bacteroides species are responsible for the majority of polysaccharide digestion occurring in the colon since they are nutritionally versatile and are able to utilize a wide variety of different carbon sources (Woodmansey 2007). Their decline could have considerable consequences for the elderly individual. E.g. reduction in amylolytic activity noted in healthy elderly population and in those treated with antibiotics.

The number of bifidobacteria species in the colon of adult subject can occur in excess of 10^{10} per gram dry weight in faeces (Finegold et al. 1983). Its decline in number and species diversity in elderly is one of the most marked changes in the elderly microbiota (Mitsuoka 1992; Gavini et al. 2001; Hopkins et al. 2001; Woodmansey et al. 2004). The reduction in bifidobacterial counts and bacterial species indicates a decline in the stability of this population in the ageing colonic ecosystem. A wide range of bifidobacterial species are found in infants and young adults, but reduced to one or two dominant species in the elderly population, which are *Bifidobacterium adolescentis* or the phenotypically similar *Bifidobacterium angulatum* and *Bifidobacterium longum* (Gavini et al. 2001; Woodmansey 2007). It has been found that the decline in species diversity is due to reduction in bifidobacterial adhesion to the intestinal mucosa of elderly individuals. This could be due to changes in the bacteria, or in the chemical composition and structure of colonic mucus (Ouwehand et al. 1999; He et al. 2001). The decline could reduce the functionality and immune responsiveness in the gut leading to increased susceptibility to GI infections observed in elderly population.

Eubacteria have been reported to be second only to bacteroides in numbers isolated from the large intestine (Woodmansey 2007). Their increase in elderly subjects may have health consequences for the host, with a possible increase in the transformation of bile acids, creating potentially harmful metabolites in the gut. One species, *Eubacterium aerofaciens*, produces moderate to severe arthritis in rats following intraperitoneal inoculation of the bacterial cell wall components (Severijnen et al. 1989). If this occurs *in vivo*, then eubacteria could, at least in part, contribute to the rise in arthritic conditions in elderly people.

4.3.5 Perturbations of Behaviour

Perturbations of behaviour induced by the central nervous system have been shown to influence the microbiota. It is now evident that behavioural disturbances such as stress can change the composition of the microbiota and lead to increased vulnerability to inflammatory stimuli in the GI tract. This could be mediated via stress-induced changes in GI physiology that alter the habitat of enteric microbiota.

Tannock and Savage (1974) studied the effect of environmental and dietary stress on the indigenous microbiota of the GI tract of mice. To induce stress, the mice were deprived of food, water and bedding for 48 hours. The stressed mice showed dramatic reductions in the population of lactobacilli while control mice had high population levels of inhabitant lactobacilli. Further, the population of fusiform-shaped bacteria in caecum and colon epithelial mucosa in stressed mice were reduced when compared to that of control. Their study also showed that the total number of anaerobes were similar in both stressed and control animals. They concluded that environmental and dietary stress markedly alters the GI microbiota in mice. It is noteworthy that their study showed that inoculation of *Salmonella typhimurium* prior to stress led to higher populations of the pathogenic bacteria than in control mice, alluding to disrupted immunity following, at least in part, disturbances to microbiota.

Several other studies have shown disruption of the microbiota by stress due to maternal separation. Bailey and Coe (1999) determined whether psychological stress, induced by maternal separation, in infant rhesus monkeys could lead to an internal environment conducive to pathogen infection. Their study showed a significant decrease in faecal bacteria, especially lactobacilli, that was evident on day 3 post separation, with a return to baseline by day 7. The decrease in the microbiota was correlated with the display of stress-indicative behaviours, but not with cortisol secretion. In addition, there was an increase in susceptibility to opportunistic bacterial infections like *Campylobacter jejuni* in infants who displayed numerous stress-indicative behaviours. These findings suggested that stress due to disruption of the mother-infant bond disturbs the microbiota and increases vulnerability to disease.

The fact that stress perturbs the microbiota and predisposes the gut to illnesses is further supported by other studies that have shown the ability of probiotics to ameliorate stress-induced changes of microbiota and the GI function (Eutamene and Bueno 2007) and to attenuate the drop in lactobacilli in maternally deprived infants (Gareau et al. 2007). Further, analysis of stool samples using 16s rRNA has revealed reduction in *Lactobacillus* species in patients with IBS, a disease characterized by an increased response to stress (Collins and Bercik 2009).

4.3.5.1 Mechanisms Behind Stress-Induced Microbiotal Disruption

The mechanisms of stress-induced microbiota disturbances following maternal separation involve increased levels of corticosterone, inflammatory cytokines, intestinal permeability, and a vulnerability of the GI tract to inflammatory stimuli. The stress could mediate these effects via cytokine-induced hyperresponsiveness of the hypothalamic–pituitary pathway (O'Mahony et al. 2009). These events could also lead to changes in epithelial cell function, mucus secretion and GI motility that directly perturb the microbiota. In addition, stress induces release of norepinephrine which is known to preferentially stimulate the growth and mucosa adherence of specific members of the microbiota (Collins and Bercik 2009). The net outcome of these events is a shift in the bacterial composition of the GI tract and its sequel of increased susceptibility of the gut to chemical and infectious inflammatory stimuli observed in stress models. The central nervous-induced microbiota disturbances are implicated in behavioural changes like depression and stress in separate human patients as well as experimental animals with IBD, IBS, Crohn's disease and ulcerative colitis. Studies in the experimental animals have shown that stress exacerbates colitis while depression increases susceptibility to inflammatory stimuli.

These events are mediated via impairment of vagal parasympathetic outflow to the gut (Collins and Bercik 2009).

4.3.6 *Gut Inflammation*

Evidence is accumulating on the effect of gut inflammation to the microbiota. It is becoming apparent that enteric inflammation alters the microbiota composition leading to disruption of colonization resistance and enhanced pathogen growth. This can be depicted by enteropathogenic bacteria, *Citrobacter rodentium* and *Salmonella enteric* spp. I serovar Typhimurium. Colonization of large intestine by these pathogens is drastically boosted by pre-existing or pathogen induced inflammation (Lupp et al. 2007; Stecher et al. 2007). According to Stecher et al. (2007), the *S. enteritidis* Typhimurium-induced inflammation dramatically alters the microbiota composition and becomes a predominant bacterial species after four days of infection. Colonization of the pathogenic *C. rodentium* in mice peaks at day seven post infection. This peak colonization correlates with a decrease in the total density of the microbiota and an increase in γ-proteobacteria as well as first manifestations of colitis symptoms (Luperchio and Schauer 2001). The γ-proteobacteria includes various prominent enteropathogens like *Salmonella* spp., *Shigella* spp., *Yersinia* spp., and *E. coli* and about 10^4–10^6 CFU per g of the microbiota (Stecher and Hardt 2008). The inflammation resolves at day 28 post infection when colonization of the pathogen declines. Similar changes in the microbiota composition are observed following *C. rodentium* infection in mice suffering from IBD as well as patients suffering from IBD or acute colitis (Lupp et al. 2007).

Several mechanisms have been hypothesized to ascertain the way inflammation disturbs the microbiota. They include food hypothesis, differential killing hypothesis, and commensal-network-disruption hypothesis.

4.3.6.1 Food Hypothesis

In a stable host-microbiota commensalism in the large intestine anaerobic fermentation of proteins and indigestible carbohydrates proceeds normally and produces essential amino acids, vitamins and SCFA. This process occurs only in those parts of the diet that cannot be processed or resorbed by the small intestine and by glycoconjugates, proteins, and cellular debris released by the mucosa. Because of this, there is scarcity of high energy nutrients in the large intestine and the microbiota uses these scarce nutrients efficiently to limit nutrient availability to any incoming pathogen. The net result is retardation in the growth of the pathogens.

In the course of inflammation, the inflamed mucosa may have changes in the nutrient mix, increases in oxygen levels, or increases in the availability of surface adhesion sites. The changes in the nutrient mix increase nutrient availability in the mucosa from inflammatory exudates (e.g. blood proteins and glycoproteins) that serve a good nutrient source for pathogens. The outcome is overgrowth of pathogenic bacteria that are known for their fast growth rates in rich media. Such bacteria include *Salmonella* spp., pathogenic *E. coli* spp., *Shigella* spp., *Citrobacter* spp., and *Vibrio cholera*.

In addition, since inflammation lowers the number of microbiota (see differential killing hypothesis), a further alteration in the substrate range leading to increased nutrient availability for enteropathogens could occur in dual ways; firstly, the decrease in total number of probiotics reduces the nutrient competition with the pathogenic bacteria; and secondly, a decrease in the microbiota fermenters avails an increased nutrient substrate range leading to increased nutrient availability for enteropathogens. The decrease in the fermenters has an additional effect of decreasing the concentrations of growth-inhibitory fermentation products like SCFA, or of toxic products such as bacteriocins; favouring further the establishment of enteropathogens. Both mechanisms for increased nutrient availability to pathogens and the decreased antibacterial effects against them, improve the pathogen growth in the inflamed gut and leads to further inflammation (Stecher and Hardt 2008).

Increased oxygen availability and the presence of additional receptors for pathogen adhesion might also enhance pathogenic colonization. It is conceivable that these conditions, especially oxygen availability, specifically foster growth of facultative anaerobic bacteria. Under these conditions it has been observed that the population size of γ-proteobacteria and other facultative anaerobic species (like certain *Lactobacillaceae*) increases following enteric infection with *Salmonella* or *Citrobacter* (Lupp et al. 2007; Stecher et al. 2007) and in IBD (Mai et al. 2006; Lupp et al. 2007). By improving the oxygen availability and providing additional receptors for pathogen adhesion, inflammation weakens colonization resistance simply by enabling fast pathogen growth.

4.3.6.2 Differential Killing Hypothesis

In the course of inflammation, inflammatory mediators are produced by cells of the immune system (macrophages, dendritic cells, and neutrophils) and the intestinal epithelial cells. Chemoattractant mediators like interleukin (IL)-8 attract the immune cells to the site of bacterial intrusion. Subsequently, the immune cells produce various antimicrobials (e.g. lysozyme, hydrolases, nitric oxide, peptides, lactoferrin, superoxide anions, hydroxyl radicals, singlet oxygen, hydrogen peroxide and halide products) that act intracellularly in the phagosomes as well as extracellularly. In addition, the intestinal mucosa expresses a diverse repertoire of α- and β-defensins that contribute to mucosa defence against pathogens (Stecher and Hardt 2008).

Since most of the antimicrobial defences are non-specific, during inflammation therefore, the luminal microbiota strains are exposed to the antimicrobial factors to the same extent as the enteropathogens. However, some other factors like differences in the binding affinity of defensins and lectins between bacterial species might cause those members of the microbiota that mediate colonization resistance to be affected whereas pathogens might resist (Peschel and Sahl 2006). This is further supported by the presence of numerous genes enhancing antimicrobial peptide-resistance and radical detoxification in the pathogens like that of *S. enteritidis* Typhimurium (Navarre et al. 2005). Thus the antimicrobial produced during inflammation may kill or inhibit growth of some members of the microbiota but not the pathogens. The affected microbiota members are the ones that would normally inhibit growth and establishment of pathogens under steady-state conditions. Hence the differential susceptibility to killing mechanisms might explain the loss of colonization resistance in the inflamed gut.

4.3.6.3 Commensal-Network-Disruption Hypothesis

This is due to disruption of commensalism as a result of loss of secondary fermenters that form the main microbiota members. Since these species are necessary for efficient growth of other microbiota species that in turn, prevent growth and establishment of pathogens under normal conditions, their exclusion favours establishment and growth of pathogens (Stecher and Hardt 2008).

Lastly, there are possibilities that changes in the overall microbiota density might lead to conditions that favour growth of pathogens such as increased nutrient availability and production of fewer inhibitory substances.

4.3.6.4 Inflammatory Bowel Disease

IBD is another example where inflammation alters the gut microbiota. Contrary to pathogen-induced inflammatory reactions, the inflammation in IBD is triggered by exaggerated immune defences directed against members of the commensal microbiota. While the exact species of microbiota that are affected remain fuzzy, there is a general mechanism favouring the growth of γ-proteobacteria member of the microbiota (Mai et al. 2006; Heimesaat et al. 2007) as well as the intestinal pathogens.

4.4 Conclusion

The intestinal microbiota plays an important role in human health only when it is properly balanced in terms of total number and composition of individual members. Bacterial colonization of the gut starts during the early days of an individual after birth and is affected even by the process and the mode of delivery. In early life, therefore, one of the major factors affecting the composition of intestinal microbiota is the mode of delivery. In this case, vaginally born infants are colonized at first by faecal and vaginal bacteria of the mother and become predominated by more beneficial microbes like Bifidobacterium species, whereas infants born through caesarean section have less beneficial microbes. Instead, they get bacteria from the hospital environment and health care workers. After infant colonization, the microbiota grows and reaches best composition during adult where optimal protection is obtained. It then changes towards decline of total protective members in elderly, leading to reduced protection and increased susceptibility to diseases. Disturbances of microbiota therefore are featured by decrease in the numbers of protective bacteria and an increase in the potentially pathogenic members. The disturbances lead to impaired functions of microbiota and may have direct or indirect effect to the health of an individual. Corrections of these disturbances by restoring the composition (e.g. consumption of probiotics) are currently a focus of alleviating disorders associated with microbiotal disturbances. However, more studies are needed to elucidate the exact microbiota species affected in various enteric disorders. The knowledge obtained from such studies will be used in designing appropriate therapeutic probiotic bacteria for alleviating specific disorders originating from disruption of the microbiota.

References

Adlerberth I, Strachan DP, Matricardi PM, Ahrné S, Orfei L, Aberg N, Perkin MR, Tripodi S, Hesselma B, Saalman R, Coates AR, Bonanno CL, Panetta V, Wold AE (2007) Gut microbiota and development of atopic eczema in 3 European birth cohorts. J Allergy Clin Immunol 120:343–350

Al Jumaili IJ, Shibley M, Lishman AH, Record CO (1984) Incidence and origin of *Clostridium difficile* in neonates. J Clin Microbiol 19:77–78

Alm JS, Swartz J, Björkstén B, Engstrand L, Engströn J, Kühn I, Lilja G, Möllby R, Norin E, Pershagen G, Reinders C, Wreiber K, Scheynius A (2002) An anthroposophic lifestyle and intestinal microflora in infancy. Pediatr Allergy Immunol 13:402–411

Attebery HR, Sutter VL, Finegold SM (1972) Effect of a partially chemically defined diet on normal human faecal flora. Am J Clin Nutr 25:1391–1398

Bailey MT, Coe CL (1999) Maternal separation disrupts the integrity of the intestinal microflora in infant rhesus monkeys. Dev Psychobiol 35:146–155

Bartosch S, Fite A, Macfarlane GT, McMurdo ME (2004) Characterization of bacterial communities in faeces from healthy elderly volunteers and hospitalized elderly patients by using real-time PCR and effects of antibiotic treatment on the faecal microbiota. Appl Environ Microbiol 70:3575–3581

Beaugerie L, Petit JC (2004) Microbial-gut interactions in health and disease. Antibiotic-associated diarrhoea. Best Pract Res Clin Gastroenterol 18:337–352

Bennet R, Nord CE (1987) Development of the faecal anaerobic microflora after caesarean section and treatment with antibiotics in newborn infants. Infection 15:332–336

Bezirtzoglou E (1997) The intestinal microflora during the first weeks of life. Anaerobe 3:173–177

Bingham SA (1999) High-meat diets and cancer risk. Proc Nutr Soc 58:243–248

Burkitt DP (1971) Epidemiology of cancer of the colon and rectum. Cancer 28:3–13

Butler LM, Sinha R, Millikan RC, Martin CF, Newman B, Gammon MD, Ammerman AS, Sandler RS (2003) Heterocyclic amines, meat intake, and association with colon cancer in a population-based study. Am J Epidemiol 157:434–445

Collins SM, Bercik P (2009) The relationship between intestinal microbiota and the central nervous system in normal gastrointestinal function and disease. Gastroenterology 136:2003–2014

Crane J, Pearce N, Shaw R, Fitzharris P, Mayers C (1994) Asthma and having siblings. BMJ 309:272

Crowther JS, Drasar BS, Goddard P, Hill MJ, Johnson K (1973) The effect of a chemically defined diet on the faecal flora and faecal steroid concentration. Gut 14:790–793

Donskey CJ, Hujer AM, Das SM, Pultz NJ, Bonomo RA, Rice LB (2003) Use of denaturing gradient gel electrophoresis for analysis of the stool microbiota of hospitalized patients. J Microbiol Meth 54:249–256

Edlund C, Nord CE (2000) Effect on the human normal microflora of oral antibiotics for treatment of urinary tract infections. J Antimicrob Chemother 46(1 Suppl):41–48

Engelbrektson A, Korzenik JR, Pittler A, Sanders ME, Klaenhammer TR, Leyer G, Kitts CL (2009) Probiotics to minimize the disruption of faecal microbiota in healthy subjects undergoing antibiotic therapy. J Med Microbiol 58:663–670

Eutamene H, Bueno L (2007) Role of probiotics in correcting abnormalities of colonic flora induced by stress. Gut 56:1495–1497

Finegold SM, Mathisen GE, George WL (1983) Changes in human intestinal flora related to the administration of antimicrobial agents. Academic Press, London, pp 355–445

Franks AH, Harmsen HJ, Raangs GC, Jansen GJ, Schut F, Welling GW (1998) Variations of bacterial populations in human faeces measured by fluorescent in situ hybridization with group-specific 16S rRNA-targeted oligonucleotide probes. Appl Environ Microbiol 64:3336–3345

Gareau MG, Jury J, MacQueen G, Sherman PM, Perdue MH (2007) Probiotic treatment of rat pups normalises corticosterone release and ameliorates colonic dysfunction induced by maternal separation. Gut 56:1522–1528

Gavini F, Cayuela C, Antoine J-M, Lecoq C, Lefebvre B, Membre J-M, Neut C (2001) Differences in the spatial distribution of bifidobacterial and enterobacterial species in human faecal microflora of three different (children, adults, elderly) age groups. Microbial Ecol Health Dis 13:40–45

Gibbs S, Surridge H, Adamson R, Cohen B, Bentham G, Reading R (2004) Atopic dermatitis and the hygiene hypothesis: a case-control study. Int J Epidemiol 33:199–207

Gorbach SL, Goldin BR (1992) Nutrition and the gastrointestinal microflora. Nutr Rev 50:378–381

Gronlund MM, Lehtonen OP, Eerola E, Kero P (1999) Faecal microflora in healthy infants born by different methods of delivery: permanent changes in intestinal flora after cesarean delivery. J Pediatr Gastroenterol Nutr 28:19–25

Guarner F, Malagelada JR (2003) Gut flora in health and disease. Lancet 361:512–519

Harmsen HJ, Wildeboer-Veloo AC, Raangs GC, Wagendorp AA, Klijn N, Bindels JG, Welling GW (2000) Analysis of intestinal microflora development in breast-fed and formula-fed infants by using molecular identification and detection methods. J Pediatr Gastroenterol Nutr 30:61–67

Hayashi H, Sakamoto M, Benno Y (2002) Phylogenetic analysis of the human gut microbiota using 16S rDNA clone libraries and strictly anaerobic culture-based methods. Microbiol Immunol 46:535–548

He F, Ouwehand AC, Isolauri E, Hosoda M, Benno Y, Salminen S (2001) Differences in composition and mucosal adhesion of bifidobacteria isolated from healthy adults and healthy seniors. Curr Microbiol 43:351–354

Heimesaat MM, Fischer A, Siegmund B, Kupz A, Niebergall J, Fuchs D, Jahn HK, Freudenberg M, Loddenkemper C, Batra A, Lehr HA, Liesenfeld O, Blaut M, Göbel UB, Schumann RR, Bereswill S (2007) Shift towards pro-inflammatory intestinal bacteria aggravates acute murine colitis via Toll-like receptors 2 and 4. PLoS One 2(7):e662

Hookman P, Barkin JS (2007) Review: *Clostridium difficile* associated disorders/diarrhoea and *Clostridium difficile colitis*: the emergence of a more virulent era. Dig Dis Sci 52:1071–1075

Hooper LV, Bry L, Falk PG, Gordon JI (1998) Host-microbial symbiosis in the mammalian intestine: exploring as internal ecosystem. Bioessays 20:336–243

Hopkins MJ, Sharp R, Macfarlane GT (2001) Age and disease related changes in intestinal bacterial populations assessed by cell culture, 16S rRNA abundance, and community cellular fatty acid profiles. Gut 48:198–205

Hughes R, Cross AJ, Pollock JRA, Bingham S (2001) Dose-dependent effect of dietary meat on endogenous colonic N-nitrosation. Carcinogenesis 22:199–202

Imase K, Takahashi M, Tanaka A, Tokunaga K, Sugano H, Tanaka M, Ishida H, Kamiya S, Takahashi S (2008) Efficacy of *Clostridium butyricum* preparation concomitantly with *Helicobacter pylori* eradication therapy in relation to changes in the intestinal microbiota. Microbiol Immunol 52:156–161

Kassinen A, Krogius-Kurikka L, Mäkivuokko H, Rinttilä T, Paulin L, Corander J, Malinen E, Apajalahti J, Palva A (2007) The faecal microbiota of irritable bowel syndrome patients differs significantly from that of healthy subjects. Gastroenterology 133:24–33

Kim KH, Fekety R, Batts DH, Brown D, Cudmore M, Silva J Jr, Waters D (1981) Isolation of *Clostridium difficile* from the environment and contacts of patients with antibiotic-associated colitis. J Infect Dis 143:42–50

Koppelman GH, Jansen DF, Schouten JP, van der Heide S, Bleecker ER, Meyers DA, Postma DS (2003) Sibling effect on atopy in children of patients with asthma. Clin Exp Allergy 33:170–175

Lay C, Sutren M, Rochet V, Saunier K, Doré J, Rigottier-Gois L (2005) Design and validation of 16S rDNA probes to enumerate members of the *Clostridium leptum* subgroup in human faecal microbiota. Environ Microbiol 7:933–946

Lee H, Pimentel M (2006) Bacteria and syndrome: the evidence forsmall intestinal bacterial overgrowth. Curr Gastroenterol Rep 8:305–311

Ley RE, Turnbaugh P, Klein S, Gordon JI (2006) Microbial ecology: human gut microbes associated with obesity. Nature 444:1022–1023

Li Y, Newton T, Weisbrodt N, Moody F (1993) Intestinal migrating myoelectric complexes in rats with acute pancreatitis and bile duct ligation. J Surg Res 55:182–187

Luperchio SA, Schauer DB (2001) Molecular pathogenesis of *Citrobacter rodentium* and transmissible murine colonic hyperplasia. Microbes Infect 3:333–340

Lupp C, Robertson ML, Wickham ME, Sekirov I, Champion OL, Gaynor EC, Finlay BB (2007) Host-mediated inflammation disrupts the intestinal microbiota and promotes the overgrowth of enterobacteriaceae. Cell Host Microbe 2:119–129

Mackie RI, Sghir A, Gaskins HR (1999) Developmental microbial ecology of the neonatal gastrointestinal tract. Am J Clin Nutr 69:1035S–1045S

Macpherson A, Khoo UY, Forgacs I, Philpott-Howard J, Bjarnason I (1996) Mucosal antibodies in inflammatory bowel disease are directed against intestinal bacteria. Gut 38:365–375

Mai V, Braden CR, Heckendorf J, Pironis B, Hirshon JM (2006) Monitoring of stool microbiota in subjects with diarrhoea indicates distortions in composition. J Clin Microbiol 44:4550–4552

Mai V, McCrary QM, Sinha R, Glei M (2009) Associations between dietary habits and body mass index with gut microbiota composition and faecal water genotoxicity: an observational study in African American and Caucasian American volunteers. Nutr J 8:49

Mäkivuokko H, Tiihonen K, Tynkkynen S, Paulin L, Rautonen N (2010) The effect of age and non-steroidal anti-inflammatory drugs on human intestinal microbiota composition. Br J Nutr 103:227–234

Malinen E, Rinttilä T, Kajander K, Mättö J, Kassinen A, Krogius L, Saarela M, Korpela R, Palva A (2005) Analysis of the faecal microbiota of irritable bowel syndrome patients and healthy controls with real-time PCR. Am J Gastroenterol 100:373–382

Mariat D, Firmesse O, Levenez F, Guimarães VD, Sokol H, Doré J, Corthier G, Furet JP (2009) The Firmicutes/Bacteroidetes ratio of the human microbiota changes with age. BMC Microbiology 9:123

Mitsuoka T (1992) Intestinal flora and aging. Nutr Rev 50:438–446

Moody F, Haley-Russell D, Muncy D (1995) Intestinal transit and bacterial translocation in obstructive pancreatitis. Dig Dis Sci 40:1798–1804

Moore WE, Moore LH (1995) Intestinal floras of populations that have a high risk of colon cancer. Appl Environ Microbiol 61:3202–3207

Moro GE, Mosca F, Miniello V, Fanaro S, Jelinek J, Stahl B, Boehm G (2003) Effects of a new mixture of prebiotics on faecal flora and stools in term infants. Acta Paediatr Suppl 91:77–79

Navarre WW, Halsey TA, Walthers D, Frye J, McClelland M, Potter JL, Kenney LJ, Gunn JS, Fang FC, Libby SJ (2005) Co-regulation of Salmonella enterica genes required for virulence and resistance to antimicrobial peptides by SlyA and PhoP/PhoQ. Mol Microbiol 56:492–508

O'Mahony SM, Marchesi JR, Scully P, Codling C, Ceolho AM, Quigley EM, Cryan JF, Dinan TG (2009) Early life stress alters behavior, immunity, and microbiota in rats: implications for irritable bowel syndrome and psychiatric illnesses. Biol Psychiatry 65:263–267

Ouwehand AC, Isolauri E, Kirjavainen PV, Salminen SJ (1999) Adhesion of four Bifidobacterium strains to human intestinal mucus from subjects in different age groups. FEMS Microbiol Lett 172:61–64

Peltonen R, Ling WH, Hanninen O, Eerola E (1992) An uncooked vegan diet shifts the profile of human faecal microflora: computerized analysis of direct stool sample gas-liquid chromatography profiles of bacterial cellular fatty acids. Appl Environ Microbiol 58:3660–3666

Penders J, Thijs C, Vink C, Stelma FF, Snijders B, Kummeling I, van den Brandt PA, Stobberingh EE (2006) Factors influencing the composition of the intestinal microbiota in early infancy. Pediatrics 118:511–521

Peschel A, Sahl HG (2006) The co-evolution of host cationic antimicrobial peptides and microbial resistance. Nat Rev Microbiol 4:529–553

Rafter J, Glinghammar B (1998) Interactions between the environment and genes in the colon. Eur J Cancer Prev 7(Suppl):S69–S74

Rastall RA (2004) Bacteria in the gut: friends and foes and how to alter the balance. J Nutr 134(Suppl):2022S–2026S

Rigottier-Gois L, Bourhis AGL, Gramet G, Rochet V, Doré J (2003) Fluorescent hybridisation combined with flow cytometry and hybridisation of total RNA to analyse the composition of microbial communities in human faeces using 16S rRNA probes. FEMS Microbiol Ecol 43:237–245

Rinne MM, Gueimonde M, Kalliomaki M, Hoppu U, Salminen SJ, Isolauri E (2005) Similar bifidogenic effects of prebiotic-supplemented partially hydrolyzed infant formula and breastfeeding on infant gut microbiota. FEMS Immunol Med Microbiol 43:59–65

Russell RM (1992) Changes in gastrointestinal function attributed to aging. Am J Clin Nutr 55(6 Suppl):1203S–1207S

Salminen S, Isolauri E (2006) Intestinal colonization, microbiota and probiotics. J Pediatr 149:S115–S120

Salminen S, Isolauri E, Onela T (1995) Gut flora in normal and disordered states. Chemotherapy 41(1 Suppl):5–15

Schneider SM, Le Gall P, Girard-Pipau F, Piche T, Pompei A, Nano JL, Hébuterne X, Rampal P (2000) Total artificial nutrition is associated with major changes in the faecal flora. Eur J Nutr 39:248–255

Seki H, Shiohara M, Matsumura T, Miyagawa N, Tanaka M, Komiyama A, Kurata S (2003) Prevention of antibiotic-associated diarrhoea in children by *Clostridium buyricum* MIYARI. Ped Internat 45:86–90

Severijnen AJ, van Kleef R, Hazenburg MP, van de Merwe J (1989) Cell wall fragments from major residents of the human intestinal flora induce chronic arthritis in rats. J Reumatol 16:1061–1068

Singh VV, Toskes PP (2004) Small bowel bacterial overgrowth: presentation, diagnosis, and treatment. Curr Treat Options Gastroenterol 7:19–28

Stecher B, Hardt WD (2008) The role of microbiota in infectious disease. Trend Microbiol 16:107–114

Stecher B, Robbiani R, Walker AW, Westendorf AM, Barthel M, Kremer M, Chaffron S, Macpherson AJ, Buer J, Parkhill J, Dougan G, von Mering C, Hardt WD (2007) Salmonella enterica serovar typhimurium exploits inflammation to compete with the intestinal microbiota. PLoS Biol 5:2177–2189

Stephen AM, Wiggins HS, Cummings JH (1987) Effect of changing transit time on colonic microbial metabolism in man. Gut 28:601–609

Tannock GW, Savage DC (1974) Influences of dietary and environmental stress on microbial populations in the murine gastrointestinal tract. Infect Immun 9:591–598

Thompson-Chagoyán OC, Maldonado J, Gil A (2005) Aetiology of inflammatory bowel disease (IBD): role of intestinal microbiota and gut-associated lymphoid tissue immune response. Clin Nutr 24:339–352

Tiihonen K, Tynkkynen S, Ouwehand A, Ahlroos T, Rautonen N (2008) The effect of ageing with and without non-steroidal anti-inflammatory drugs on gastrointestinal microbiology and immunology. Br J Nutr 100:130–137 van Felius ID, Akkermans LM, Bosscha K, Verheem A, Harmsen W, Visser MR, Gooszen HG (2003) Interdigestive small bowel motility and duodenal bacterial overgrowth in experimental acute pancreatitis. Neurogastroenterol Motil 15:267–276

van Vliet MJ, Tissing WJE, Dun CAJ, Meessen NEL, Kamps WA, de Bont ESJM, Harmsen HJM (2009) Chemotherapy treatment in pediatric patients with acute myeloid leukemia receiving antimicrobial prophylaxis leads to a relative increase of colonization with potentially pathogenic bacteria in the gut. Clin Infect Dis 49:262–270

Warny M, Pepin J, Fang A, Killgore G, Thompson A, Brazier J, Frost E, McDonald LC (2005) Toxin production by an emerging strain of Clostridium difficile associated with outbreaks of severe disease in North America and Europe. Lancet 366:1079–1084

Wiström J, Norrby SR, Myhre EB, Eriksson S, Granström G, Lagergren L, Englund G, Nord CE, Svenungsson B (2001) Frequency of antibiotic-associated diarrhoea in 2462 antibiotic-treated hospitalized patients: a prospective study. J Antimicrob Chemother 47:43–50

Wlodarska M, Finlay BB (2010) Host immune response to antibiotic perturbation of the microbiota. Mucosal Immuol 3:100–103

Wollowski I, Rechkemmer G, Pool-Zobel BL (2001) Protective role of probiotics and prebiotics in colon cancer. Am J Clin Nutr 73(Suppl):S451–445

Woodmansey EJ (2007) Intestinal bacteria and ageing. J Appl Microbiol 102:1178–1186

Woodmansey EJ, McMurdo ME, Macfarlane GT, Macfarlane S (2004) Comparison of compositions and metabolic activities of faecal microbiotas in young adults and in non-antibiotic-treated elderly subjects. Appl Env Microbiol 70:6113–6122

Young VB, Schmidt TM (2004) Antibiotic-associated diarrhoea accompanied by large-scale alterations in the composition of the faecal microbiota. J Clin Microbiol 42:1203–1206

Chapter 5
The Gut Microbiota, Probiotics and Infectious Disease

Cormac G. M. Gahan, Gerald C. O'Sullivan and J. Kevin Collins

5.1 Introduction

The human gastrointestinal (GI) tract represents a vast microbial ecosystem with a high degree of complexity and inter-individual diversity. The large intestine contains an estimated 10^{11}–10^{12} bacteria per gram of contents and it is thought that total bacterial cells outnumber human cells by at least an order of magnitude (O'Sullivan 2005; O'Hara and Shanahan 2006). These bacteria comprise between 1,000 and 1,150 different species including 160 species which are common between individuals and may provide core functions that are essential for the health of the host (Qin et al. 2010). In this context the human individual has been proposed as a 'superorganism' in which the individual co-exists with their unique microbiota (Wilson and Sober 1989; Zilber-Rosenberg and Rosenberg 2008). The gut microbiota therefore comprise a 'virtual organ' that carries out significant metabolic, digestive and immunoregulatory roles that benefit the host (O'Sullivan 2005; O'Hara and Shanahan 2006).

An important role of the microbiota is to provide direct protection against infectious disease, a function referred to as 'colonisation resistance'. Perturbations in the gut microbiota (for example following antibiotic treatment) can lead to increased susceptibility to food-borne infections as well as infection with *Clostridium difficile*. The importance of the microbiota is highlighted through studies in germ-free mice (mice raised under sterile conditions) which exhibit increased susceptibility to GI pathogens (O'Hara and Shanahan 2006). In addition, human intervention studies using live probiotic cultures have indicated efficacy in preventing or ameliorating GI infections (O'Sullivan et al. 2005; Penner et al. 2005). Presumably this is through the transient restoration of colonization resistance in these individuals.

G. C. O'Sullivan (✉)
Cork Cancer Research Centre, University College Cork, Cork, Ireland
e-mail: geraldc@iol.ie

G. C. O'Sullivan
Department of Surgery, Mercy University Hospital, Grenville Place, Cork, Ireland

J. J. Malago et al. (eds.), *Probiotic Bacteria and Enteric Infections*,
DOI 10.1007/978-94-007-0386-5_5, © Springer Science+Business Media B.V. 2011

The mechanisms by which colonization resistance is maintained by the host microbiota are unclear. However, recent work is beginning to shed light upon this phenomenon. A significant body of evidence suggests that the microbiota function to reduce local inflammation in the gut; the inflammation favours growth of the pathogen at the expense of the microbiota (Stecher and Hardt 2008). During inflammation growth of the pathogen may be stimulated by release of cellular debris (the *food hypothesis*) and pathogens are better equipped than commensal microorganisms for resisting the effects of local immune responses (the *differential killing hypothesis*). In addition the microbiota directly communicates with epithelial cells through the production of specific metabolites which regulate local defences (including production of defensins) (Stecher and Hardt 2008). Commensal organisms also produce specific antimicrobial factors (for example bacteriocins) as well as acidic end products of metabolism which are inhibitory to invading pathogens (Corr et al. 2009). Finally there is good evidence to suggest that the presence of the microbiota can reduce the expression of specific virulence factors by pathogens in the GI tract (Corr et al. 2009). This review will consider the role of the microbiota in protection against infectious disease and the possible mechanisms that underpin this activity. The use of live probiotic cultures in the treatment or prevention of GI infections is also discussed.

5.2 Loss of Colonization Resistance in Germ-Free Mice

Germ-free mice are raised under sterile conditions and therefore provide a useful model with which to study the role of the microbiota in infectious disease. Comparisons of germ-free and conventionally colonized mice reveal a major role for the gut microbiota in local and systemic lymphoid organogenesis and colonization resistance (Macpherson and Harris 2004; Sekirov and Finlay 2009). Germ-free mice have smaller Peyer's patches, fewer intraepithelial lymphocytes and poor digestive enzyme activity all of which may influence susceptibility to GI infection (O'Sullivan 2005; O'Hara and Shanahan 2006). Furthermore commensal microorganisms are known to produce metabolic products (e.g. butyrate and lithocholic acid) that induce production of anti-microbial peptides (defensins) in the host GI tract (Kida et al. 2006; Termen et al. 2008). In general, interactions between the commensal microbiota and host toll-like receptors (TLRs) serve to modulate local immune responses, maintain epithelial cell homeostasis and promote intestinal repair following injury and inflammation (O'Hara and Shanahan 2006; Sekirov and Finlay 2009). In the absence of a normal intestinal microbiota these functions are dysregulated in germ-free mice.

A number of studies show that germ-free mice are susceptible to a variety of intestinal pathogens. Germ-free mice exhibit a significant increase in susceptibility to oral infection with *Listeria monocytogenes, Salmonella enterica* subspecies Typhimurium, pathogenic *E. coli* and *Clostridium difficile* (Inagaki et al. 1996; Nardi et al. 1991; Taguchi et al. 2002; Wilson et al. 1986). Studies demonstrate that monocolonization of germ-free mice before infection is sufficient to protect these animals against *S.* Typhimurium, *Shigella flexneri, E. coli* and *L. monocytogenes* infection (Filho-Lima et al. 2000; Hudault et al. 1997; Rodrigues et al. 2000; Vieira et al. 2008). This suggests that the microbiota alone can mediate colonization resistance against acute infections. However, it is difficult to determine the exact mechanism by which the pathogens are impeded in these studies although direct antagonism of the pathogen and immunomodulation (e.g. to increase secretory antibody production) have been suggested (Hudault et al. 1997; Nardi et al. 1991; Rodrigues et al. 2000). Overall these studies support clinical evaluations in humans and indicate that probiotic interventions have significant potential to limit GI infections.

5.3 Antibiotic-Induced Changes to the Microbiota

Antibiotic administration in both humans and animals is known to affect the normal GI microbiota and to lower resistance to GI pathogens. This is clearly evident in *C. difficile* infections where infection occurs in patients undergoing treatment with broad spectrum antibiotics (notably clindamycin or ampicillin) (Wilcox 2003). Analysis of a human microbiota-associated mouse model indicates that broad spectrum antibiotic treatment does not quantitatively modify the total microbiota. Rather specific groups of microorganisms increased (*Bacteroides-Porphyromonas-Prevotella* group) whilst other microorganisms decreased (*Clostridium coccoides-Eubacterium rectale* group). This resultant imbalance may increase susceptibility to *C. difficile* colonization (Barc et al. 2004). Khoruts et al. (2009) recently used a 16s DNA sequencing approach to demonstrate that the microbiota of an elderly patient with *C. difficile* infection was disrupted in components of the Firmicutes and Bacteroides groups. Following microbial transplantation from a healthy donor the microbiota of the patient returned to a normal balance of bacterial constituents resembling that of the donor. This correlated with a resolution of the patients symptoms (Khoruts et al. 2009).

Dethlefsen et al. (2008) have recently carried out deep sequencing of 16s DNA in the microbiome of three human subjects before and after treatment with ciprofloxacin, a broad spectrum antibiotic known to have an association with *C. difficile* infections (Deshpande et al. 2008). In this study approximately one third of bacterial taxa in the gut were affected by the antibiotic treatment. Whilst this greatly reduced the overall diversity of the gut microbiota, subjects did not report any tangible symptomatic effects, indicating functional redundancy of the components of the microbiota. Generally the constituents of the microbiota returned to normal (pre-antibiotic) levels within 4 weeks of cessation of the antibiotic. However, in some cases individual bacterial taxa could not be detected even after 6 months following antibiotic treatment (Dethlefsen et al. 2008). This study provides a potential 'map' of the microbiota as it exists prior to and following antibiotic treatment.

Perturbation of the intestinal microbiota in mice through oral administration of vancomycin and streptomycin significantly increases susceptibility to oral *S.* Typhimurium infection (Sekirov et al. 2008). Disruption of the murine microbiota prior to infection led to increased colonization by the pathogen and an increase in the severity of inflammatory symptoms (Sekirov et al. 2008). Antibiotic administration before infection also resulted in higher levels of shedding and potential for transmission of the pathogen suggesting a role for the microbiota in controlling disease spread between individuals (Lawley et al. 2008).

Antibiotic-mediated disruption of the gut microbiota leads to a transient reduction in innate immune parameters similar to those seen in germ-free animals. In particular, antibiotic treatment of mice leads to downregulation of local expression of Reg3g, a C-type lectin (an antimicrobial compound active against Gram positive bacteria) (Brandl et al. 2008). This renders mice susceptible to local infection with vancomycin resistant enterococci (VRE). This effect can be reversed through oral administration of lipopolysaccharide (the ligand for TLR4) which re-induces Reg3g and reinstates local resistance to VRE (Brandl et al. 2008).

5.4 Evidence of Colonization Resistance from Human Probiotic Intervention Studies

A 2004 Cochrane meta-analysis reviewed the available evidence for the use of probiotics in diarrhoeal illness and determined that probiotics appear to be a useful adjunct to rehydration therapy in treating acute, infectious diarrhoea in adults and children (Allen et al. 2004). There was no evidence that poor study design had led to an overestimation of the benefits of probiotics. The analysis investigated a variety of probiotic strains including the yeast *Saccharomyces boulardii* and the well-studied bacterium *Lactobacillus rhamnosus* GG and concluded that a variety of probiotics reduced infectious diarrhoea in children and adults in various settings. The authors suggest that a mechanism common to most probiotics such as 'colonization resistance' is most likely responsible as the activity is mediated against a variety of gut pathogens (Allen et al. 2004).

Another meta-analysis reviewed the outcomes from 12 randomised controlled clinical trials examining probiotic efficacy against traveller's diarrhoea (McFarland 2007). Most of these infections are thought to be of bacterial aetiology with pathogenic enterobacteriaceae being the most likely cause. In this setting *S. boulardii* was shown to be effective in preventing onset of diarrhoea especially when administered at the highest dose. Efficacy was more pronounced in some destinations (North Africa) compared to others (South America) and was ineffective in travellers to India. Similarly *L. rhamnosus* GG was more effective in travellers to a specific region in Turkey (Alanya) but was less effective in tourists visiting another region (Marmaris). *L. rhamnosis* GG was also effective in a separate trial monitoring visitors to the USA. Analysis of trial data clearly indicated the overall efficacy of probiotics in preventing traveller's diarrhoea but demonstrated variations with respect to destinations visited (McFarland 2007).

There is good evidence from double-blind placebo-controlled randomized studies that probiotic administration can reduce the duration of rotaviral diarrhoea in children when administered alongside typical rehydration therapies. In one large study involving 291 neonatal patients *L. rhamnosus* GG administration significantly shortened the duration of symptoms compared to patients administered the placebo (Guandalini et al. 2000). Another study in older children showed that *L. reuteri* SD was also effective in reducing the duration of symptoms compared to children receiving the placebo (Shornikova et al. 1997). Other studies showed that *L. rhamnosus* GG is effective in reducing duration of symptoms but is not effective as a prophylactic in prevention of rotavirus infection (Mastretta et al. 2002; Szajewska et al. 2001).

A recent thorough, evidence-based review of clinical trials suggests that both antibiotic-associated diarrhoea and *C. difficile* infection may respond well to probiotic interventions (McFarland 2009). A mixture of *L. casei* DN114001, *S. thermophilus* and *L. bulgaricus* was effective in reducing antibiotic-associated diarrhoea in hospitalized patients relative to those receiving the placebo (Hickson et al. 2007). In another study 269 children taking antibiotics for ear or respiratory tract infections were randomized to receive either a placebo or *S. boulardii*. Over a short follow-up period (two weeks) the frequency of diarrhoea in the probiotic group was significantly less than in the placebo group (Kotowska et al. 2005). McFarland (2009) concludes that whilst most meta-analyses indicate that probiotics are effective for preventing antibiotic-associated diarrhoea, not all are effective and that effects are strain specific.

A meta-analysis of six randomized controlled trials using probiotics in association with standard antibiotic treatments for *C. difficile* revealed that probiotics significantly reduced the risk of recurrent *C. difficile* infections (McFarland 2006). The most comprehensive study remains a 1994 analysis of 124 adult *C. difficile* patients on varied doses of vancomycin or metronidazole who were randomized to receive either *S. boulardii* probiotic or a placebo in addition to their antibiotic therapy (McFarland et al. 1994). In this study 15/57 (26.3%) of patients receiving the probiotic suffered relapses of *C. difficile* infection compared to 30/67 (44.8%) in the control group (McFarland et al. 1994). Relatively few studies have examined the prevention of *C. difficile* primary infections by probiotic administrations. A recent analysis of a probiotic mix of *L. casei, L. bulgaricus* and *S. thermophilus* indicated that 0/56 (0%) out of the probiotic group developed *C. difficile* infection over the course of the study compared to 9/53 (17%) of the control group (Hickson et al. 2007). It is suggested that further trials are required with higher numbers of participants in order to determine the optimal probiotic strains and doses to enhance both prophlyaxis and treatment of *C. difficile* infections (McFarland 2009).

Gotteland et al. (2006) have carried out a systematic review of clinical trial data to examine the usefulness of probiotic interventions in controlling gastric colonization by *Helicobacter pylori*. Many of these trails utilize *Lactobacillus* species as the predominant probiotic strain as members of this genus survive relatively well in the stomach and can be detected in gastric biopsies following oral administration in human volunteers (Gotteland et al. 2006; Valeur et al. 2004). Many of these strains have also been shown to be effective in mouse models in the prevention of *H. pylori* colonization and in eliminating the pathogen in infected mice (Johnson-Henry et al. 2004; Kabir et al. 1997). In one set of human clinical studies probiotics were used in conjunction with triple therapy for the elimination of *H. pylori*. Gotteland et al. (2006) conclude that probiotic supplementation decreases the side effects of triple therapy and improves drug compliance. In some studies (but not all) eradication rates were improved (Gotteland et al. 2006). In other studies probiotics were utilized as the sole treatment. The double blind placebo-controlled trials demonstrated probiotic efficacy (primarily of *L. johnsonii*) in reducing urea breath test results consistent with a reduction of *H. pylori* colonization (Gotteland et al. 2006). However, whilst probiotics may act as a potential adjunct to therapy total eradication of the pathogen is most effective following triple therapy (Gotteland et al. 2005, 2006).

Overall meta-analyses and systematic reviews of clinical trial data appear to indicate that administration of probiotics can improve colonization resistance in the host to limit the effects of GI pathogens. However, clearly more work is needed in this area in order to match particular probiotic strains to particular pathogens and to establish the appropriate probiotic dose and delivery matrix needed to prevent or limit colonization by specific pathogens. It has been suggested that further double-blind placebo-controlled randomized trials with greater numbers of volunteers, specified strains of probiotics, specified doses and with appropriate controls are needed to generate consistent data that will support health claims surrounding probiotic food products (Farnworth 2008).

5.5 Mechanistic Basis of Colonization Resistance

The mechanisms that underpin the establishment of colonization resistance by the gut microbiota or probiotic strains are currently unclear. However, they are thought to comprise immunomodulation, direct antagonism, enhancement of epithelial barrier function and direct inhibition of virulence factor expression in specific pathogens Fig. 5.1. The evidence supporting these proposed mechanisms is discussed below. Research in this area is essential in order to underpin more applied research in the area of probiotics, to support health claims and even to drive the development of bacteria-free medicines for GI administration (the 'bugs-to-drugs' concept) (Shanahan 2010).

5.5.1 Regulation of Inflammation

It is clear that administration of probiotic commensal stains to mice can reduce inflammatory responses to intestinal pathogens such as *S.* Typhimurium (O'Mahony et al. 2008; Sekirov and Finlay 2009). *In vitro* studies indicate that the presence of commensal strains added to cell monolayers during infection by various pathogens (including *L. monocytogenes* and *S.* Typhimurium) can reduce production of proinflammatory cytokines and chemokines and increase secretion of anti-inflammatory mediators (such as IL-10) (Corr et al. 2007a; O'Hara et al. 2006; Silva et al. 2004). Furthermore human intervention studies indicate that specific probiotic strains can reduce the symptoms of irritable bowel syndrome (IBS) (O'Mahony et al. 2005) and may have potential in the maintenance of treatment of inflammatory bowel disease (IBD) (Hedin et al. 2007).

The mechanisms by which the microbiota elicit these effects are beginning to become established. Through examination of mice expressing luciferase as a marker of NF-κB expression, O'Mahony et al. (2008) have recently demonstrated significant reduction in NF-κB activation during *Salmonella* infection mediated through oral administration of a potential probiotic strain, *B. infantis* (O'Mahony et al. 2008). Others have demonstrated that a commensal bacterium, *Bacteroides thetaiotaomicron* reduces inflammation normally induced through *Salmonella*-TLR5 interactions (Kelly et al. 2004). The mechanism underpinning this anti-inflammatory response was dependent upon peroxisome proliferator activated receptor- γ (PPAR-γ)-mediated inhibition of NF-κB and was directly induced by *B. thetaiotaomicron* (Kelly et al. 2004).

An emerging model suggests that pathogens elicit inflammatory responses as a survival mechanism to disrupt the normal microbiota in order to overcome colonization resistance (Stecher and Hardt 2008). *Citrobacter rodentium* infection in mice results in significant perturbations of the murine gut microbiota which resemble the changes observed following chemical induction of inflammation (Lupp et al. 2007). Similarly inflammation is a natural feature of *S.* Typhimurium infection in mice and is associated with an ability of the pathogen to out-compete the resident microbiota (Stecher et al. 2007; Woo et al. 2008). During *S.* Typhimurium infection in mice the microbiota is significantly disrupted and this has been attributed to induction of inflammation rather than a direct result of pathogen-microbiota interactions (Stecher et al. 2008). Indeed *S.* Typhimurium mutants that fail to induce inflammation are incapable of survival in the gut and are outcompeted by the resident microbiota (Stecher et al. 2007).

The induction of inflammation is thought likely to increase the availability of nutrients available to invading pathogens (*the food hypothesis*) (Stecher and Hardt 2008). It is postulated that high energy glycoproteins released from mucin, epithelial cells and inflammatory immune cells are more readily available in the inflamed gut. Since enteropathogens (and members of the γ-proteobacteria group in general) such as *Salmonella, Citrobacter* and *E. coli* are capable of fast growth rates under optimal growth conditions they may be well placed to take advantage of this increase in nutrient availability (Stecher and Hardt 2008). This hypothesis is supported by the fact that in murine models of inflammation there is a significant increase in the γ-proteobacteria group (Heimesaat et al. 2007; Lupp et al. 2007).

Whilst induction of inflammation will increase bacterial killing in the gut it is likely that pathogens show greater resistance to this killing effect than members of the normal microbiota (*the differential killing hypothesis*) (Stecher and Hardt 2008). Many pathogens (including *S.* Typhimurium) have evolved mechanisms to resist killing by invading phagocytes and so may be capable of survival during a localised inflammatory response. It is thought that the commensal microbiota may suffer a degree of 'collateral damage' during pathogen-induced inflammation (Stecher and Hardt 2008). Clearly the normal microbiota or probiotic strains play a significant role in tempering inflammation which may benefit the community structure within the gut and prevent colonization by pathogens.

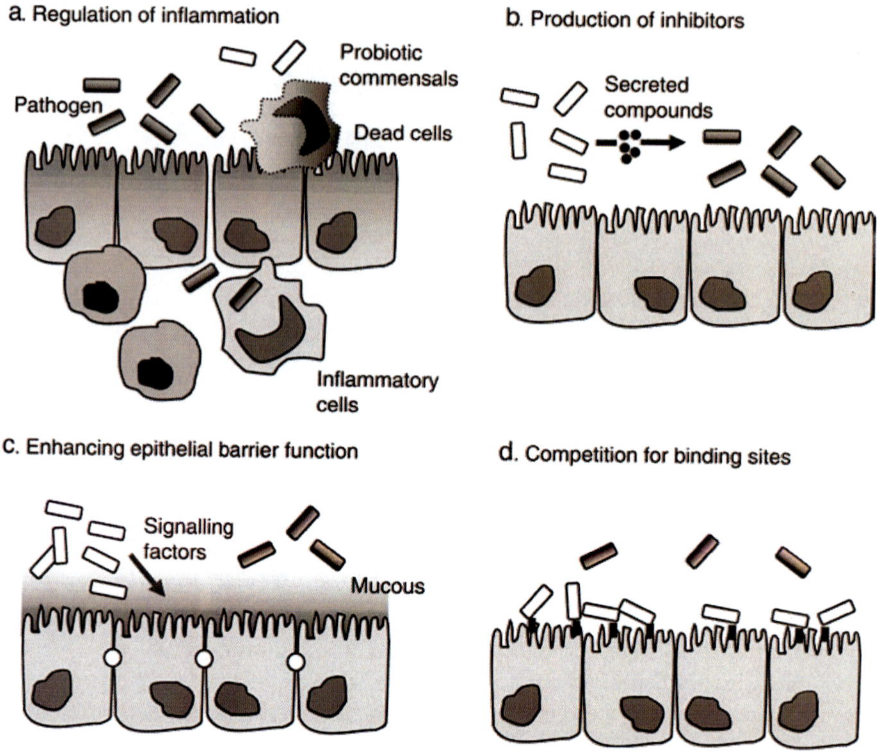

Fig. 5.1 The mechanisms by which probiotic commensal microorganisms impede gastrointestinal pathogens. **a** Inflammation is thought to promote the release of dead cells and nutrients into the lumen of the gastrointestinal tract favouring growth of the γ-proteobacteria (including Gram negative pathogens) (*the food hypothesis*). Gram negative pathogens are also well adapted to survive within inflammatory cells which invade the site of infection (*the differential killing hypothesis*). Specific probiotic commensal microorganisms can modulate inflammation and thereby favour growth of commensal organisms. **b** Probiotic commensal microorganisms can produce inhibitory compounds which down-regulate virulence factor expression by pathogens or which directly kill or impede pathogens in the GI tract. **c** Probiotic commensal microorganisms can directly signal to enterocytes in the GI tract using compounds such as short chain fatty acids, to enhance barrier function. Specific probiotic commensal strains are also capable of inducing mucous production by enterocytes, a mechanism that is directly inhibitory to infection by enteropathogenic *E. coli*. **d** It is thought that probiotic commensal strains can compete with pathogens for binding sites on gastrointestinal enterocytes

5.5.2 Production of Acid and Secretion of Inhibitory Substances

Lactobacillus and *Bifidobacterium* spp. are capable of producing organic acids through normal metabolic processes. Indeed production of both lactic acid and proteinaceous substances by *Lactobacillus* and *Bifidobacterium* spp. has been shown to impede infection of human intestinal cells by enterohemorrhagic *E. coli* O157:H7 (Gopal et al. 2001). The authors found that pre-treatment of *E. coli* O157:H7 with concentrated cell-free culture supernatants from the commensal strains significantly reduced numbers of culturable *E. coli* and the invasiveness of the pathogen (Gopal et al. 2001).

Asahara et al. (2004) demonstrated that selected *Bifidobacterium* species, including *B. breve* strain Yakult, display anti-infectious activity against Shiga toxin-producing *E. coli* (STEC) O157:H7 in mice. A dramatic decrease in bodyweight and subsequent death was observed in placebo-fed mice, while bodyweight was maintained and no fatalities were observed in *B. breve*-fed mice. The authors proposed that this effect was due to production of acetic acid by *B. breve* and lowering of intestinal pH, which had the combined effect of inhibiting Shiga-like toxin (Stx) production (Asahara et al. 2004).

The probiotic *E. coli* strain Nissle 1917 interferes with invasion of human embryonic intestinal epithelial INT407 cells by *S.* Typhimurium via secretion of inhibitory substances (Altenhoefer et al. 2004). This phenomenon was shown through the separation of the probiotic from the pathogen and host cells by a semi-permeable membrane. The membrane allowed the passage of small molecules but prevented direct contact between the probiotic and the pathogen (Altenhoefer et al. 2004). A similar transwell chamber system was used to demonstrate that lactobacilli are capable of inhibiting *Listeria monocytogenes* invasion of C2Bbe1 epithelial cells in the absence of direct contact, through secretion of proteinaceous molecule(s) (Corr et al. 2007a). However, the nature of the proteinaceous agent(s) was not identified.

Pridmore et al. (2008) examined the production of hydrogen peroxide by the human gut commensal *Lactobacillus johnsonii* NCC533. Through *in silico* analysis of the genome of *Lactobacillus johnsonii* NCC533 they identified the means by which hydrogen peroxide is synthesized. Furthermore they demonstrated that hydrogen peroxide was produced by NCC533 *in vitro* at levels that were inhibitory for *S.* Typhimurium (Pridmore et al. 2008). Further studies analyzing isogenic mutants which fail to produce hydrogen peroxide will be informative in determining whether this phenomenon is significant in mediating colonization resistance in mice.

Bacteriocins are compounds with potential anti-microbial activity synthesized by many bacterial species, including many commensal bacteria (Cotter et al. 2005). Bacteriocins have long been suggested as potential mediators of anti-pathogen activities and colonization resistance (Gotteland et al. 2006). However, systematic analysis of the *in vivo* role of bacteriocins is limited. Bacteriocins have been shown to be necessary for oral bacterial replacement therapy which employs long-term oral colonisation by non-cariogenic variants of *S. mutans* (Tagg and Dierksen 2003). The bacteriocin lacticin 3147 has also demonstrated efficacy in a veterinary setting against Gram positive pathogens that cause bovine mastitis (Klostermann et al. 2009).

Another bacteriocin that has been investigated for *in vivo* efficacy is Abp118, a two component bacteriocin produced by *L. salivarius* UCC118 (Flynn et al. 2002). Molecular genetic analysis of the properties of Abp118 and the sequencing of the genome of *L. salivarius* UCC118 enabled the creation of stable mutants which fail to produce the bacteriocin (Claesson et al. 2006; Corr et al. 2007b; Flynn et al. 2002). Wild-type *L. salivarius* UCC118 clearly reduced the severity of oral *L. monocytogenes* infections in mice whereas mice were not protected by the Abp118 null mutant strain. This demonstrated that bacteriocin production is the primary mediator of protection in the oral listeriosis model. Furthermore, *L. salivarius* UCC118 did not offer any protection when mice were infected with a strain of *L. monocytogenes* expressing the cognate Abp118 bacteriocin immunity protein AbpIM. This is the first work to link GI colonization resistance to production of a bacteriocin using a molecular genetic approach (Corr et al. 2007b).

5.5.3 Inhibition of Virulence Factor Expression

Recent research indicates that members of the gut microbiota have the capacity to influence gene expression in specific pathogens. Cell-free supernatants of *L. acidophilus* inhibit quorum sensing and expression of the LEE virulence genes responsible for inducing attaching and effacing lesions in *E. coli* O157:H7 cell infections, but do not affect production of shiga toxin in this strain (Medellin-Pena et al. 2007). A recent extension of this work indicates that the effect is also seen *in vivo* in the mouse model where attachment of *E. coli* O157:H7 is impeded by the *L. acidophilus* La-5 strain (Medellin-Pena and Griffiths 2009).

Whole genome microarray analyses have been used to measure global gene expression in *E. coli* O157:H7 following co-incubation with *Lactobacillus rhamnosus GG* (LGG). This work indicated that LGG co-incubation lowers expression of the *stx* genes encoding shiga toxin production in *E. coli* O157:H7 (Carey et al. 2008). Further work indicated that a variety of *Lactobacillus, Pediococcus*, and *Bifidobacterium* strains can repress *stxA* expression in this model system, suggesting a global mechanism by which the microbiota could impede virulence factor expression (Carey et al. 2008).

Another recent study examined a variety of potential probiotic strains for ability to inhibit the ureolytic pathogen *Yersinia enterocolitica* (Lavermicocca et al. 2008). One probiotic strain, *L. plantarum* ITM21B, was capable of inhibiting urease activity in the pathogen (Lavermicocca et al. 2008). Ryan et al. (2009) have shown that *L. salivarius* (though not other species of *Lactobacillus*) significantly downregulates expression of CAG virulence genes in *H. pylori*. This correlates with a *Lactobacillus*-mediated reduction in inflammatory cytokine production by host cells following *H. pylori* infection and may suggest a mechanism for localized control of inflammation by probiotic commensal strains (Ryan et al. 2009). Future studies will be needed to uncover the regulatory mechanisms that determine signalling between pathogens and commensals. It is possible that further investigations of this nature will identify compounds with therapeutic potential that downregulate virulence gene expression in specific pathogens.

5.5.4 Epithelial Barrier Function and Probiotic Signaling

A key mechanism by which probiotics may exert colonization resistance is *via* induction of enhanced barrier function within the epithelial monolayer mediated by elevated mucous secretion or conformational changes (Mack et al. 1999; Ng et al. 2009). DNA microarray analysis has been used to determine the effect of the probiotic commensal *E. coli* strain Nissle 1917 upon host gene expression profiles during infection of T84 epithelial cells with enteropathogenic *E. coli* (Zyrek et al. 2007). The Nissle 1917 strain was seen to restore barrier function to the monolayer and more than 300 host genes exhibited altered expression following addition of the commensal strain. Changes included enhanced expression and distribution of zonula occludens-2 (ZO-2), a tight-junction protein (Zyrek et al. 2007). Other studies have demonstrated that VSL#3 probiotic bacteria prevent the redistribution of zonula occludens-1 (ZO-1) that is associated with *Salmonella dublin* infection of cell monolayers. The probiotic mixture also stabilized the overall barrier function of the monolayer (Otte and Podolsky 2004).

Pre-treatment of epithelial monolayers with probiotic-commensal bacteria reduces epithelial injury following exposure to *E. coli* O157:H7 and *E. coli* O127:H6 by preventing the pathogen-induced drop in transepithelial resistance (TER), a measure of barrier integrity (Sherman et al. 2005). These probiotics also reduced the number of attaching and effacing lesions formed in response to *E. coli* O157:H7 a property mediated by alpha-actinin. In this study, viable lactic acid-producing bacteria were necessary to mediate the observed effects (Sherman et al. 2005). Similarly, *S. thermophilus* and *L. acidophilus* enhanced general cytoskeletal and tight junction architechture in epithelial cells through enhanced actinin and occludin expression (Resta-Lenert and Barrett 2003). In another recent study lactobacilli minimized F-actin rearrangements and morphological alterations in the cell monolayers in Hep-2 cell monolayers when applied prior to infection with enterohaemorrhagic *E. coli* (EHEC) (Hugo et al. 2008). These studies collectively indicate that live probiotic commensal bacteria are capable of influencing cytoskeletal arrangements and modulating barrier activity of epithelial cells in culture.

The exact molecular mechanisms by which probiotics stimulate alterations in epithelial cell function are not yet fully understood. A number of studies have shown that specific metabolites produced by probiotic strains can improve epithelial and mucosal barrier function (Madsen et al. 2001; Qin et al. 2005). These include production of short-chain fatty-acids (SCFAs) as by-product of microbial fermentation, such as butyrate which induces epithelial cell differentiation and increases barrier integrity (Cook and Sellin 1998). Recent work by Ewaschuk et al. (2008) demonstrated that abiotic supernatant from *Bifidobacteria infantis* cultures was capable of increasing TER and enhancing expression of occludin and ZO-1 in T84 cells *in vitro*. The conditioned medium was also capable of reducing permeability in the colons of mice. The active agents were shown to be peptides released by *B. infantis*. The results are significant as they indicate that biologically active peptides from this probiotic commensal strain are capable of retaining *in vivo* activity when administered orally to mice (Ewaschuk et al. 2008).

Another physiological change induced by probiotics in host cell monolayers is the induction or overexpression of mucin (Mack et al. 2003; Mattar et al. 2002). GI tract mucins are large, carbohydrate-rich, high-molecular-weight glycoproteins which are the major components of mucous (Mattar et al. 2002). Mucin forms a physicochemical barrier which protects epithelial cells from chemical, enzymatic, mechanical and microbial damage, and limits microbial adherence and subsequent invasion (Mack et al. 2003). A number of potential mucin encoding genes have been identified in humans, and of these MUC2 and MUC3 encode the predominant ileo-colonic mucins (Mack et al. 2003). Adherence of selected *Lactobacillus* strains to the human intestinal HT29 epithelial-cell line induces up-regulation of mucin gene expression, and correlates with increased host cell secretion of MUC3 (Mack et al. 2003). *L. plantarum* 299v and *L. rhamnosus* GG inhibit the adherence of entero-pathogenic *E. coli* to HT29 intestinal epithelial cells *via* induction or overexpression of mucin (Mack et al. 1999). In an *in vitro* Caco-2 cell model, *L. casei* LGG up-regulates expression of both MUC2 mRNA and protein and reduces bacterial translocation of the intestinal epithelium (Mattar et al. 2001, 2002). Thus, increased expression of intestinal mucin in response to lactobacilli mediates inhibition of adherence of pathogens to intestinal cells. However, analysis of this phenomenon directly using *in vivo* infection models has not yet been undertaken.

Collado et al. (2008) have recently shown that specific probiotic strains have the capacity to prevent adhesion of the opportunistic pathogen *Enterobacter sakazakii* (now *Cronobacter sakazakii*) to immobilized human mucous. These studies indicate that specific probiotic strains have the capacity to competitively exclude pathogens from human mucous potentially as a mechanism for preventing transient colonisation of the mucous layer by GI pathogens (Collado et al. 2008).

Probiotic commensal strains also have the potential to induce the release of defensins from epithelial cells. These small peptides/proteins are active against bacteria, fungi and viruses and are also known to stabilise gut barrier function (Furrie et al. 2005). *E. coli* Nissle 1917 induces human beta defensin-2 (hBD-2) gene expression in Caco-2 intestinal epithelial cells *in vitro* (Wehkamp et al. 2004). Cellular expression of defensin genes was enhanced through NF-κB and AP-1 signalling pathways (Wehkamp et al. 2004). Recently, several *Lactobacillus* strains and VSL#3 probiotic mixture were found to induce human beta defensin 2 (hBD-2) gene expression in Caco-2 cells (Schlee et al. 2008). This was also dependent upon NF-κB and AP-1 signalling pathways as well as MAPK extracellular regulated kinase (ERK 1/2), p38 and c-Jun N-terminal kinase (JNK), although to varying degrees (Schlee et al. 2008). Another recent study demonstrated enhanced expression of the host antibacterial peptide LL-37 in response to bacterial butyrate production. The study showed that this process was mediated by upregulation of the CAMP gene (encoding LL-37) via the PU.1 transcriptional regulator (Termen et al. 2008).

5.5.5 Competition for Pathogen Binding Sites

The competitive blocking of pathogen binding sites on host cells has been proposed to represent a potential mechanism for mediating colonization resistance (O'Sullivan et al. 2005; Sherman et al. 2005; Tsai et al. 2005). However, this phenomenon is difficult to investigate and direct proof is limited. In one study, BALB/c mice were orally administered *Lactobacillus* strains LAP5 or LF33 originally isolated from swine and poultry for 7 consecutive days before oral challenge with *S. enterica* serovar Typhimurium (Tsai et al. 2005). Numbers of *Salmonella* invading internal organs of probiotic-fed mice were significantly lower than placebo controls, and it was thought that the adhesiveness of the *Lactobacillus* cells to the mouse intestinal epithelium was an important factor in antagonism against *Salmonella* invasion. However this inference was based upon *in vitro* assessment of adherence to intestinal cell lines and was not proven *in vivo* (Tsai et al. 2005). A more recent study investigated Spa pili formation in *L. rhamnosus* GG and determined that binding of the bacterium to human mucous could be blocked by antibodies to SpaC (Kankainen et al. 2009). As many pathogens use specialized mechanisms to bind to mucous it is possible that probiotic strains (such as *L. rhamnosus* GG) could play a role in impeding this interaction. Future studies which investigate exclusion of pathogens from specific binding sites are awaited with interest.

5.6 Conclusions

There is a significant body of evidence to support the role for normal microbiota in preventing colonization by pathogens (colonization resistance). Clearly the gut flora of an individual is as unique as a fingerprint (Eckburg et al. 2005). However, key bacterial genera must be represented in order to maintain GI homeostasis and to provide colonization resistance. When this microbial balance is disrupted (such as through antibiotic treatment or inflammation) then the individual may be at increased risk of infection by gut pathogens (such as *C. difficile*). There is good evidence that intervention by administration of live probiotic cultures may bolster colonization resistance. However, not all probiotics are effective against all pathogens and much more research is needed to match specific commensal strains to specific disease states. Future studies with larger cohorts of volunteers and accurate/representative placebo controls will be required in order to truly establish a role for probiotics in the prevention of infectious disease in otherwise healthy subjects. In addition, establishing the molecular mechanisms that underpin colonization resistance both by the normal microbiota and by probiotics will support health claims provided for specific interventions and may lead to bacteria-free molecules with therapeutic properties.

References

Allen SJ, Okoko B, Martinez E, Gregorio G, Dans LF (2004) Probiotics for treating infectious diarrhoea. Cochrane Database Syst Rev. doi:10.1002/14651858 (CD003048)

Altenhoefer A, Oswald S, Sonnenborn U, Enders C, Schulze J, Hacker J, Oelschlaeger TA (2004) The probiotic *Escherichia coli* strain Nissle 1917 interferes with invasion of human intestinal epithelial cells by different enteroinvasive bacterial pathogens. FEMS Immunol Med Microbiol 40:223–229

Asahara T, Shimizu K, Nomoto K, Hamabata T, Ozawa A, Takeda Y (2004) Probiotic bifidobacteria protect mice from lethal infection with Shiga toxin-producing *Escherichia coli* O157:H7. Infect Immun 72:2240–2247

Barc MC, Bourlioux F, Rigottier-Gois L, Charrin-Sarnel C, Janoir C, Boureau H, Dore J, Collignon A (2004) Effect of amoxicillin-clavulanic acid on human faecal flora in a gnotobiotic mouse model assessed with fluorescence hybridization using group-specific 16S rRNA probes in combination with flow cytometry. Antimicrob Agents Chemother 48:1365–1368

Brandl K, Plitas G, Mihu CN, Ubeda C, Jia T, Fleisher M, Schnabl B, DeMatteo RP, Pamer EG (2008) Vancomycin-resistant enterococci exploit antibiotic-induced innate immune deficits. Nature 455:804–807

Carey CM, Kostrzynska M, Ojha S, Thompson S (2008) The effect of probiotics and organic acids on Shiga-toxin 2 gene expression in enterohemorrhagic *Escherichia coli* O157:H7. J Microbiol Methods 73:125–132

Claesson MJ, Li Y, Leahy S et al. (2006) Multireplicon genome architecture of *Lactobacillus salivarius*. Proc Natl Acad Sci U S A 103:6718–6723

Collado MC, Isolauri E, Salminen S (2008) Specific probiotic strains and their combinations counteract adhesion of *Enterobacter sakazakii* to intestinal mucus. FEMS Microbiol Lett 285:58–64

Cook SI, Sellin JH (1998) Review article: short chain fatty acids in health and disease. Aliment Pharmacol Ther 12:499–507

Corr SC, Gahan CG, Hill C (2007a) Impact of selected Lactobacillus and Bifidobacterium species on Listeria monocytogenes infection and the mucosal immune response. FEMS Immunol Med Microbiol 50:380–388

Corr SC, Li Y, Riedel CU, O'Toole PW, Hill C, Gahan CG (2007b) Bacteriocin production as a mechanism for the antiinfective activity of *Lactobacillus salivarius* UCC118. Proc Natl Acad Sci U S A 104:7617–7621

Corr SC, Hill C, Gahan CG (2009) Understanding the mechanisms by which probiotics inhibit gastrointestinal pathogens (Chap. 1). Adv Food Nutr Res 56:1–15

Cotter PD, Hill C, Ross RP (2005) Bacteriocins: developing innate immunity for food. Nat Rev Microbiol 3:777–788

Deshpande A, Pant C, Jain A, Fraser TG, Rolston DD (2008) Do fluoroquinolones predispose patients to *Clostridium difficile* associated disease? A review of the evidence. Curr Med Res Opin 24:329–333

Dethlefsen L, Huse S, Sogin ML, Relman DA (2008) The pervasive effects of an antibiotic on the human gut microbiota, as revealed by deep 16S rRNA sequencing. PLoS Biol 6:e280

Eckburg PB, Bik EM, Bernstein CN, Purdom E, Dethlefsen L, Sargent M, Gill SR, Nelson KE, Relman DA (2005) Diversity of the human intestinal microbial flora. Science 308:1635–1638

Ewaschuk JB, Diaz H, Meddings L, Diederichs B, Dmytrash A, Backer J, Looijer-van Langen M, Madsen KL (2008) Secreted bioactive factors from *Bifidobacterium infantis* enhance epithelial cell barrier function. Am J Physiol Gastrointest Liver Physiol 295:G1025–G1034

Farnworth ER (2008) The evidence to support health claims for probiotics. J Nutr 138:1250S–1254S

Filho-Lima JV, Vieira EC, Nicoli JR (2000) Antagonistic effect of *Lactobacillus acidophilus*, *Saccharomyces boulardii* and *Escherichia coli* combinations against experimental infections with *Shigella flexneri* and *Salmonella enteritidis* subsp. typhimurium in gnotobiotic mice. J Appl Microbiol 88:365–370

Flynn S, van Sinderen D, Thornton GM, Holo H, Nes IF, Collins JK (2002) Characterization of the genetic locus responsible for the production of ABP-118, a novel bacteriocin produced by the probiotic bacterium *Lactobacillus salivarius* subsp. salivarius UCC118. Microbiology 148:973–984

Furrie E, Macfarlane S, Kennedy A, Cummings JH, Walsh SV, O'Neil DA, Macfarlane GT (2005) Synbiotic therapy (Bifidobacterium longum/Synergy 1) initiates resolution of inflammation in patients with active ulcerative colitis: a randomised controlled pilot trial. Gut 54:242–249

Gopal PK, Prasad J, Smart J, Gill HS (2001) In vitro adherence properties of Lactobacillus rhamnosus DR20 and *Bifidobacterium lactis* DR10 strains and their antagonistic activity against an enterotoxigenic *Escherichia coli*. Int J Food Microbiol 67:207–216

Gotteland M, Poliak L, Cruchet S, Brunser O (2005) Effect of regular ingestion of *Saccharomyces boulardii* plus inulin or Lactobacillus acidophilus LB in children colonized by *Helicobacter pylori*. Acta Paediatr 94:1747–1751

Gotteland M, Brunser O, Cruchet S (2006) Systematic review: are probiotics useful in controlling gastric colonization by *Helicobacter pylori*? Aliment Pharmacol Ther 23:1077–1086

Guandalini S, Pensabene L, Zikri MA et al. (2000) Lactobacillus GG administered in oral rehydration solution to children with acute diarrhoea: a multicenter European trial. J Pediatr Gastroenterol Nutr 30:54–60

Hedin C, Whelan K, Lindsay JO (2007) Evidence for the use of probiotics and prebiotics in inflammatory bowel disease: a review of clinical trials. Proc Nutr Soc 66:307–315

Heimesaat MM, Fischer A, Siegmund B et al. (2007) Shift towards pro-inflammatory intestinal bacteria aggravates acute murine colitis via toll-like receptors 2 and 4. PLoS One 2:e662

Hickson M, D'Souza AL, Muthu N, Rogers TR, Want S, Rajkumar C, Bulpitt CJ (2007) Use of probiotic Lactobacillus preparation to prevent diarrhoea associated with antibiotics: randomised double blind placebo controlled trial. BMJ 335:80

Hudault S, Liévin V, Bernet-Camard MF, Servin AL (1997) Antagonistic activity exerted in vitro and in vivo by *Lactobacillus casei* (strain GG) against *Salmonella typhimurium* C5 infection. Appl Environ Microbiol 63:513–518

Hugo AA, Kakisu E, De Antoni GL, Perez PF (2008) Lactobacilli antagonize biological effects of enterohaemorrhagic *Escherichia coli* in vitro. Lett Appl Microbiol 46:613–619

Inagaki H, Suzuki T, Nomoto K, Yoshikai Y (1996) Increased susceptibility to primary infection with Listeria monocytogenes in germfree mice may be due to lack of accumulation of L-selectin+ CD44+ T cells in sites of inflammation. Infect Immun 64:3280–3287

Johnson-Henry KC, Mitchell DJ, Avitzur Y, Galindo-Mata E, Jones NL, Sherman PM (2004) Probiotics reduce bacterial colonization and gastric inflammation in H. pylori-infected mice. Dig Dis Sci 49:1095–1102

Kabir AM, Aiba Y, Takagi A, Kamiya S, Miwa T, Koga Y (1997) Prevention of *Helicobacter pylori* infection by lactobacilli in a gnotobiotic murine model. Gut 41:49–55

Kankainen M, Paulin L, Tynkkynen S et al. (2009) Comparative genomic analysis of *Lactobacillus rhamnosus* GG reveals pili containing a human- mucus binding protein. Proc Natl Acad Sci U S A 106:17193–17198

Kelly D, Campbell JI, King TP, Grant G, Jansson EA, Coutts AG, Pettersson S, Conway S (2004) Commensal anaerobic gut bacteria attenuate inflammation by regulating nuclear-cytoplasmic shuttling of PPAR-gamma and RelA. Nat Immunol 5:104–112

Khoruts A, Dicksved J, Jansson JK, Sadowsky MJ (2009) Changes in the composition of the human faecal microbiome after bacteriotherapy for recurrent Clostridium difficile-associated diarrhoea. J Clin Gastroenterol 44(5):354–360

Kida Y, Shimizu T, Kuwano K (2006) Sodium butyrate up-regulates cathelicidin gene expression via activator protein-1 and histone acetylation at the promoter region in a human lung epithelial cell line, EBC-1. Mol Immunol 43:1972–1981

Klostermann K, Crispie F, Flynn J, Meaney WJ, Paul Ross R, Hill C (2009) Efficacy of a teat dip containing the bacteriocin lacticin 3147 to eliminate Gram-positive pathogens associated with bovine mastitis. J Dairy Res 77(2):231–238

Kotowska M, Albrecht P, Szajewska H (2005) Saccharomyces boulardii in the prevention of antibiotic-associated diarrhoea in children: a randomized double-blind placebo-controlled trial. Aliment Pharmacol Ther 21:583–590

Lavermicocca P, Valerio F, Lonigro SL, Di Leo A, Visconti A (2008) Antagonistic activity of potential probiotic Lactobacilli against the ureolytic pathogen *Yersinia enterocolitica*. Curr Microbiol 56:175–181

Lawley TD, Bouley DM, Hoy YE, Gerke C, Relman DA, Monack DM (2008) Host transmission of *Salmonella enterica* serovar Typhimurium is controlled by virulence factors and indigenous intestinal microbiota. Infect Immun 76:403–416

Lupp C, Robertson ML, Wickham ME, Sekirov I, Champion OL, Gaynor EC, Finlay BB (2007) Host-mediated inflammation disrupts the intestinal microbiota and promotes the overgrowth of Enterobacteriaceae. Cell Host Microbe 2:204

Mack DR, Michail S, Wei S, McDougall L, Hollingsworth MA (1999) Probiotics inhibit enteropathogenic *E. coli* adherence in vitro by inducing intestinal mucin gene expression. Am J Physiol 276:G941–G950

Mack DR, Ahrne S, Hyde L, Wei S, Hollingsworth MA (2003) Extracellular MUC3 mucin secretion follows adherence of Lactobacillus strains to intestinal epithelial cells in vitro. Gut 52:827–833

Macpherson AJ, Harris NL (2004) Interactions between commensal intestinal bacteria and the immune system. Nat Rev Immunol 4:478–485

Madsen K, Cornish A, Soper P, McKaigney C, Jijon H, Yachimec C, Doyle J, Jewell L, De Simone C (2001) Probiotic bacteria enhance murine and human intestinal epithelial barrier function. Gastroenterology 121:580–591

Mastretta E, Longo P, Laccisaglia A, Balbo L, Russo R, Mazzaccara A, Gianino P (2002) Effect of Lactobacillus GG and breast-feeding in the prevention of rotavirus nosocomial infection. J Pediatr Gastroenterol Nutr 35:527–531

Mattar AF, Drongowski RA, Coran AG, Harmon CM (2001) Effect of probiotics on enterocyte bacterial translocation in vitro. Pediatr Surg Int 17:265–268

Mattar AF, Teitelbaum DH, Drongowski RA, Yongyi F, Harmon CM, Coran AG (2002) Probiotics up-regulate MUC-2 mucin gene expression in a Caco-2 cell-culture model. Pediatr Surg Int 18:586–590

McFarland LV (2006) Meta-analysis of probiotics for the prevention of antibiotic associated diarrhoea and the treatment of *Clostridium difficile* disease. Am J Gastroenterol 101:812–822

McFarland LV (2007) Meta-analysis of probiotics for the prevention of traveler's diarrhoea. Travel Med Infect Dis 5:97–105

McFarland LV (2009) Evidence-based review of probiotics for antibiotic-associated diarrhoea and *Clostridium difficile* infections. Anaerobe 15:274–280

McFarland LV, Surawicz CM, Greenberg RN et al. (1994) A randomized placebo-controlled trial of Saccharomyces boulardii in combination with standard antibiotics for *Clostridium difficile* disease. JAMA 271:1913–1918

Medellin-Pena MJ, Griffiths MW (2009) Effect of molecules secreted by *Lactobacillus acidophilus* strain La-5 on *Escherichia coli* O157:H7 colonization. Appl Environ Microbiol 75:1165–1172

Medellin-Pena MJ, Wang H, Johnson R, Anand S, Griffiths MW (2007) Probiotics affect virulence-related gene expression in *Escherichia coli* O157:H7. Appl Environ Microbiol 73:4259–4267

Nardi RM, Vieira EC, Crocco-Afonso LC, Silva ME, Bambirra EA, Andrade AM, Nicoli JR (1991) Bacteriological and immunological aspects of conventional and germfree mice infected with *Salmonella typhimurium*. Rev Latinoam Microbiol 33:239–243

Ng SC, Hart AL, Kamm MA, Stagg AJ, Knight SC (2009) Mechanisms of action of probiotics: recent advances. Inflamm Bowel Dis 15:300–310

O'Hara AM, Shanahan F (2006) The gut flora as a forgotten organ. EMBO Rep 7:688–693

O'Hara AM, O'Regan P, Fanning A, O'Mahony C, Macsharry J, Lyons A, Bienenstock J, O'Mahony L, Shanahan F (2006) Functional modulation of human intestinal epithelial cell responses by *Bifidobacterium infantis* and *Lactobacillus salivarius*. Immunology 118:202–215

O'Mahony L, McCarthy J, Kelly P et al. (2005) Lactobacillus and bifidobacterium in irritable bowel syndrome: symptom responses and relationship to cytokine profiles. Gastroenterology 128:541–551

O'Mahony C, Scully P, O'Mahony D et al. (2008) Commensal-induced regulatory T cells mediate protection against pathogen-stimulated NF-kappaB activation. PLoS Pathog 4:e1000112

O'Sullivan GC, Kelly P, O'Halloran S, Collins C, Collins JK, Dunne C, Shanahan F (2005) Probiotics: an emerging therapy. Curr Pharm Des 11:3–10

Otte JM, Podolsky DK (2004) Functional modulation of enterocytes by gram-positive and gram-negative microorganisms. Am J Physiol Gastrointest Liver Physiol 286:G613–G626

Penner R, Fedorak RN, Madsen KL (2005) Probiotics and nutraceuticals: non-medicinal treatments of gastrointestinal diseases. Curr Opin Pharmacol 5:596–603

Pridmore RD, Pittet AC, Praplan F, Cavadini C (2008) Hydrogen peroxide production by *Lactobacillus johnsonii* NCC 533 and its role in anti-Salmonella activity. FEMS Microbiol Lett 283:210–215

Qin HL, Shen TY, Gao ZG, Fan XB, Hang XM, Jiang YQ, Zhang HZ (2005). Effect of lactobacillus on the gut microflora and barrier function of the rats with abdominal infection. World J Gastroenterol 11:2591–2596

Qin J, Li R, Raes J et al. (2010) A human gut microbial gene catalogue established by metagenomic sequencing. Nature 464:59–65

Resta-Lenert S, Barrett KE (2003) Live probiotics protect intestinal epithelial cells from the effects of infection with enteroinvasive *Escherichia coli* (EIEC). Gut 52:988–997

Rodrigues AC, Cara DC, Fretez SH, Cunha FQ, Vieira EC, Nicoli JR, Vieira LQ (2000) *Saccharomyces boulardii* stimulates sIgA production and the phagocytic system of gnotobiotic mice. J Appl Microbiol 89:404–414

Ryan KA, O'Hara AM, van Pijkeren JP, Douillard FP, O'Toole PW (2009) Lactobacillus salivarius modulates cytokine induction and virulence factor gene expression in *Helicobacter pylori*. J Med Microbiol 58:996–1005

Schlee M, Harder J, Koten B, Stange EF, Wehkamp J, Fellermann K (2008) Probiotic lactobacilli and VSL#3 induce enterocyte beta-defensin 2. Clin Exp Immunol 151:528–535

Sekirov I, Finlay BB (2009) The role of the intestinal microbiota in enteric infection. J Physiol 587:4159–4167

Sekirov I, Tam NM, Jogova M, Robertson ML, Li Y, Lupp C, Finlay BB (2008) Antibiotic-induced perturbations of the intestinal microbiota alter host susceptibility to enteric infection. Infect Immun 76:4726–4736

Shanahan F (2010) Gut microbes: from bugs to drugs. Am J Gastroenterol 105:275–279

Sherman PM, Johnson-Henry KC, Yeung HP, Ngo PS, Goulet J, Tompkins TA (2005) Probiotics reduce enterohemorrhagic *Escherichia coli* O157:H7- and enteropathogenic *E. coli* O127:H6-induced changes in polarized T84 epithelial cell monolayers by reducing bacterial adhesion and cytoskeletal rearrangements. Infect Immun 73:5183–5188

Shornikova AV, Casas IA, Isolauri E, Mykkanen H, Vesikari T (1997) *Lactobacillus reuteri* as a therapeutic agent in acute diarrhoea in young children. J Pediatr Gastroenterol Nutr 24:399–404

Silva AM, Barbosa FH, Duarte R, Vieira LQ, Arantes RM, Nicoli JR (2004) Effect of *Bifidobacterium longum* ingestion on experimental salmonellosis in mice. J Appl Microbiol 97:29–37

Stecher B, Hardt WD (2008) The role of microbiota in infectious disease. Trends Microbiol 16:107–114

Stecher B, Robbiani R, Walker AW et al. (2007) Salmonella enterica serovar typhimurium exploits inflammation to compete with the intestinal microbiota. PLoS Biol 5:2177–2189

Stecher B, Barthel M, Schlumberger MC, Haberli L, Rabsch W, Kremer M, Hardt WD (2008) Motility allows *S. typhimurium* to benefit from the mucosal defence. Cell Microbiol 10:1166–1180

Szajewska H, Kotowska M, Mrukowicz JZ, Armanska M, Mikolajczyk W (2001) Efficacy of Lactobacillus GG in prevention of nosocomial diarrhoea in infants. J Pediatr 138:361–365

Tagg JR, Dierksen KP (2003) Bacterial replacement therapy: adapting 'germ warfare' to infection prevention. Trends Biotechnol 21:217–223

Taguchi H, Takahashi M, Yamaguchi H, Osaki T, Komatsu A, Fujioka Y, Kamiya S (2002) Experimental infection of germ-free mice with hyper-toxigenic enterohaemorrhagic *Escherichia coli* O157:H7, strain 6. J Med Microbiol 51:336–343

Termen S, Tollin M, Rodriguez E, Sveinsdottir SH, Johannesson B, Cederlund A, Sjovall J, Agerberth B, Gudmundsson GH (2008) PU.1 and bacterial metabolites regulate the human gene CAMP encoding antimicrobial peptide LL-37 in colon epithelial cells. Mol Immunol 45:3947–3955

Tsai CC, Hsih HY, Chiu HH, Lai YY, Liu JH, Yu B, Tsen HY (2005) Antagonistic activity against Salmonella infection in vitro and in vivo for two Lactobacillus strains from swine and poultry. Int J Food Microbiol 102:185–194

Valeur N, Engel P, Carbajal N, Connolly E, Ladefoged K (2004) Colonization and immunomodulation by *Lactobacillus reuteri* ATCC 55730 in the human gastrointestinal tract. Appl Environ Microbiol 70:1176–1181

Vieira LQ, dos Santos LM, Neumann E, da Silva AP, Moura LN, Nicoli JR (2008) Probiotics protect mice against experimental infections. J Clin Gastroenterol 42(3 Suppl, Pt 2):S168–S169

Wehkamp J, Harder J, Wehkamp K et al. (2004) NF-kappaB- and AP-1-mediated induction of human beta defensin-2 in intestinal epithelial cells by *Escherichia coli* Nissle 1917: a novel effect of a probiotic bacterium. Infect Immun 72:5750–5758

Wilcox MH (2003) Gastrointestinal disorders and the critically ill. *Clostridium difficile* infection and pseudomembranous colitis. Best Pract Res Clin Gastroenterol 17:475–493

Wilson DS, Sober E (1989) Reviving the superorganism. J Theor Biol 136:337–356

Wilson KH, Sheagren JN, Freter R, Weatherbee L, Lyerly D (1986) Gnotobiotic models for study of the microbial ecology of *Clostridium difficile* and *Escherichia coli*. J Infect Dis 153:547–551

Woo H, Okamoto S, Guiney D, Gunn JS, Fierer J (2008) A model of Salmonella colitis with features of diarrhoea in SLC11A1 wild-type mice. PLoS One 3:e1603

Zilber-Rosenberg I, Rosenberg E (2008) Role of microorganisms in the evolution of animals and plants: the hologenome theory of evolution. FEMS Microbiol Rev 32:723–735

Zyrek AA, Cichon C, Helms S, Enders C, Sonnenborn U, Schmidt MA (2007) Molecular mechanisms underlying the probiotic effects of *Escherichia coli* Nissle 1917 involve ZO-2 and PKCzeta redistribution resulting in tight junction and epithelial barrier repair. Cell Microbiol 9:804–816

Part IV
Application of Molecular Biology and -omics of Probiotics in Enteric Protection

Chapter 6
Application of Molecular Biology and Genomics of Probiotics for Enteric Cytoprotection

Saloomeh Moslehi-Jenabian, Dennis Sandris Nielsen and Lene Jespersen

6.1 Introduction

The intestinal microbiota plays an essential role in host nutrition, intestinal cell proliferation and differentiation, development of the immune system and acquired responses to pathogens. Alterations in the composition of the intestinal microbiota have recently been linked to various diseases, including inflammatory bowel disease, allergy and diabetes type II (Guarner and Malagelada 2003; Larsen et al. 2010; Lomax and Calder 2009). Probiotics are among the variable indigenous constituents of the gut mictobiota. There are various evidences for different beneficial functions of probiotics and the mechanisms underlying these health effects include both microbe-microbe and microbe-host interactions. Nevertheless, the molecular basis of these mechanisms is still largely unknown. However, recent modern molecular biology based –omics technologies (genomics, proteomics and metabolomics), allowing simultaneous analysis of huge numbers of genes, proteins or metabolites, have revealed insights into understanding the molecular basis for these health promoting activities and increased our knowledge concerning the roles of probiotics in microbe–microbe and host–microbe interactions. The microbial genomic content reflects metabolism, physiology, biosynthetic capabilities of the microorganism, and its ability to adapt to varying conditions and environments. Hence, genome analysis of probiotics will help us to understand their metabolic processes and functionality in human health and well-being. Beside the scientific importance, it will provide a way to improve functional foods, which attracts the interest of the industry and consumers. Consequently, it is of significant concern to exploit the recent studies on the molecular details of the interaction of probiotics with the human host and other microbes. This chapter provides an overview of current progresses in molecular and genomic technologies of probiotics to elucidate the role of these microorganisms in human health and well being. Emphasis will be on the model probiotic

S. Moslehi-Jenabian (✉)
Department of Food Science, Food Microbiology, Faculty of Life Sciences,
University of Copenhagen, Rolighedsvej 30, 1958 Frederiksberg, Denmark

J. J. Malago et al. (eds.), *Probiotic Bacteria and Enteric Infections*,
DOI 10.1007/978-94-007-0386-5_6,

bacteria *Lactobacillus* spp. and *Bifidobacterium* spp., which are phylogenetically distant relatives with different features. When relevant, references will be made to the probiotic yeast *Saccharomyces cerevisiae* var. *boulardii* (van der Aa Kühle and Jespersen 2003), which is widely used as a therapeutic agent.

6.2 Functional Genomics

Functional genomic analyses including whole genome sequencing, genome data mining and comparative genomics have been useful in understanding the influence of genetic content, organization, function and regulation on gut and probiotic functionality as well as to identify the differences and similarities between probiotics since many of the probiotic features are species and even strain dependent. Functional genomic analysis is therefore essential to understand the cellular physiology, metabolic pathways, sensing and signalling in order to clarify mechanisms underlying the probiotic functions of these microorganisms (Klaenhammer et al. 2002). In addition, genomic tools to investigate the gene regulatory networks are important in order to analyse the response of microorganisms to different environmental conditions, especially, the gut-related environmental stresses.

Various studies have investigated the molecular response of probiotics using *in vitro* models mimicking the gut and intestinal environment, for instance acid and bile stress response and tolerance. In many cases, the genes and proteins identified encompass the general stress proteins like GroEL, GroES and DnaK (Frees et al. 2003; Lim et al. 2000; Weiss and Jespersen 2010), and functions related to maintenance of the cell-envelope integrity due to the destructive effect of bile on the cell wall (Bron et al. 2006). It has been shown that these responses are controlled by different regulators that are involved in control of the general stress response (Ferreira et al. 2001, 2003). *In vitro* models are useful for investigating the response of the microorganism to a specific intestinal stress. However, investigation of the full response of a given microorganism will only be achieved using *in vivo* approaches. Therefore, some functional genomic approaches have focused on the study of genetic responses of microorganisms *in vivo* with the goal of identifying bacterial genes that are important during residence in the gut.

Three main strategies have been developed for the identification of genes that are highly expressed *in vivo*, as compared with laboratory conditions: (1) (recombination-based) *in vivo* expression technology ((R-)IVET), (2) signature-tagged mutagenesis (STM), and (3) selective capture of transcribed sequences (SCOTS). These *in vivo* gene identification strategies have been applied for investigation of important genes in bacterial pathogenesis (Mahan et al. 2000). In addition, IVET has recently been employed to identify genes potentially influencing the probiotic functionality in both *Lactobacillus reuteri* 100-23 (Walter et al. 2003) and *Lactobacillus plantarum* WCFS1 (Bron et al. 2004). This approach allows identification of promoter elements that are expressed during *in vivo* transit of probiotic cultures, and reveals the corresponding genes driven by these promoters.

The DNA microarray technique is a functional genomic approach enabling monitoring the global transcriptional response at the time of sampling and can be used to elucidate the genomic expression of gut-related bacteria in the intestinal tract (Azcarate-Peril et al. 2004; Denou et al. 2007). This approach together with real-time PCR, can be used for quantitative analysis of the transcriptional response of the cells under conditions of interest, e.g., cells that are located at specific intestinal sites (Tao et al. 2006).

Targeted insertional mutagenesis is another alternative to study the gene regions that are presumed to be involved in probiotic traits, and thereby a number of gene regions have been characterized and functionally correlated to important phenotypes (Azcarate-Peril et al. 2004; Velez et al. 2007).

Thus far, functional genomic analyses have revealed a number of interesting features that are generally considered to be important for the roles of probiotics in enteric cytoprotection and health.

6.3 Genes and Molecules Involved in Adaptation of Probiotics to the Gut Niche

Tolerance of probiotics to the stress conditions of the intestinal environment and their adaptation to the gut niche play significant roles in the functionality of probiotics. Different genomic studies have demonstrated the genetic adaptation and metabolic activity of *Lactobacillus* spp. or *Bifidobacterium* spp. in the intestinal environment, which will be discussed in detail in the following sections.

6.3.1 Genes and Molecules Involved in Stress Adaptation

Genes encoding acid resistance responses are essential in tolerance of probiotics to intestinal stress. As an example induction of putative heat shock proteins, i.e., DnaK, DnaJ, GrpE, GroES and GroEL, in acid adapted cells (exposure of cells to sub-lethal adaptive acid conditions) has been shown in *Lactobacillus acidophilus* CRL 639 (Lorca et al. 2002). Recently, a transcriptomic study has shown the expression of stress related genes GroEL, DnaK and ClpP in *L. acidophilus* NCFM after exposure to gastric juice following passage through an *in vitro* gastrointestinal tract model (Weiss and Jespersen 2010). In *L. acidophilus*, the *atp* operon is an acid inducible operon containing 8 genes encoding the various subunits of the F_1F_0-ATPase, a multimeric enzyme either synthesizing ATP using protons or conversely expulse protons out of the cell with the energy provided by ATP hydrolysis. Acidic stress induces expression of the *atp* operon accompanied by an increase in the activity of the membrane-bound enzyme, which results in active expulsion of protons out of the cell and maintenance of cytoplasmic pH under acidic environmental conditions (Kullen and Klaenhammer 1999). Further studies have shown the presence of four loci contributing to acid resistance in the *L. acidophilus* NCFM genome. The role of the four loci in acid tolerance was proved by insertional mutagenesis in these regions, which resulted in acid sensitive derivatives (Azcarate-Peril et al. 2004). A two-component regulatory system has been found in *L. acidophilus* NCFM playing a role in acid resistance (Azcarate-Peril et al. 2005). Insertional mutagenesis of this two-component regulatory system resulted in an acid sensitive mutant. Whole-genome microarray analysis of the mutant showed that expression of 80 genes including two oligopeptide-transport systems, other components of the proteolytic enzyme system, and a *luxS* homolog was affected by the mutation. The gene *luxS* is involved in AI-2 mediated interspecies quorum sensing (cell-to-cell communication) among bacteria (Federle and Bassler 2003). A transcriptomic study has shown that the *luxS* gene is induced by acidic stress in *L. acidophilus* NCFM and *Lactobacillus rhamnosus* GG and plays a role in the acid stress response in these probiotics. It was observed that in both species, the *luxS* gene was transiently up-regulated after acidic shock (pH 4.0). Acid adaptation of cells attenuated the transcription of the *luxS* gene. Thus, this gene might be important in not only the survival of *Lactobacillus* spp. during the passage through the gastrointestinal tract, but also in the cell-to-cell communication among bacteria in the intestinal microbiota (Moslehi-Jenabian et al. 2009). Genome wide expression analysis experiments using microarrays have revealed that in *L. reuteri* ATCC 55730, the *clpL* chaperone gene (encoding an ATPase with chaperone activity) was involved in the early response to severe acidic shock. This was validated by mutation in *clpL* and the mutant was significantly more sensitive to acidic stress compared to the wild type (Wall et al. 2007).

Genes involved in the tolerance to bile salts are also important for survival of probiotics after passage through the gastrointestinal tract. DNA micro-array analysis of the global transcriptional response of *L. plantarum* WCFS1 against bile revealed 12 bile-responsive gene clusters. Seven of the identified bile-responsive genes and gene clusters encoded typical stress-related functions, including glutathione reductase and glutamate decarboxylase, involved in oxidative and acid stress defence, respectively. Besides, 14 bile-responsive genes and gene clusters were detected that encoded proteins located in the cell envelope, including the *dlt* operon and the F_1F_0 ATPase. The induction of a high number of genes encoding cell envelope functions show the significant effect of bile salts on the integrity and/or functionality of the cytoplasmic membrane and cell wall (Bron et al. 2006). Genes encoding bile salt hydrolases (*bsh*) have been identified in intestinal *Lactobacillus* spp., i.e., *L. acidophilus* NCFM (McAuliffe et al. 2005), *Lactobacillus johnsonii* 100-100 (Elkins et al. 2001) and *L. plantarum* WCFS1 (Lambert et al. 2008a), which shows the ecological adaptation of these species to the intestine and the importance of this trait for *Lactobacillus* spp. in order to colonize the lower gastrointestinal tract.

As for *Lactobacillus* spp., *Bifidobacterium* spp. have developed a system that attempt to maintain their cytoplasmic pH near neutral under acidic stress. In this respect, the proton-translocating ATPase (F_1F_0-ATPase) plays an important role and is encoded by the *atp* operon including nine genes. This multi-subunit enzyme is essential for growth of *Bifidobacterium* spp. under acidic conditions (Ventura et al. 2004). It has been shown that bile induces expression of the F_1F_0-ATPase and increases the membrane-bound H^+-ATPase activity in *Bifidobacterium animalis*. Comparison of *B. animalis* IPLA 4549 and a mutant with acquired resistance to bile (*B. animalis* 4549dOx) has shown that the bile-resistance mutant was able to tolerate bile by increasing the intracellular ATP reserve and by inducing proton pumping by the F_1F_0-ATPase (Sanchez et al. 2006). Genes encoding bile salt hydrolases (*bsh*) have been detected in *Bifidobacterium longum* BB536 (Shuhaimi et al. 2001), *Bifidobacterium bifidum* ATCC 11863 (Kim et al. 2004), *Bifidobacterium adolescentis* ATCC 15705 (Kim et al. 2005) and a bile tolerant strain of *B. animalis* subsp. *lactis* KL612 (Kim and Lee 2008). In a recent study, two putative multidrug resistance (MDR) transporter genes, i.e. the BL0920 gene from *B. longum* subsp. *longum* NCC2705 and its homolog, Bbr0838 gene, from *Bifidobacterium breve* UCC2003, were induced after exposure to sub-inhibitory concentrations of bile. The expression of the BL0920 gene in *Escherichia coli* conferred resistance to bile, which was probably mediated by active efflux from the cells. This study represents the first identified bifidobacterial bile efflux pump (Gueimonde et al. 2009). Molecular analysis of *B. longum* NCC2705 cells grown in the intestinal tract of mice revealed that different genes and proteins are expressed in the cells for adaptation of *B. longum* to intestinal stress. Among these, EF-Tu (related to the retention or attachment), bile salt hydrolase and stress proteins which protect *B. longum* against the action of bile salts and other destructive components of the gastrointestinal tract have been identified. In addition, it has been found that intestinal growth triggered phosphorylation of LuxS protein (the active form of LuxS) that possibly play a key role in the regulation of quorum sensing between microorganisms of intestinal microbiota (Yuan et al. 2008).

6.3.2 Genes and Molecules Involved in Nutritional Adaptation

The complete sequencing of several *Lactobacillus* spp. genomes has revealed a considerable degree of auxotrophy for amino acids and other cellular components. To compensate for these auxotrophies, *Lactobacillus* spp. have been shown to encode multiple genes for transport and uptake of macromolecules and metabolism of complex carbohydrates (Pfeiler and Klaenhammer 2007). Due to their auxotrophy, *Lactobacillus* spp. will predominantly be present in the ileum, which is a nutritional richer environment than e.g. the colon. Comparing the genome sequence of intestinal isolates of *Lactobacillus* spp. with food isolates indicates a strong degree of niche adaptation. As an example, *Lactobacillus helveticus* DPC 4571, a cheese starter culture, has additional genes for fatty acid biosynthesis and specific amino-acid metabolism, but remarkably fewer cell-surface proteins and phosphoenolpyruvate phosphotransferase systems for sugar utilization compared to *L. acidophilus* NCFM, which is a closely related species well adapted to the intestine. In addition, no functional mucus-binding proteins or transporters for complex carbohydrates are encoded by the *L. helveticus* DPC 4571 genome, indicating adaptation to the milk environment. Whereas *L. acidophilus* that is adapted to the gut ecological niche, contains functional gene sets such as mucus-binding and cell surface proteins and enzyme complexes that are absent from *L. helveticus* DPC 4571 (Altermann et al. 2005; Callanan et al. 2008), emphasizing the importance of these gene sets for gut adaptation and probiotic functionality. The genes encoding the mucus-binding or cell surface proteins found in the genome of intestinal *Lactobacillus* spp. are predicted to produce secreted proteins such as the S-layer proteins, which are maintained at the cell envelope via either covalent interactions affected by the sortase enzyme or electrostatic interactions, and interact with human intestinal compounds such as extracellular matrix proteins and mucus (Åvall-Jääskeläinen and Palva 2005). These extracellular proteins are essential not only in the interaction of probiotics with host cells or tissues, but also in degradation of complex extracellular carbon sources and have a prominent role in the adaptation to environmental changes and intestinal persistence (Boekhorst et al. 2006; Buck et al. 2005). Analysis of the predicted extracellular proteins of *L. plantarum* WCFS1 has revealed that at least 12 proteins are predicted to be directly involved in adherence to host components like collagen and mucin, and about 30 extracellular enzymes, mainly hydrolases and transglycosylases, predicted to be involved in substrate degradation by *L. plantarum* WCFS1 to maintain the growth in different environmental niches (Boekhorst et al. 2006).

In vivo studies using an IVET strategy based on the *in vivo* selection of an antibiotic-resistant phenotype have shown induction of 3 *in vivo* induced genes that are highly expressed in *L. reuteri* 100-23 during intestinal colonisation in *Lactobacillus*-free mice (Walter et al. 2003). In another study using a recombinase-based-IVET approach in *L. plantarum* WCFS1, 72 different genes were induced during passage through the gastrointestinal tract of conventional mice. Most of these genes were related to carbon and amino-acid metabolism and stress response (Bron et al. 2004). The homologues of many of these genes have been found in intestinal pathogens and associated with survival and adaptation to the gut environment. Whole genome transcriptional profiling of *L. plantarum* during colonization in the cecum of germ-free mice showed up-regulation of genes involved in carbohydrate transport and metabolism, compared with *in vitro* growth conditions. Indeed, the mouse diet had an essential impact on the *in situ* transcriptome of *L. plantarum* WCFS1 (Marco et al. 2009). Similar studies have shown transcription of metabolic genes in *Lactobacillus casei* DN-114 001 (Oozeer et al. 2005) and in *L. johnsonii* NCC533 (Denou et al. 2007, 2008) as adaptation to the environmental conditions in the murine intestine. In the latter species the expression of different sets of genes was observed to depend on its location in the mouse intestine (Denou et al. 2007, 2008). Some *Lactobacillus* spp. can utilize fructo-oligosaccharides which are known as prebiotics (non-digestible oligosaccharides which stimulate growth and/or metabolic activity of probiotics in the host intestine) and thereby interact metabolically with host and other microbes. *L. acidophilus* NCFM metabolise fructo-oligosaccharides by inducing the transcription of a specific transport and degradation system (Barrangou et al. 2003). Similarly, *L. plantarum* WCFS1 have a specific gene expression pattern when exposed to fructo-oligosaccharides, even though it is only able to degrade the short chains of these compounds (Saulnier et al. 2007).

Contrary to *Lactobacillus* spp., *Bifidobacterium* spp. are autotrophic and are therefore well adapted to growth in an environment with low concentrations of some growth substrates such as the human colon (Ventura et al. 2007a). This property gives them an ecological advantage in the intestinal ecosystem. *Bifidobacterium* genomics has demonstrated their relative broad autotrophy for amino acids, nucleotides, vitamins, and cofactors and has verified their ability to degrade and utilize complex carbohydrates (Schell et al. 2002). Gene clusters coding for complex sugar degradation pathways are abundant in bifidobacterial genomes (Ventura et al. 2007b) and preliminary intestinal transcriptomic studies have shown expression of bifidobacterial genes including oligosaccharide metabolism and vitamin production in the human infant gut (Klaassens et al. 2009). *Bifidobacterium* spp. are also able to hydrolyse different types of fructo-oligosaccharides (prebiotics) and the operons for fructo-oligosaccharide metabolism, specific transporters and hydrolases for oligosaccharides have been identified in the bifidobacterial genome (Gonzalez et al. 2008; Ryan et al. 2005).

6.4 Genes and Molecules Involved in Interaction of Probiotics with Enteropathogens and Gut Microbial Symbionts

Interaction of probiotics with enteropathogens in the intestinal tract involves different mechanisms, including nutrient-based interactions, competition for specific adhesion sites (competitive exclusion) and production of antimicrobial compounds.

6.4.1 Genes and Molecules Involved in Nutrient-Based Interactions

Nutrient-based interactions between probiotic bacteria and other members of the gut microbiota has been proved using germ-free mice models colonized by *Bacteroides thetaiotaomicron* ATCC 29148 (a prominent component of the adult human gut microbiota), *B. longum* NCC2705 and *L. casei* DN-114 001 or combinations of these microorganisms. Whole genome transcriptional profiling of all bacterial species as well as the intestinal epithelium showed that presence of *B. longum* triggered an expansion in the diversity of polysaccharides targeted for degradation by *B. thetaiotaomicron* (e.g., mannose- and xylose-containing glycans), and induced host genes involved in innate immunity. Presence of *L. casei* in this model resulted in an expanded capacity of *B. thetaiotaomicron* to metabolize polysaccharides and increased expression of genes for inorganic ion transport and metabolism, the same results as those observed by *B. longum*. This model showed how a resident symbiont and a probiotic species adapt their substrate utilization in response to each other (Sonnenburg et al. 2006). Indeed, it has been proposed that depletion of iron by *Bifidobacterium* spp. which is an essential nutrient for many intestinal pathogens (but not for *Lactobacillus* spp.) could be an important factor in the protective effect of *Bifidobacterium* spp. against pathogens in the gut (Kot and Bezkorovainy 1993).

6.4.2 Genes and Molecules Involved in Competitive Exclusion

One of the beneficial roles of probiotics is competition with enteropathogens to adhere to intestinal mucus or competitive exclusion. Therefore, the capacity of probiotic bacteria to adhere to the intestinal mucosa is an important factor for competitive exclusion. Different molecular methods including comparative genomics have revealed a number of genes involved in the adhesion of probiotic *Lactobacillus* spp. to the intestinal tract, such as genes encoding mucus-binding proteins (Altermann et al. 2005; Buck et al. 2005), surface layer proteins (Buck et al. 2005; van Pijkeren et al. 2006), fibronectin-binding proteins (Altermann et al. 2005; Buck et al. 2005), fimbrae (Pridmore et al. 2004), EPS clusters (Altermann et al. 2005; Pridmore et al. 2004), mucus-binding pilli (Kankainen et al. 2009) and mannose-specific adhesion proteins (Pretzer et al. 2005).

Multiple copies of genes encoding mucus-binding proteins have been found in different *Lactobacillus* spp. The predicted mucus-binding proteins are unusually large proteins representing the largest open reading frames (ORFs) in the genome, with relatively low amino acid identity offering considerable sequence variability within surface proteins which are supposed to have important roles in mucus binding (Altermann et al. 2005; Pridmore et al. 2004). Inactivation of genes encoding a mucus-binding protein, a fibronectin-binding protein and a surface layer protein in *L. acidophilus* had a great impact on adherence to intestinal Caco-2 epithelial cells. The adhesion ability was reduced significantly in the mucus-binding protein mutant (65%), the fibronectin-binding protein mutant (76%), and the surface layer protein mutant (84%). However, the decreased adhesion ability in the latter mutant was due to the loss of multiple surface proteins that may be embedded in the S-layer. This study showed that in *L. acidophilus* NCFM multiple cell surface proteins individually have a role in the ability of organism to attach to intestinal cells (Buck et al. 2005). Recently, a transcriptomic study using an *in vitro* gastrointestinal tract model has shown up-regulation of the genes encoding mucin binding protein and fibronectin-binding protein in *L. acidophilus* NCFM after exposure to duodenal juice and bile (Weiss and Jespersen 2010). The important role of mucus-binding pilli in the adhesion ability has been proved in *L. rhamnosus* GG. Comparative genomics of this probiotic bacterium with a starter culture strain *L. rhamnosus* LC705 (exhibiting reduced binding to mucus) revealed one genomic island in *L. rhamnosus* GG which was not present in the other strain and contained 3 pilli encoding genes (spaCBA). Molecular analysis showed that the spaC gene is involved in the adherence of strain *L. rhamnosus* GG to human intestinal mucus and presence of this gene is crucial for the interaction between *Lactobacillus* spp. and host tissues offering a likely explanation of the longer persistence of *L. rhamnosus* GG in the intestinal tract compared to other *L. rhamnosus* strains (Kankainen et al. 2009). Furthermore, a mannose-specific adhesin gene (*msa*) which encodes a sortase-dependent cell surface protein has been found in *L. plantarum* WCFS1 (Pretzer et al. 2005). In a pig model, a *msa* knock-out mutant of *L. plantarum* 299v exhibited decreased association with intestinal epithelia and increased jejunal fluid absorption. The wild-type *L. plantarum* 299v induced expression of the gene encoding pancreatitis-associated protein, a protein with proposed bactericidal properties but this feature was not observed in the *msa* mutant that suggests a role for the *msa* gene in the induction of host responses in the pig intestine (Gross et al. 2008). Adhesion to mannose residues is a likely mechanism behind various bacterium-host interactions. Presence of mannose specific adhesin genes and mannose-specific binding properties has been observed in different pathogens such as *E. coli* and *Salmonella enterica* serovar Typhimurium and is the basis for competitive exclusion by the potent probiotic yeast *S. cerevisiae* var. *boulardii* that have mannose containing polysaccharides in the cell wall (Moslehi-Jenabian et al. 2010). *S. cerevisiae* var. *boulardii* prevents bacterial adherence and translocation in the intestinal epithelial cells, due to the capacity of cell wall to bind enteropathogens. The *S. cerevisiae* var. *boulardii* cell wall has been shown to bind enterohaemorrhagic *E. coli* and *S. enterica* serovar Typhimurium (Gedek 1999).

The genome sequence of *B. longum* and other *Bifidobacterium* spp. contain predicted glycoprotein-binding fimbriae and mucus and fibronectin-binding proteins that could be involved in the bacterial adhesion to the intestinal tract (Klaassens et al. 2009; Schell et al. 2002). *B. adolescentis* BB-119 binds to type V collagen at galactose chains as target site via its two cell surface proteins with molecular masses of 36 kDa and 52 kDa and lectin-like activity (Mukai et al. 1997). It has been shown that several species of *Bifidobacterium* produce a compound in the growth media which inhibits binding of enterotoxic *E. coli*-expressing colonization factor antigen II to gangliotetraosylceramice (asialo GMT1 or GA1), a common bacterium-binding structure (Fujiwara et al. 1997).

6.4.3 Genes and Molecules Involved in Production of Antimicrobial Compounds

Probiotics are able to interact with enteropathogens by production of bacteriocins (antimicrobial peptides). Bacteriocins are a heterogeneous family of small, heat stable peptides with antimicrobial activity against closely related bacteria (Cotter et al. 2005). Numerous studies have shown the production of various bacteriocins by probiotics with antimicrobial effect against enteropathogens (Corr et al. 2007; Todorov and Dicks 2004; Zamfir et al. 2007). However, in most of these studies, it was not proved that the bacteriocin production was the main reason for inhibitory effect against pathogens by the probiotics. Nevertheless, bacteriocin-based interaction of probiotics and enteropathogens have been proved for *L. salivarius* UCC118 which has the ability to eliminate *Listeria monocytogenes* EGDe and LO28 from a mouse model due to the production of the broad spectrum bacteriocin Abp118 (also known as salivaricin) (Corr et al. 2007). It was observed that a bacteriocin-negative derivative of *L. salivarius* UCC118 was not able to protect mice against listerial infection. On the other hand, *L. salivarius* UCC118 could not protect the mice against infection with a *L. monocytogenes* derivative expressing the bacteriocin-immunity protein (Corr et al. 2007). This study demonstrates precisely the importance of bacteriocin production by probiotics for the protection against enteropathogens. In addition to bacteriocins, production of lactic acid and H_2O_2 has also been shown to be important measures used by *Lactobacillus* spp. against enteropathogens (De Keersmaecker et al. 2006; Pridmore et al. 2008).

In addition to inhibiting enteropathogens by production of antimicrobial compounds, it has been shown that the probiotic yeast *S. cerevisiae* var. *boulardii* produces two proteins of 54 and 120 kDa being responsible for degradation or neutralisation of bacterial toxins. The 54 kDa protein is a serine protease that decreases the enterotoxic and cytotoxic activities of *Clostridium difficile* by proteolysis of *C. difficile* toxin A and inhibits binding of the toxin to its brush border membrane receptor. *In vivo* studies have shown that oral administration of *S. cerevisiae* var. *boulardii* or its supernatant decreases toxin A-induced intestinal secretion and permeability due to activity of this enzyme (Castagliuolo et al. 1996, 1999; Pothoulakis et al. 1993). The 120 kDa protein has no proteolytic activity but competes specifically with the chloride secretion stimulated by the toxins of *Vibrio cholera* by reducing the cyclic adenosine monophosphate (cAMP) in the intestinal cells (Czerucka et al. 1994; Czerucka and Rampal 1999). Both *S. cerevisiae* var. *boulardii* and *S. cerevisiae* W303 have the ability to protect Fisher rats against cholera toxin (Brandão et al. 1998). *S. cerevisiae* var. *boulardii* also synthesizes a protein phosphatase that dephosphorylates endotoxins such as lipopolysaccharides of *E. coli* 055B5 and inactivates its cytotoxic effects (Buts et al. 2006).

6.5 Genes and Molecules Involved in Interaction of Probiotics with Host

Probiotic-host interactions that benefit the host can be investigated by genome mining and molecular analysis of the bacterial proteins or macromolecules, which might be involved. Probiotics interact with host and confer beneficial effects by means of different mechanisms including metabolic interactions, modulation of mucosal barrier function and modulation of the innate and adaptive immune system.

6.5.1 Genes and Molecules Involved in Metabolic Interactions

Probiotics interact metabolically with the host by modifying the nutritive function of the epithelium. For example, expression of the *ldh* gene encoding lactate hydrogenase by *Lactobacillus* spp. after entrance to the gastrointestinal tract and production of lactate by these bacteria, that later could be converted to butyric acid by the activity of some of the colon bacteria such as *Eubacterium hallii* (Duncan et al. 2004; Marco et al. 2007; Oozeer et al. 2005). The production of butyrate is important due to its beneficial effect on the gut epithelium. Butyrate is an important source of energy for the colonic mucosal cells, and it has been suggested to be essential for the maintenance of the colonic epithelium (Hamer et al. 2008).

Bile salt hydrolase activity and metabolism of bile salts is another positive effect of probiotics on the host positively influencing host physiology due to its role in biological processes in the host, i.e. in serum cholesterol lowering. Bile salts have antimicrobial and emulsification properties and probiotics by metabolizing these compounds might affect the amount of fat absorbed by the body. Bile salt deconjugation is the obligatory first reaction in further oxidation and dehydroxylation steps of bile salts by intestinal bacteria, and it includes the production of secondary bile salts, which negatively affects the host by being involved in formation of gallstones and colon cancer. On the other hand, bile salt deconjugation plays a role in mucin production and excretion in the intestinal lumen, and this could affect the nutritional environment encountered by the intestinal microbiota (Lambert et al. 2008b). Besides, bile acids act as local signalling molecules that regulate innate immunity (Inagaki et al. 2006), while re-absorbed bile salts act as signalling molecules involved in regulation of systemic endocrine functions (Watanabe et al. 2006).

6.5.2 Genes and Molecules Involved in Modulation of Mucosal Barrier Function

Probiotics preserve the barrier function by different mechanisms such as induction of mucin secretion (Mack et al. 2003), up-regulation of cytoprotective heat shock proteins (Petrof et al. 2004; Tao et al. 2006), enhancement of tight-junction functions (Klingberg et al. 2005; Seth et al. 2008) and modulation of epithelial cell apoptosis (Yan et al. 2007). Some of the signalling pathways involved in these mechanisms have been identified; however, the probiotic effector molecules and the genes encoding them are mostly unidentified.

Induction of mucin secretion is one of the mechanisms by which probiotics strengthen the intestinal barrier functionality. This mechanism is dependent on direct adhesion of probiotics to the epithelial cells as it has been shown by losses in the ability of mucin induction followed by spontaneous mutation in the *adh* gene (involved in adhesion) in *L. plantarum* 299v (Mack et al. 2003).

Increase in the level of inflammatory cytokines and apoptosis of intestinal epithelial cell lead to disruption of epithelial integrity. It has been indicated that *L. rhamnosus* GG prevents cytokine-induced apoptosis in human and mouse intestinal epithelial cells by regulating signalling pathways, i.e., by activation of the anti-apoptotic Akt/protein kinase B and inhibition of activation of the pro-apoptotic p38/mitogen-activated protein kinase by tumor necrosis factor-alpha (TNF-α), interleukin-1 alpha (IL-1 α), or gamma-interferon (IFN-γ) (Yan and Polk 2002). Two secreted proteins (p75 and p40) have been found in the spent culture of this probiotic bacterium, which activate Akt that has inhibitory effects on cytokine-induced apoptosis and loss of intestinal epithelial cells. Thereby these two proteins promote cell growth in human and mouse colon epithelial cells and cultured mouse colon explants (Yan et al. 2007). Intestinal epithelial tight junction is structured by distribution of different specific proteins such as occludin, zonula occludens (ZO-1, ZO-2, and ZO-3), claudins, E-cadherin, beta-catenin and junctional adhesion molecules (Anderson and VanItallie 1995). Hydrogen peroxide induces the re-distribution of these proteins and cause disruption of tight junctions. Secretory proteins of *L. rhamnosus* GG have been shown to protect intestinal epithelial tight junctions and the barrier function from hydrogen peroxide-induced damages by preserving the distribution of occludin, zonula occludens (ZO-1), E-cadherin, and beta-catenin in the intercellular junctions by a protein kinase C (PKC)- and mitogen-activated protein (MAP) kinase-dependent mechanism (Seth et al. 2008). In addition, an acid and heat stable low-molecular-weight peptide has been found in the spent culture of *L. rhamnosus* GG that induce expression of heat shock proteins (Hsp25 and Hsp72) in intestinal epithelial cells in a time- and concentration-dependent manner (Tao et al. 2006). DNA microarray experiments showed that Hsp72 is one of the genes most highly up-regulated in response to exposure to *L. rhamnosus* GG spent culture. Real-time PCR and electrophoretic mobility shift assays indicated that the *L. rhamnosus* GG spent culture modulates the activity of certain signalling pathways in intestinal epithelial cells by activating MAP kinases. In addition, functional studies suggested that treatment of gut epithelial cells with *L. rhamnosus* GG spent culture protects them from oxidative stress, possibly by preserving cytoskeletal integrity. Inhibition of nuclear factor-kappaB (NF-κB) and induction of heat shock proteins in colonic epithelial cells through proteasome inhibition has also been observed after exposure of the epithelial cells to spent culture of the probiotic mixture VSL#3 (*L. casei, L. plantarum, L. acidophilus, L. delbrueckii* subsp. *bulgaricus, B. longum, Bifidobacterium infantis, B. breve* and *Streptococcus salivarius* subsp. *thermophilus*) (Petrof et al. 2004). Investigation of individual strains of VSL#3 showed that spent culture of *B. infantis* had the highest effect on increasing the TER compared with spent cultures of other probiotic strains in the mixture. *B. infantis* spent culture decreased claudin-2, and increased ZO-1 and occludin expression in T84 cells, which was mediated by changes in MAP kinases. Besides, *B. infantis* spent culture inhibited reduction of TER induced by TNF-α and IFN-γ and re-distribution of tight junction proteins. In addition, oral administration of spent culture reduced colonic permeability in mice (Ewaschuk et al. 2008). These results may account for the anti-inflammatory and cytoprotective effects reported for probiotics and the mechanism of microbial-epithelial interaction. However, more research is needed to identify the unknown factor(s) in spent culture of various probiotics, which exert the protective effects on intestinal epithelial cells mediated by multiple signalling pathways.

Anti-inflammatory effects and lowering the proinflammatory response has also been shown for *S. cerevisiae* var. *boulardii* upon exposure to enteropathogens (Chen et al. 2006; van der Aa Kühle et al. 2005). Production of products with anti-inflammatory effect has also been shown by *S. cerevisiae* var. *boulardii*. This yeast produces a soluble factor (<10 kDa) that exerts anti-inflammatory effects after stimulation with *C. difficile*-toxin A, by reducing secretion of IL-8 in human colonocytes and activation of extracellular signal-regulated protein kinases 1 and 2 (ERK1/2) in both human colonocytes and murine ileal loops (Chen et al. 2006). Besides, *S. cerevisiae* var. *boulardii* produces a low molecular weight soluble factor (<1 kDa) which blocks NF-κB activation and NF-κB-mediated IL-8 gene expression in intestinal epithelial cells and monocytes (Sougioultzis et al. 2006).

6.5.3 Genes and Molecules Involved in Modulation of the Innate and Adaptive Immune System

Modulation of the innate and adaptive immune system is another beneficial effect of probiotics on human health. Modulation of the immune system by *Lactobacillus* spp. is mostly ascribed to priming immunoregulatory responses via modulation of dendritic cell (DC) function and induction of regulatory T cells (Rook and Brunet 2005). Numerous studies have shown the immunomodulating effects of probiotics. However, only a few probiotic effector molecules have been identified in this context. One of the probiotic effector molecules which has been identified based on genomics of *B. longum* NCC2705 is a gene encoding a homologue of the eukaryotic-type serine protease inhibitor (serpin) (Schell et al. 2002). Some of the members of the serpin family have the capacity to suppress inflammatory responses through inhibition of human neutrophil elastase (Ivanov et al. 2006).

The ability of probiotic *Lactobacillus* spp. to prime DCs to drive development of regulatory T-cell varies among different species and depends on their recognition of the C-type lectin receptor DC-SIGN (DC-specific intercellular adhesion molecule 3-grabbing nonintegrin), which has an important role in regulatory T-cell stimulation (Smits et al. 2005). It has been shown that mutation in a surface layer protein (SlpA) in *L. acidophilus* NCFM resulted in significant reduction in binding capacity to DC-SIGN. The protein SlpA is the first identified probiotic bacterial DC-SIGN ligand that is functionally involved in the modulation of DC and T cell functions (Konstantinov et al. 2008).

Teichoic acids, and especially lipoteichoic acids (LTA), are one of the major immunostimulatory components of pathogenic Gram-positive bacteria (Morath et al. 2001). Transcriptional profiling of *L. plantarum* WCFS1 during colonization of the cecum of germ-free mice have shown that a set of bacterial genes encoding cell surface-related functions were differentially regulated compared with *in vitro* growth conditions. Genes encoding the biosynthetic pathway responsible for addition of D-alanyl substituent to LTA were down regulated, while the expression levels of teichoic acid biosynthetic genes remained virtually unchanged and it has been proposed that *L. plantarum* WCFS1 modifies its gene expression *in vivo* and minimize the levels of D-alanylated LTA present on the cell surface (Marco et al. 2009). The importance of teichoic acid composition on the cell wall of probiotic *Lactobacillus* spp. in their immunomodulatory effect has been studied using a mutant of *L. plantarum* NCIMB8826 (*dlt*) which was modified in the teichoic acid biosynthesis pathway and presented much less D-Ala in its teichoic acids compared to the wild type strain. This mutation positively influenced the Toll-like receptor 2 (TLR-2)-dependent immunomodulatory properties of *L. plantarum* NCIMB8826 (Grangette et al. 2005). A considerably lower secretion of proinflammatory cytokines by peripheral blood mononuclear cells and monocytes in parallel with a significant increase in IL-10 production was observed after stimulation with the mutant as compared to the parental strain. In addition, the mutant was significantly more protective in a murine colitis model compared to its wild type counterpart. These studies demonstrated that composition of teichoic acids has a great impact on the immunomodulatory effect of *L. plantarum* and this probiotic modifies its gene expression *in vivo* in a way that will increase its immunomodulatory effect. However, it should be kept in mind that *dlt* mutants of other strains of *Lactobacillus* spp. have shown other behaviour. As an example *dlt* mutations in *L. rhamnosus* GG resulted in unaltered immunomodulation and the mutant showed lower survival under intestinal conditions (Velez et al. 2007). Similar results were found for a *L. reuteri* 100-23 *dlt* mutant (Walter et al. 2007). These studies confirm that the immunomodulatory effects of probiotic *Lactobacillus* spp. are strongly species and even strain dependent.

Investigation of the immunomodulatory effect of probiotic preparation VSL#3 has shown that VSL#3 is a strong inducer of IL-10 by DCs from blood and intestinal tissue, and prevents generation of Th1 cells. However, individual strains within VSL#3 presented different immunomodulatory effects on DCs and bifidobacteria strains (*B. longum, B. infantis* and *B. breve*) offered the highest anti-inflammatory effects. Interaction of cell wall components of these *Bifidobacterium* spp. with human intestinal lamina propria mononuclear cells, whole blood, or an enriched blood dendritic cell population showed that *Bifidobacterium* spp. up-regulate IL-10 production by DCs and decrease IFN-γ production by T cells (Hart et al. 2004). In a similar study using a murine macrophage-like cell line, *B. adolescentis* and *B. longum* induced higher secretion of a proinflammatory cytokine IL-12 and TNF-α, compared to *B. bifidum, B. breve,* and *B. infantis*, whereas *B. adolescentis* did not stimulate the production of anti-inflammatory IL-10 as the other tested bacteria, showing that this bacterium is less capable of down-regulating the inflammatory response in macrophage-like cell line (He et al. 2002). However, the signalling pathways and genes involved in these immunomodulatory effects have not yet been identified.

6.6 Conclusion and Future Perspectives

Recent functional genomic analyses and molecular studies have identified some of the genes and molecules offering the health benefits of probiotics. These studies have proved the role of both stress response genes and genes involved in adaptation to new ecological environments being important for the functionality of the probiotics. In addition, different genes have been found to be involved in adhesion of probiotics to intestinal epithelial cells and production of antimicrobial products, which might be crucial for competitive exclusion and interaction with enteropathogens. More specifically, different genes have been recognized to be involved in mucin secretion, in regulation of the different signalling pathways resulting in anti-inflammatory effects, and in strengthening the epithelial tight junctions, which have protective effects on intestinal epithelial functionality. However, only a limited number of genes have been identified in this regard and additional studies are necessary to uncover all genes involved and to clarify the specific mechanisms at the molecular level. Several studies have detected some genes that are involved in modulating the immune response toward development of T-regulatory cells, but the exact mechanisms and genes are still missing. Altogether, genomics and molecular studies have been proved to be useful to unravel the genes and molecules involved in probiotic functionality and to recognize the regions in the genome that might be interesting for further investigation to identify the exact mechanisms involved in the beneficial effects of probiotics on human health. In this context, effective *in vitro* and *in vivo* models combined with omic approaches and assisted by mathematical models that help exploiting the complex information obtained will facilitate identification of the precise mechanisms by which probiotic microorganisms confer their health benefits. However, considering the biodiversity of these microorganisms in the gut microbiota and the fact that their mode of action is species and even strain dependent, we are still in the beginning of this field of research. Therefore, it is crucial to understand the human gut microbiome to get a comprehensive view of the genetic diversity of the gut residents and the variations in their molecular characteristics and host interactions. This will allow us to find more efficient probiotics and to develop new functional food products and therapeutic agents with improved quality.

References

Altermann E, Russell WM, Azcarate-Peril MA, Barrangou R, Buck BL, McAuliffe O, Souther N, Dobson A, Duong T, Callanan M, Lick S, Hamrick A, Cano R, Klaenhammer TR (2005) Complete genome sequence of the probiotic lactic acid bacterium *Lactobacillus acidophilus* NCFM. Proc Natl Acad Sci U S A 102(11):3906–3912

Anderson JM, VanItallie CM (1995) Tight junctions and the molecular-basis for regulation of paracellular permeability. Am J Physiol-Gastrointest Liver Physiol 32(4):G467–G475

Åvall-Jääskeläinen S, Palva A (2005) *Lactobacillus* surface layers and their applications. FEMS Microbiol Rev 29(3):511–529

Azcarate-Peril MA, Altermann E, Hoover-Fitzula RL, Cano RJ, Klaenhammer TR (2004) Identification and inactivation of genetic loci involved with *Lactobacillus acidophilus* acid tolerance. Appl Environ Microbiol 70(9):5315–5322

Azcarate-Peril MA, McAuliffe O, Altermann E, Lick S, Russell WM, Klaenhammer TR (2005) Microarray analysis of a two-component regulatory system involved in acid resistance and proteolytic activity in *Lactobacillus acidophilus*. Appl Environ Microbiol 71(10):5794–5804

Barrangou R, Altermann E, Hutkins R, Cano R, Klaenhammer TR (2003) Functional and comparative genomic analyses of an operon involved in fructooligosaccharide utilization by *Lactobacillus acidophilus*. Proc Natl Acad Sci U S A 100(15):8957–8962

Boekhorst J, Wels M, Kleerebezem M, Siezen RJ (2006) The predicted secretome of *Lactobacillus plantarum* WCFS1 sheds light on interactions with its environment. Microbiology 152:3175–3183

Brandão RL, Castro IM, Bambirra EA, Amaral SC, Fietto LG, Tropia MJ, Neves MJ, Dos Santos RG, Gomes NC, Nicoli JR (1998) Intracellular signal triggered by cholera toxin in *Saccharomyces boulardii* and *Saccharomyces cerevisiae.* Appl Environ Microbiol 64(2):564–568

Bron PA, Grangette C, Mercenier A, de Vos WM, Kleerebezem M (2004) Identification of *Lactobacillus plantarum* genes that are induced in the gastrointestinal tract of mice. J Bacteriol 186(17):5721–5729

Bron PA, Molenaar D, Vos WM, Kleerebezem M (2006) DNA micro-array-based identification of bile-responsive genes in *Lactobacillus plantarum.* J Appl Microbiol 100(4):728–738

Buck BL, Altermann E, Svingerud T, Klaenhammer TR (2005) Functional analysis of putative adhesion factors in *Lactobacillus acidophilus* NCFM. Appl Environ Microbiol 71(12):8344–8351

Buts JP, Dekeyser N, Stilmant C, Delem E, Smets F, Sokal E (2006) *Saccharomyces boulardii* produces in rat small intestine a novel protein phosphatase that inhibits *Escherichia coli* endotoxin by dephosphorylation. Pediatr Res 60(1):24–29

Callanan M, Kaleta P, O'Callaghan J, O'Sullivan O, Jordan K, McAuliffe O, Sangrador-Vegas A, Slattery L, Fitzgerald GF, Beresford T, Ross RP (2008) Genome sequence of lactobacillus helveticus, an organism distinguished by selective gene loss and insertion sequence element expansion. J Bacteriol 190(2):727–735

Castagliuolo I, LaMont JT, Nikulasson ST, Pothoulakis C (1996) *Saccharomyces boulardii* protease inhibits *Clostridium difficile* toxin A effects in the rat ileum. Infect Immun 64(12):5225–5232

Castagliuolo I, Riegler MF, Valenick L, LaMont JT, Pothoulakis C (1999) *Saccharomyces boulardii* protease inhibits the effects of *Clostridium difficile* toxins A and B in human colonic mucosa. Infect Immun 67(1):302–307

Chen X, Kokkotou EG, Mustafa N, Bhaskar KR, Sougioultzis S, O'Brien M, Pothoulakis C, Kelly CP (2006) *Saccharomyces boulardii* inhibits ERK1/2 mitogen-activated protein kinase activation both *in vitro* and *in vivo* and protects against *Clostridium difficile* toxin A-induced enteritis. J Biol Chem 281(34):24449–24454

Corr SC, Li Y, Riedel CU, O'Toole PW, Hill C, Gahan CGM (2007) Bacteriocin production as a mechanism for the antfinfective activity of *Lactobacillus salivarius* UCC118. Proc Natl Acad Sci U S A 104(18):7617–7621

Cotter PD, Hill C, Ross RP (2005) Bacteriocins: developing innate immunity for food. Nat Rev Microbiol 3(10):777–788

Czerucka D, Rampal P (1999) Effect of *Saccharomyces boulardii* on cAMP- and Ca^{2+} -dependent Cl- secretion in T84 cells. Dig Dis Sci 44(11):2359–2368

Czerucka D, Roux I, Rampal P (1994) *Saccharomyces boulardii* inhibits secretagogue-mediated adenosine 3′,5′-cyclic monophosphate induction in intestinal cells. Gastroenterology 106(1):65–72

De Keersmaecker SCJ, Verhoeven TLA, Desair J, Marchal K, Vanderleyden J, Nagy I (2006) Strong antimicrobial activity of *Lactobacillus rhamnosus* GG against *Salmonella typhimurium* is due to accumulation of lactic acid. FEMS Microbiol Lett 259(1):89–96

Denou E, Berger B, Barretto C, Panoff JM, Arigoni F, Brussow H (2007) Gene expression of commensal *Lactobacillus johnsonii* strain NCC533 during *in vitro* growth and in the murine gut. J Bacteriol 189(22):8109–8119

Denou E, Pridmore RD, Berger B, Panoff JM, Arigoni F, Brussow H (2008) Identification of genes associated with the long-gut-persistence phenotype of the probiotic *Lactobacillus johnsonii* strain NCC533 using a combination of genomics and transcriptome analysis. J Bacteriol 190(9):3161–3168

Duncan SH, Louis P, Flint HJ (2004) Lactate-utilizing bacteria, isolated from human faeces, that produce butyrate as a major fermentation product. Appl Environ Microbiol 70(10):5810–5817

Elkins CA, Moser SA, Savage DC (2001) Genes encoding bile salt hydrolases and conjugated bile salt transporters in *Lactobacillus johnsonii* 100-100 and other *Lactobacillus* species. Microbiology 147:3403–3412
Ewaschuk JB, Diaz H, Meddings L, Diederichs B, Dmytrash A, Backer J, Langen MLV, Madsen KL (2008) Secreted bioactive factors from Bifidobacterium infantis enhance epithelial cell barrier function. Am J Physiol-Gastrointest Liver Physiol 295(5):G1025–G1034
Federle MJ, Bassler BL (2003) Interspecies communication in bacteria. J Clin Invest 112(9):1291–1299
Ferreira A, O'Byrne CP, Boor KJ (2001) Role of sigma(B) in heat, ethanol, acid, and oxidative stress resistance and during carbon starvation in *Listeria monocytogenes*. Appl Environ Microbiol 67(10):4454–4457
Ferreira A, Sue D, O'Byrne CP, Boor KJ (2003) Role of *Listeria monocytogenes* sigma(B) in survival of lethal acidic conditions and in the acquired acid tolerance response. Appl Environ Microbiol 69(5):2692–2698
Frees D, Vogensen FK, Ingmer H (2003) Identification of proteins induced at low pH in *Lactococcus lactis*. Int J Food Microbiol 87(3):293–300
Fujiwara S, Hashiba H, Hirota T, Forstner JF (1997) Proteinaceous factor(s) in culture supernatant fluids of bifidobacteria which prevents the binding of enterotoxigenic *Escherichia coli* to gangliotetraosylceramide. Appl Environ Microbiol 63(2):506–512
Gedek BR (1999) Adherence of *Escherichia coli* serogroup 0 157 and the *Salmonella Typhimurium* mutant DT 104 to the surface of *Saccharomyces boulardii*. Mycoses 42(4):261–264
Gonzalez R, Klaassens ES, Malinen E, de Vos WM, Vaughan EE (2008) Differential transcriptional response of *Bifidobacterium longum* to human milk, formula milk, and galactooligosaccharide. Appl Environ Microbiol 74(15):4686–4694
Grangette C, Nutten S, Palumbo E, Morath S, Hermann C, Dewulf J, Pot B, Hartung T, Hols P, Mercenier A (2005) Enhanced anti inflammatory capacity of a *Lactobacillus plantarum* mutant synthesizing modified teichoic acids. Proc Natl Acad Sci U S A 102(29):10321–10326
Gross G, van der Meulen J, Snel J, van der Meer R, Kleerebezem M, Niewold TA, Hulst MM, Smits MA (2008) Mannose-specific interaction of *Lactobacillus plantarum* with porcine jejunal epithelium. FEMS Immunol Med Microbiol 54(2):215–223
Guarner F, Malagelada JR (2003) Gut flora in health and disease. Lancet 361(9356):512–519
Gueimonde M, Garrigues C, van Sinderen D, de los Reyes-Gavilan CG, Margolles A (2009) Bile-inducible efflux transporter from *Bifidobacterium longum* NCC(2705) conferring bile resistance. Appl Environ Microbiol 75(10):3153–3160
Hamer HM, Jonkers D, Venema K, Vanhoutvin S, Troost FJ, Brummer RJ (2008) Review article: the role of butyrate on colonic function. Aliment Pharmacol Thera 27(2):104–119
Hart AL, Lammers K, Brigidi P, Vitali B, Rizzello F, Gionchetti P, Campieri M, Kamm MA, Knight SC, Stagg AJ (2004) Modulation of human dendritic cell phenotype and function by probiotic bacteria. Gut 53(11):1602–1609
He F, Morita H, Ouwehand AC, Hosoda M, Hiramatsu M, Kurisaki J, Isolauri E, Benno Y, Salminen S (2002) Stimulation of the secretion of pro-inflammatory cytokines by *Bifidobacterium* strains. Microbiol Immunol 46(11):781–785
Inagaki T, Moschetta A, Lee YK, Peng L, Zhao GX, Downes M, Yu RT, Shelton JM, Richardson JA, Repa JJ, Mangelsdorf DJ, Kliewer SA (2006) Regulation of antibacterial defense in the small intestine by the nuclear bile acid receptor. Proc Natl Acad Sci U S A 103(10):3920–3925
Ivanov D, Emonet C, Foata F, Affolter M, Delley M, Fisseha M, Blum-Sperisen S, Kochhar S, Arigoni F (2006) A serpin from the gut bacterium *Bifidobacterium longum* inhibits eukaryotic elastase-like serine proteases. J Biol Chem 281(25):17246–17252
Kankainen M, Paulin L, Tynkkynen S, von Ossowski I, Reunanen J, Partanen P, Satokari R, Vesterlund S, Hendrickx APA, Lebeer S, De Keersmaecker SCJ, Vanderleyden J, Hamalainen T, Laukkanen S, Salovuori N, Ritari J, Alatalo E, Korpela R, Mattila-Sandholm T, Lassig A, Hatakka K, Kinnunen KT, Karjalainen H, Saxelin M, Laakso K, Surakka A, Palva A, Salusjarvi T, Auvinen P, de Vos WM (2009) Comparative genomic analysis of *Lactobacillus rhamno-*

sus GG reveals pili containing a human-mucus binding protein. Proc Natl Acad Sci U S A 106(40):17193–17198
Kim GB, Lee BH (2008) Genetic analysis of a bile salt hydrolase in Bifidobacterium animalis subsp lactis KL612. J Appl Microbiol 105(3):778–790
Kim GB, Miyamoto CM, Meighen EA, Lee BH (2004) Cloning and characterization of the bile salt hydrolase genes (bsh) from Bifidobacterium bifidum strains. Appl Environ Microbiol 70(9):5603–5612
Kim GB, Brochet M, Lee BH (2005) Cloning and characterization of a bile salt hydrolase (bsh) from Bifidobacterium adolescentis. Biotechnol Lett 27(12):817–822
Klaassens ES, Boesten RJ, Haarman M, Knol J, Schuren FH, Vaughan EE, de Vos WM (2009) Mixed-species genomic microarray analysis of faecal samples reveals differential transcriptional responses of Bifidobacteria in breast- and formula-fed infants. Appl Environ Microbiol 75(9):2668–2676
Klaenhammer T, Altermann E, Arigoni F, Bolotin A, Breidt F, Broadbent J, Cano R, Chaillou S, Deutscher J, Gasson M, van de GM, Guzzo J, Hartke A, Hawkins T, Hols P, Hutkins R, Kleerebezem M, Kok J, Kuipers O, Lubbers M, Maguin E, McKay L, Mills D, Nauta A, Overbeek R, Pel H, Pridmore D, Saier M, van SD, Sorokin A, Steele J, O'Sullivan D, de VW, Weimer B, Zagorec M, Siezen R (2002) Discovering lactic acid bacteria by genomics. Antonie Van Leeuwenhoek 82(1–4):29–58
Klingberg TD, Pedersen MH, Cencic A, Budde BB (2005) Application of measurements of transepithelial electrical resistance of intestinal epithelial cell monolayers to evaluate probiotic activity. Appl Environ Microbiol 71(11):7528–7530
Konstantinov SR, Smidt H, de Vos WM, Bruijns SCM, Singh SK, Valence F, Molle D, Lortal S, Altermann E, Klaenhammer TR, van Kooyk Y (2008) S layer protein A of *Lactobacillus acidophilus* NCFM regulates immature dendritic cell and T cell functions. Proc Natl Acad Sci U S A 105(49):19474–19479
Kot E, Bezkorovainy A (1993) Effects of Mg^{2+} and Ca^{2+} on Fe^{2+} uptake by Bifidobacterium thermophilum. Int J Biochem 25(7):1029–1033
Kullen MJ, Klaenhammer TR (1999) Identification of the pH-inducible, proton-translocating F_1F_0-ATPase (atpBEFHAGDC) operon of Lactobacillus acidophilus by differential display: gene structure, cloning and characterization. Mol Microbiol 33(6):1152–1161
Lambert JM, Bongers RS, de Vos WM, Kleerebezem M (2008a) Functional analysis of four bile salt hydrolase and penicillin acylase family members in *Lactobacillus plantarum* WCFS1. Appl Environ Microbiol 74(15):4719–4726
Lambert JM, Siezen RJ, de Vos WM, Kleerebezem M (2008b) Improved annotation of conjugated bile acid hydrolase superfamily members in Gram-positive bacteria. Microbiology 154:2492–2500
Larsen N, Vogensen FK, van den Berg FWJ, Nielsen DS, Andreasen AS, Pedersen BK, bu Al-Soud W, Sorensen SJ, Hansen LH, Jakobsen M (2010) Gut microbiota in human adults with type 2 diabetes differs from non-diabetic adults. Plos One 5(2):e9085
Lim EM, Ehrlich SD, Maguin E (2000) Identification of stress-inducible proteins in *Lactobacillus delbrueckii* subsp. *bulgaricus*. Electrophoresis 21(12):2557–2561
Lomax AR, Calder PC (2009) Probiotics, immune function, infection and inflammation: a review of the evidence from studies conducted in humans. Curr Pharm Des 15(13):1428–1518
Lorca GL, De Valdez GF, Ljungh A (2002) Characterization of the protein-synthesis dependent adaptive acid tolerance response in *Lactobacillus acidophilus*. J Mol Microbiol Biotechnol 4(6):525–532
Mack DR, Ahrne S, Hyde L, Wei S, Hollingsworth MA (2003) Extracellular MUC3 mucin secretion follows adherence of *Lactobacillus* strains to intestinal epithelial cells *in vitro*. Gut 52(6):827–833
Mahan MJ, Heithoff DM, Sinsheimer RL, Low DA (2000) Assessment of bacterial pathogenesis by analysis of gene expression in the host. Ann Rev Genet 34:139–164
Marco ML, Bongers RS, de Vos WM, Kleerebezem M (2007) Spatial and temporal expression of *Lactobacillus plantarum* genes in the gastrointestinal tracts of mice. Appl Environ Microbiol 73(1):124–132

Marco ML, Peters THF, Bongers RS, Molenaar D, van Hemert S, Sonnenburg JL, Gordon JI, Kleerebezem M (2009) Lifestyle of *Lactobacillus plantarum* in the mouse caecum. Environ Microbiol 11(10):2747–2757

McAuliffe O, Cano RJ, Klaenhammer TR (2005) Genetic analysis of two bile salt hydrolase activities in *Lactobacillus acidophilus* NCFM. Appl Environ Microbiol 71(8):4925–4929

Morath S, Geyer A, Hartung T (2001) Structure-function relationship of cytokine induction by lipoteichoic acid from *Staphylococcus aureus*. J Exp Med 193(3):393–397

Moslehi-Jenabian S, Gori K, Jespersen L (2009) AI-2 signalling is induced by acidic shock in probiotic strains of *Lactobacillus* spp. Int J Food Microbiol 135(3):295–302

Moslehi-Jenabian S, Lindegaard L, Jespersen L (2010) Beneficial effects of probiotic and food borne yeasts on human health. Nutrients 2(4):449–473

Mukai T, Toba T, Ohori H (1997) Collagen binding of *Bifidobacterium adolescentis*. Curr Microbiol 34(5):326–331

Oozeer R, Furet JP, Goupil-Feuillerat N, Anba J, Mengaud J, Corthier G (2005) Differential activities of four *Lactobacillus casei* promoters during bacterial transit through the gastrointestinal tracts of human-microbiota-associated mice. Appl Environ Microbiol 71(3):1356–1363

Petrof EO, Kojima K, Ropeleski MJ, Musch MW, Tao Y, De Simone C, Chang EB (2004) Probiotics inhibit nuclear factor-kappa B and induce heat shock proteins in colonic epithelial cells through proteasome inhibition. Gastroenterology 127(5):1474–1487

Pfeiler EA, Klaenhammer TR (2007) The genomics of lactic acid bacteria. Trends Microbiol 15(12):546–553

Pothoulakis C, Kelly CP, Joshi MA, Gao N, O'Keane CJ, Castagliuolo I, LaMont JT (1993) *Saccharomyces boulardii* inhibits *Clostridium difficile* toxin A binding and enterotoxicity in rat ileum. Gastroenterology 104(4):1108–1115

Pretzer G, Snel J, Molenaar D, Wiersma A, Bron PA, Lambert J, de Vos WM, van der Meer R, Smits MA, Kleerebezem M (2005) Biodiversity-based identification and functional characterization of the mannose-specific adhesin of *Lactobacillus plantarum*. J Bacteriol 187(17):6128–6136

Pridmore RD, Berger B, Desiere F, Vilanova D, Barretto C, Pittet AC, Zwahlen MC, Rouvet M, Altermann E, Barrangou R, Mollet B, Mercenier A, Klaenhammer T, Arigoni F, Schell MA (2004) The genome sequence of the probiotic intestinal bacterium *Lactobacillus johnsonii* NCC 533. Proc Natl Acad Sci U S A 101(8):2512–2517

Pridmore RD, Pittet AC, Praplan F, Cavadini C (2008) Hydrogen peroxide production by *Lactobacillus johnsonii* NCC 533 and its role in anti-*Salmonella* activity. FEMS Microbiol Lett 283(2):210–215

Rook GAW, Brunet LR (2005) Microbes, immunoregulation, and the gut. Gut 54(3):317–320

Ryan SM, Fitzgerald GF, van Sinderen D (2005) Transcriptional regulation and characterization of a novel beta-fructofuranosidase-encoding gene from *Bifidobacterium breve* UCC2003. Appl Environ Microbiol 71(7):3475–3482

Sanchez B, Reyes-Gavilan CGD, Margolles A (2006) The F_1F_0-ATPase of *Bifidobacterium animalis* is involved in bile tolerance. Environ Microbiol 8(10):1825–1833

Saulnier DAA, Molenaar D, de Vos WA, Gibson GR, Kolida S (2007) Identification of prebiotic fructooligosaccharide metabolism in *Lactobacillus plantarum* WCFS1 through microarrays. Appl Environ Microbiol 73(6):1753–1765

Schell MA, Karmirantzou M, Snel B, Vilanova D, Berger B, Pessi G, Zwahlen MC, Desiere F, Bork P, Delley M, Pridmore RD, Arigoni F (2002) The genome sequence of *Bifidobacterium longum* reflects its adaptation to the human gastrointestinal tract. Proc Natl Acad Sci U S A 99(22):14422–14427

Seth A, Yan F, Polk DB, Rao RK (2008) Probiotics ameliorate the hydrogen peroxide-induced epithelial barrier disruption by a PKC- and MAP kinase-dependent mechanism. Am J Physiol Gastrointest Liver Physiol 294(4):G1060–G1069

Shuhaimi M, Ali AM, Saleh NM, Yazid AM (2001) Cloning and sequence analysis of bile salt hydrolase (bsh) gene from *Bifidobacterium longum*. Biotechnol Lett 23(21):1775–1780

Smits HH, Engering A, van der Kleij D, de Jong EC, Schipper K, van Capel TMM, Zaat BAJ, Yazdanbakhsh M, Wierenga EA, van Kooyk Y, Kapsenberg ML (2005) Selective probiotic bacteria induce IL-10-producing regulatory T cells *in vitro* by modulating dendritic cell function through dendritic cell-specific intercellular adhesion molecule 3-grabbing nonintegrin. J Allergy Clin Immunol 115(6):1260–1267

Sonnenburg JL, Chen CTL, Gordon JI (2006) Genomic and metabolic studies of the impact of probiotics on a model gut symbiont and host. Plos Biol 4(12):2213–2226

Sougioultzis S, Simeonidis S, Bhaskar KR, Chen XH, Anton PM, Keates S, Pothoulakis C, Kelly CP (2006) *Saccharonzyces boulardii* produces a soluble anti-inflammatory factor that inhibits NF-kappa B-mediated IL-8 gene expression. Biochem Biophys Res Commun 343(1):69–76

Tao Y, Drabik KA, Waypa TS, Musch MW, Alverdy JC, Schneewind O, Chang EB, Petrof EO (2006) Soluble factors from *Lactobacillus* GG activate MAPKs and induce cytoprotective heat shock proteins in intestinal epithelial cells. Am J Physiol-Cell Physiol 290(4):C1018–C1030

Todorov SD, Dicks LMT (2004) Effect of medium components on bacteriocin production by *Lactobacillus pentosus* ST151BR, a strain isolated from beer produced by the fermentation of maize, barley and soy flour. World J Microbiol Biotechnol 20(6):643–650

van der Aa Kühle A, Jespersen L (2003) The taxonomic position of *Saccharomyces boulardii* as evaluated by sequence analysis of the D1/D2 domain of 26S rDNA, the ITS1-5.8S rDNA-ITS2 region and the mitochondrial cytochrome-c oxidase II gene. Syst Appl Microbiol 26(4):564–571

van der Aa Kühle A, Skovgaard K, Jespersen L (2005) *In vitro* screening of probiotic properties of *Saccharomyces cerevisiae* var. boulardii and food-borne *Saccharomyces cerevisiae* strains. Int J Food Microbiol 101(1):29–39

van Pijkeren JP, Canchaya C, Ryan KA, Li Y, Claesson MJ, Sheil B, Steidler L, O'Mahony L, Fitzgerald GF, van Sinderen D, O'Toole PW (2006) Comparative and functional analysis of sortase-dependent proteins in the predicted secretome of *Lactobacillus salivarius* UCC118. Appl Environ Microbiol 72(6):4143–4153

Velez MP, Verhoeven TLA, Draing C, Von Aulock S, Pfitzenmaier M, Geyer A, Lambrichts I, Grangette C, Pot B, Vanderleyden J, De Keersmaecker SCJ (2007) Functional analysis of D-alanylation of lipoteichoic acid in the probiotic strain *Lactobacillus rhamnosus* GG. Appl Environ Microbiol 73(11):3595–3604

Ventura M, Canchaya C, van Sinderen D, Fitzgerald GF, Zink R (2004) *Bifidobacterium lactis* DSM 10140: identification of the atp (atpBEFHAGDC) operon and analysis of its genetic structure, characteristics, and phylogeny. Appl Environ Microbiol 70(5):3110–3121

Ventura M, Canchaya C, Fitzgerald GF, Gupta RS, van Sinderen D (2007a) Genomics as a means to understand bacterial phylogeny and ecological adaptation: the case of bifidobacteria. Antonie Van Leeuwenhoek 91(4):351–372

Ventura M, Canchaya C, Tauch A, Chandra G, Fitzgerald GF, Chater KF, van Sinderen D (2007b) Genomics of *actinobacteria*: tracing the evolutionary history of an ancient phylura. Microbiol Mol Biol Rev 71(3):495–548

Wall T, Bath M, Britton RA, Jonsson H, Versalovic J, Roos S (2007) The early response to acid shock in *Lactobacillus reuteri* involves the ClpL chaperone and a putative cell wall-altering esterase. Appl Environ Microbiol 73(12):3924–3935

Walter J, Heng NC, Hammes WP, Loach DM, Tannock GW, Hertel C (2003) Identification of *Lactobacillus reuteri* genes specifically induced in the mouse gastrointestinal tract. Appl Environ Microbiol. 69(4):2044–2051

Walter J, Loach DM, Alqumber M, Rockel C, Hermann C, Pfitzenmaier M, Tannock GW (2007) D-alanyl ester depletion of teichoic acids in *Lactobacillus reuteri* 100-23 results in impaired colonization of the mouse gastrointestinal tract. Environ Microbiol 9(7):1750–1760

Watanabe M, Houten SM, Mataki C, Christoffolete MA, Kim BW, Sato H, Messaddeq N, Harney JW, Ezaki O, Kodama T, Schoonjans K, Bianco AC, Auwerx J (2006) Bile acids induce energy expenditure by promoting intracellular thyroid hormone activation. Nature 439(7075):484–489

Weiss GM, Jespersen L (2010) Transcriptional analysis of genes associated with stress and adhesion in *Lactobacillus acidophilus* NCFM during the passage through an *in vitro* gastrointestinal tract model. J Mol Microbiol Biotechnol 18(4):206–214

Yan F, Polk DB (2002) Probiotic bacterium prevents cytokine-induced apoptosis in intestinal epithelial cells. J Biol Chem 277(52):50959–50965

Yan F, Cao H, Cover TL, Whitehead R, Washington MK, Polk DB (2007) Soluble proteins produced by probiotic bacteria regulate intestinal epithelial cell survival and growth. Gastroenterology 132(2):562–575

Yuan J, Wang B, Sun ZK, Bo X, Yuan X, He X, Zhao HQ, Du XY, Wang F, Jiang Z, Zhang L, Jia LL, Wang YF, Wei KH, Wang J, Zhang XM, Sun YS, Huang LY, Zeng M (2008) Analysis of host-inducing proteome changes in *Bifidobacterium longum* NCC2705 grown *in vivo*. J Proteome Res 7(1):375–385

Zamfir M, Brezeanu A, de Vuyst L (2007) Bactericidal effect of acidophilin 801, a bacteriocin produced by *Lactobacillus acidophilus* IBB 801. Roman Biotechnol Lett 12(6):3521–3531

Chapter 7
Application of Probiotic Proteomics in Enteric Cytoprotection

Hans Christian Beck, Søren Feddersen and Jørgen Petersen

7.1 Introduction

Proteomics is a systematic analysis of all proteins expressed by a genome and involves identification of proteins in living organisms and determination of their functions in a physiological context. Proteomics aims at quantifying the expression level of the complete protein complement (the proteome) in a cell at any given time. While proteomics research was initially focused on two-dimensional gel electrophoresis (2-DE) for protein separation and identification, it now refers to any procedure that characterizes the function of large sets of proteins. The recent developments of sensitive, high resolution, high speed scanning instruments and alternative dissociation methods such as electron transfer dissociation, have made mass spectrometry an indispensible tool in proteomics research and has allowed a more complete characterization of protein isoforms and post-translational modifications. Additionally, the introduction of stable isotope labeling techniques such as stable isotope labeling by amino acids in cell culture (SILAC) and isobaric tag for relative and absolute quantification (iTRAQ) has transformed mass spectrometry-based proteomics from a rather descriptive technique into a tool for the un-biased measurement of dynamic changes in protein expression, in specific post-translational modifications and in protein interactions. Even though this technology has become an indispensible tool in molecular biology research and despite the general acceptance that proteins are the major cause of probiotic treats such as adhesion and immune stimulation, proteomics has, however, only to a limited extent been applied in the field of probiotics. For example, an important criterion for the selection of probiotic strains is their ability to adhere to epithelial cells and the intestinal mucosa thereby enhancing the intestinal persistence of probiotics during the initial contact with the mucosa (Servin and Coconnier 2003). The adhesive properties form the basis for a transient colonization in the gastrointestinal tract making the probiotic bacteria able to exert positive health effects such as stimulation of the immune sys-

H. C. Beck (✉)
The Proteomics Group, Danish Technological Institute, Holbergsvej 10, 6000 Kolding, Denmark
e-mail: hcb@teknologisk.dk

J. J. Malago et al. (eds.), *Probiotic Bacteria and Enteric Infections*,
DOI 10.1007/978-94-007-0386-5_7,

tem (Oelschlaeger 2010), reduction of cholesterol levels (Naruszewicz et al. 2002), and lowering of increased blood pressure (Brunser et al. 2007). Furthermore, by adherence to epithelial cells, probiotic bacteria likely exclude pathogenic microorganisms from binding and thereby prevent translocation (Mangell et al. 2006). These features are important treats of probiotics and it is, although the molecular mechanisms are not fully understood, generally accepted that proteins located on the outer bacterial surface or secreted by the bacterium play an important role in these host-probiotic interactions. In fact, several proteins involved in adhesion and immune stimulation have been identified and characterized in probiotic strains (Granato et al. 2004, 1999; Rojas et al. 2002; Roos and Jonsson 2002; Bergonzelli et al. 2006). Furthermore, an increasing number of reports indicate that proteins known for their intracellular functions such as involvement in glycolysis are located on the outer surface and contribute to the adhesive properties of probiotic bacteria (Beck et al. 2009; Granato et al. 2004; Bergonzelli et al. 2006; Izquierdo et al. 2009; Kinoshita et al. 2008). Despite this only a few systematic proteomic studies have explored the potential of bacterial species and sub-species as probiotics. Here, we provide an update on "state-of-the-art" proteomics technologies applied in biological research and also on the proteomic methods applied in probiotics.

7.2 State-of-the-Art Proteomics Technologies

The proteome *per se* refers to the proteins expressed by a genome at a given time point. The information stored in the genome is more or less static while the proteome provides an overall view of the cell machinery at a given time point. By studying the proteome of a cell under various conditions information on its dynamics is obtained.

The recent developments in mass spectrometry and bioinformatics together with sample preparation methods have made mass spectrometry-based proteomics an indispensible tool in molecular biology research. Decades ago the proteome was studied by 2-DE, first separating proteins according to their pI followed by separation according to protein size. This made it possible to resolve hundreds of proteins and quantify changes in protein expression levels between samples, although identification of these proteins was hampered by the low throughput and the general limitations of the Edman protein sequencing technology. The development of mass spectrometry ionization techniques such as matrix-assisted laser desorption (MALDI) and electro spray ionization (ESI) in the 1990s allowed large scale protein identification and characterization and almost completely replaced Edman sequencing. The recent developments in nano-flow high performance liquid chromatography (nHPLC), high resolution fast scanning hybrid mass spectrometers and stable isotope labeling techniques such as iTRAQ and SILAC have enabled the identification and characterization of thousands of proteins in a single experiment. Thus, in today's proteomic laboratories it is now possible to robustly separate complex protein mixtures with high resolution, extract the proteins of interest and interrogate them with mass spectrometry, and then ultimately search protein databases using mass spectral data to identify proteins with high confidence.

7.3 Generic Proteomics Strategies

Samples for proteome analysis are prepared by a variety of techniques depending on nature and type of sample and the experiment to be performed. This is typically the first step of any proteomic experiment and aims to reduce the complexity of the sample (=enrich for relevant sub-proteome). Numerous strategies have been developed for sample preparation and include isolation of cell organelles such as nuclei or membranes prior to protein extraction or enrichment for proteins carrying specific post-translational modifications such as acetylations (Kocher and Superti-Furga 2007; Thingholm et al. 2009; Frisk et al. 1998; Cordwell and Thingholm 2010).

After sample preparation the proteins are prepared for nano-LC MSMS analysis by proteolytic digestion. Most often, trypsin that specifically cleaves proteins at the C-termini of arginine and lysine is applied in proteomics experiment. In case of discrete proteins isolated by 2-DE the peptides are often analysed by MALDI-TOF or MALDI TOF-TOF producing a peptide mass map and, in case of MALDI TOF-TOF analysis, is combined with amino acid sequences from a few selected peptides. The recent introduction of hybride analyser instruments such as quadrupole time of flight (q-TOF) and iontrap-orbitrap (LTQ orbitrap) instruments combined with novel fragmentation techniques such as electron transfer dissociation provide sensitive, high-resolution (more than 100,000), high mass-accuracy (<1 ppm), high dynamic range (>5,000) and fast scanning (Olsen et al. 2009). In combination with robust nHPLC systems these features allow the identification and quantification of thousands of proteins in a single experiment and have completely out-competed the 2-DE–MALDI TOF and TOF-TOF strategy in terms of workload, number of proteins identified and quantified, and post-translational modifications characterized in a single experiment.

7.4 Quantitative Proteomics

Prerequisite for quantitative proteomics is the incorporation of stable isotopes into the peptides to be measured. Even though several approaches exist the most popular methods used rely on the SILAC (Boschiroli et al. 2001) or the iTRAQ (Ross et al. 2004) methods (Fig. 7.1). Principally, in SILAC heavy isotope labelled amino acids (e.g. C-13 arginine) replace normal amino acids (C-12 arginine) in the cell cultures to be investigated. When cells are growing the heavy amino acids are incorporated into all proteins. Protein extracts from labelled and non-labelled cell cultures are then mixed in a 1:1 ratio, digested and analysed by mass spectrometry (Fig. 7.1). Upon digestion (e.g. with trypsin) all arginine-containing peptides from the heavy

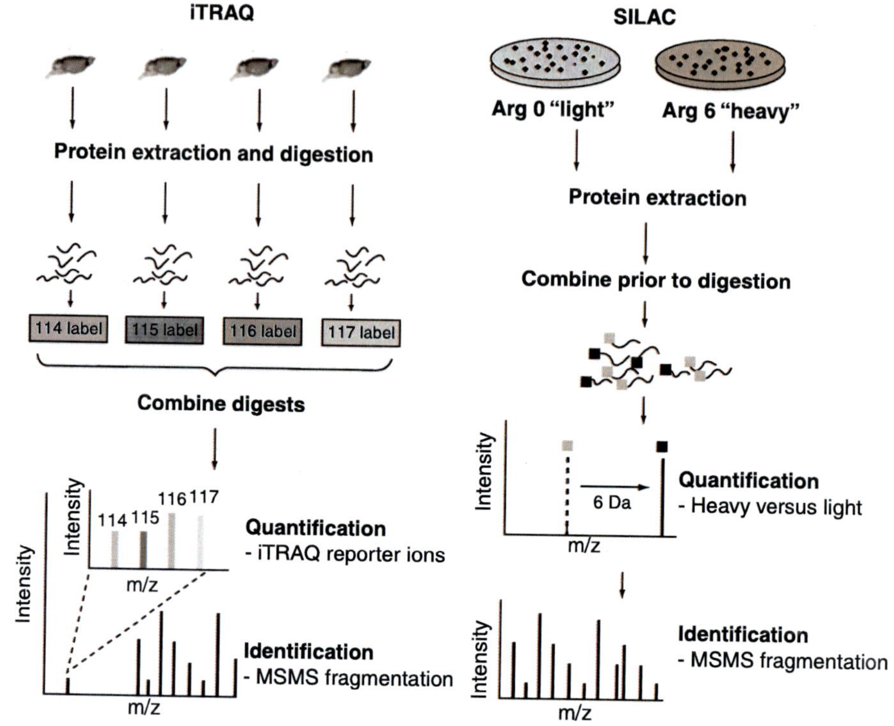

Fig. 7.1 Principle of the iTRAQ and SILAC labeling methods in proteomics

labelled cell culture will be 6 Da heavier than their chemical identical counterparts from the culture cultivated in parallel with normal amino acids. Thus, chemically identical peptides can be differentiated due to their mass differences and the relative intensities of the measured peptides reflect the abundance ratios for the two proteins.

Another "gold standard" labeling method in proteomics is the iTRAQ labeling method; this method is based on the covalent labeling of the N-terminal of peptides from tryptic digest with varying tags (Fig. 7.1). From two and up to eight different samples or samples extracted from cell cultures maintained at eight different conditions can be labeled. Samples are pooled in equal ratios, fractionated by nano liquid chromatography and analyzed by MSMS. The mixture of tagged peptides labeled with different tags cannot be distinguished by mass (m/z value), but upon fragmentation peptides with identical m/z values release reporter ions (fragment ions) of different masses enabling the simultaneous sequencing and relative quantification of labeled peptides.

Other more rarely used stable isotope labeling techniques in proteomics include the ^{16}O–^{18}O labeling strategy (Bantscheff et al. 2004) and dimethyl labeling (Hsu et al. 2003); whereas dimethyl labeling is based on the reductive amination of the N-terminus and ε-amino groups of lysine the ^{16}O–^{18}O labeling strategy relies on the enzymatic incorporation of ^{18}O into the C-terminus of the resulting peptides upon proteolytic cleavage by serine proteases such as trypsin.

7.5 Proteomics Methods Applied in Probiotics

The interaction between probiotic bacteria and the human intestine is mediated by proteins exposed on the outer surface of the bacterium. Therefore these proteins are of outmost importance when evaluating possible probiotic properties of a certain bacteria. Surface proteins can be divided into covalently and non-covalently bound proteins. Covalent bound proteins include lipoproteins that are bound to lipids of the cytoplasmatic membrane via an N-terminal lipobox sequence (LXXC) (Sutcliffe and Harrington 2002) and also proteins with a C-termal PPXTG motif that gets covalently attached to the S-layer (peptidoglycan) by enzymes that recognize, cleave and attach the protein to the S-layer (Navarre and Schneewind 1999). Non-covalently bound surface proteins include transmembrane proteins, proteins with specific C-terminal repeats or Glycine-tryptophan (GW) motifs that attach to the lipoteichoic or teichoic acids that are part of the cell wall matrix and S-layer proteins that binds to lipoteichoic acids or polysaccharides via a specific SL domain (Bath et al. 2005). Due to the complexity of the membrane and the different mechanism by which proteins associate with the membrane proteomic analysis of this part of the cell is particular challenging.

In the literature, a number of proteomics strategies have been described for the characterization of bacteria including lactic acid bacteria used for the production and fermentation of food products. Only minor part of these studies, however, includes the study of biologically active proteins secreted into the surroundings or displayed on the outer bacterial surface from probiotic strains. Briefly, these studies include a PBS shaving of loosely bound proteins from the outer surface of the bacterial cell (Beck et al. 2009), isolation of the S-layer proteins using lysozyme treatment and LiCl extraction (Izquierdo et al. 2009), and a proteolytic digestion of intact cells using a "shave and conqueror" strategy.

Beck et al. (2009) used a gentle PBS wash of whole cells of the probiotic *Lactobacillus plantarum* 299v strain to isolate the loosely bound surface-associated proteins of this bacterium. A protein solubilizing agent was added in order to increase the yield of this procedure. This method is very robust but the major drawback of this method is that it only targets the non-covalently bound proteins that are attached to the bacterial outer surface. The extracted proteins were then separated by SDS-PAGE and protein bands were excised, trypsinated and analysed by capillary LC-MSMS. The identified proteins are summarized in Table 7.1.

Another method was used by Izpuierdo et al. that analysed the adhesive proteome of probiotic lactobacillus strains by comparing the expression of the outer surface proteome for a panel of probiotic lactobacillus strains with varying adhesion properties (Izquierdo et al. 2009). Cells of three different strains of *lactobacilli* were treated with lysozyme in order to release proteins associated with the outer cell sur-

Table 7.1 Summary of proteins associated to the outer cell surface of probiotic bacteria and identified by proteomics technologies

Functional category	Putative function	Accession number[a]	Study[b]	Probiotic function
Cell envelope biogenesis	Muramidase	gi|28379158	A, C	?
	Extracellular protein, putative gamma-D-glutamate-meso-diaminopimelate muropeptidase	gi|28378774	A, C	?
	Extracellular protein—homologeous to aggregation promoting factor	gi|28379317	A, C	?
	Extracellular protein—homologeous to putative gamma-D-glutamate-meso-diaminopimelate muropeptidase	gi|28379752	A, C	?
	Extracellular protein (putative chitin-binding protein)	gi|28271221	A	?
	Extracellular protein—homologeous to aggregation promoting factor-like surface protein	gi|28270057	A, C	Adhesion
Central metabolism	Phosphopyruvate hydratase/enolase	gi|28377645	A	Adhesion
	Glucose-6-phosphate isomerase	gi|28379049	A, C	?
	Phosphoglycerate kinase	gi|28377643	A, C	Adhesion
	Glyceraldehyde 3-phosphate dehydrogenase	gi|28377642	A, B, C	Adhesion
	Fructose-bisphosphate aldolase	gi|28377250	A	?
	Triosephosphate isomerase	gi|28377644	A, C	Adhesion, Competitive exclusion
	D-lactate dehydrogenase	gi|28378688	B	Adhesion
	Glutathione reductase	gi|28379636	B	Adhesion
	6-Phosphogluconate dehydrogenase	gi|28378252	B	Adhesion
	Citrate lyase, beta chain	gi|28377896	B	Adhesion
	Conjugated bile acid hydrolase	gi|149533	B	Adhesion
	Cystein aminopeptidase	gi|2837779	B	Adhesion
	Probable phosphoketolase 1	gi|28379168	B	Adhesion
	Phosphoglycerate mutase	gi|28379560	A, C	?
Protein folding	GroEL	gi|28377591	A, B, C	Adhesion, immuno-modulation
	GroES co-chaperonine gi|28377590	A, B	?	
Stress response	Heat shock protein DnaK	gi|28378660	A, B, C	Adhesion, immuno-modulation
	Heat shock protein GrpE	gi|28378661	A	?
	Cold shock protein CspC	gi|28377809	B	Adhesion

Table 7.1 (continued)

Functional category	Putative function	Accession number[a]	Study[b]	Probiotic function
	Stress induced DNA binding protein	gi|28379536	A, B, C	Adhesion
Transcription and translation	Elongation factor Tu	gi|28378740	A	Adhesion, Competitive exclusion
	Elongation factor TS	gi|28378685	A	?
	Trigger factor	gi|28378739	A, B, C	Adhesion
	Ribosomal protein L6	gi|28377847	A	?
	Ribosomal protein L12/L7	gi|28377498	A, C	Immunomodulation
	DNA-directed DNA polymerase III, beta chain	gi|28376976	B	?
	DNA-binding protein II	gi|28378532	A	?
	Ribosomal protein L30	gi|28377850	A	?
	Glyceryl-tRNA synthetase alpha chain	gi28378608	B	Adhesion
	Bifunctional GMP synthetase/ glutamine amidotransferase protein	gi28377744	B	Adhesion
	Seryl-tRNA synthetase 2	gi|28377821	B	Adhesion
	Glycyl-tRNA synthetase alpha chain	gi|28378608	B	Adhesion
	Ribosomal protein L29	gi|28377840	A	?
Unknown	Hypothetical protein lp_2253—homologeous to DUF28 Superfamily	gi|28378848	A	?
	Hypothetical protein lp_1872	gi|28378527	B	Adhesion
	Hypothetical protein lp_1708	gi|28378396	A, C	?
	Hypothetical protein lp_2809	gi|28379289	A	?
	Extracellular protein—homologeous to hypothetical protein LVIS_0893	gi|28378760	A	?
	S-ribosylhomocysteinase	gi|28377630	B	Adhesion

[a] Accession numbers from the nrNCBI database for *L. plantarum* WCFS1

[b] References: A: (Beck et al. 2009), B: (Izquierdo et al. 2009), C: (Beck and Petersen 2009)

face. The resulting cell wall extracts were subjected to differential proteome analysis using 2-DE, tryptic digestion and chip-LC-q-TOF MSMS analysis. The analysed protein spots were identified using database search. Of the analysed lactobacillus strains 21 differentially expressed spots (Table 7.1) were identified that correlated with the adhesive properties of the studied strains. The differential analysis was performed using image analysis.

As stated above, the investigation of the outer cell surface-associated proteins of microorganisms such as probiotics posses several challenges. The traditional methods for the isolation of membrane or surface-associated proteins target specific groups of proteins such as peptidoglycan-bound proteins, loosely bound proteins or

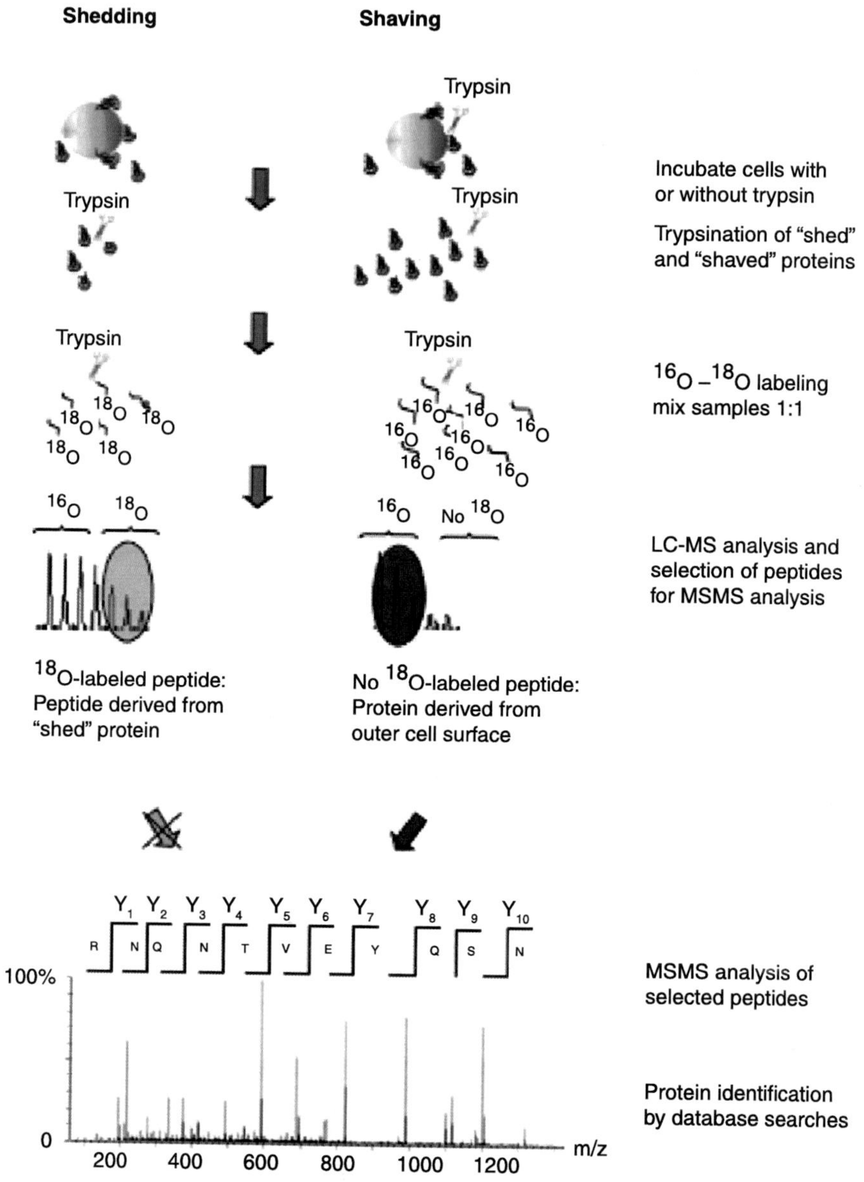

Fig. 7.2 A proteomic strategy for the identification of outer surface-associated proteome of probiotics

hydrophobic proteins that are embedded in the plasma membrane. The overall challenge is to isolate the membrane or surface-associated proteins without contaminating the sample with unwanted proteins, i.e. avoiding the contamination of "true" extracellular proteins with "true" intracellular proteins due to cell lysis caused by enzymatic activity or harsh conditions like 5 M LiCl treatment. In order to circumvent this challenge we introduced a proteomic method based on proteolytic shaving of the outer cell surface followed by stable isotope labeling and LC-MSMS analysis (Fig. 7.2) (Beck and Petersen 2009).

The cell culture is split into two and incubated with or without trypsin for a very short period of time in order to minimize contamination caused by cell lysis. After "shaving" the outer cell surface with a protease such as trypsin the cells are removed from both samples by centrifugation. The supernatants containing either "background" protein released by shedding and cell lysis or proteins released from the outer surface by proteolytic shaving were supplemented with trypsin and incubated over night for complete proteolytic digestion. The resulting peptides were then isotopically labeled using the ^{16}O–^{18}O labeling strategy (Bantscheff et al. 2004), the two samples were then mixed and subsequently analysed by nHPLC-MS. Tryptic peptides from lysis- or shed-derived proteins can then be discriminated from "true" extracellular proteins via their relative intensities in the LC-MS analysis. Ideally, the peptides from lysis-derived proteins will display a 1:1 ratio whereas peptides derived from true extracellular proteins will display relative intensities that deviate from a 1:1 ratio. An outline of this strategy is presented in Fig. 7.2.

7.6 The Outer Surface Proteome of Probiotic Bacteria

Usually, the *in vitro* tests important for the selection of probiotics are related to the investigation of the survival, growth, and adhesion of the probiotic bacterium and are carried out in conditions simulating physiological conditions, such as gastric, pancreatic and intestinal environments. Other health-promoting features of probiotics are very difficult to assess by *in vitro* tests. These include anti-inflammatory (Baroja et al. 2007), anti-carcinogenic (Commane et al. 2005), and cholesterol-lowering (Larkin et al. 2009) effects, and also the prevention of allergies (Prescott and Bjorksten 2007). The evaluation of these effects most often requires *in vivo* studies and cannot directly be assessed by proteomics or any other analytical *in vitro* technique. Thus, at this stage functional proteomics has been limited to a single study that analyse the "adhesive" proteome of the outer bacterial surface. In one study the outer surface proteome of three *Lactobacillus* strains; *L. platarum* WHE 92, *L. plantarum* 299v, and *L. plantarum* CECT 4185 were analysed in order to identify protein profiles involved in adhesion using mucin as a model probiotic property (Izquierdo et al. 2009). Protein extracts from the selected strains were analyzed by 2D PAGE and the differentially expressed proteins across the strains with different binding properties were used as a measure for proteins involved in adhesion. A total of 21 proteins were found to be over expressed in the cell wall proteome of the highly adhesive *L. plantarum* WHE 92 strain, indicating their involvement in the adhesion process (Table 7.1).

Six of these proteins overlapped the findings in two other proteomic studies assessing the outer surface proteome of the probiotic bacteria *L. plantarum* 299v and *L. plantarum* WCFS1 (Beck et al. 2009; Beck and Petersen 2009). Among these the elongation factor EF-Tu, GroES, GroEL chaperonin, the molecular chaperone DnaK and also glyceraldehydes-3-phosphate dehydrogenase (GAP-DH) are proteins that previously have been associated with adhesion properties (Table 7.1).

The most abundant protein present on the outer surface in these studies was the glycolytic enzyme GAPDH. This protein was also highly expressed in the cell wall proteome of several other lactobacillus strains (Kinoshita et al. 2008; Antikainen et al. 2007; Madsen et al. 2005) and it was found that its association with the outer bacterial surface was pH-dependent as a pH close to the pI of the protein favoured the association to the outer cell surface. This pH-dependent specific binding to Caco-2 (derived from a human colorectal carcinoma) cell surfaces was demonstrated not only for GAPDH, but also for Enolase (Eno), and phosphoglycerate kinase (PGK). Caco-2 cells express several markers distinct of normal small intestinal villi and has proven excellent as *in vitro* model for probiotic adhesion and pathogenic invasion (Granato et al. 2004; Lehto and Salminen 1997).

Another aspect that has to be considered is how GAPDH and also Eno, and PGK are attached to the outer surface of the bacterial cell. Cloning of these enzymes revealed no binding motifs such as the C-terminal LPXTG motif that is typical for covalently attached proteins. Nor did the genes corresponding to the identified proteins encode signal sequences that could explain their extracellular location (Madsen et al. 2005). In a recent study Antikeinen et al., demonstrated that GAPDH and Eno is anchored on the outer cell surface of *Lb. crispatus* via the interaction with lipoteichoic acid (LTA) (Antikainen et al. 2007). This binding was due to the interaction of the negatively charged carboxylic acid of LTA with the positively charged amino groups of GAPDH and Eno at low pH. pI analyses of GAPDH, Eno, and PGK revealed pI's of 5.30, 4.90, and 5.04, respectively. In fact, these enzymes associate to the outer cell surface at pH 5 but are released into the growth medium at neutral and slightly alkaline pH. This demonstrates that the process of adhesion to the host is a pH-dependent process involving several proteinaceus compounds.

The Protein EF-Tu was identified as a component of the outer surface of *Lactobacillus johnsonii* NCC 533 (La1) where it mediates a pH-dependent binding to epithelial cells and human gut mucin (Granato et al. 2004). Similar findings were also done with the heat shock protein GroEL from the same *Lactobacillus* strain, that also exhibited a pH-dependent binding to epithelial cells with a more efficient binding at acidic pH (Bergonzelli et al. 2006). DnaK was also identified from *M. tuberculosis* where it was demonstrated to bind to and activate human plasminogen to generate plasmin and is known as an immuno-dominant antigen for both humoral and cell-mediated immune response in humans and infected animals (Bartow and McMurray 1997) indicating a role of this protein in the immunomodulatory properties of *L. plantarum* 299v (Schultz et al. 2002; Herias et al. 1999).

7.7 Clinical Applications of Probiotics

Hitherto, no single protein has been applied for the cure and protection of various gastrointestinal disorders but a variety of potential application of "whole cell" therapy has emerged. These include probiotic-based therapies against inflammatory bowel disease (IBD), irritable bowel syndrome (IBS), gluten intolerance, gastroenteritis, *Heliobacter pylori* infections, and colon cancer (Yan and Polk 2010). Mechanisms of actions of probiotics include properties such as increased enzyme production, maintenance of the host microbial balance in the gastrointestinal tract (GIT) via the production of substances such as proteins that compete with pathogens for adherence to the GIT thereby promoting intestinal epithelial cell survival, and regulating immune responses by enhancing the innate immunity and preventing pathogen-induced inflammation reactions. In fact, these mechanisms of actions are most probably mediated by proteins and other molecules derived from probiotic microorganisms and they act via the stimulation and regulation of host signalling pathways such as the MAPK (mitogen-activated protein kinase), NF-κB (nuclear factor kappa B), and PI3K (phosphatidylinositol-3′-kinase) signalling pathways (Vanderpool et al. 2008).

For example, the intestinal electrolyte absorption is regulated through the stimulation of the apical chloride/hydroxyl exchange activity by a PI3K-dependent mechanism mediated by outer cell surface-located and/or soluble molecules secreted by the probiotic bacteria *Lactobacillus acidophilus* (Borthakur et al. 2008). In another study it was shown that surface proteins from *L. casei* prevented inflammation through the regulation of the secretion of several factors from the intestinal epithelial cells. In addition outer surface-associated proteins from *L. casei* inhibited TNF-induced T-cell chemokine interferon (IFN)-inducible (IP-10) secretion by inhibiting vesicular pathways, followed by degradation of this pro-inflammatory chemokine (Hoermannsperger et al. 2009).

Other examples of the prevention of cytokine and chemical-induced epithelial apoptosis and disruption of barrier function include the protective roles in cytokine-induced intestinal epithelial cell apoptosis of two novel proteins secreted by *Lactobacillus rhamnosus* GG. The proteins, p40 and p75, act via the activation of the anti-apoptotic PI3K/Akt signalling pathways *in vitro* and in colon organ culture (Yan et al. 2007). More examples of probiotic effects of secreted or outer surface-associated proteins include the suppression of NF-κB signalling to limit excessive inflammation by soluble factors released by probiotic bacteria such as *Bifidobacterium* and *Lactobacillus* ssp and others (Spehlmann and Eckmann 2009; Heuvelin et al. 2009; Petrof et al. 2009).

7.8 Conclusions

Despite the fact that mass spectrometry-based proteomics has become an indispensable tool in molecular biology research this technology has only to a very limited extent been applied in the identification and characterization of proteins involved in probiotic effects such as adhesion and immunomodulation. Three proteomic studies exploiting the outer surface-associated proteome from five probiotic lactobacillus strains using different experimental procedures identified six common proteins that in previous studies were shown to be involved in the adhesion to the host. These results demonstrate that proteomics has a potential as a screening technology in the search for new probiotic strains with specific properties. Thus, future research should focus on the correlation of the outer-surface-associated proteome of probiotic micro organisms with specific probiotic treats and also on the development of mass spectrometry-based proteomic methods for the high-throughput screening of potential probiotic strains.

References

Antikainen J, Kuparinen V, Lahteenmaki K, Korhonen TK (2007) pH-dependent association of enolase and glyceraldehyde-3-phosphate dehydrogenase of *Lactobacillus crispatus* with the cell wall and lipoteichoic acids. J Bacteriol 189:4539–4543

Bantscheff M, Dumpelfeld B, Kuster B (2004) Femtomol sensitivity post-digest (18)O labeling for relative quantification of differential protein complex composition. Rapid Commun Mass Spectrom 18:869–876

Bartow RA, Mcmurray DN (1997) Cellular and humoral immune responses to mycobacterial stress proteins in experimental pulmonary tuberculosis. Tuber Lung Dis 78:185–193

Bath K, Roos S, Wall T, Jonsson H (2005) The cell surface of *Lactobacillus reuteri* ATCC 55730 highlighted by identification of 126 extracellular proteins from the genome sequence. FEMS Microbiol Lett 253:75–82

Beck HC, Petersen J (2009) A proteomic strategy for the analysis of the outer cell surface proteome of probiotic bacteria. 18th International mass spectrometry conference, Bremen, Germany, 30 Aug–04 Sept. Poster presentation

Beck HC, Madsen SM, Glenting J, Petersen J, Israelsen H, Norrelykke MR, Antonsson M, Hansen AM (2009) Proteomic analysis of cell surface-associated proteins from probiotic *Lactobacillus plantarum*. FEMS Microbiol Lett 297:61–66

Bergonzelli GE, Granato D, Pridmore RD, Marvin-Guy LF, Donnicola D, Corthesy-Theulaz IE (2006) GroEL of *Lactobacillus johnsonii* La1 (NCC 533) is cell surface associated: potential role in interactions with the host and the gastric pathogen *Helicobacter pylori*. Infect Immun 74:425–434

Borthakur A, Gill RK, Tyagi S, Koutsouris A, Alrefai WA, Hecht GA, Ramaswamy K, Dudeja PK (2008) The probiotic *Lactobacillus acidophilus* stimulates chloride/hydroxyl exchange activity in human intestinal epithelial cells. J Nutr 138:1355–1359

Boschiroli ML, Foulongne, V, O'callaghan D (2001) Brucellosis: a worldwide zoonosis. Curr Opin Microbiol 4:58–64

Brunser O, Gotteland M, Cruchet S (2007) Functional fermented milk products. Nestle Nutr Workshop Ser Pediatr Progr 60:235–247; discussion 247–250

Commane D, Hughes R, Shortt C, Rowland I (2005) The potential mechanisms involved in the anti-carcinogenic action of probiotics. Mutat Res 591:276–289

Cordwell SJ, Thingholm TE (2010) Technologies for plasma membrane proteomics. Proteomics 10:611–627

Frisk A, Ison CA, Lagergard T (1998) GroEL heat shock protein of *Haemophilus ducreyi*: association with cell surface and capacity to bind to eukaryotic cells. Infect Immun 66:1252–1257

Granato D, Perotti F, Masserey I, Rouvet M, Golliard M, Servin A, Brassart D (1999) Cell surface-associated lipoteichoic acid acts as an adhesion factor for attachment of *Lacto-*

bacillus johnsonii La1 to human enterocyte-like Caco-2 cells. Appl Environ Microbiol 65:1071–1077

Granato D, Bergonzelli GE, Pridmore RD, Marvin L, Rouvet M, Corthesy-Theulaz IE (2004) Cell surface-associated elongation factor Tu mediates the attachment of *Lactobacillus johnsonii* NCC533 (La1) to human intestinal cells and mucins. Infect Immun 72:2160–2169

Herias MV, Hessle C, Telemo E, Midtvedt T, Hanson LA, Wold AE (1999) Immunomodulatory effects of *Lactobacillus plantarum* colonizing the intestine of gnotobiotic rats. Clin Exp Immunol 116:283–290

Heuvelin E, Lebreton C, Grangette C, Pot B, Cerf-Bensussan N, Heyman M (2009) Mechanisms involved in alleviation of intestinal inflammation by *Bifidobacterium breve* soluble factors. PLoS One 4:e5184

Hoermannsperger G, Clavel T, Hoffmann M, Reiff C, Kelly D, Loh G, Blaut M, Holzlwimmer G, Laschinger M, Haller D (2009) Post-translational inhibition of IP-10 secretion in IEC by probiotic bacteria: impact on chronic inflammation. PLoS One 4:e4365

Hsu JL, Huang SY, Chow NH, Chen SH (2003) Stable-isotope dimethyl labeling for quantitative proteomics. Anal Chem 75:6843–6852

Izquierdo E, Horvatovich P, Marchioni E, Aoude-Werner D, Sanz Y, Ennahar S (2009) 2-DE and MS analysis of key proteins in the adhesion of Lactobacillus plantarum, a first step toward early selection of probiotics based on bacterial biomarkers. Electrophoresis 30:949–956

Kinoshita H, Uchida H, Kawai Y, Kawasaki T, Wakahara N, Matsuo H, Watanabe M, Kitazawa H, Ohnuma S, Miura K, Horii A, Saito T (2008) Cell surface *Lactobacillus plantarum* LA 318 glyceraldehyde-3-phosphate dehydrogenase (GAPDH) adheres to human colonic mucin. J Appl Microbiol 104:1667–1674

Kocher T, Superti-Furga G (2007) Mass spectrometry-based functional proteomics: from molecular machines to protein networks. Nat Methods 4:807–815

Larkin TA, Astheimer LB, Price WE (2009) Dietary combination of soy with a probiotic or prebiotic food significantly reduces total and LDL cholesterol in mildly hypercholesterolaemic subjects. Eur J Clin Nutr 63:238–245

Lehto EM, Salminen SJ (1997) Inhibition of *Salmonella typhimurium* adhesion to Caco-2 cell cultures by Lactobacillus strain GG spent culture supernate: only a pH effect? FEMS Immunol Med Microbiol 18:125–132

Baroja ML, Kirjavainen PV, Hekmat S, Reid G (2007) Anti-inflammatory effects of probiotic yogurt in inflammatory bowel disease patients. Clin Exp Immunol 149:470–479

Madsen SM, Glenting J, Vrang A, Ravn P, Riemann HK, Israelsen H, Norrelykke MR, Hansen AM, Antonsson M, Ahrne S, Beck HC (2005) Cell surface-associated glycolytic enzymes from Lactobacillus plantarum 299v mediate adhesion to human epithelial cells and extracellular matrix proteins. J Biotech 118:S142–S142

Mangell P, Lennernas P, Wang M, Olsson C, Ahrne S, Molin G, Thorlacius H, Jeppsson B (2006) Adhesive capability of *Lactobacillus plantarum* 299v is important for preventing bacterial translocation in endotoxemic rats. Apmis 114:611–618

Naruszewicz M, Johansson ML, Zapolska-Downar D, Bukowska H (2002) Effect of *Lactobacillus plantarum* 299v on cardiovascular disease risk factors in smokers. Am J Clin Nutr 76:1249–1255

Navarre WW, Schneewind O (1999) Surface proteins of gram-positive bacteria and mechanisms of their targeting to the cell wall envelope. Microbiol Mol Biol Rev 63:174–229

Oelschlaeger TA (2010) Mechanisms of probiotic actions—a review. Int J Med Microbiol 300(1):57–62

Olsen JV, Schwartz JC, Griep-Raming J, Nielsen ML, Damoc E, Denisov E, Lange O, Remes P, Taylor D, Splendore M, Wouters ER, Senko M, Makarov A, Mann M, Horning S (2009) A dual pressure linear ion trap Orbitrap instrument with very high sequencing speed. Mol Cell Proteomics 8:2759–2769

Petrof EO, Claud EC, Sun J, Abramova T, Guo Y, Waypa TS, He SM, Nakagawa Y, Chang EB (2009) Bacteria-free solution derived from *Lactobacillus plantarum* inhibits multiple NF-kappaB pathways and inhibits proteasome function. Inflamm Bowel Dis 15:1537–1547

Prescott SL, Bjorksten B (2007) Probiotics for the prevention or treatment of allergic diseases. J Allergy Clin Immunol 120:255–262
Rojas M, Ascencio F, Conway PL (2002) Purification and characterization of a surface protein from *Lactobacillus fermentum* 104R that binds to porcine small intestinal mucus and gastric mucin. Appl Environ Microbiol 68:2330–2336
Roos S, Jonsson H (2002) A high-molecular-mass cell-surface protein from *Lactobacillus reuteri* 1063 adheres to mucus components. Microbiology 148:433–442
Ross PL, Huang YN, Marchese JN, Williamson B, Parker K, Hattan S, Khainovski N, Pillai S, Dey S, Daniels S, Purkayastha S, Juhasz P, Martin S, Bartlet-Jones M, He F, Jacobson A, Pappin DJ (2004) Multiplexed protein quantitation in *Saccharomyces cerevisiae* using amine-reactive isobaric tagging reagents. Mol Cell Proteomics 3:1154–1169
Schultz M, Veltkamp C, Dieleman LA, Grenther WB, Wyrick PB, Tonkonogy SL, Sartor RB (2002) *Lactobacillus plantarum* 299V in the treatment and prevention of spontaneous colitis in interleukin-10-deficient mice. Inflamm Bowel Dis 8:71–80
Servin AL, Coconnier M.-H. (2003) Adhesion of probiotic strains to the intestinal mucosa and interaction with pathogens. Best Pract Res Clinl Gastroenterol 17:741–754
Spehlmann ME, Eckmann L (2009) Nuclear factor-kappa B in intestinal protection and destruction. Curr Opin Gastroenterol 25:92–99
Sutcliffe IC, Harrington DJ (2002) Pattern searches for the identification of putative lipoprotein genes in Gram-positive bacterial genomes. Microbiology 148:2065–2077
Thingholm TE, Jensen ON, Larsen MR (2009) Analytical strategies for phosphoproteomics. Proteomics 9:1451–1468
Vanderpool C, Yan F, Polk DB (2008) Mechanisms of probiotic action: implications for therapeutic applications in inflammatory bowel diseases. Inflamm Bowel Dis 14:1585–1596
Yan F, Polk DB (2010) Probiotics: progress toward novel therapies for intestinal diseases. Curr Opin Gastroenterol 26:95–101
Yan F, Cao H, Cover TL, Whitehead R, Washington MK, Polk DB (2007) Soluble proteins produced by probiotic bacteria regulate intestinal epithelial cell survival and growth. Gastroenterology 132:562–575

Chapter 8
Promoting Gut Health with Probiotic Metabolomics

Sebastiano Collino, François-Pierre J. Martin, Sunil Kochhar and Serge Rezzi

8.1 Introduction

Modern nutrition research has promoted the use of metabolomics techniques to gain a deeper understanding of the interactions between nutrition, health, and physiological processes (Rezzi et al. 2007). Metabolomics is a well established analytical approach for the analysis of physiological regulatory processes via the quantitative measurement of dynamic metabolic changes (metabolic profiling) of living systems in response to genetic modifications or physiological stimuli, including diet and nutrients. Nowadays, metabolomic applications have evolved towards deciphering the cellular and molecular processes in response to different individual dietary modulations, predicting health and disease outcomes. For example, it recently showed the full versatility of this science as a diagnostic tool during active and quiescent ulcerative colitis (UC), providing new insights into patho-physiological mechanisms (Bjerrum et al. 2009). Indeed, by the global study of low molecular weight compounds (<1,500 Da) in biofluids (mostly plasma/serum and urine) and tissues, metabolomics assures the characterization of individual metabolic phenotypes, or metabotypes (Rezzi et al. 2008; Nicholson et al. 1984). Metabolomics employs mainly two analytical techniques based on ^{1}H nuclear magnetic resonance (NMR) spectroscopy and mass spectrometry coupled to gas or liquid chromatography (GC/MS and LC/MS). NMR spectroscopy offers the unique prospect to holistically profile hundreds of metabolites with no prior selection with minimal sample preparation. Interestingly, this technique is not only used for the profiling of biological fluids (liquid state NMR), but it is also today commonly employed for the study of metabolic profiling of intact tissue biopsies, using High Resolution Magic Angle Spinning NMR (HR-MAS). MS methods are also commonly employed for global and targeted profiling and are inherently more sensitive, but require a more comprehensive sample preparation (Bruce et al. 2009) with separation of the metabolite

S. Rezzi (✉)
BioAnalytical Science, Metabolomics & Biomarkers, Nestlé Research Center,
P.O. Box 44, 1000 Lausanne 26, Switzerland
e-mail: serge rezzi@rdls nestle.com

J. J. Malago et al. (eds.), *Probiotic Bacteria and Enteric Infections*,
DOI 10.1007/978-94-007-0386-5_8,

components using either gas chromatography (GC) or liquid chromatography (LC). The acquired spectral profiles reflect the metabolic state of an individual that is expressed as a result of genes, environment, lifestyle, and diet. Indeed, the obtained metabolic profiles encapsulate information on the metabolic activity of the gut microbiota that co-evolved with complex organisms, and which represent a major determinant in nutrition and health (Martin et al. 2009b; Rezzi et al. 2007; Nicholson and Lindon 2008; Nicholson et al. 2005). Both methods are then comprehensively used to generate high density data, from which meaningful biological information are recovered using advanced statistical tools (Montoliu et al. 2009; Eriksson et al. 2004).

8.2 The Role of Gut Microbes in Determining Gastrointestinal Functional Ecology and Influencing Mammalian Metabolic Phenotypes

The human body contains *circa* 10 trillion parenchymal cells, but the gut contains approximately 100 trillion commensal, symbiotic and pathogenic organisms, with about 17 families, 50 genera, 400–500 different species and an undefined number of subspecies, which vary in community composition through life according to lifestyle and nutrition (Xu et al. 2007; Beale 2002; Tannock 2005). The gut microbiota consortium also communicates with surrounding tissues to stimulate development of a host environment that fosters implantation of the microbiota and contribute to host nutrition by generating simplified carbohydrates, short chain fatty acids (SCFA), amino acids, and vitamins. With the recent Nobel Prize awarded to B. Marshall (Marshall 2003) and J.R. Warren (Warren 2000) for their discovery of the bacterium *Helicobacter pylori* and its role in gastritis and peptic ulcer disease, an appropriate consideration has been given to individual human gut microbial activities and their roles in the future of personalized health-care. Their work highlighted the potential of *H. pylori* infection to trigger an ulcer by stimulating increased acid production in the stomach, leading to damage to the stomach and/or intestinal lining. There is a growing awareness of the importance of the variation in the gut microbiota as a factor that influences human and animal health, with major mammalian metabolic processes and inter-organ cross-talks being strongly connected to the microbial activity, which in turn may have long-term health consequences to the host. Recent applications of top-down system biology approach revealed the depth and width of the long range effects of the gut microbiota modulation in complex organisms, resulting in modulation of host lipid, carbohydrate and amino acid metabolism at a panorganismal scale (Martin et al. 2009d; Nicholson et al. 2005). Wikoff et al. (2009) have also shown the influence of the microbiome and specific bacteria on mammalian plasma biochemistry. Their application of MS-based metabolic approach revealed significant differences among circulating molecules in blood plasma from germ-free mice and conventional mice. Their work provided additional evidence that the specific metabolic activities of a single gut bacterial species can provide the host with new biochemical compounds in sufficient amount to be detected in the general blood circulation.

Martin et al. exemplified how the gut microbial modulation of the gastrointestinal system and extensive microbial-mammalian co-metabolism may fine-tune host metabolic processes and may induce metabolic deregulations (Martin et al. 2007a). The authors inoculated germ free mice with a model of simplified human microbiota (HBM) or made them acquire a normal mouse microbiome by exposing them to conventional mice. The authors described profound metabolic differences between normal and "re-conventionalised" mice on the one hand and HBM mice on the other hand, at both systemic and tissue-specific metabolic changes across a wide range of metabolic pathways and metabolite classes. In particular, the authors highlighted the different bacterial modulation of the bile acid metabolism and enterohepatic cycle, with consequent effects on the absorption of dietary fat and concomitant to lipid accumulation in the livers of HBM animals. In this case, gut bacteria exert modulation over the host metabolism via reprocessing of signalling molecules, i.e. bile acids. As such, bile acids may be an example of transgenomic mechanism of quorum sensing (Nicholson et al. 2005) whereby microbial cells communicate with each other and disperse their metabolic functions thus behaving like a multi-cellular organism (Martin et al. 2007a).

Competition between microbes and their mammalian hosts has been co-evolving for millions of years, and the ability of commensal bacteria to protect the host from microbial-induced disease is well established (Reid 2001). Mammalian-microbial symbiosis play a strong role in the metabolism of exogenous and endogenous compounds having a direct influence in the aetiology of several diseases such as Crohn's disease (Gupta et al. 2000), irritable bowel syndrome (IBS) (Martin et al. 2009c; Bergonzelli et al. 2005), obesity (Ley et al. 2006), cardiovascular disease (Pereira and Gibson 2002), and gastrointestinal cancers (Dunne 2001). Moreover, the host exerts also controls over the microbiome community structure, a process that has evolved to attain specific benefits ranging from protection, nutrition to physiology (Nicholson et al. 2005). Indeed, the gut microbial community contains multiple cell types providing an extended genome interacting with a number of important mammalian metabolic regulatory functions. For instance, gut microbes perform multiple digestive and metabolic functions for the host, whilst interacting extensively with the host through metabolic exchange and co-metabolism of substrates. Such interactions between metabolomes, i.e. the complete sets of small-molecule metabolites to be found within a single organism, are poorly understood, but are important in understanding the development and expression of many disease processes such as ulcerative colitis, or pouchitis (Sartor 2004).

The interactions between a mammalian host and its enteric microbiota have been studied for decades by applying a large range of biomolecular techniques in experimental animals (Tannock 2005; Hooper and Gordon 2001). The current metabolomic revolution offers an unprecedented opportunity to identify the molecular foundations of these relationships so that we can understand how they contribute to our normal physiology and how they can be exploited to develop new therapeutic strategies. In the past, conventional techniques have demonstrated that the complex microbial community differs in composition along the length (with an increasing gradient of indigenous microbes from the stomach to the colon) and across the diameter of the gastrointestinal tract, and is comprised of both rapidly transiting and relatively persistent components (Dunne 2001). The gut symbionts communicate with surrounding tissues to shape a host environment that fosters their implantation and persistence (Hooper et al. 2002). In addition, the modulation of host-cell pathways by bacteria is reported as a determinant factor in the aetiology of disease and onset of homeostasis deregulation (Bhavsar et al. 2007). ^{1}H HR-MAS NMR spectroscopy was recently used to characterize the biochemical components of intact epithelial tissues from the different gut compartments along the human gastrointestinal tract (GIT) including the antrum, duodenum, jejunum, ileum, and transverse colon (Wang et al. 2007). Region-specific metabolic variations were observed relating to different physiological properties in each intestinal tissue, including functional variations in energy metabolism, osmoregulation, gut microbial activity, and oxidative protection. Furthermore, as assessed by Martin et al. (2009a), symbiotic gut microbes can have a significant influence on host health and disease aetiology through a region-specific modulation of gastrointestinal functions. The author analyzed the effects of inoculating germfree mice with a model of simplified human baby microbiota (HBM) on the biochemical composition of intact intestinal tissues (duodenum, jejunum, ileum, proximal and distal colon) using ^{1}H HR-MAS NMR spectroscopy. The HBM tissue metabolite profiles with those from conventional and conventionalized mice were compared. Each topographical intestinal region showed a specific metabolic profile that was altered differentially by the various microbiomes, especially for osmolytes. The duodenum and jejunum were characterized with higher concentrations of amino acids, which illustrate the essential role of the small intestine in catabolizing up to 50% of the dietary amino acids for energy production, intestinal *de novo* synthesis and maintenance of mucosa (Wu 1998). Notably, conventional mice had higher levels of amino acids in the jejunum, whereas HBM animals showed higher amino acid levels in the duodenum. These microbial-dependent variations indicate that the type of gut bacteria may have important implications for the utilization of dietary proteins and amino acids and their subsequent availability to extra-intestinal tissues. The ileum is regarded as the major site for absorption of luminal bile acids and emulsified dietary lipids, and the observation of higher concentrations of phosphocholine, glycerophospholipids and glutathione, taurine and betaine was consistent with the role of the ileum in lipid and bile acid metabolism. Intriguingly, the jejunum of HBM colonized mice showed metabolic similarities to the ileal profile, marked by higher levels of glutathione (GSH), and lower concentrations of its precursors in the γ-glutamyl cycle when compared to conventional animals, which indicated different protective mechanisms against gut inflammation, mucosal damage, and oxidative stress. Since the jejunum was recently described as a primary location for digested fat storage (Robertson et al. 2003; Backhed et al. 2004) and that colonization of germfree mice with HBM was associated with higher enterohepatic recirculation of bile acids, and of dietary fat absorption and re-circulation (Martin et al. 2007a), changes in jejunal GSH may be suggestive of a different mechanism of protection against oxidative damage in the intestinal mucosa related to intestinal lipid absorption and processing in ex-germfree animals. Such an alteration of GSH metabolism may have inferred effects on the regulation of protein and prostaglandin metabolism and intestinal immune responses (Martin et al. 2009c).

Moreover, metabolic profiling, coupled with the metagenomic study of a novel antibiotic model, illustrated the close inter-relationship between the host and microbial "metabotypes". The evaluation of the relative contribution of the gut microbiota to the host metabolism was also assessed by investigating the effects of the antibiotic vancomycin (Yap et al. 2008), which impact specifically microbial metabolism since it does not have systemic absorption. The host's metabotype was characterized using ^{1}H NMR spectroscopy of urine and faecal extract samples. Time-course changes in the gut microbiotal community after administration of vancomycin were monitored using 16S rRNA gene PCR and denaturing gradient gel electrophoresis analysis and showed a strong effect on several species, mostly within the Firmicutes. Vancomycin treatment was associated with faecal excretion of uracil, amino acids and SCFA, highlighting the contribution of the gut microbiota to the production and metabolism of these dietary compounds. Clear differences in gut microbial communities between control and antibiotic-treated mice were observed in the current study. Reduced urinary excretion of gut microbial co-metabolites phenylacetylglycine and hippurate was also observed. Moreover, the lower faecal concentration of amino acids and SCFAs could be explained by the antibiotic induced disruption of the gut bacterial ecology. This in turn could cause a dysregulation of the absorptive abilities of the intestinal border.

Metabolomics has therefore proven to be a well established approach to characterize gut microbiome modulations of the intestinal biochemistry and physiology. Such technique is permitting today the *in vivo* analysis of the gut microbiome, revealing their protective mechanisms, creating new avenues of research that could have significant clinical applications. Indeed, the 'gut origin of sepsis' concept (Kinross et al. 2009) describes the role of the intestine in the development of sepsis and the post-operative Multi Organ Dysfunction Syndrome (MODS).

Overall, the metabolomic methods described here are looking for general and specific aspects of the gut microbiome revealing alteration in species that are responsible for changes in the host's physiological regulatory processes, probing the understanding of the microbial-mammalian metabolic axis.

8.3 Metabolomics: A Unique Non-invasive Tool to Measure the Contribution of Gut Microbial Metabolism to Gut Disorders

The colon is one of the most populated sections of the gastrointestinal tract, with high density of bacterial cells up to 10^{12} per milliliter. The measurement of the gut microbial populations and metabolism is generally confined to faecal samples, which represent only a limited range of microbial species and activities within the colon, in part due to the elevated colonic absorption of bacterial metabolites (Berg and Savage 1972). Yet, faecal analysis provides essential information on the impact of microbial activities on specific biological processes, including bile acids, fatty acids, N-nitroso compounds, and heterocyclic amines metabolism. The metabolic composition of faecal extracts provides a window for elucidating the complex metabolic interplay between mammals and their intestinal ecosystems, and these metabolite profiles can yield information on a range of gut disorders. The systematic characterization of faecal composition in humans, mice, and rats, together with the evaluation of inherent compositional variability, provided an important milestone for biomarker discovery in faeces, for both diagnostic and therapeutic surveillance applications (Saric et al. 2008). The metabolites present in aqueous faecal extracts of humans, mice and rats were characterized using high-resolution ^{1}-H NMR spectroscopy coupled with multivariate pattern recognition techniques. Although many faecal metabolites were common across the examined biological systems, such as SCFA and branched chain amino acids, each species generated a unique profile. Relatively higher levels of uracil, hypoxanthine, phenylacetate, glucose, glycine, and tyrosine were present in the rat, with beta-alanine being unique to the rat, and glycerol and malonate being unique to the human. Human faecal extracts showed a greater interindividual variation than the two rodent species, reflecting the natural genetic and environmental diversity in human populations. Faecal composition in healthy mice was found to change over time, which might be explained by altered gut microbial presence or activity.

Marchesi et al. (2007) described the first characterization of faecal extracts of IBD (inflammatory bowel disease) patients using metabolomics as a diagnostic tool for differentiating patients affected with UC and Crohn's disease (CD), hence improving disease surveillance. The faecal extracts of both CD and UC patients were characterized by reduced levels of butyrate, acetate, methylamine, and trimethylamine in comparison with a control population, suggesting changes in the gut microbial community. Also, elevated quantities of amino acids were present in the faeces from both disease groups, implying malabsorption caused by the inflammatory disease or an element of protein losing enteropathy. Metabolic differences in faecal profiles were more marked in the CD group in comparison with the control group, indicating that the inflammation caused by CD is more extensive in comparison with UC and involves the whole intestine. Furthermore, glycerol resonances were a dominant feature of faecal spectra from patients with CD but were present in much lower intensity in the control and UC groups. This work illustrates the potential of metabolomics as a novel diagnostic technique for further disease recognition and surveillance.

8.4 Metabolomics: New Research Venues to Decipher Probiotics' Effects on the Mammalian Gut Microbial Metabolic Cross Talk

Nowadays, functional foods are commonly used to modulate the composition of the gut microbiota contributing to the maintenance of the host health and prevention of diseases. One of the methods currently used to modulate the balance of intestinal microbiota is based on oral administration of probiotics. Probiotics are mostly administered through functional foods (i.e., fermented milks, yogurts) as dairy products. *Lactobacillus* and *Bifidobacterium* are the most produced probiotic strains worldwide. Probiotic supplementation aims at replacing or balancing the number of potentially harmful *E. coli* and *Clostidria* in the intestine by enriching the population of good bacteria that produce carbohydrate with little proteolytic activity.

The use of probiotics as treatment for a broad number of gastrointestinal and other infectious diseases has gained wide interest over the last few years, but little is known about the underlying mechanism of their action on the gut microbiota (Forestier et al. 2001). A very long list of beneficial effects arising from changes to the intestinal milieu has been proposed such as increased resistance to infectious disease (Berg 1983), decreased duration of diarrhoea (Brassart and Schiffrin 1997), reduction in blood pressure (Cahil et al. 1997), and reduced serum cholesterol concentration (Berg and Savage 1972). It is the impact of probiotics on the composition of the intestinal microbiota, nevertheless, that forms the basis for the probiotic concept. The presence of probiotic strains in the intestinal tract is dependent on their consumption and they may never become a true component of the intestinal ecosystem in the majority of the consumers (Tannock 2005). There are several reports of disease prevention or immune function enhancement resulting from oral administration of certain members of the intestinal microbiota. Much of the protection attributed to probiotics may result from colonization resistance (Alander et al. 1999; Mercenier and Grangette 1999). For instance, the intestinal bacteria populations are altered during diarrhoea. Specifically, numbers of *Bifidobacteria, Lactobacillus*, and *Bacteroides* are decreased and *Coliform* numbers increase during episodes of diarrhoea (Bercik et al. 2005). Supplementation of the microbiota with probiotic organisms may therefore protect from diarrhoea. In addition, probiotics may strengthen the epithelial barrier, thereby preventing pathogenic translocation from the epithelium (Campieri and Gionchetti 1999). Indirect evidence of probiotic enhancement of the innate immune system has also been reported including increased macrophage phagocytosis, epithelial cell defence functions and stimulation of lymphocyte activation and antibody production (Jijon et al. 2004; Christensen et al. 2002; Madsen et al. 2001). It has been shown *in vitro* and *in vivo* that *L. paracasei* is able to shape the intestinal physiology, aiding digestion and absorption of nutrients (Annuk et al. 2003; Sarker et al. 2005). Moreover, *L. paracasei* has been shown to produce antagonist metabolites and antioxidants to stimulate cells of the immune system *in vitro* (Ibnou-Zekri et al. 2003), to induce a homeostatic microenvironment that favours tolerance rather than allergy in gnotobiotic mice, and to contribute to the normalization of gastrointestinal disorders (Verdu et al. 2004; Prioult et al. 2003).

Metabolomics is nowadays a powerful tool used to underlie the mechanistic effects of probiotics on the microbial ecology and on human health. One approach to aid in the understanding of the interactions between micro-organisms and their hosts is to utilize a simplified mammalian ecosystem in the absence of bacteria and to evaluate the effects on host functions after colonizing the gut with a single microbial species (Falk et al. 1998). 1-HR-MAS techniques are able to generate specific metabolic signatures reflecting the structure and function of intact tissue biopsies and provide unique opportunity to explore specific bacterial metabolic effect onto gut structure and functions. In a recent study, Martin et al. employed such an approach to generate a region-specific metabolic mapping of the gut tract of a gnotobiotic mouse, in which they assessed the effects of *Lactobacillus paracasei* (Martin et al. 2007b). A total of 24 female gnotobiotic mice were divided into three groups: a control group supplemented with water and two groups supplemented with either live *L. paracasei* or a gamma-irradiated equivalent. Variations in relative concentrations of amino acids, anti-oxidant, and creatine were observed relating to different physiological properties in each intestinal tissue. Metabolic characteristics of lipogenesis and fat storage were observed in the jejunum and colon. Colonization with live *L. paracasei* induced region-dependent changes in the metabolic profiles of all intestinal tissues, except for the colon, consistent with modulation of intestinal digestion, absorption of nutrients, energy metabolism, lipid synthesis and protective functions. These signals acted as reference profiles with which to compare changes in response to gut microbiota manipulation at the tissue level as demonstrated by ingestion of a bacterial probiotic. Contrary to the effects induced by live *L. paracasei*, no changes were seen with supplementation of irradiated-killed bacteria. The results presented here demonstrated the importance of the transgenomic interactions between *L. paracasei* and the host, at the metabolic level to modulate the gut functions.

Additionally, a novel top-down level system biological approach of the host response to probiotic intervention was also revealed, where amino acids metabolism, methylamines, and SCFA pathways appeared to be altered. Here, the transgenomic metabolic effects of exposure to *Lactobacillus paracei* or *Lactobacillus rhamnous* probiotics were measured by a combined NMR and UPLC-MS metabolomic approach (Martin et al. 2008). Biofluid and tissue analysis of germ free mice model, colonized with a simplified model of human microbiota, revealed that probiotic exposure exerted microbiome modification, resulting in altered hepatic lipid metabolism, lowered plasma lipoprotein level, and stimulated glycolysis.

In a follow-up study, Martin et al. employed a similar strategy to monitor biochemical changes across the major compartments of the intestinal tract in response to the inoculation of germfree mice with different microbiomes and subsequent nutritional intervention with probiotics and synbiotics (combination of probiotics and prebiotic) (Martin et al. 2009a). A prebiotic is a "non-digestible food ingredient that beneficially affects the host by selectively stimulating the growth and/or activity of one or a limited number of bacteria in the colon that have the potential to improve host health" (Gibson and Roberfroid 1995). Colon-specific osmo-regulatory functions for balancing luminal and intracellular milieus were highlighted through higher levels of two osmolytes, *myo*-inositol and *scyllo*-inositol. HBM mice showed a reduction in colon levels of *myo*-inositol and taurine, another strong osmoprotectant, which suggested an alteration of muscular and osmo-regulatory functions. Proximal and distal colon tissues were differentiated in the different microbiome models by the concentrations of bacterial products, those being the highest in conventional animals. Moreover, supplementing HBM mice with probiotics or synbiotics resulted in different metabolic signatures in the proximal and distal colon, which shows how dietary modulation of bacterial activity may alter host intestinal biochemistry.

In another selected mouse model, characterized by simplified gut microbiota mouse system, the depth of the gut microbiome modulations of host biochemistry were also uncovered by modulating the gut functional ecology with pro-, pre-, and synbiotics (Martin et al. 2009a). The dietary modulation of the gut microbota at a panorganismal scale was measured using a multicompartmental modeling approach with metabolic imprints from 10 tissue/fluid samples. Induced microbial changes affected host lipids, carbohydrates, and amino acids metabolism in every major mammalian biological analyzed compartment. Selectively, probiotic supplementation reduced hepatic glycogen, glutamine, and adrenal ascorbate with intrinsic effects on energy homeostasis, antioxidation, and stereogenesis.

These studies show the breadth and the depth of gut microbiome modulations of the intestinal metabolite pathways and indicate that the type of gut microbiota may be a key factor in the determination of the intestinal homeostasis, osmo-protection, motility and calorie recovery from the diet.

8.5 Probiotic Action and Therapeutic Surveillance with Metabonomics

Recent evidence has shown that the influence of the gut microorganisms might be more important in the progression of human diseases than previously suspected (Guarner and Malagelada 2003). For instance, the intestinal microbiota may contribute to gut dysfunctions (Madden and Hunter 2002) such as IBS (MacDonald and Monteleone 2005) by affecting the immune system, or by directly changing muscular excitatory neurotransmission (Husebye et al. 2001). The importance and a renewal of interest in the impact of symbiont microbes on mammalian health has led to recent publications showing the association of particular strains of microbe with various disorders of the gastrointestinal tract, including IBS, IBD, or certain cancers (Dunne 2001; Holmes and Nicholson 2005). For these reasons, much attention today in metabolomics applications are also given to decipher the mechanism of actions of probiotics to promote gut health, restore gastrointestinal homeostasis and immune response, but also to prevent and monitor allergies and inflammatory incidences. For instance, the effects of probiotics intervention on global lipidomic profile in humans indicated that *Lactobacillus rhamnosus* strain GG intervention leads to a perceived decrease levels of lysophosphatidylcholines (LysoGPCho), sphingomyelins and several glycerophosphatidylcholines, while triacylglycerols displayed a small increase during the intervention period (Kekkonen et al. 2008). Interestingly, for example, the small picked up trend in LysoGPCho levels had been related to IBD (Minami et al. 1994; Haapamaki et al. 1999) where, specifically, *Lactobacillus rhamnosus* is well known to normalise gut permeability (Isolauri et al. 1993), improving mucosal integrity and epithelial cell survival (Lam et al. 2007; Yan et al. 2007).

It is well known that one of the most common functional digestive disorders affecting around 20% of the industrialized adult population is IBS (Loftus 2004). IBS generally involves abdominal pain and bowel habit disturbance, with changes in stool frequency and consistency. The aetiology, although multi-factorial, remains largely undefined. Factors implicated in the onset and development of IBS include perturbation of the gut microbiota precipitated by infection, diet, and genetic predisposition (Barbara et al. 2002; Rodriguez and Ruigomez 1999) but, it is not clear whether the altered microbiota is a cause or a consequence of the gut dysfunction. In affected patients, abnormalities in colonic microbiota have recently been suggested (Haller 2006) with significant reduction in the amount of intraluminar *Bifidobacteria* or/and *Lactobacilli.*

The model of *Trichinella spiralis* (*T. spiralis*) infection in NIH Swiss mice has a number of similarities to post-infective IBS and allows direct investigations of the effects of IBS on intestinal structure and function (Verdu et al. 2004). The model is reminiscent of observations in humans, which show the persistence of neuromotor dysfunctions after recovery from induced gastroenteritis or abnormal gut motility in patients in remission from UC (Barbara et al. 1997). There is persistence of the increased contractile response of jejunal longitudinal muscle after resolution of the mucosal inflammation. During the infection, the physical properties of the muscle are modified and may reflect tissue oedema and changes in extracellular matrix components. These immune mediated changes in intestinal physiology are associated with worm expulsion (Khan and Collins 2006). The altered physiology is maintained, subsequently by cyclooxygenase-2 (COX-2), transforming growth factor (TGF-β1) and prostaglandin (PGE_2) production in the neuromuscular layers of the gut. Thus, modulation of the immune response and its consequences may attenuate the development of the persistent alteration in gut physiology. A metabolomic strategy had been used to understand the underlying processes of IBS via the analysis of the metabolic profiles of plasma, jejunal wall and myenteric plexus-longitudinal muscle (LMM) tissues obtained from *T. spiralis* infected mice (Martin et al. 2006). In particular, LMM is the most external segment of the gut wall, containing the myenteric plexus in association with the longitudinal muscle layers and the connective tissue and enables new possibilities for the study of the underlying mechanisms of IBS. In such a post-infective model, manipulation of host-microbiota interactions with probiotic supplementation offers a promising approach to assess the efficacy of therapeutic interventions (Sartor 2004). For example, treating *T. spiralis* infected mice with the probiotic *L. paracasei* in its spent culture medium (SCM) and the SCM devoid of live bacteria attenuated gut muscle hyper-contractility and normalized the immune markers suggesting amelioration of the inflammatory response. Consistent biomarkers in plasma and intestinal tissue profiles of mice infected with *T. spiralis* were identified with a NMR-based metabolomic strategy, enhancing the current knowledge concerning the pathophysiology of the infection. One of the main effects of the infection was a modification of energy metabolism to meet the energy requirements due to the intestinal muscular hyper-contractility. This was shown with an increased energy metabolism (lactate, citrate, alanine), fat mobilization (acetoacetate, 3-D-hydroxybutyrate, lipoproteins) and a disruption of amino acid metabolism due to increased protein breakdown as a result of the experienced intestinal hypercontractility. Furthermore, changes in taurine, glycerophosphocholine, creatine and choline in jejunal LMM and plasma metabolic profiles could be correlated to the modification of the intestinal physiology, i.e. hypertrophy and disrupted jejunal functions. Then, several biomarkers (glutamine, lysine, methionine) known to play crucial role in modulating immune response had their levels changed due to infection which could be related to the late immune response previously reported (Verdu et al. 2004). In addition, the jejunal wall tissues showed variations in the levels of gut microbiota-related metabolites that were correlated to the disruption of the gut microbiota that was observed in IBS. The metabolic profiling approach also confirmed the role of *L. paracasei* probiotics in normalising the muscular activity and the disturbed energy metabolism as evidenced by decreased glycogenesis and elevated lipid breakdown in comparison with untreated *T. spiralis* infected mice. This metabolomic approach further showed that *L. paracasei* improved the gut microbiota metabolism related to the intestinal dysfunction in infected mice.

Even though *L. paracasei* free-SCM efficiently normalized gut contractility, it induced limited effects on plasma and jejunal wall metabolic profiles. In particular, no metabolic effects on microbiota related compounds (such as amino acids or SCFA), which corroborate previous findings that SCM had no measurable effects on small intestinal microbiota (Verdu et al. 2004). The results suggested that *L. paracasei* and its metabolites excreted in its SCM may regulate gut contractility by additional host-bacterial interactions specific to the probiotic and that do not require normalising gut microbiota or its metabolism. LMM tissues, with respect to both treatments, were shown to be metabolically identical to the infected samples despite reported modulation of protein and gene expression related to immune responses. Metabolomic analysis suggested that even if treatments induced a normalisation of muscular activity, the changes in muscular morphology and disturbed metabolic profiles, a consequence of hypertrophy and hyperplasia, were not resolved. Hence, supplementation with specific probiotics appeared to offer considerable promise in the treatment of post-infective IBS by modulating host metabolism and host-microbiota interactions. This work not only confirmed the potential of metabolic profiling for disease and therapeutic surveillance, but it also created a framework for surveying for the modulation of mammalian-microbiota interactions. In addition, it aided at the understanding of the mechanisms of microbe-mammalian metabolic interactions related to the processes of host diseases, which can be further followed using conventional molecular and microbiological methods.

While today there is no direct experimental evidence for cancer suppression in human subjects as a result of the consumption of probiotic cultures in fermented or unfermented dairy products, current investigations are shading new light into possible mechanisms of inhibition. Over 940,000 cases of colorectal cancers (CRC) are diagnosed each year making it one of the most spread cancer type (Midgley and Kerr 1999). CRC is responsible for approximately 529,000 deaths worldwide each year, with 60% of these deaths occurring in developed countries (Goldberg 2005). Numerous studies reported that the general population of intestinal bacteria is associated with initiation of colon cancer via the production of carcinogens, co-carcinogens, or pro-carcinogens (O'Keefe 2008). Current scientific evidence, from animal studies, is displaying today that supplementation of probiotics might (Shahani and Ayebo 1980) reduce tumour occurrence and precancerous lesion in the colon. Early ^{1}H NMR spectroscopy study demonstrated significant increases in taurine, choline-containing compounds and lipid resonances in malignant colon mucosa (Moreno et al. 1993). More recent proteomic work (Bi et al. 2006) has indicated that alterations of metabolic pathways such as purine and pyrimidine metabolism, glycolysis, gluconeogenesis, glucoronic acid synthesis and tricarboxylic acid cycle also occur in CRC. With the aid of metabolomics techniques such alterations in different metabolic pathways are expected to result in associated changes in metabolic profiles for future identification of targets for measuring the efficacy of therapeutic and/or probiotic manipulations. Chan et al. (2009) compared the metabolic profile of biopsied colorectal tumours and their matched normal mucosae obtained from 31 CRC patients using high-resolution magic angle spinning–NMR and GCMS and obtained 31 marker metabolites that distinguish normal from malignant samples and further colon from rectal cancers. The majority of these were associated with expected metabolic perturbations in CRC including elevated tissue hypoxia, glycolysis, nucleotide biosynthesis, lipid metabolism, inflammation and steroid metabolism. Moreover, the application of NMR metabolic profiling to faecal water extracts has interesting diagnostic potential for detecting CRC (Monleon et al. 2009). NMR metabolic profiles of faecal water extracts from patients with CRC and healthy individuals were obtained with possible differences between them. Low concentration of SCFA (acetate and butyrate) appeared to be the most prominent markers. Concentrations of proline and cysteine, which are major components of most colonic epithelium mucus glycoproteins, also displayed significant changes in samples from CRC. In addition, recent studies on the rat rat-azoxymethane model (Le Leu et al. 2005) displayed that resistant starch and *Bifidobacterium lactis* significantly protected against the development of CRC beyond the benefit of either agent alone.

The era of modern nutrition unfolds a greater understanding of how foods and dietary bioactive components, especially probiotics, influence human diseases and health. This scientific challenge is critical for the development of tailored approaches to reduce disease burden.

8.6 Conclusion

Recent advances in immunology, metabolomics, and microbial ecology have shown that the contribution of the intestinal microbiota to the overall health status of the host has been so far underestimated. It is therefore of crucial importance for nutrition to understand and to metabolically characterize the interactive molecular processes between the host and its microbiome.

By opening a direct biochemical window into the regulation of physiological and regulatory processes at the system level, metabolomics provides a unique approach for the identification of biomarkers capable of providing better understanding on the molecular mechanisms between nutrition and health. Moreover, metabolomics appears to be a favourable approach for assessing gut microbiota metabolism through the analysis of biofluids and tissues.

As displayed in this chapter, we indeed have now the microbial inventories and genetic blueprints to begin tackling intestinal microbial ecology at an unprecedented level of detail. Through the rigorous characterization of interactions between the diet and the microbiota, metabolomics is providing new ventures for modulating the microbiota towards the improvement of human health. The prospective of preventing the progression of human diseases, by normalizing their effects, by specific nutritional intervention programs, such as probiotic supplementation, could benefit from the applications of metabolomics for therapeutic surveillance and assessment of treatment efficacy.

Undeniably, metabolic knowledge of the functional ecology of the gut microbiota will bring the design of novel and enhanced functional foods aiming at promoting health and delaying onsets of metabolic disruption towards pathological conditions.

References

Alander M, Satokari R, Korpela R, Saxelin M, Vilpponen-Salmela T, Mattila-Sandholm T, von Wright A (1999) Persistence of colonization of human colonic mucosa by a probiotic strain, Lactobacillus rhamnosus GG, after oral consumption. Appl Environ Microbiol 65(1):351–354

Annuk H, Shchepetova J, Kullisaar T, Songisepp E, Zilmer M, Mikelsaar M (2003) Characterization of intestinal lactobacilli as putative probiotic candidates. J Appl Microbiol 94:403–412

Backhed F, Ding H, Wang T, Hooper LV, Koh GY, Nagy A, Semenkovich CF, Gordon JI (2004) The gut microbiota as an environmental factor that regulates fat storage. Proc Natl Acad Sci U S A 101:15718–15723

Barbara G, Vallance BA, Collins SM (1997) Persistent intestinal neuromuscular dysfunction after acute nematode infection in mice. Gastroenterology 113(4):1224–1232

Barbara G, De GR, Stanghellini V, Cremon C, Corinaldesi R (2002) A role for inflammation in irritable bowel syndrome? Gut 51(Suppl 1):141–144

Beale B (2002) Probiotics: their tiny worlds are under scrutiny. Scientist 16:20–22

Bercik P, Verdu EF, Collins SM (2005) Is irritable bowel syndrome a low-grade inflammatory bowel disease? Gastroenterol Clin North Am 34(2):235–239

Berg RG (1983) Host immune response to antigens of the indigenous intestinal flora. In: Hentges DJ (ed) Human intestinal microflora in health and disease. Academic Press, New York, pp 101–126

Berg RG, Savage DC (1972) Immunological responses and micro-organisms indigenous to the gastrointestinal tract. Am J Clin Nutr 25:1364–1371

Bergonzelli GE, Blum S, Brussow H, Corthesy-Theulaz I (2005) Probiotics as a treatment strategy for gastrointestinal diseases? Digestion 72(1):57–68

Bhavsar AP, Guttman JA, Finlay BB (2007) Manipulation of host-cell pathways by bacterial pathogens. Nature 449:827–834

Bi X, Lin Q, Foo TW, Joshi S, You T, Shen HM, Ong CN, Cheah PY, Eu KW, Hew CL (2006) Proteomic analysis of colorectal cancer reveals alterations in metabolic pathways: mechanism of tumorigenesis. Mol Cell Proteomics 5(6):1119–1130

Bjerrum JT, Nielsen OH, Hao F, Tang H, Nicholson JK, Wang Y, Olsen J (2009) Metabonomics in ulcerative colitis: diagnostics, biomarker identification, and insight into the pathophysiology. J Proteome Res 9:1535–3907

Brassart T, Schiffrin EJ (1997) The use of probiotics to reinforce mucosal defence mechanisms. Trends Food Sci Technol 8:321–326

Bruce SJ, Tavazzi I, Parisod V, Rezzi S, Kochhar S, Guy PA (2009) Investigation of human blood plasma sample preparation for performing metabolomics using ultrahigh performance liquid chromatography/mass spectrometry. Anal Chem 81(9):3285–3296

Cahill RJ, Foltz CJ, Fox JG, Dangler CA, Powrie F, Schauer DB (1997) Inflammatorybowel disease: an immunity-mediated condition triggered by bacterial infection with *Helicobacter hepaticus*. Infect Immun 65:3126–3131

Campieri M, Gionchetti P (1999) Probiotics in inflammatory bowel disease: new insight to pathogenesis or a possible therapeutic alternative? Gastroenterology 116(5):1246–1249

Chan EC, Koh PK, Mal M, Cheah PY, Eu KW, Backshall A, Cavill R, Nicholson JK, Keun HC (2009) Metabolic profiling of human colorectal cancer using high-resolution magic angle spinning nuclear magnetic resonance (HR-MAS NMR) spectroscopy and gas chromatography mass spectrometry (GC/MS). J Proteome Res 8(1):352–361

Christensen HR, Frokiaer H, Pestka JJ (2002) Lactobacilli differentially modulate expression of cytokines and maturation surface markers in murine dendritic cells. J Immunol 168(1):171–178

Dunne C (2001) Adaptation of bacteria to the intestinal niche: probiotics and gut disorder. Inflamm Bowel Dis 7(2):136–145

Eriksson L, Antti H, Gottfries J, Holmes E, Johansson E, Lindgren F, Long I, Lundstedt T, Trygg J, Wold S (2004) Using chemometrics for navigating in the large data sets of genomics, proteomics, and metabonomics (gpm). Anal Bioanal Chem 380(3):419–429

Falk PG, Hooper LV, Midtvedt T, Gordon JI (1998) Creating and maintaining the gastrointestinal ecosystem: what we know and need to know from gnotobiology. Microbiol Mol Biol Rev 62(4):1157–1170

Forestier C, de Champs C, Vatoux C, Joly B (2001) Probiotic activities of Lactobacillus casei rhamnosus: in vitro adherence to intestinal cells and antimicrobial properties. Res Microbiol 152(2):167–173

Gibson GR, Roberfroid MB (1995) Dietary modulation of the human colonic microbiota: introducing the concept of prebiotics. J Nutr 125(6):1401–1412

Goldberg RM (2005) Advances in the treatment of metastatic colorectal cancer. Oncologist 10(Suppl 3):40–48

Guarner F, Malagelada JR (2003) Gut flora in health and disease. Lancet 361(9356):512–519

Gupta P, Andrew H, Kirschner BS, Guandalini S (2000) Is lactobacillus GG helpful in children with Crohn's disease? Results of a preliminary, open-label study. J Pediatr Gastroenterol Nutr 31(4):453–457

Haapamaki MM, Gronroos JM, Nurmi H, Irjala K, Alanen KA, Nevalainen TJ (1999) Phospholipase A2 in serum and colonic mucosa in ulcerative colitis. Scand J Clin Lab Invest 59(4):279–287

Haller D (2006) Intestinal epithelial cell signalling and host-derived negative regulators under chronic inflammation: to be or not to be activated determines the balance towards commensal bacteria. Neurogastroenterol Motil 18(3):184–199

Holmes E, Nicholson J (2005) Variation in gut microbiota strongly influences individual rodent phenotypes. Toxicol Sci 87(1):1–2

Hooper LV, Gordon JI (2001) Commensal host-bacterial relationships in the gut. Science 292:1115–1118

Hooper LV, Midtvedt T, Gordon JI (2002) How host-microbial interactions shape the nutrient environment of the mammalian intestine. Ann Rev Nutr 22:283–307

Husebye E, Hellstrom PM, Sundler F, Chen J, Midtvedt T (2001) Influence of microbial species on small intestinal myoelectric activity and transit in germ-free rats. Am J Physiol Gastrointest Liver Physiol 280(3):G368–G380

Ibnou-Zekri N, Blum S, Schiffrin EJ, von der Weid T (2003) Divergent patterns of colonization and immune response elicited from two intestinal Lactobacillus strains that display similar properties in vitro. Infect Immun 71(1):428–436

Isolauri E, Majamaa H, Arvola T, Rantala I, Virtanen E, Arvilommi H (1993) Lactobacilluscasei strain GG reverses increased intestinal permeability induced by cow milk in suckling rats. Gastroenterology 105(6):1643–1650

Jijon H, Backer J, Diaz H, Yeung H, Thiel D, McKaigney C, De Simone C, Madsen K (2004) DNA from probiotic bacteria modulates murine and human epithelial and immune function. Gastroenterology 126(5):1358–1373

Kekkonen RA, Sysi-Aho M, Seppanen-Laakso T, Julkunen I, Vapaatalo H, Oresic M, Korpela R (2008) Effect of probiotic Lactobacillus rhamnosus GG intervention on global serum lipidomic profiles in healthy adults. World J Gastroenterol 14(20):3188–3194

Khan WI, Collins SM (2006) Gut motor function: immunological control in enteric infection and inflammation. Clin Exp Immunol 143(3):389–397

Kinross J, von Roon AC, Penney N, Holmes E, Silk D, Nicholson JK, Darzi A (2009) The gut microbiota as a target for improved surgical outcome and improved patient care. Curr Pharm Des 15(13):1537–1545

Lam EK, Tai EK, Koo MW, Wong HP, Wu WK, Yu L, So WH, Woo PC, Cho CH (2007) Enhancement of gastric mucosal integrity by Lactobacillus rhamnosus GG. Life Sci 80(23):2128–2136

Le Leu RK, Brown IL, Hu Y, Bird AR, Jackson M, Esterman A, Young GP (2005) A synbiotic combination of resistant starch and Bifidobacterium lactis facilitates apoptotic deletion of carcinogen-damaged cells in rat colon. J Nutr 135(5):996–1001

Ley R, Turnbaugh P, Klein S, Gordon J (2006) Microbial ecology: human gut microbes associated with obesity. Nature 444(7122):1022–1023

Loftus EV Jr (2004) Clinical epidemiology of inflammatory bowel disease: incidence, prevalence, and environmental influences. Gastroenterology 126(6):1504–1517

MacDonald T, Monteleone G (2005) Immunity, inflammation, and allergy in the gut. Science 307(5717):1920–1925

Madden JA, Hunter JO (2002) A review of the role of the gut microflora in irritable bowel syndrome and the effects of probiotics. Br J Nutr 88(Suppl 1):S67–S72

Madsen K, Cornish A, Soper P, McKaigney C, Jijon H, Yachimec C, Doyle J, Jewell L, De Simone C (2001) Probiotic bacteria enhance murine and human intestinal epithelial barrier function. Gastroenterology 121(3):580–591

Marchesi JR, Holmes E, Khan F, Kochhar S, Scanlan P, Shanahan F, Wilson ID, Wang Y (2007) Rapid andnoninvasive metabonomic characterization of inflammatory bowel disease. J Proteome Res 6:546–551

Marshall B (2003) Helicobacter pylori: past, present and future. Keio J Med 52(2):80–85

Martin FP, Verdu EF, Wang Y, Dumas ME, Yap IK, Cloarec O, Bergonzelli GE, Corthesy-Theulaz I, Kochhar S, Holmes E, Lindon JC, Collins SM, Nicholson JK (2006) Transgenomic metabolic interactions in a mouse disease model: interactions of Trichinella spiralis infection with dietary Lactobacillus paracasei supplementation. J Proteome Res 5(9):2185–2193

Martin FP, Dumas ME, Wang Y, Legido-Quigley C, Yap IK, Tang H, Zirah S, Murphy GM, Cloarec O, Lindon JC, Sprenger N, Fay LB, Kochhar S, van Bladeren P, Holmes E, Nicholson JK (2007a) A top-down systems biology view of microbiome-mammalian metabolic interactions in a mouse model. Mol Sys Biol 3:112–118

Martin FP, Wang Y, Sprenger N, Holmes E, Lindon JC, Kochhar S, Nicholson JK (2007b) Effects of probiotic Lactobacillus paracasei treatment on the host gut tissue metabolic profiles probed via magic-angle-spinning NMR spectroscopy. J Proteome Res 6(4):1471–1481

Martin FP, Wang Y, Sprenger N, Yap IK, Rezzi S, Ramadan Z, Pere-Trepat E, Rochat F, Cherbut C, van Bladeren P, Fay LB, Kochhar S, Lindon JC, Holmes E, Nicholson JK (2008) Top-down systems biology integration of conditional prebiotic modulated transgenomic interactions in a humanized microbiome mouse model. Mol Syst Biol 4:205

Martin FP, Wang Y, Yap IK, Sprenger N, Lindon JC, Rezzi S, Kochhar S, Holmes E, Nicholson JK (2009a) Topographical variation in murine intestinal metabolic profiles in relation to microbiome speciation and functional ecological activity. J Proteome Res 8(7):3464–3474
Martin FP, Rezzi S, Pere-Trepat E, Kamlage B, Collino S, Leibold E, Kastler J, Rein D, Fay LB, Kochhar S (2009b) Metabolic effects of dark chocolate consumption on energy, gut microbiota, and stress-related metabolism in free-living subjects. J Proteome Res 8(12):5568–5579
Martin FP, Rezzi S, Montoliu I, Philippe D, Tornier L, Messlik A, Holzlwimmer G, Baur P, Quintanilla-Fend L, Loh G, Blaut M, Blum S, Kochhar S, Haller D (2009c) Metabolic assessment of gradual development of moderate experimental colitis in IL-10 deficient mice. J Proteome Res 8(5):2376–2387
Martin FP, Sprenger N, Yap IK, Wang Y, Bibiloni R, Rochat F, Rezzi S, Cherbut C, Kochhar S, Lindon JC, Holmes E, Nicholson JK (2009d) Panorganismal gut microbiome-host metabolic crosstalk. J Proteome Res 8(4):2090–2105
Mercenier A, Grangette C (1999) Immunity and probiotics, nutrition and health collection. John Libbey, Paris, pp 9–44
Midgley R, Kerr D (1999) Colorectal cancer. Lancet 353(9150):391–399
Minami T, Tojo H, Shinomura Y, Matsuzawa Y, Okamoto M (1994) Increased group II phospholipase A2 in colonic mucosa of patients with Crohn's disease and ulcerative colitis. Gut 35(11):1593–1598
Monleon D, Morales JM, Barrasa A, Lopez JA, Vazquez C, Celda B (2009) Metabolite profiling of faecal water extracts from human colorectal cancer. NMR Biomed 22(3):342–348
Montoliu I, Martin FP, Collino S, Rezzi S, Kochhar S (2009) Multivariate modeling strategy for intercompartmental analysis of tissue and plasma 1H NMR spectrotypes. J Proteome Res 8(5):2397–2406
Moreno A, Rey M, Montane JM, Alonso J, Arus C (1993) 1H NMR spectroscopy of colon tumors and normal mucosal biopsies; elevated taurine levels and reduced polyethyleneglycol absorption in tumors may have diagnostic significance. NMR Biomed 6(2):111–118
Nicholson JK, Lindon JC (2008) Systems biology: Metabonomics. Nature 455:1054–1056
Nicholson JK, Sadler PJ, Bales JR, Juul SM, MacLeod AF, Sonksen PH (1984) Monitoring metabolic disease by proton NMR of urine. Lancet 2(8405):751–752
Nicholson JK, Holmes E, Wilson ID (2005) Gut microorganisms, mammalian metabolism and personalized health care. Nat Rev Microbiol 3(5):431–438
O'Keefe SJ (2008) Nutrition and colonic health: the critical role of the microbiota. Curr Opin Gastroenterol 24:51–58
Pereira DI, Gibson GR (2002) Effects of consumption of probiotics and prebiotics on serum lipid levels in humans. Crit Rev Biochem Mol Biol 37(4):259–281
Prioult G, Fliss I, Pecquet S (2003) Effect of probiotic bacteria on induction and maintenance of oral tolerance to beta-lactoglobulin in gnotobiotic mice. Clin Diagn Lab Immunol 10(5):787–792
Reid G (2001) Can bacterial interference prevent infection? Trends Microbiol 9:424–428
Rezzi S, Ramadan Z, Fay LB, Kochhar S (2007) Nutritional metabonomics: applications and perspectives. J Proteome Res 6(2):513–525
Rezzi S, Vera FA, Martin FP, Wang S, Lawler D, Kochhar S (2008) Automated SPE-RP-HPLC fractionation of biofluids combined to off-line NMR spectroscopy for biomarker identification in metabonomics. J Chromatogr B Anal Technol Biomed Life Sci 871(2):271–278
Robertson MD, Parkes M, Warren BF, Ferguson DJ, Jackson KG, Jewell DP, Frayn KN (2003) Mobilisation of enterocyte fat stores by oral glucose in humans. Gut 52:834–839
Rodriguez LA, Ruigomez A (1999) Increased risk of irritable bowel syndrome after bacterial gastroenteritis: cohort study. Br Med J 318(7183):565–566
Saric J, Wang Y, Li J, Coen M, Utzinger J, Marchesi JR, Keiser J, Veselkov K, Lindon JC, Nicholson JK, Holmes E (2008) Species variation in the faecal metabolome gives insight into differential gastrointestinal function. J Proteome Res 7(1):352–360

Sarker SA, Sultana S, Fuchs GJ, Alam NH, Azim T, Brussow H, Hammarstrom L (2005) Lactobacillus paracasei strain ST11 has no effect on rotavirus but ameliorates the outcome of nonrotavirus diarrhoea in children from Bangladesh. Pediatrics 116(2):e221–e228

Sartor RB (2004) Therapeutic manipulation of the enteric microflora in inflammatory bowel diseases: antibiotics, probiotics, and prebiotics. Gastroenterology 126(6):1620–1633

Shahani KM, Ayebo AD (1980) Role of dietary lactobacilli in gastrointestinal microecology. Am J Clin Nutr 33(Suppl 11):2448–2457

Tannock GW (2005) Commentary: Remembrance of microbes past. Int J Epidemiol 34(1):13–15

Verdu EF, Bercik P, Bergonzelli GE, Huang XX, Blennerhasset P, Rochat F, Fiaux M, Mansourian R, Corthesy-Theulaz I, Collins SM (2004) Lactobacillus paracasei normalizes muscle hypercontractility in a murine model of postinfective gut dysfunction. Gastroenterology 127(3):826–837

Wang Y, Holmes E, Comelli EM, Fotopoulos G, Dorta G, Tang H, Rantalainen MJ, Lindon JC, Corthesy-Theulaz IE, Fay LB, Kochhar S, Nicholson JK (2007) Topographical variation in metabolic signatures of human gastrointestinal biopsies revealed by high-resolution magic-angle spinning (1)H NMR spectroscopy. J Proteome Res 6(10):3944–3951

Warren JR (2000) Gastric pathology associated with Helicobacter pylori. Gastroenterol Clin North Am 29(3):705–751

Wikoff WR, Anfora AT, Liu J, Schultz PG, Lesley SA, Peters EC, Siuzdak G (2009) Metabolomics analysis reveals large effects of gut microflora on mammalian blood metabolites. Proc Natl Acad Sci U S A 106(10):3698–3703

Wu G (1998) Intestinal mucosal amino acid catabolism. J Nutr 128:1249–1252

Xu J, Mahowald MA, Ley RE, Lozupone CA, Hamady M, Martens EC, Henrissat B, Coutinho PM, Minx P, Latreille P, Cordum H, Van Brunt A, Kim K, Fulton RS, Fulton LA, Clifton SW, Wilson RK, Knight RD, Gordon JI (2007) Evolution of symbiotic bacteria in the distal human intestine. PLoS Biol 5:e156

Yan F, Cao H, Cover TL, Whitehead R, Washington MK, Polk DB (2007) Soluble proteins produced by probiotic bacteria regulate intestinal epithelial cell survival and growth. Gastroenterology 132(2):562–575

Yap IK, Li JV, Saric J, Martin FP, Davies H, Wang Y, Wilson ID, Nicholson JK, Utzinger J, Marchesi JR, Holmes E (2008) Metabonomic and microbiological analysis of the dynamic effect of vancomycin-induced gut microbiota modification in the mouse. J Proteome Res 7(9):3718–3728

Part V
Probiotics, Gut Immunology and Enteric Protection

Chapter 9
Microbiota, Probiotics and Natural Immunity of the Gut

Eduardo Jorge Schiffrin and Anne Donnet-Hughes

9.1 The Intestinal Immune Organ

The symbiotic relationship between the host and its microbiota is testimony to the effectiveness of the intestinal immune system in discerning which microbes are potentially pathogenic. However, tolerance to the microbiota is not due to immunological ignorance. Innate and adaptive components of the immune system work in concert to prevent exaggerated tissue-damaging responses to the microbiota but effective defence against invading pathogens.

Innate immunity has been classically defined as a rapid, non-specific immune response with identical characteristics each time that the same type of pathogen is encountered. The cellular participants in this response form the host's first line of defence and include short-lived neutrophils, macrophages, dendritic cells (DC), natural killer and epithelial cells. The capacity of these cells to sense environmental antigens depends on their expression of a limited repertoire of evolutionarily inherited receptors which do not recognise specific antigens but broadly shared molecular motifs. Although these cells are traditionally considered to lack immunological memory, they are instrumental in coordinating subsequent antigen-specific responses.

The genes encoding antigen receptors on B and T cells are assembled from different gene segments through the recombination-activating gene (RAG)-protein. Somatic recombination (Schatz et al. 1992) results in a diverse repertoire of receptors that allows most antigenic determinants to be recognised by adaptive immune receptors. Primary antigen stimulation of the adaptive immune response promotes the maturation of the antigen receptors. Further diversity in B cells is achieved through somatic hypermutation. In both cases, there is increased affinity for specific ligands. The long-lived, antigen-specific memory B and T lymphocytes that are then generated respond faster and more efficiently upon successive encounters of the specific pathogen (Kelly et al. 2005).

E. J. Schiffrin (✉)
Nestlé Nutrition, R & D, Nestec Ltd, 22 Avenue Reller, 1800 Vevey, Switzerland
e-mail: eduardo.schiffrin@nestle.com

J. J. Malago et al. (eds.), *Probiotic Bacteria and Enteric Infections*,
DOI 10.1007/978-94-007-0386-5_9,

9.2 Pattern Recognition Receptors in the Innate Immune Response

Microbially associated molecular patterns (MAMPs or "infectious non-self") (Janeway 1989; Akira et al. 2006; Medzhitov 2007) or endogenous "danger signals" derived from stressed, damaged or infected tissue (Matzinger 1994; Kabelitz and Medzhitov 2007) can stimulate innate inflammatory defences and subsequently initiate adaptive immune responses through activation and maturation of antigen presenting cells such as DC. They achieve this by interacting with pattern recognition receptors (PRR), the best characterised of which are the family of Toll-like receptors (TLR). These germ-line encoded, transmembrane molecules are expressed on the plasma membrane or in intracellular endosomal compartments. Upon ligation, they dimerise and induce cell signalling through a series of adaptor molecules such as myeloid differentiation factor 88 (MyD88), TIR domain-containing adaptor protein inducing IFN-β (TRIF) and interferon regulatory factor (IRF). The end result is NF-κB activation or the production of type I interferons (Kawai and Akira 2006). However, the precise nature of the transduced signal is dependent on the type of ligand and the cellular location of the TLR (Barton and Kagan 2009).

Humans have ten TLRs, all of which mediate cellular signalling through a common TIR (for Toll/interleukin (IL)-1 receptor) domain. They are classified according to their cellular location and their ligands. TLRs 1, 2, 4, 5, 6 and 10 are found on the cell surface (Kawai and Akira 2006; Barton and Kagan 2009). Of these, TLRs 1, 2, 4, 6 and 10 are involved in recognition of bacterial lipids whilst TLR5 interacts with proteins of bacteria and parasites. TLRs 3, 7, 8 and 9 are localised in endosomal compartments where they recognise viral and bacterial nucleic acids (Rakoff-Nahoum and Medzhitov 2008).

In addition to the transmembrane PRR, there are intracellular receptors which recognise bacteria and virus in the cytosol (Medzhitov 2007; Martinon and Tschopp 2004; Meylan et al. 2006). A large family of cytosolic PRR, the nucleotide binding oligomerization domain (Nod)-like receptor (NLR) family (Fritz et al. 2006) has various subfamilies based on N-terminus motifs such as a caspase activation and recruitment domain, CARD (NOD subfamily), a pyrin domain protein (NACHT-LRR-PYD-containing proteins, the NALP subfamily) (Meylan et al. 2006) or a BIR (baculoviral inhibitor-of-apoptosis protein repeat) domain (the NAIP subfamily) (Tig and Davis 2005).

Nod1 and Nod2 sense bacterial molecules produced during the synthesis, degradation and remodelling of peptidoglycan, a widely expressed component of bacterial cell walls. Nod2 reacts with muramyl dipeptide (MDP) and as such, is a sensor for most bacteria, whereas Nod1 recognises peptidoglycan fragments containing γ-D-glutamyl-mesodiaminopimelic acid, from mainly Gram-negative bacteria (Strober et al. 2006). In both cases, ligand binding leads to NF-κB activation.

The NALP and NAIP subfamilies are involved in the inflammatory response mediated by IL-1β, IL-18 and IL-33 (Meylan et al. 2006) and are particularly important in sensing bacterial components that have gained access to the intracellular compartment. These NLRs form part of a multiprotein complex known as the inflammasome which upon interaction with its ligands, terminates in activation of caspases (1, 4 and 5 in humans) and the release of pro-inflammatory cytokines from their precursors (Martinon and Tschopp 2004).

Intracellular sensors of virus involve the helicases retinoic-acid-inducible gene I (RIG-I) and melanoma differentiation-associated gene 5 (MDA5), which detect viral RNA in the cytosol and DNA-dependent activator of interferon-regulatory factors (DAI) which detects viral DNA (Yoneyama et al. 2004; Unterholzner and Bowie 2008). Engagement of these receptors induces expression of type I interferons and pro-inflammatory cytokines. Together with TLR 3, 7 and 9 in the endosome, these receptors are central to the anti-viral immune response. Indeed, the natural killer (NK) cells which are key players in anti-viral responses, are activated via viral nucleic acids that are bound to intracellular PRR or by plasmacytoid DC via TLRs 3, 7 and 9. The resultant anti-viral effector functions include apoptosis of the virally infected cells and increased IFN-γ production (Lanier 2005).

Ligands for PRR are common to pathogens and symbiotic microorganisms and yet, the intestine initiates a protective response to the former while allowing an important and complex microbiota to establish itself at the intestinal surface without any detrimental effect on homeostasis. Discrimination between pathogens and commensals clearly involves other mechanisms. The expression and engagement of PRR in different cellular compartments, as well the presence of additional "danger signals" from stressed or damaged tissues may be important determining factors. Certainly, most pathogens express a number of virulence determinants such as adherence to and invasion of host cells, resistance to antimicrobial proteins and the production of toxins. An important virulence trait of pathogens like *Salmonella* and *Shigella* is their capacity to access the intracellular environment where they interact with PRR such as NALP, NAIP or RIG-I.

Several PRR recognise endogenous molecules as well as microbial components. For example, high-mobility group box 1 (HMGB1), a terminal mediator of inflammation, is released from injured tissues and upon binding to TLR4 amplifies innate immune activation (Scaffidi et al. 2002). In addition, NALP and NAIP in the cytosol can bind endogenous molecules such as crystals of monosodium urate that are also expressed during cellular stress or damage (Pasare and Medzhitov 2005). Thus, PRR of the innate immune system are used in the surveillance of the extracellular space, endolysosomal compartments and the cytoplasm for signs of, not only infection, but tissue injury.

9.3 The Role of Innate Immunity in Gut Homeostasis

The intestinal epithelium is a single layer of cells which line the intestine and provide the first sensory line of defence and active sampling of resident bacteria, pathogens and other antigens. The villus epithelium of the small intestine is composed of several epithelial cell types. The most predominant cells are enterocytes which sample the luminal environment and secrete cytokines and chemokines which programme immune cells in the underlying tissue; goblet cells which secrete mucins; enteroendocrine cells which produce peptide hormones and Paneth cells which express antimicrobial peptides in the villus crypt. In addition, M cells overlying Peyer's patches and other lymphoid aggregates actively sample luminal antigens and present them to immune cells in the lymphoid follicles. The integrity of the mucosal barrier depends on the rate of epithelial renewal as well as on the expression of intercellular tight junction proteins which control permeability of the barrier.

While innate recognition of low numbers of bacteria in blood or tissues triggers first an inflammatory response and thereafter, an adaptive immune response, substantially higher quantities of bacteria reside in the gut lumen without initiating any detrimental inflammatory response (Stack et al. 2009). This homeostasis is achieved through robust cellular and molecular mechanisms which are involved in intestinal barrier function, cytoprotection and innate immunity. For example, reinforcement of the barrier is provided by a mucous gel layer which contains highly glycosylated mucin proteins that are produced by goblet cells and give the gel its viscoelastic properties (Corfield et al. 2000). This mucous gel covers the luminal surface of the gastrointestinal tract and limits microbial interaction with host tissues (Johansson et al. 2008). In addition, enterocytes and Paneth cells express a number of antimicrobial proteins which protect against pathogens and limit invasion of indigenous microbes (Mukherjee et al. 2008). By confining the majority of microbes to the luminal side of the intestinal barrier, the host not only benefits from the colonisation resistance function of the microbiota against pathogens, but also ensures that the systemic immune system does not interact extensively with the indigenous microbes (Hooper 2009). However, the intestinal epithelium is not completely impermeable to microbial components. Indeed, it appears that a "supervised" interaction between microbial components and host cells is vital for the development and maturation of the intestinal barrier and its associated immune system.

Studies comparing conventionalised and germ-free animals have shown that the microbiota influences gut barrier development and integrity at various levels. A microbial presence in the intestine promotes the development of intestinal lymphoid tissues (Cebra et al. 1998); decreases epithelial cell cycle time and increases proliferation in villus crypts (Alam et al. 1994); influences the thickness, compactness, mucin composition and structure of the intestinal mucous layer (Szentkuti et al. 1990; Comelli et al. 2008), as well as intestinal permeability (Yang et al. 2002). Although the presence or absence of an intestinal microbiota does not have a major influence on the production of antimicrobial defensins (Pütsep et al. 2000), the microbiota significantly increases Paneth cell production of other antimicrobial peptides (Ismail and Hooper 2005; Cash et al. 2006) such as cryptdin-related sequence (CRS) peptides and the C-type lectin, RegIIIγ (Karlsson et al. 2008). In addition, the microbiota induces a cytoprotective response of epithelial cells with concomitant expression of heat-shock proteins (Arvans et al. 2005).

Innate responses initiated through PRR recognition of microbial components play a critical role in maintaining intestinal homeostasis. However, there is only a low expression of PRR on the apical membrane of epithelial cells (Cario et al. 2002). Furthermore, the microbiota is sequestered in the intestinal lumen and has no ready access to PRR that are expressed on the basolateral membranes or intracellular compartments of sentinel cells. Clearly, some microbial components must succeed in breaching the epithelial barrier. Indeed, in the healthy intestine there is a low but continuous, physiological uptake of bacteria by M cells of the follicle-associated epithelium (FAE) (Man et al. 2004) and by lamina propria DC which modulate the expression of tight junction proteins on epithelial cells and thereby, allow their dendrites to pass through the epithelium and sample bacteria from the gut lumen (Rescigno et al. 2001; Rescigno 2003). As a precautionary measure, intestinal macrophages may eliminate translocating bacteria through multiple antimicrobial mechanisms (Medzhitov 2007).

The net result of the above processes is production of secretory IgA (sIgA), which further limits bacterial penetration across the barrier (MacPherson and Uhr 2004). Intraluminal injection of TLR2 and TLR4 ligands enhances M cell transport of microparticles and induces migration of subepithelial DCs into the FAE (Chabot et al. 2006), thus suggesting that TLR activation regulates the gate-keeping functions of the FAE. However, there is also some evidence that such activation on intestinal epithelial cells contributes to microbial sampling. Studies in wild-type and mutant mice have shown a novel mechanism by which enterocyte TLR4 expression mediates phagocytosis and translocation of bacteria across the intestinal barrier (Neal et al. 2006). It is of relevance that although TLR4 expression on intestinal epithelial cells is considered to be low, it is increased following exposure to pro-inflammatory cytokines (Abreu et al. 2003) and can be found at higher levels in the mucosa of inflammatory bowel patients (Cario and Podolsky 2000).

Studies in mice lacking TLR2, TLR4, or MyD88, an adaptor molecule for TLR signalling provide some of the most convincing evidence that such receptors are involved in maintaining gut homeostasis and protecting against injury. Such mice are more susceptible to dextran sodium sulphate-induced colitis and exhibit reduced intestinal epithelial cell proliferation, increased bacterial translocation, decreased mucous production and/or deficient intestinal repair mechanisms (Fukata et al. 2005; Rakoff-Nahoum et al. 2004). TLR activation also contributes to epithelial cytoprotective responses and the production of heat shock proteins 25 and 72 (Rakoff-Nahoum et al. 2004; Ismail and Hooper 2005; Arvans et al. 2005).

While some antimicrobial peptides are constitutively produced, there is evidence that PRR engagement by microbial ligands is involved in the production of others (Hooper et al. 2003; Kobayashi et al. 2005). Paneth cells can detect the presence of abnormally high bacterial counts and increase their production of anti-microbial products accordingly. They sense enteric bacteria via cell-autonomous MyD88-dependent TLR activation (Vaishnava et al. 2008) or via NOD-1 and 2 activation (Voss et al. 2006). Subsequent production of antimicrobial peptides then limits translocation and dissemination of microbes across the mucosal barrier (Vaishnava et al. 2008).

9.3.1 *Maintaining Immune Cell Homeostasis to Avoid Inflammation and Modulation of the Adaptive Immune Response*

One mechanism of immune homeostasis depends on the active modulation of innate immune signalling in epithelial cells by symbiotic microorganisms. Several studies support an active suppression of pro-inflammatory signalling pathways in the epithelial cell. It has been reported that attenuated *Salmonella* species can prevent epithelial innate responses by retaining the transcription factor NF-κB in the cytoplasm (Neish et al. 2000). Another symbiont, *Bacteroides thetaiotamicron,* inhibits NF-κB gene activation by promoting nuclear export of the NF-κB p65 subunit in a peroxisome proliferator-activated receptor -γ (PPAR-γ) dependent mechanism (Kelly et al. 2004). In addition, non-pathogenic components of the microbiota and certain probiotic strains are able to elicit a specific cytokine response from intestinal epithelial cells. While some Gram negative bacteria induce secretion of pro-inflammatory molecules, specific lactobacilli can stimulate production of homeostatic cytokines such as TGF-β which then contribute to the conditioning of the mucosal environment (Haller et al. 2000a).

The intestinal lamina propria and Peyer's patches contain a rich network of DCs which play a critical role in the decision to tolerate or to mount an immune response to non-self or danger signals. The choice depends on the type and amplitude of the microbial challenge as well as intestinal conditioning of the DCs via micro-environmental signals such as retinoic acid (Wang et al. 2009) and TGF-β (Coombes and Powrie 2008; Rescigno 2009). A distinct epithelial response to bacterial components (Haller et al. 2000a) is critical in the conditioning of both DCs and macrophages (Haller et al. 2002). Lamina propria DC subsets have distinct origins and functions (Bogunovic et al. 2009; Varol et al. 2009). While phagocytosis of microbes by epithelial cell-conditioned CD103+DCs promotes the differentiation of Forkhead box (Fox) p3+ T regulatory cells with specificity for commensals and their components (Sun et al. 2007), CD103−DC exposed to flagellated invasive bacteria give rise to Th17 and Th1 cells with pro-inflammatory capacity (Coombes and Powrie 2008). Thus, by regulating DC function, epithelial cells not only participate in barrier function, they preserve gut homeostasis (Fig. 9.1). In both cases, innate and adaptive immune responses are involved (Iliev el al. 2009).

9.4 Innate and Adaptive Immune Response, Functional Continuity and Common Functional Characteristics

Although innate and adaptive immune responses have different functional characteristics, it is important to underscore their multiple interactions and functional similarities.

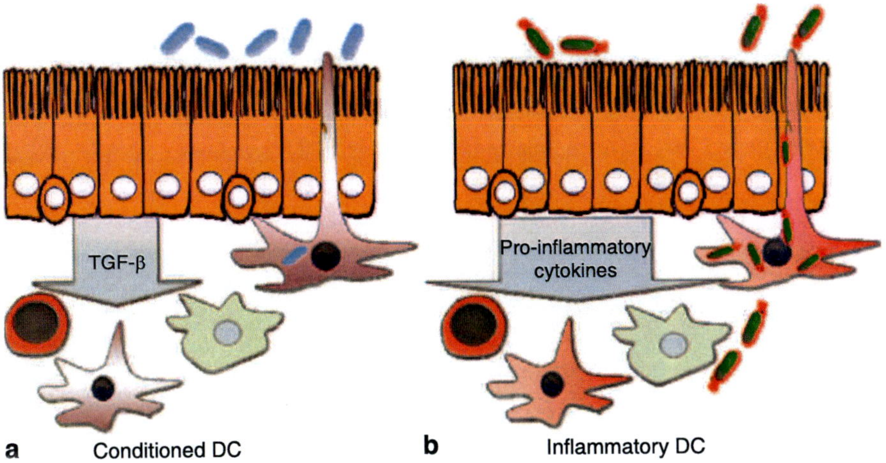

Fig. 9.1 Innate decisions in intestinal homeostasis and defence. In the steady-state, epithelial cell-derived factors condition dendritic cells (DC) in the intestinal lamina propria (**a**) but during pathogenic attack, there is increased production of pro-inflammatory cytokines by epithelial cells and/or lamina propria that have been activated by invading microbes (**b**). This results in an inflammatory DC phenotype

Epithelial cells and DCs are true sentinels and are among the first cellular points-of-contact for antigens at the mucosal interface. The innate responses generated by these cells participate in the induction of adaptive responses. Adaptive immune cells expressing random specificities also recognise antigens but they are not exposed to antigenic characteristics that alert the host of danger (Medzhitov 2007). Their immune response is dependent on the type and location of preceding PRR-mediated innate signals. Thus, microbial recognition by PRR on antigen presenting cells such as DCs governs antigen-specific recognition by lymphocyte receptors. The temporal relationship between the antigen and MAMP signalling in DCs determines the type of immune reaction that is induced in T or B cells (Banchereau and Steinman 1998).

There is increasing evidence that some lymphoid populations such as B1 cells, marginal zone B cells, γδ-T cells and invariant Natural Killer T cells (iNKT) can behave like innate immune cells. Using specific receptors such as immunoglobulin molecules or the T cell receptor (TCR), they can mount effector functions cytolysis or cytokine production upon first encounter with antigen. However, the diversity of antigen receptors is restricted and their specificities are limited to a predefined repertoire of ligands (Bendelac et al. 2001). In addition, innate-like lymphocytes have a skewed distribution in immune compartments where it is more likely to find pre-defined specific ligands. Interestingly, B1 cell antigen receptors also recognise MAMPs. These cells are localised in the peritoneal and pleural cavities and express IgM antibodies that are specific for bacterial polysaccharides (Berland and Wortis 2002).

Some T cells, such as iNKT, also behave like innate cells. These cells express a semi-invariant TCR composed of a conserved alpha chain with a limited repertoire of beta chains. CD1 molecules bind both endogenous lipid and foreign, microbial lipid molecules (Cohen et al. 2009) and during pathogen challenge, they help promote iNKT cells.

Interestingly, some innate cells may have memory. Certain NK cell subsets expressing germ-line encoded invariant receptors that recognise cytomegalovirus (CMV) proliferate and mount a more efficient response upon a second encounter with CMV. Moreover, in contrast to naïve NK cells, these "memory" NK cells can also be stimulated by other environmental signals (Lanier and Sun 2009).

Adaptive immune cells with memory have the capacity to mount an energetic response towards specific antigens but can also respond to environmental signals independently of the presence of the antigen. For example, memory B cells can secrete antibody and IL-15 when stimulated with CpG containing DNA in the absence of specific antigens. Indeed, recent studies have demonstrated that purified B and T lymphocytes express several TLRs and respond directly to TLR ligands (Pasare and Medzhitov 2005; Ruprecht and Lanzavecchia 2006). Furthermore, in the absence of the specific antigen, memory T cells can secrete IFN-γ in response to IL-12 and IL-18 produced by DCs. This may also occur *in vivo*. Mice injected with polyinosinic:polycytidylic acid or endotoxin have activated memory T cells that secrete IFN-γ (Lanier and Sun 2009).

Thus, both memory cells of the adaptive immune system (B, T lymphocytes) and antigen-primed innate cells such as CMV-reactive NK cells can be activated in a bystander manner by environmental signals. This new paradigm adds another level of complexity in our understanding of the mechanisms underlying inflammatory responses at the intestinal mucosa.

9.5 Impaired Innate Immunity in the Gut as a Pathogenic Factor

Dysregulation of host-microbiota homeostasis can lead to chronic or acute pathogenic states such as inflammatory bowel disease (IBD).

9.5.1 Chronic Disease

A basic defect in IBD is incapacity of innate mechanisms of defence to control bacterial density at the intestinal mucosa. This leads to greater numbers of mucosa-associated bacteria. Certainly, the many disease susceptibility genes that have been recently identified suggest that host cell innate immune mechanisms cannot clear bacteria (Fukata et al. 2005) in extracellular and intracellular compartments (Xavier and Podolsky 2007).

NOD2 was the first gene associated with Crohn's disease susceptibility (Hugot et al. 2001). This gene codes for a cytosolic PRR that recognises bacterial peptidoglycans. Several roles for this PRR in disease susceptibility have been postulated. The NOD2 mutations seen in Crohn's disease result in different cellular aspects but one of the most important is diminished NF-κB activation. This may be related to lower defensin production by Paneth cells and lower cytoprotective function at the epithelial layer. As NOD2 activation has been shown to inhibit TLR2 signalling (Xavier and Podolsky 2007), it is also possible that the pathogenetic variants expressed on Crohn's patients fail to modulate pro-inflammatory TLR2 activated pathways. However, it has recently been reported that NOD-2 mutant proteins, unlike their wild type counterparts, actively suppress IL-10 production in a peptidoglycan- and NF-κB-independent mechanism (Noguchi et al. 2009).

9.5.2 Acute Disease

Intestinal exposure to stressors can damage intestinal barrier function and cause an increased intestinal permeability to bacteria. In addition to the increased load of bacteria which challenge epithelial and lamina propria cells, cell injury directly triggers a response of the host cells. Together, these changes alter innate regulation of gut homeostasis.

A frequent injury of the gut is the ischemia-reperfusion damage that is seen in critically ill patients with varied pathologies. Hypoxia is increasingly recognised as an important determinant of cell dysfunction. It promotes cellular activation of the transcription factor, hypoxia inducible factor 1 α (HIF-1α) (Jantsch et al. 2008). When this molecule is expressed on sentinel cells of the gut, pro-inflammatory responses are amplified and the innate immune tolerance that preserves gut homeostasis, is disturbed (Yang et al. 2002).

In different critically ill patients, intestinal blood deprivation can be due to a vascular stress response that has redirected blood flow to other organs or to the use of vasoactive drugs for maintaining blood pressure in shocked patients. In either case, the intestinal environment becomes a source of inflammatory signals that amplify the systemic inflammatory response and contribute to organ failure.

9.6 Modulation of the Microbiota to Improve Gut Defences, Modulate Innate Reactivity and Deliver Health Benefits

9.6.1 Gut Microbiota and Probiotics to Enhance Innate Immune Defences

Within the complexity of the gut microbiota, bacteria have been classified according to the type of interaction that they have with the host. While symbiotic bacteria are those with a health-promoting function, commensals are permanent residents of the microbiota which entertain rather neutral interactions -neither good nor bad- with the host. A third category includes bacteria with the potential to induce pathology. The term "probiotic" is used to define dietary micro-organisms that behave like symbiotic microbes and impart health benefits to the host.

Different methods, ranging from *in vitro* experimental models to human clinical studies, have been utilised to study the biological activities of probiotics. Germ-free and gnotobiotic animal models have proven to be particularly important experimental tools for studying the biological effects of the microbiota on the host. Indeed, formerly germ-free animals which have been colonised with a known, generally simple, bacterial inoculum allow a well-controlled evaluation of the genetic and functional consequences of bacterial colonisation.

In broad terms, colonisation by the microbiota modulates expression of host genes that participate in diverse and fundamental physiological functions such as intestinal proliferation and absorptive function, mucosal barrier fortification and intestinal maturation (Stappenbeck et al. 1998; Hooper et al. 2001).

Bacteroides thetaiotamicron colonisation of germ-free mice has been studied by many researchers because it is a prominent member of both the mouse and human microbiota and because it has symbiotic activities. In the gnotobiotic monocolonised mouse, it induces a response that is indicative of improved host defences at the intestinal interface. More specifically, it increases the number of IgA-producing B1 cells and their expression of the polymeric immunoglobulin receptor that transports IgA across the epithelial layer. In addition, colonisation by this Bacteroides increases expression of the CRP-ductin gene that encodes a mucous layer component (MUCLIN) and a receptor for the trefoil peptides that are involved in mucosal barrier integrity and epithelial restitution respectively. It also increases intestinal expression of the bactericidal protein angiogenin (Hooper et al. 2001). *Bacteroides thetaiotamicron* has also been shown to increase Paneth cell expression of the antimicrobial peptide, regenerating islet derived 3γ (REG3γ) (Cash et al. 2006). As the antibacterial activity of this peptide is directed toward some Gram+ microorganisms, it is likely that *Bacteroides thetaiotamicron* directs gut innate responses in such a way that they preserve their own environment and in so doing, induce host protection. Simultaneous induction of the various genes may avert massive colonisation of the mucosal surface as well as bacteria translocation across the epithelial barrier.

Another model frequently used to study the protective effects of specific microbes involves pathogenic challenge of germ-free or conventional animals in the presence of probiotics. Studies in such animals have demonstrated protective effects of probiotics against enteropathogens such as *Salmonella* (Silva et al. 2004; Sprinz et al. 1961) and *Escherichia coli* O157:H7 (Asahara et al. 2004).

The mechanisms of the increased susceptibility to infectious agents in germ-free or poorly colonised hosts probably involve a number of different components (Round and Mazmanian 2009). Several reports point to inadequate T cell function and low antibody production, particularly of secretory IgA.

Improved protection by symbiotic or probiotic bacteria against pathogens may be due to their stimulation of innate defences. Indeed, a study in humans has demonstrated that a fermented milk containing *Lactobacillus johnsonii* La1 significantly enhances both the phagocytic activity and the respiratory burst of peripheral blood leukocytes (Donnet-Hughes et al. 1999; Schiffrin et al. 1997).

As the microbiota is considered to interact with immunocompetent cells in defined compartments of the mucosal immune system, potential probiotic organisms have been studied in co-culture with different immune cell populations. Such studies have shown that while *Lactobacillus johnsonii* La1 strongly induces cellular production of IFN-γ and IL-12, *Escherichia coli* and lipopolysaccharide preferentially induce IL-10. The cells activated by *L. johnsonii* were NK cells that required the presence and the contact of monocyte-derived macrophages (Haller et al. 2000b). Although the majority of NK cells are located in the liver and the peripheral blood, they are also home to the intestinal mucosa where they play a key role in innate responses against tumours, viral infections and intestinal pathogens.

The molecular basis of gut microbiota-host interactions has been extensively described in recent years and there is a general acceptance that innate immune function is a central player in bacterial sensing and in minimising inflammatory damage in the intestinal microenvironment (Stack et al. 2009; Rakoff-Nahoum 2004). Although cellular and pre-clinical immunomodulatory activities have been attributed to probiotic organisms, clinical evidence supporting a protective role of probiotics against infectious processes is lacking. Nevertheless, some studies have demonstrated beneficial effects of well-selected probiotics in children suffering from acute infectious diarrhoea. There is good evidence that administration of *Lactobacillus rhamnosus* GG (LGG) reduces the duration of acute diarrhoeal episodes in children, particularly when they are of viral aetiology (Guandalini et al. 2000). Probiotics are also effective in preventing acute viral diarrhoea in a clinical setting. The administration of *Bifidobacterium lactis* and *Streptococcus thermophilus* in a placebo controlled study conducted in hospitalised children has been shown to lower the incidence of diarrhoea and of viral shedding (Saavedra et al. 1994). Another study in young Bangladeshi children with acute watery diarrhoea examined the protective effects of *Lactobacillus paracasei* (ST11) or a placebo. Interestingly, in children with non-rotavirus diarrhoea, there was a significantly positive effect of ST11 in reducing the daily intake of an oral rehydration solution and in reducing stool output and frequency. In contrast, no protective effect was observed in rotavirus-infected children (Sarker et al. 2005).

9.6.2 Gut Microbiota and Probiotics to Promote Immune Intestinal Homeostasis

An almost persistent feature of symbiotic bacteria is their capacity to prevent an inflammatory response in the host and thereby, avoid the type of response which ordinarily eliminates "non-self" microbial cells. In fact, this property is probably at the very heart of the symbiotic relationship with the host (Round and Mazmanian 2009) and is a key for promoting homeostasis at the mucosal level.

The selection of probiotics to restore the gut homeostasis in IBD (Sartor 2004; Gionchetti et al. 2003) or systemic conditions of altered immune reactivity such as allergy (Isolauri et al. 2000) or autoimmunity (Ljungberg et al. 2006) has been based on very general patterns of bacterially induced responses in host cells *in vitro* or in the animal host.

The capacity to stimulate production of homeostatic mediators in different immune cell types *in vitro* has been a criterion used by many laboratories to select probiotics. Intestinal epithelial cultures are frequently used in such studies. Indeed, *Lactobacilli* that can promote TGF-β gene activation and do not induce a pro-inflammatory response in epithelial cell cultures have been associated with potentially homeostatic function at the gut level (Haller et al. 2000b, 2002).

The importance of the epithelial cell-DC tandem to interact with commensal microbiota and support mucosal homeostasis was mentioned earlier. A population of epithelial cell-conditioned CD103+ DCs plays a central role in the development of intestinal FOXP3+ T regulatory cells (Sun et al. 2007) in a TGF-β and retinoic acid-dependent manner (Coombes and Powrie 2008). Thus, innate anti-inflammatory responses at the epithelial cell level can greatly contribute to the initiation of tolerogenic adaptive immune responses that are based on the capacity of CD103+ DCs to induce peripheral T regulatory cells from naive T cells (von der Weid et al. 2001; O'Mahony et al. 2008). Intestinal DCs can also interact directly with luminal bacterial via cytoplasmic prolongations that pass between epithelial cell tight-junctions (Rescigno et al. 2001). Studies examining the interaction of DCs with specific probiotics have shown that bacteria such as *Lactobacillus rhamnosus* support an immature DC phenotype and promote T regulatory development from naive CD4+ T cells (Foligne et al. 2007).

Some experimental models support the contention that some probiotics, and even attenuated pathogens, can promote homeostatic epithelial responses and avoid exaggerated inflammation. *In vivo* demonstration of homeostatic activities may depend on a combination of epithelial responses—a sort of innate immune tolerance with concomitant induction of T regulatory cell differentiation. In fact, it is almost impossible to make a distinction between the modulation of innate or adaptive immune response *in vivo* (Neish et al. 2000; Kelly et al. 2004).

Many studies have reported the benefits of probiotics in the modulation of intestinal inflammation. Several different animal studies, in particular, have shown potential benefits of probiotics in IBD (Sartor 2004). Such benefits have subsequently been confirmed in very specific human conditions (Gionchetti et al. 2003) such as pouchitis. Although the underlying mechanism is not totally understood, an increased production of the homeostatic cytokines TGF-β and IL-10 has been reported following the administration of a blend of probiotic bacteria.

There is now a general consensus that components of the microbiota contribute to the pathogenesis of IBD. True pathogens and virulent components of the normal microbiota components have both been implicated. More recently, it has also been shown that the dominant phylum *Firmicutes* and the *Faecalibacterium prausnitzii* species in particular, are present in low numbers in the microbiota of IBD patients. It is of great interest therefore, that oral administration of either live *F. prausnitzii* or its metabolites reduces the severity of experimental colitis and tends to correct the dysbiosis. *In vitro* experiments suggest that these effects are due, at least in part, to an inhibition of NF-κB dependent gene activation (Sokol et al. 2008).

9.7 Conclusion

The intestinal innate immune system is responsible to maintain homeostasis at the gut level. The permanent control of DC differentiation, level of activation and function by the commensal microbiota-epithelial cell cross talk help to generate the tolerogenic DC that promote peripheral T regulatory cell differentiation and B1 cell secretory IgA production. Together these responses prevent a pathogenic chronic inflammation observed in immune-related gut disorders like inflammatory bowel.

References

Abreu MT, Thomas LS, Arnold ET, Lukasek K, Michelsen KS, Arditi M (2003) TLR signaling at the intestinal epithelial interface. J Endotoxin Res 9:322–330

Akira S, Uematsu S, Takeuchi O (2006) Pathogen recognition and innate immunity. Cell 124:783–801

Alam M, Midtvedt T, Uribe A (1994) Differential cell kinetics in the ileum and colon of germfree rats. Scand J Gastroenterol 29:445–451

Arvans DL, Vavricka SR, Ren H, Musch MW, Kang L, Rocha FG, Lucioni A, Turner JR, Alverdy J, Chang EB (2005) Luminal bacterial flora determines physiological expression of intestinal epithelial cytoprotective heat shock proteins 25 and 72. Am J Physiol Gastrointest Liver Physiol 288:G696–G704

Asahara T, Shimizu K, Nomoto K, Hamabata T, Ozawa A, Takeda Y (2004) Probiotic bifidobacteria protect mice from lethal infection with Shiga toxin-producing *Escherichia coli* O157:H7. Infect Immun 72:2240–2247

Banchereau J, Steinman RM (1998) Dendritic cells and the control of immunity. Nature 392:245–252

Barton GM, Kagan JC (2009) A cell biological view of Toll-like receptor function: regulation through compartmentalization. Nat Rev Immunol 9:535–542

Bendelac A, Bonneville M, Kearney JF (2001) Auto-reactivity by design: innate B and T lymphocytes. Nat Rev Immunol 1:177–186

Berland R, Wortis HH (2002) Origins and functions of B-1 cells with notes on the role of CD5. Ann Rev Immunol 20:253–300

Bogunovic M, Ginhoux F, Helft J, Shang L, Hashimoto D, Greter M, Liu K, Jakubzick C, Ingersoll MA, Leboeuf M, Stanley ER, Nussenzweig M, Lira SA, Randolph GJ, Merad M (2009) Origin of the lamina propria dendritic cell network. Immunity 31:513–525

Cario E, Podolsky DK (2000) Differential alteration in intestinal epithelial cell expression of Toll-like receptor 3 (TLR3) and TLR4 in inflammatory bowel disease. Infect Immun 68:7010–7017

Cario E, Brown D, McKee M, Lynch-Devaney K, Gerken G, Podolsky DK (2002) Commensal-associated molecular patterns induce selective toll-like receptor-trafficking from apical membrane to cytoplasmic compartments in polarized intestinal epithelium. Am J Pathol 160:165–173

Cash HL, Whitman CV, Behrendt CL, Hooper LV (2006) Symbiotic bacteria direct expression of an intestinal bactericidal lectin. Science 313:1126–1130

Cebra JJ, Periwal SB, Lee G, Lee F, Shroff KE (1998) Development and maintenance of the gut-associated lymphoid tissue (GALT): the roles of enteric bacteria and viruses. Dev Immunol 6:13–18

Cohen NR, Garg S, Brenner MB (2009) Antigen presentation by CD1 lipids, T cells, and NKT cells in microbial immunity. Adv Immunol 102:1–94

Comelli EM, Simmering R, Faure M, Donnicola D, Mansourian R, Rochat F, Corthesy-Theulaz I, Cherbut C (2008) Multifaceted transcriptional regulation of the murine intestinal mucus layer by endogenous microbiota. Genomics 91:70–77

Coombes JL, Powrie F (2008) Dendritic cells in intestinal immune regulation. Nat Rev Immunol 8:435–446

Corfield AP, Myerscough N, Longman R, Sylvester P, Arul S, Pignatelli M (2000) Mucins and mucosal protection in the gastrointestinal tract: new prospects for mucins in the pathology of gastrointestinal disease. Gut 47:589–594

Chabot S, Wagner JS, Farrant S, Neutra MR (2006) TLRs regulate the gatekeeping functions of the intestinal follicle-associated epithelium. J Immunol 176:4275–4283

Donnet-Hughes A, Rochat F, Serrant P, Aeschlimann JM, Schiffrin EJ (1999) Modulation of non-specific mechanisms of defense by lactic acid bacteria: effective dose. J Dairy Sci 82:863–869

Fritz JH, Ferrero RL, Philpott DJ, Girardin SE (2006) Nod-like proteins in immunity, inflammation and disease. Nat Immunol 7:1250–1257

Foligne B, Zoumpopoulou G, Dewulf J, Ben Younes A, Chareyre F, Sirard JC, Grangette C (2007) A key role of dendritic cells in probiotic functionality. PLoS One 2(3):e313. doi:10.1371/journal.pone.0000313

Fukata M, Michelsen KS, Eri R, Thomas LS, Hu B, Lukasek K, Nast CC, Lechago J, Xu R, Naiki Y, Soliman A, Arditi M, Abreu MT (2005) Toll-like receptor-4 is required for intestinal response to epithelial injury and limiting bacterial translocation in a murine model of acute colitis. Am J Physiol Gastrointest Liver Physiol 288:G1055–G1065

Gionchetti P, Rizzello F, Helwig U, Venturi A, Lammers KM, Brigidi P, Vitali B, Poggioli G, Miglioli M, Campieri M (2003) Prophylaxis of pouchitis onset with probiotic therapy: a double-blind, placebo-controlled trial. Gastroenterology 124:1202–1209

Guandalini S, Pensabene L, Zikri MA, Dias JA, Casali LG, Hoekstra H, Kolacek S, Massar K, Micetic-Turk D, Papadopoulou A, de Sousa JS, Sandhu B, Szajewska H, Weizman Z (2000) Lactobacillus GG administered in oral rehydration solution to children with acute diarrhoea: a multicenter European trial. J Pediatr Gastroenterol Nutr 30:54–60

Haller D, Bode C, Hammes W, Pfeifer A, Schiffrin EJ, Blum S (2000a) Non-pathogenic bacteria elicit a differential cytokine response by intestinal epithelial cell/leucocyte co-cultures. Gut 47:79–87

Haller D, Blum S, Bode C, Hammes WP, Schiffrin EJ (2000b) Activation of human peripheral blood mononuclear cells by nonpathogenic bacteria in vitro: evidence of NK cells as primary targets. Infect Immun 68:752–759

Haller D, Serrant P, Peruisseau G, Bode C, Hammes WP, Schiffrin EJ, Blum S (2002) IL-10 producing CD14 low monocytes inhibit lymphocyte-dependent activation of intestinal epithelial cells by commensal bacteria. Microbiol Immunol 46:195–205

Hooper LV (2009) Do symbiotic bacteria subvert host immunity? Nat Rev Microbiol 7:367–374

Hooper LV, Wong MH, Thelin A, Hansson L, Falk PG, Jeffrey I, Gordon JI (2001) Molecular analysis of commensal host-microbial relationships in the intestine. Science 291:881–884

Hooper LV, Stappenbeck TS, Hong CV, Gordon JI (2003) Angiogenins: a new class of microbicidal proteins involved in innate immunity. Nat Immunol 4:269–273

Hugot JP, Chamaillard M, Zouali H, Lesage S, Cézard JP, Belaiche J, Almer S, Tysk C, O'Morain CA, Gassull M, Binder V, Finkel Y, Cortot A, Modigliani R, Laurent-Puig P, Gower-Rousseau C, Macry J, Colombel JF, Sahbatou M, Thomas G (2001) Association of NOD2 leucine-rich repeat variants with susceptibility to Crohn's disease. Nature 411:599–603

Iliev ID, Mileti E, Matteoli G, Chieppa M, Rescigno M (2009) Intestinal epithelial cells promote colitis-protective regulatory T-cell differentiation through dendritic cell conditioning. Mucosal Immunol 2:340–350

Ismail AS, Hooper LV (2005) Epithelial cells and their neighbors. IV. Bacterial contributions to intestinal epithelial barrier integrity. Am J Physiol Gastrointest Liver Physiol 289:G779–G784

Isolauri E, Arvola T, Sütas Y, Moilanen E, Salminen S (2000) Probiotics in the management of atopic eczema. Clin Exp Allergy 30:1604–1610

Janeway CA Jr (1989) Approacing the asymptote? Evolution and revolution in immunology. Cold Spring Harb Symp Quant Biol 54(Pt 1):1–13

Jantsch J, Chakravortty D, Turza N, Prechtel AT, Buchholz B, Gerlach RG, Volke M, Gläsner J, Warnecke C, Wiesener MS, Eckardt KU, Steinkasserer A, Hensel M, Willam C (2008) Hypoxia and hypoxia-inducible factor-1 alpha modulate lipopolysaccharide-induced dendritic cell activation and function. J Immunol 180:4697–4705

Johansson ME, Phillipson M, Petersson J, Velcich A, Holm L, Hansson GC (2008) The inner of the two Muc2 mucin-dependent mucus layers in colon is devoid of bacteria. Proc Natl Acad Sci U S A 105:15064–15069

Kabelitz D, Medzhitov R (2007) Innate immunity- cross-talk with adaptive immunity through pattern recognition receptors and cytokines. Curr Opin Immunol 19:1–3

Karlsson J, Pütsep K, Chu H, Kays RJ, Bevins CL, Andersson M (2008) Regional variations in Paneth cell antimicrobial peptide expression along the mouse intestinal tract. BMC Immunol 9:37

Kawai T, Akira S (2006) TLR signalling. Cell Death Diff 13:816–825

Kelly D, Campbell JI, King TP, Grant G, Jansson EA, Coutts AG, Pettersson S, Conway S (2004) Commensal anaerobic gut bacteria attenuate inflammation by regulating nuclear-cytoplasmic shuttling of PPAR-gamma and RelA. Nat Immunol 5:104–112

Kelly DF, Pollard AJ, Moxon ER (2005) Immunological memory. JAMA 294:3019–3023

Kobayashi KS, Chamaillard M, Ogura Y, Henegariu O, Inohara N, Nuñez G, Flavell RA (2005) Nod2-dependent regulation of innate and adaptive immunity in the intestinal tract. Science 307:731–734

Lanier LL (2005) NK cell recognition. Ann Rev Immunol 23:225–274

Lanier LL, Sun J (2009) Do the terms innate and adaptive immunity create conceptual barriers? Nat Rev Immunol 9:302–303

Ljungberg M, Korpela R, Ilonen J, Ludvigsson J, Vaarala O (2006) Probiotics for the prevention of beta cell autoimmunity in children at genetic risk of type 1 diabetes-the PRODIA study. Ann N Y Acad Sci 1079:360–364

MacPherson AJ, Uhr T (2004) Induction of protective IgA by intestinal dendritic cells carrying commensal bacteria. Science 303:1662–1665

Man AL, Prieto-Garcia ME, Nicoletti C (2004) Improving M cell mediated transport across mucosal barriers: do certain bacteria hold the keys? Immunology 113:15–22

Martinon F, Tschopp J (2004) Inflammatory caspases: linking an intracellular innate immune system to autoinflammatory diseases. Cell 117:561–574

Matzinger P (1994) Tolerance, danger, and the extended family. Ann Rev Immunol 12:991–1045

Medzhitov R (2007) Recognition of mciroorganisms and activation of the immune response. Nature 449:819–826

Meylan E, Tschopp J, Karin M (2006) Intracellular pattern recognition receptors in the host response. Nature 442:39–44

Mukherjee S, Vaishnava S, Hooper LV (2008) Multi-layered regulation of intestinal antimicrobial defense. Cell Mol Life Sci 65:3019–3027

Neal MD, Leaphart C, Levy R, Prince J, Billiar TR, Watkins S, Li J, Cetin S, Ford H, Schreiber A, Hackam DJ (2006) Enterocyte TLR4 mediates phagocytosis and translocation of bacteria across the intestinal barrier. J Immunol 176:3070–3079

Neish AS, Gewirtz AT, Zeng H, Young AN, Hobert ME, Karmali V, Rao AS, Madara JL (2000) Prokaryotic regulation of epithelial responses by inhibition of IkappaB-alpha ubiquitination. Science 289:1560–1563

Noguchi E, Homma Y, Kang X, Netea MG, Ma X (2009) A Crohn's disease-associated NOD2 mutation suppresses transcription of human IL10 by inhibiting activity of the nuclear ribonucleoprotein hnRNP-A1. Nat Immunol 10(May):471–479

O'Mahony C, Scully P, O'Mahony D, Murphy S, O'Brien F, Lyons A, Sherlock G, MacSharry J, Kiely B, Shanahan F, O'Mahony L (2008) Commensal-induced regulatory T cells mediate protection against pathogen-stimulated NF-κB activation. PLoS Pathog 4:e1000112

Pasare C, Medzhitov R (2005) Control of B-cell responses by Toll-like receptors. Nature 436:364–368

Pütsep K, Axelsson LG, Boman A, Midtvedt T, Normark S, Boman HG, Andersson M (2000) Germ-free and colonized mice generate the same products from enteric prodefensins. J Biol Chem 275:40478–40482

Rakoff-Nahoum S, Medzhitov R (2008) Innate immune recognition of the indigenous microbial flora. Mucosal Immunol 1(1 Suppl):S10–S14

Rakoff-Nahoum S, Paglino J, Eslami-Varzaneh F, Edberg S, Medzhitov R (2004) Recognition of commensal microflora by Toll-like receptors is required for intestinal homeostasis. Cell 118:229–241

Rescigno M (2003) Identification of a new mechanism for bacterial uptake at mucosal surfaces, which is mediated by dendritic cells. Pathol Biol (Paris) 51:69–70

Rescigno M (2009) Before they were gut dendritic cells. Immunity 31:454–456

Rescigno M, Rotta G, Valzasina B, Ricciardi-Castagnoli P (2001) Dendritic cells shuttle microbes across gut epithelial monolayers. Immunobiology 204:572–581

Round JL, Mazmanian SK (2009) The gut microbiota shapes intestinal immune responses during health and disease. Nat Rev Immunol 9:313–323

Ruprecht CR, Lanzaveccjia A (2006) Toll-like receptor stimulation as a third signal required for activation of human naive B cells. Eur J Immunol 36:810–816

Saavedra JM, Bauman NA, Oung I, Perman JA, Yolken RH (1994) Feeding of *Bifidobacterium bifidum* and *Streptococcus thermophilus* to infants in hospital for prevention of diarrhoea and shedding of rotavirus. Lancet 344:1046–1049

Sarker SA, Sultana S, Fuchs GJ, Alam NH, Azim T, Brüssow H, Hammarström L (2005) *Lactobacillus paracasei* strain ST11 has no effect on rotavirus but ameliorates the outcome of nonrotavirus diarrhoea in children from Bangladesh. Pediatrics 116(2):e221–e228

Sartor RB (2004) Therapeutic manipulation of the enteric microflora in inflammatory bowel diseases: antibiotics, probiotics, and prebiotics. Gastroenterology 126:1620–1633

Scaffidi P, Misteli T, Bianchi ME (2002) Release of chromatin protein HMGB1 by necrotic cells triggers inflammation. Nature 418:191–195

Schatz DG, Oettinger MA, Schlissel MS (1992) V(D) recombination, molecular biology and regulation. Ann Rev Immunol 10:359–383

Schiffrin EJ, Brassart D, Servin AL, Rochat F, Donnet-Hughes A (1997) Immune modulation of blood leukocytes in humans by lactic acid bacteria: criteria for strain selection. Am J Clin Nutr 66:515S–520S

Silva AM, Barbosa FH, Duarte R, Vieira LQ, Arantes RM, Nicoli JR (2004) Effect of *Bifidobacterium longum* ingestion on experimental salmonellosis in mice. J Appl Microbiol 97:29–37

Sokol H, Pigneur B, Watterlot L, Lakhdari O, Bermúdez-Humarán LG, Gratadoux JJ, Blugeon S, Bridonneau C, Furet JP, Corthier G, Grangette C, Vasquez N, Pochart P, Trugnan G, Thomas G, Blottière HM, Doré J, Marteau P, Seksik P, Langella P (2008) *Faecalibacterium prausnitzii* is an anti-inflammatory commensal bacterium identified by gut microbiota analysis of Crohn disease patients. Proc Natl Acad Sci U S A 105:16731–16736

Sprinz H, Kundel DW, Dammin GJ, Horowitz RE, Schneider H, Formal SB (1961) The response of the germ-free guinea pig to oral bacterial challenge with *Escherichia coli* and *Shigella flexneri*. With special reference to lymphatic tissue and the intestinal tract. Am J Pathol 39:681–695

Stack E, Hapfelmeir S, Stecher B, Velykoredko Y, Stoel M, Lawson MAE, Geuking MB, Beutler B, Tedder TF, Hardt W-D, Bercik P, Verdu E, McCoy KD, Macpherson AJ (2009) Innate and adaptive immunity cooperate flexibly to maintain host-microbiota mutualism. Science 325:617–620

Stappenbeck TS, Wong MH, Saam JR, Mysorekar IU, Gordon JI (1998) Notes from some crypt watchers: regulation of renewal in the mouse intestinal epithelium. Curr Opin Cell Biol 10:702–709

Strober W, Murray PJ, Kitani A, Watanabe T (2006) Signalling pathways and molecular interactions of NOD1 and NOD2. Nat Rev Immunol 6:9–20
Sun CM, Hall JA, Blank RB, Bouladoux N, Oukka M, Mora JR, Belkaid Y (2007) Small intestine lamina propria dendritic cells promote de novo generation of Foxp3 T reg cells via retinoic acid. J Exp Med 204:1775–1785
Szentkuti L, Riedesel H, Enss ML, Gaertner K, Von Engelhardt W (1990) Pre-epithelial mucus layer in the colon of conventional and germ-free rats. Histochem J 22:491–497
Tig JP, Davis BK (2005) CATERPILLER: a novel gene family important in immunity, cell death, and diseases. Ann Rev Immunol 23:387–414
Unterholzner L, Bowie AG (2008) The interplay between viruses and innate immune signaling: recent insights and therapeutic opportunities. Biochem Pharmacol 75:589–602
Vaishnava S, Behrendt CL, Ismail AS, Eckmann L, Hooper LV (2008) Paneth cells directly sense gut commensals and maintain homeostasis at the intestinal host-microbial interface. Proc Natl Acad Sci U S A 105:20858–20863
Varol C, Vallon-Eberhard A, Elinav E, Aychek T, Shapira Y, Luche H, Fehling HJ, Hardt WD, Shakhar G, Jung S (2009) Intestinal lamina propria dendritic cell subsets have different origin and functions. Immunity 31(3):502–512 (Sept 18)
von der Weid T, Bulliard C, Schiffrin EJ (2001) Induction by a lactic acid bacterium of a population of CD4(+) T cells with low proliferative capacity that produce transforming growth factor beta and interleukin-10. Clin Diagn Lab Immunol 8:695–701
Voss E, Wehkamp J, Wehkamp K, Stange EF, Schröder JM, Harder J (2006) NOD2/CARD15 mediates induction of the antimicrobial peptide human beta-defensin-2. J Biol Chem 281:2005–20011
Wang J, Huizinga TW, Toes RE (2009) De novo generation and enhanced suppression of human CD4+CD25+ regulatory T cells by retinoic acid. J Immunol 183:4119–4126
Xavier RJ, Podolsky DK (2007) Unravelling the pathogenesis of inflammatory bowel disease. Nature 448:427–434
Yang R, Gallo DJ, Baust JJ, Watkins SK, Delude RL, Fink MP (2002) Effect of hemorrhagic shock on gut barrier function and expression of stress-related genes in normal and gnotobiotic mice. Am J Physiol Regul Integr Comp Physiol 283:R1263–R1274
Yoneyama M, Kikuchi M, Natsukawa T, Shinobu N, Imaizumi T, Miyagishi M, Taira K, Akira S, Fujita T (2004) The RNA helicase RIG-I has an essential function in double-stranded RNA-induced innate antiviral responses. Nat Immunol 5:730–737

Chapter 10
Interaction of Probiotics with the Acquired Immune System in the Gut

Rossana D'Arienzo, Kathryne B. Schwartz and Mauro Rossi

10.1 Introduction

The mucosa of the intestinal tract is exposed to a vast number of orally acquired environmental antigens, as well as a large community of commensal bacteria. The intestinal epithelium, mucosal immune system, and gut microbiota coordinate a network of immunological and non-immunological activities that are responsible for counteracting foreign pathogens, while simultaneously tolerating microbiota. As long as the bacteria are kept outside the physical barrier of the intestines, inflammation is not initiated. This phenomenon is essentially due to the fact that gut epithelial cell are sealed to one another by tight junctions that control the paracellular route of solutes and actively distinguish between invasive and noninvasive bacteria. Antimicrobial peptides released by enterocytes also help control the translocation and propagation of microbes. Meanwhile, the epithelium and gut-associated lymphoid tissue (GALT) develop complex immune responses to varied stimuli in order to ensure pathogen elimination.

The gut microbiota and probiotics influence both the development and regulation of immune and non-immune defences, as demonstrated by a study that investigated the impact of a germ-free environment in mice (Tlaskalová-Hogenová et al. 2004). The authors observed modifications in both localized intestinal and systemic levels of immunity, specifically immature GALT, decreased immunoglobulin levels in the sera, and smaller spleens. Data shows, however, that the stimulatory effects of probiotics, not limited to innate immunity alone, also influence pre-established adaptive immunity. The following chapter will focus primarily on these probiotic mechanisms and GALT interactions that serve to regulate the host's adaptive immune responses. In addition, the possible applications of probiotics in the modulation of ongoing immune responses will also be reviewed.

M. Rossi (✉)
Institute of Food Sciences, CNR, via Roma 64, 83100 Avellino, Italy
Tel.: +39-0825-299391
Fax: +39-0825-299104
e-mail: mrossi@isa.cnr.it

J. J. Malago et al. (eds.), *Probiotic Bacteria and Enteric Infections*,
DOI 10.1007/978-94-007-0386-5_10,

10.2 GALT Components Involved in Adaptive Immunity

The intestines are home to the largest amount of lymphoid tissue in the body, commonly referred to as the GALT. The anatomical compartments of the GALT are divided into two categories: effector sites, including the intra-epithelial lymphocytes (IEL) and lamina propria (LP), and inductive sites, such as the mesenteric lymph nodes (MLN), Peyer's patches (PP), isolated lymphoid follicles (ILF), and cryptopatches. IEL represent a distinct class of T lymphocytes that are mainly involved in innate mechanisms found scattered throughout the intestinal epithelial layer. Some IEL display cytotoxic activity, whereas others exert a regulatory function that contributes to the maintenance of oral and peripheral tolerance (Locke et al. 2006). The LP, however, is the main effector site of adaptive immunity and, as such, contains a large proportion of the body's lymphocytes, including immunoglobulin A (IgA) producing plasma cells, and αβ-TCR $CD4^+$ and $CD8^+$ effector T cells (Johansson-Lindbom and Agace 2007). T regulatory (Treg) cell subsets, important for the induction of tolerance, have also been identified at this site (Izcue et al. 2009). The LP also hosts the largest population of tissue resident macrophages and dendritic cells (DC) that are professional antigen-presenting cells (APC) originating from circulating monocytes (Schenk and Mueller 2008).

Meanwhile, several studies have evidenced that MLN are crucial for oral tolerance (Mowat 2003; Spahn et al. 2002; Worbs et al. 2006). APC that are present in the LP migrate to the MLN after encountering antigens, thereby enabling recognition and activation of naïve T cells (Kunkel et al. 2003). The induction of adaptive immunity, as well as oral tolerance, in the MLN is greatly dependent on the migration of antigen-loaded DC. PP, structures consisting of randomly distributed lymphoid aggregates, represent the principal inductive sites within the intestinal mucosa (Newberry and Lorenz 2005). The follicle-associated epithelium of PP contains highly specialized epithelial microfold (M) cells that function as active transporters of dietary or microbial antigens from the gut lumen to the sub-epithelial dome region. Once in this region, an active or tolerogenic immune response can be established (Jump and Levine 2002; Suzuki et al. 2008).

ILF are similar to PP in many ways, therefore giving them the capacity to function as inductive sites for mucosal immunity as well (Lorenz and Newberry 2004). These structures are barrel-shaped, single-domed, mononuclear cell aggregates located throughout the surface of the small intestine (Cesta 2006), the development of which requires the presence of lumen antigens (Hamada et al. 2002; Lorenz et al. 2003). Lastly, cryptopatches are small clusters of immature lymphocyte-like cells, surrounded by DC and stromal cells (Kanamori et al. 1996), found predominantly in the small intestine adjacent to the intestinal crypts (Newberry and Lorenz 2005). Based on laboratory investigations, cryptopatches appear to be involved in the organization of intestinal lymphoid structures, particularly the formation of ILF (Eberl 2005).

10.3 Regulatory Elements of Immune Tolerance in GALT

Activated plasma cells in the lamina propria are crucial for adaptive immune responses triggered by lumen antigens, because they promote large-scale production of IgA that is then secreted into the gut lumen (Suzuki and Fagarasan 2008). Coating of bacteria by IgA in the gut lumen facilitates bacterial translocation through M cells in the follicle-associated epithelium of PP and ILF and contributes to mucosal homeostasis by enabling antigen sampling by DC in the subepithelial dome (Rey et al. 2004). Additionally, activated T lymphocytes launch a range of responses targeted towards innocuous antigens and pathogens. $CD4^+$ Treg cells are responsible for controlling these responses in order to avoid inappropriate inflammation in the gut. Treg cells suppress the proliferation and activity of nearly all types of immune cells (Carrier et al. 2007). $CD4^+$ Treg subsets are noticeable in naturally-occurring (Sakaguchi et al. 2006; Uhlig et al. 2006) and induced $CD25^+Foxp3^+$ Treg cells (Coombes et al. 2007; Sun et al. 2007), TGF-β^+Foxp3$^+$ TH3 cells (Faria and Weiner 2005), and IL-10^+Foxp3$^-$ Tr1 cells (Roncarolo et al. 2006). $Foxp3^+CD4^+$ Treg cells, in particular, represent 20–40% of $CD4^+$ T cells in the LP, highlighting the importance of this subset in the control of intestinal immune responses (Ivanov et al. 2007). This point is further emphasized by the fact that induced $CD25^+Foxp3^+$ Treg cells emerged in the intestinal immune system in clear correlation with the appearance of commensal bacteria and dietary antigens in the gut (Coombes et al. 2007).

A few different mechanisms have been proposed to explain the suppressive function of $Foxp3^+CD4^+$ Treg cells, one of which is supported by indications that IL-10 secreted by these cells is mainly involved in blocking effector T cell responses to microbial antigens (Li and Flavell 2008). A study conducted by Kuhn et al. (1993) found that IL-10 deficiency in mice results in the development of a lethal form of colitis. Interestingly, when mice were kept under sterile conditions, however, colitis did not develop, nor were there any signs of immune system activation (Sellon et al. 1998).

Along with IL-10, TGF-β is a crucial cytokine for effective maintenance of tolerance. TGF-β performs this function by influencing the development and balance of effector and regulatory immune cells within the system (Li and Flavell 2008). Furthermore, TGF-β appears particularly important for maintenance of Foxp3 expression in peripheral Treg cells (Marie et al. 2005). Recent evidences suggest that the development of inflammatory TH17 cells, located almost exclusively in the LP of the small intestine, and that of $Foxp3^+$ Treg cells are linked. This is because both cell types require TGF-β to achieve proper differentiation. At low concentrations, TGF-β acts synergistically with IL-6 and IL-21 to promote TH17 differentiation, yet activates $Foxp3^+$ Treg cells at high concentrations (Zhou et al. 2008).

Recently, the importance of another cytokine, IL-2, was considered regarding the suppression of TH17 polarization, and thus regulating the balance between regulatory and inflammatory immune cells (Laurence et al. 2007). More specifically, IL-2 promotes STAT5-mediated suppression of IL-17 production while concurrently enabling Treg expansion and survival, thereby limiting local inflammation. Furthermore, production of $Foxp3^+$ Treg with suppressive activity from mice can be stimulated *in vitro* with IL-2 and TGF-β exposure. Interestingly, the addition of retinoic acid (RA) can further enhance this production process (Mucida et al. 2007), demonstrating the unique role of RA in the mediation of a reciprocal interplay between a generation of Foxp3$^+$ Treg and inflammatory TH17 cells.

Finally, mechanisms have also been proposed for the way in which Treg directly target APC (Shevach 2009). One such hypothesized explanation is supported in large part by the fact that constitutive surface expression of cytotoxic T lymphocyte antigen 4 (CTLA-4) enables the interaction with co-stimulatory molecules CD80 and CD86 on APCs. Consequently, this communication causes down-regulated expression of CD80 and CD86, as well as limits development of adaptive immune responses (Shevach 2009; Onishi et al. 2008). This knowledge is evidenced by a study conducted with CTLA-4 knockout mice. It was observed that the absence of CTLA-4 in these mice resulted in the development of multi-organ inflammation and premature death (Read et al. 2006; Wing et al. 2008).

10.4 Probiotics and Induction of Tolerance

Probiotic strains are capable of regulating the magnitude of immune activation in response to pathogens, or other harmful antigens, in order to prevent the development of excessive inflammation. It has been shown that several probiotic strains possess the ability to induce IL-10 production in epithelial cells both *in vitro* and *in vivo* (Medina et al. 2007; Ménard et al. 2008). In light of these facts, it seems likely that they may also play a role in lymphocyte differentiation (Forchielli and Walker 2005), a process mediated via modulation of TLR-signaling that is essential for a balanced conversion of naïve lymphocytes into effector T cells or regulatory Treg cells (Baumgart and Carding 2007) (Fig. 10.1). The proposed effects of probiotic bacteria on lymphocyte differentiation are supported by the fact that certain probiotic strains were shown to induce cytokine patterns in DC different from those created by pathogenic bacteria. *Lactobacillus* and *Bifidobacterium*, induced IL-10 in MLN-derived DC, while *Salmonella* induced Th1-polarizing cytokines under the same experimental conditions (O'Mahony et al. 2006). IL-10 production and CD83 expression, a maturation surface marker, were induced in cord blood DC by *B. bifidum, B. longum,* and *B. pseudocatenulatum* strains, thus provoking a shift of the immune response toward a Th2 profile (Young et al. 2004). Moreover, TLR2 interactions with culture supernatant from *B. breve* BbC50SN promoted DC maturation by up-regulating CD83 expression, as well as that of CD86 and HLA-DR (Hoarau et al. 2006). Again, the maturation of DC was associated with IL-10 production. Interestingly, the probiotic mixture VSL#3 was also found to influence DC maturation (Hart et al. 2004). VSL#3 contains eight different bacterial strains (four lactobacilli, three bifidobacteria, and one *Streptococcus thermophilus* subsp. *Salivarius*) and has been shown beneficial in the treatment of some inflammatory bowel diseases (IBD). In a study authored by Hart et al., VSL#3 was discovered to be a potent activator of IL-10 production by DC isolated from blood and intestinal LP. Moreover, VSL#3 also exhibited the ability to diminish proinflammatory effects of LPS by decreasing LPS-induced IL-12 production while simultaneously maintaining IL-10 production. Interestingly, individual bacterial strains within VSL#3 displayed distinct immunomodulatory effects on DC, with the most marked anti-inflammatory effects resulting from bifidobacteria strains. It is important to bear in mind, however, that different strains of the same species may also differentially polarize the phenotype of the immune response. For example, some strains of *B. longum* were shown to induce high levels of IL-4 and IL-10 (Th2-like response) in splenocytes, while others promoted a Th1-like phenotype characterized by IFN-γ and TNF-α secretions (Ménard et al. 2008). Therefore, further work is still necessary to obtain a full understanding of the molecular mechanisms underlying this marked specificity in cognate strains. Overall, the findings presented by the previously described studies clearly denote the value of preliminary tests for screening specific probiotic strains for the ability to change the outcome of the immune response in order to alleviate intestinal pathologies.

Meanwhile, the modulatory activity of probiotics on DC produces 'natural adjuvant effects' on intestinal Treg cells (Fig. 10.1). Smits et al. (2005) found that some strains of *L. reuteri* and *L. casei* can induce IL-10-producing Treg cells *in vitro* by modulating DC through the use of specific adhesion molecules. Recently, syngenic transfer of DC from mice treated with *L. salivarius* Ls33 or *L. rhamnosus* Lr32, but

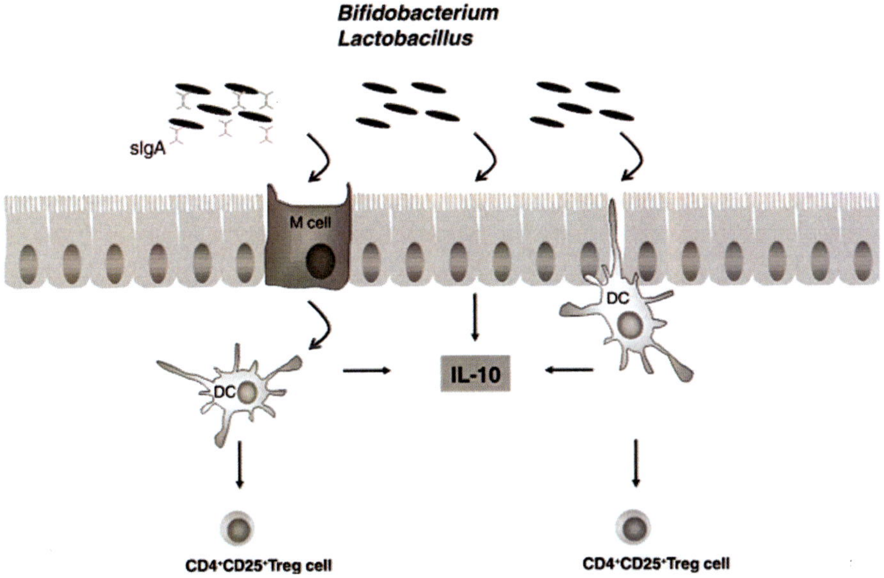

Fig. 10.1 Tolerance induction by probiotics. Probiotics can directly interact with TLR expressed on the surface of enterocytes and DC sampling the intestinal lumen, inducing the anti-inflammatory IL-10. This activity has been observed also following coating of bacteria with IgA that facilitates bacterial translocation through M cells in the follicle-associated epithelium of PP and ILF, enabling antigen sampling by dendritic cells in the subepithelial dome. In addition, interaction of probiotics with DC produce 'natural adjuvant effects' on intestinal Treg, with the specific induction of $CD4^{+}CD25^{+}$ Treg cells

not *L. acidophilus* NCFM, protected against 2, 4, 6-trinitrobenzenesulfonic acid (TNBS)-induced colitis, an effect associated with the induction of CD4⁺CD25⁺ Treg cells (Foligne et al. 2007). In agreement with the above results, administration of *Bifidobacterium infantis* 35,624 in mice that were subsequently infected with *Salmonella typhimurium* revealed an inhibition of infection that was correlated with reduced NF-kappaB activity (O'Mahony et al. 2008). Notably, NF-kappaB inhibitory activity also resulted from adoptive transfer of $CD4^+CD25^+$ T cells isolated from *B. infantis*-fed mice. Complimenting this data, a recent study featuring oral administration of the probiotic strain *Lactobacillus casei* DN-114 001 in a mouse model of colitis, induced by effector/memory CD8(+) T cells, showed that this treatment reduced the magnitude of the disease (Hacini-Rachinel et al. 2009). Again, the beneficial effect produced by the described treatment was found to be mediated by increased activity of $Foxp3^+$ $CD4^+$ Treg cells derived from colon LP.

10.5 Probiotics and Enhancement of Adaptive Immunity in the GALT

In addition to regulatory activities, probiotic bacteria can also contribute to the effector T cell response toward pathogens by enhancing induction of pro-inflammatory cytokines (Cross 2002) (Fig. 10.2). In particular, several *Lactobacillus* strains

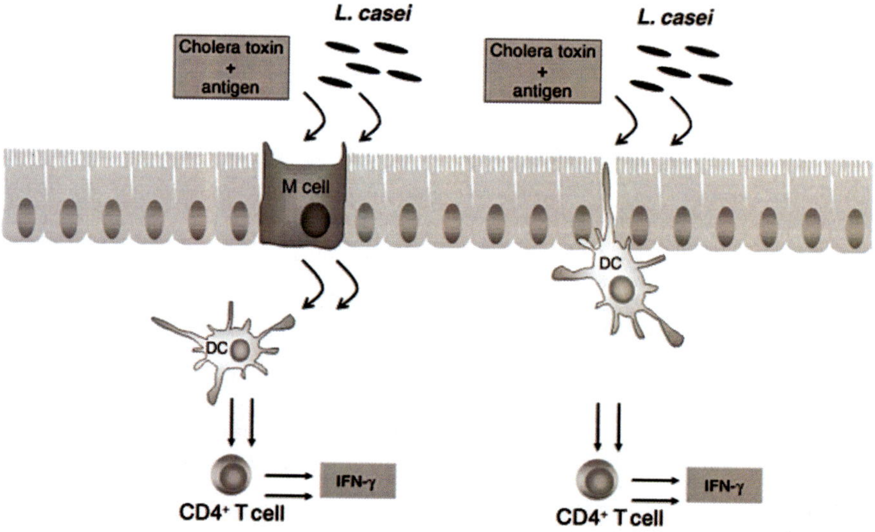

Fig. 10.2 Ability of probiotics to enhance the antigen specific adaptive immunity into the GALT. Probiotic bacteria can give a contribution to the effector T cell response toward ongoing immune response by enhancing the induction of pro-inflammatory cytokines. Among the tested strains, *L. casei* has been shown to exert such effects. acting by exhacerbating the T cell responses toward a Th1 phenotype

administered orally were able to efficiently improve the response toward parenterally immunized antigens (Maassen et al. 2000); most importantly, the adjuvanticity of probiotic strains were associated with extremely reduced levels of intrinsic immunogenicity (Maassen et al. 2003). Out of all of the strains tested, *L. casei* had the greatest impact on the adaptive response, and together with dextran, promoted a significant increase of NK cell activities in healthy humans (Ogawa et al. 2006). In a series of murine models, oral administration of *L. casei* resulted in suppression of allergic responses (Kim et al. 2005; Shida et al. 2002), an event associated with *in vitro* inhibition of IL-4 and IL-5 secretions together with activation of IFN-γ secretion by stimulated splenocytes (Shida et al. 1998). It was recently discovered that induction of intestinal hypersensitivity towards gliadin in DQ8 transgenic mice, a model of celiac disease, could be strongly enhanced by the administration of an oral challenge with *L. casei* (D'Arienzo et al. 2008), although the stronger IFN-γ production induced by this treatment was not accompanied by any sign of enteropathy. This property of *L. casei* was confirmed in another study that demonstrated the development of protective immunity against *Salmonella* following administration of recombinant *L. casei* expressing the flagellar antigen (Kajikawa et al. 2007). Together, these results serve to strongly emphasize the potential application of *L. casei* as a safe vaccine adjuvant that acts mainly by promoting T cell responses toward a Th1 phenotype.

In addition to the influence of *L. casei*, other probiotics were found to improve the humoral response by enhancing antigen presentation to B lymphocytes, therefore increasing secretory IgA production against pathogens (Yasui et al. 1999). As a result of this knowledge, the co-adjuvant activity of *Lactobacillus fermentum* (CECT5716) was recently tested in human clinical trials. These studies showed that oral administration of CECT5716 strengthened the systemic effectiveness of an anti-influenza vaccine by increasing both the T-helper type 1 response and the production of virus-neutralizing antibodies (Olivares et al. 2007). Based on the aforementioned data, the possible use of probiotics to change the outcome of an antibody response seems a promising proposition. To this end, oral bacteriotherapy featuring probiotics could prove instrumental in the management of IgE-mediated allergic diseases. The administration of *L. casei* Shirota (LcS) was recently investigated, revealing its ability to reduce seasonal allergic rhinitis (SAR) (Ivory et al. 2008). Specifically, Ivory et al. observed increased levels of specific IgG and decreased levels of IgE in the patients treated with probiotics. These results were accompanied by a significant reduction of *in vitro* IL-5, IL-6, and IFN-γ production promoted by PBMC.

Yet more recently, the effects of VSL#3 preparations were studied in Parj1-sensitized mice. Results showed the ability of VSL#3 to shift, *in vitro*, the allergen-specific cytokine production activated in splenocytes towards a Treg/Th0 profile (Mastrangeli et al. 2009). Interestingly, this prophylactic treatment with VSL#3 *in vivo* caused a significant reduction of serum specific IgG. These results suggest that the probiotic preparations do not only encourage primary responses towards a Treg/Th0-type profile but can also modify an established Th2 response.

10.6 New Applications of Probiotics for GALT Immunomodulation

The development of genetically modified bacterial strains co-expressing antigens and cytokines seems to be one of the most promising paths towards the establishment of new therapeutic applications of probiotics targeting GALT immune responses. A pioneering study by Steidler et al. (1998) using intranasal administration of *L. lactis* expressing tetanus toxin fragment C (TTFC) and either mIL-2 or mIL-6 demonstrated the soundness of this claim. The authors' results showed significantly higher antibody titres resulting from the modified strains with mIL-2 or mIL-6 than those produced by the parental strain expressing TTFC alone (Steidler et al. 1998). More recently, another study utilized intranasal co-administration of *L. lactis* secreting mIL-12 alongside with *L. lactis* engineered to produce the E7 antigen of human papillomavirus type 16 (Bermudez-Humaran et al. 2003). The described treatment produced enhanced levels of IL-2 and IFN-γ following *in vitro* stimulation of splenocytes isolated from immunized mice. On the other hand, an intragastric treatment using a form of *L. lactis* that produces immunomodulatory cytokine IL-10 was effective in reducing colitis in various mouse models (Steidler et al. 2000), indicating that the acidic pH of the stomach did not hinder local IL-10 delivery in the gut by means of probiotic strain. These findings have forged the way for the advancement of a safe and effective therapy for GALT in the form of dietary supplementation with IL-10. A key characteristic of this treatment, largely contributing to its level of accomplishment, is the way in which it limits systemic exposure to IL-10, thus avoiding the possible shortcomings associated with parenteral IL-10 therapy. Accordingly, a Phase 1 clinical trial was designed to evaluate the effects of oral administration of IL-10- producing *Lactobacillus* in patients suffering from colitis. Analysis of the results revealed a beneficial impact of the treatment on disease activity with only minor side effects (Braat et al. 2006).

Likewise, strategies to prevent or cure food allergy are also likely to involve elevation of IL-10 levels in the gut in order to cause induction of antigen-specific regulatory T cells (Treg) and IgA antibodies. In fact, the administration of IL-10-secreting *L. lactis* to mice prior to sensitization with β-lactoglobulin (BLG) showed a significantly inhibited frequency of anaphylaxis (Frossard et al. 2007). Closer analysis of the results revealed that the pre-treatment had functioned to prevent BLG-specific IgE and IgG1 antibody production in serum and increase BLG-specific secretory IgA antibodies in faeces. In addition, pretreated mice exhibited an elevated number of IL-10 secreting T cells in PP, similar to those conditions seen in actively tolerized mice.

Another possible avenue to use engineered probiotics is to reinduce oral tolerance to auto-antigens. To this end, the oral delivery of *L. lactis* expressing recombinant ovalbumin (OVA) was investigated. The treatment produced decreased antigen-specific delayed-type hypersensitivity (DTH) in transgenic mice carrying OVA-specific T cell receptors (TcR) and up-regulated antigen-specific IL-10 secretion in both the intestinal and systemic compartments (Huibregtse et al. 2007). The resulting therapeutic effect appeared to be mediated by induction of $CD4^{+}CD25^{+}Foxp3^{+}$ Treg cells that function by means of a TGF-β-dependent mechanism.

A second promising application of engineered probiotics can be appreciated with consideration of the local production and secretion of monoclonal antibodies (mAb). mAb often represent the first available tools to block identified targets in different pathologies. Classical mAb therapy usually requires systemic infusions of large amounts of purified mAb, a method that is unfortunately often associated with collateral effects. In light of this fact, an orally administered form of mAb would be an ideal alternative to the classic therapies, because it would likely avoid, if not completely diminish, the side effects there were previously observed. In effect, one report using *L. lactis* bacteria engineered to secrete anti-TNF antibody fragments has already demonstrated the feasibility and efficacy of this proposition (Vandenbroucke et al. 2006). In this study, beneficial effects were observed in both DSS-induced colitis and established enterocolitis in IL-10-/- mice following oral administration of the anti-TNF antibody-secreting *L. lactis*. The authors noted that the administered anti-TNF antibodies were regularly found in association with LP cell surfaces in areas of the mucosa containing inflammatory infiltrates. Interestingly, no measurable levels of antibodies were detected within systemic circulation. Given the whole of this data, the oral treatment featured in this study shows a lot of promise for the realization of a more effective means of managing colitis, without the adverse effects characteristic of traditional anti-TNF therapies.

10.7 Concluding Remarks

Even though humans have been using probiotics for many years, only recently did scientific evidence arise to support the value of these bacteria in the regulation of mucosal immunity. Over the course of this text, several potential medical applications of specific probiotic strains have been discussed, including their role in increasing defences against acute infections by acting as vaccine adjuvants, improving chronic inflammatory bowel conditions, and inducing tolerance to harmless environmental antigens, such as those triggering food allergies. More important, however, is the newly improved understanding of the mechanisms orchestrating how these microorganisms behave, that could expose new strategies for selecting and designing useful probiotic strains.

Moreover, the innovation of genetic engineering applied to the study of probiotics could ultimately result in the creation of completely new bacterial strains that carry elements derived from an eukaryotic background. The capacity for immunomodulation as provided by this proposed procedure could be expanded even further by discriminatory selection of the host organisms. It is important to keep in mind, however, that the physiological impact of probiotics is much subtler than expected on different diseases/conditions and still largely controversial in regards to healthy individuals. Nevertheless, if designed with adequate attention placed on biological safety, the development of specialized probiotic strains has great potential to revolutionize therapeutic applications of microorganisms, especially those specified to target GALT immune responses and topical immunomodulation.

References

Baumgart DC, Carding SR (2007) Inflammatory bowel disease: cause and immunobiology. Lancet 369:1627–1640

Bermudez-Humaran LG, Langella P, Cortes-Perez NG, Gruss A, Tamez-Guerra RS, Oliveira SC, Saucedo-Cardenas O, Montes de Oca-Luna R, Le Loir Y (2003) Intranasal immunization with recombinant Lactococcus lactis secreting murine interleukin-12 enhances antigen-specific Th1 cytokine production. Infect Immun 71(4):1887–1896

Braat H, Rottiers P, Hommes DW, Huyghebaert N, Remaut E, Remon JP, van Deventer SJ, Neirynck S, Peppelenbosch MP, Steidler L (2006) A phase I trial with transgenic bacteria expressing interleukin-10 in Crohn's disease. Clin Gastroenterol Hepatol 4:754–759

Carrier Y, Yuan J, Kuchroo VK, Weiner HL (2007) Th3 cells in peripheral tolerance. II. TGF-beta-transgenic Th3 cells rescue IL-2-deficient mice from autoimmunity. J Immunol 178:172–178

Cesta MF (2006) Normal structure, function, and histology of mucosa-associated lymphoid tissue. Toxicol Pathol 34:599–608

Coombes JL, Siddiqui KR, Arancibia-Carcamo CV, Hall J, Sun CM, Belkaid Y, Powrie F (2007) A functionally specialized population of mucosal CD103+ DC induces Foxp3+ regulatory T cells via a TGF-beta and retinoic acid-dependent mechanism. J Exp Med 204:1757–1764

Cross M (2002) Microbes versus microbes: immune signals generated by probiotic lactobacilli and their role in protection against microbial pathogens. FEMS Inmunol Medical Microbiol 34:245–253

D'Arienzo R, Maurano F, Luongo D, Mazzarella G, Stefanile R, Troncone R, Auricchio S, Ricca E, David C, Rossi M (2008) Adjuvant effect of *Lactobacillus casei* in a mouse model of gluten sensitivity. Immunol Lett 119:78–83

Eberl G (2005) Inducible lymphoid tissues in the adult gut: recapitulation of a fetal developmental pathway? Nat Rev Immunol 5:413–420

Faria AM, Weiner HL (2005) Oral tolerance. Immunol Rev 206:232–259

Foligne B, Zoumpopoulou G, Dewulf J, Ben Younes A, Chareyre F, Sirard JC, Pot B, Grangette CA (2007) A key role of dendritic cells in probiotic functionality. PLoS One 2(3):e313

Forchielli ML, Walker WA (2005) The role of gut-associated lymphoid tissues and mucosal defence. Br J Nutr 93(Suppl 1):S41–S48

Frossard CP, Steidler L, Eigenmann PA (2007) Oral administration of an IL-10-secreting Lactococcus lactis strain prevents food-induced IgE sensitization. J Allergy Clin Immunol 119(4):952–959

Hacini-Rachinel F, Nancey S, Boschetti G, Sardi F, Doucet-Ladevaze R, Durand PY, Flourié B, Kaiserlian D (2009) CD4+ T cells and Lactobacillus casei control relapsing colitis mediated by CD8+ T cells. J Immunol 183(9):5477–5486

Hamada H, Hiroi T, Nishiyama Y, Takahashi H, Masunaga Y, Hachimura S, Kaminogawa S, Takahashi-Iwanaga H, Iwanaga T, Kiyono H, Yamamoto H, Ishikawa H (2002) Identification of multiple isolated lymphoid follicles on the antimesenteric wall of the mouse small intestine. J Immunol 168:57–64

Hart AL, Lammers K, Brigidi P, Vitali B, Rizzello F, Gionchetti P, Campieri M, Kamm MA, Knight SC, Stagg AJ (2004) Modulation of human dendritic cell phenotype and function by probiotic bacteria. Gut 53:1602–1609

Hoarau C, Lagaraine C, Martin L, Velge-Roussel F, Lebranchu Y (2006) Supernatant of *Bifidobacterium breve* induces dendritic cell maturation, activation, and survival through a Toll-like receptor 2 pathway. J Allergy Clin Immunol 117:696–702

Huibregtse IL, Snoeck V, de Creus A, Braat H, De Jong EC, Van Deventer SJ, Rottiers P (2007) Induction of ovalbumin-specific tolerance by oral administration of Lactococcus lactis secreting ovalbumin. Gastroenterology 133(2):517–528

Ivanov II, Zhou L, Littman DR (2007) Transcriptional regulation of Th17 cell differentiation. Semin Immunol 19:409–417

Ivory K, Chambers SJ, Pin C, Prieto E, Arqués JL, Nicoletti C (2008) Oral delivery of *Lactobacillus casei* Shirota modifies allergen-induced immune responses in allergic rhinitis. Clin Exp Allergy 38(8):1282–1289

Izcue A, Coombes JL, Powrie F (2009) Regulatory lymphocytes and intestinal inflammation. Ann Rev Immunol 27:313–338

Johansson-Lindbom B, Agace WW (2007) Generation of gut-homing T cells and their localization to the small intestinal mucosa. Immunol Rev 215:226–242

Jump RL, Levine AD (2002) Murine Peyer's patches favor development of an IL-10-secreting, regulatory T cell population. J Immunol 168:6113–6119

Kajikawa A, Satoh E, Leer RJ, Yamamoto S, Igimi S (2007) Intragastric immunization with recombinant *Lactobacillus casei* expressing flagellar antigen confers antibody-independent protective immunity against *Salmonella enterica* serovar Enteritidis. Vaccine 25:3599–3605

Kanamori Y, Ishimaru K, Nanno M, Maki K, Ikuta K, Nariuchi H, Ishikawa H (1996) Identification of novel lymphoid tissues in murine intestinal mucosa where clusters of ckit+ IL-7R +Thy1+ lympho-hemopoietic progenitors develop. J Exp Med 184:1449–1459

Kim H, Kwack K, Kim DY, Ji GE (2005) Oral probiotic bacterial administration suppressed allergic responses in an ovalbumin-induced allergy mouse model. FEMS Immunol Med Microbiol 45(2):259–267

Kuhn R, Lohler J, Rennick D, Rajewsky K, Muller W (1993) Interleukin-10-deficient mice develop chronic enterocolitis. Cell 75:263–274

Kunkel D, Kirchhoff D, Nishikawa S, Radbruch A, Scheffold A (2003) Visualization of peptide presentation following oral application of antigen in normal and Peyer's patches deficient mice. Eur J Immunol 33:1292–1301

Laurence A, Tato CM, Davidson TS, Kanno Y, Chen Z, Yao Z, Blank RB, Meylan F, Siegel R, Hennighausen L, Shevach EM, O'Shea J (2007) Interleukin-2 signaling via STAT5 constrains T helper 17 cell generation. Immunity 26:371–381

Li MO, Flavell RA (2008) Contextual regulation of inflammation: a duet by transforming growth factor-beta and interleukin-10. Immunity 28:468–476

Locke NR, Stankovic S, Funda DP, Harrison LC (2006) TCR gamma delta intraepithelial lymphocytes are required for self-tolerance. J Immunol 176:6553–6559

Lorenz RG, Newberry RD (2004) Isolated lymphoid follicles can function as sites for induction of mucosal immune responses. Ann N Y Acad Sci 1029:44–57

Lorenz RG, Chaplin DD, McDonald KG, McDonough JS, Newberry RD (2003) Isolated lymphoid follicle formation is inducible and dependent upon lymphotoxinsufficient B lymphocytes, lymphotoxin beta receptor, and TNF receptor I function. J Immunol 170:5475–5482

Maassen CB, van Holten-Neelen C, Balk F, den Bak-Glashouwer MJ, Leer RJ, Laman JD, Boersma WJ, Claassen E (2000) Strain-dependent induction of cytokine profiles in the gut by orally administered *Lactobacillus* strains. Vaccine 18:2613–2623

Maassen CB, Boersma WJ, Van Holten-Neelen C, Claassen E, Laman JD (2003) Growth phase of orally administered Lactobacillus strains differentially affects IgG1/IgG2a ratio for soluble antigens: implications for vaccine development. Vaccine 21(21–22):2751–2757

Marie JC, Letterio JJ, Gavin M, Rudensky AY (2005) TGF-beta1 maintains suppressor function and Foxp3 expression in CD4+CD25+ regulatory T cells. J Exp Med 201:1061–1067

Mastrangeli G, Corinti S, Butteroni C, Afferni C, Bonura A, Boirivant M, Colombo P, Di Felice G (2009) Effects of live and inactivated VSL#3 probiotic preparations in the modulation of in vitro and in vivo allergen-induced Th2 responses. Int Arch Allergy Immunol 150(2):133–143

Medina M, Izquierdo E, Ennahar S, Sanz Y (2007) Differential immunomodulatory properties of *Bifidobacterium longum* strains: relevance to probiotic selection and clinical applications. Clin Exp Immunol 150:531–538

Ménard O, Batel MJ, Gaboriau-Routhiau V, Waligora-Dupriet AJ (2008) Gnotobiotic mouse immune response induced by *Bifidobacterium* sp. strains isolated from infants. Appl Environ Microbiol 74:660–666

Mowat AM (2003) Anatomical basis of tolerance and immunity to intestinal antigens. Nat Rev Immunol 3:331–341

Mucida D, Park Y, Kim G, Turovskaya O, Scott I, Kronenberg M, Cheroutre H (2007) Reciprocal TH17 and regulatory T cell differentiation mediated by retinoic acid. Science 317:256–260

Newberry RD, Lorenz RG (2005) Organizing a mucosal defense. Immunol Rev 206:6–21

Ogawa T, Asai Y, Tamai R, Makimura Y, Sakamoto H, Hashikawa S, Yasuda K (2006) Natural killer cell activities of synbiotic *Lactobacillus casei* ssp. *casei* in conjunction with dextran. Clin Exp Immunol 143(1):103–109

Olivares M, Díaz-Ropero MP, Sierra S, Lara-Villoslada F, Fonollá J, Navas M, Rodríguez JM, Xaus J (2007) Oral intake of Lactobacillus fermentum CECT5716 enhances the effects of influenza vaccination. Nutrition 23(3):254–260

O'Mahony L, O'Callaghan L, McCarthy J, Shilling D, Scully P, Sibartie S, Kavanagh E, Kirwan WO, Redmond HP, Collins JK, Sanan F (2006) Differential cytokine response from dendritic cells to commensal and pathogenic bacteria in different lymphoid compartments in humans. Am J Physiol Gastrointest Liver Physiol 290:G839–G845

O'Mahony C, Scully P, O'Mahony D, Murphy S, O'Brien F, Lyons A, Sherlock G, MacSharry J, Kiely B, Shanahan F, O'Mahony L (2008) Commensal-induced regulatory Tcells mediate protection against pathogen-stimulated NF-kappaB activation. PLoS Pathog 4(8):e1000112

Onishi Y, Fehervari Z, Yamaguchi T, Sakaguchi S (2008) Foxp3+ natural regulatory T cells preferentially form aggregates on dendritic cells in vitro and actively inhibit their maturation. Proc Natl Acad Sci U S A 105:10113–10118

Read S, Greenwald R, Izcue A, Robinson N, Mandelbrot D, Francisco L, Sharpe AH, Powrie F (2006) Blockade of CTLA-4 on CD4+CD25+ regulatory T cells abrogates their function in vivo. J Immunol 177:4376–4383

Rey J, Garin N, Spertini F, Corthesy B (2004) Targeting of secretory IgA to Peyer's patch dendritic and T cells after transport by intestinal M cells. J Immunol 172:3026–3033

Roncarolo MG, Gregori S, Battaglia M, Bacchetta R, Fleischhauer K, Levings MK (2006) Interleukin-10-secreting type 1 regulatory T cells in rodents and humans. Immunol Rev 212:28–50

Sakaguchi S, Ono M, Setoguchi R, Yagi H, Hori S, Fehervari Z, Shimizu J, Takahashi T, Nomura T (2006) Foxp3+ CD25+ CD4+ natural regulatory T cells in dominant self tolerance and autoimmune disease. Immunol Rev 212:8–27

Schenk M, Mueller C (2008) The mucosal immune system at the gastrointestinal barrier. Best Pract Res Clin Gastroenterol 22:391–409

Sellon RK, Tonkonogy S, Schultz M, Dieleman LA, Grenther W, Balish E, Rennick DM, Sartor RB (1998) Resident enteric bacteria are necessary for development of spontaneous colitis and immune system activation in interleukin-10-deficient mice. Infect Immun 66:5224–5231

Shevach EM (2009) Mechanisms of foxp3+ T regulatory cell-mediated suppression. Immunity 30:636–645

Shida K, Makino K, Morishita A, Takamizawa K, Hachimura S, Ametani A, Sato T, Kumagai Y, Habu S, Kaminogawa S (1998) *Lactobacillus casei* inhibits antigen-induced IgE secretion through regulation of cytokine production in murine splenocyte cultures. Int Arch Allergy Immunol 115(4):278–287

Shida K, Takahashi R, Iwadate E, Takamizawa K, Yasui H, Sato T, Habu S, Hachimura S, Kaminogawa S (2002) *Lactobacillus casei* strain Shirota suppresses serum immunoglobulin E and immunoglobulin G1 responses and systemic anaphylaxis in a food allergy model. Clin Exp Allergy 32(4):563–570

Smits HH, Engering A, van der Kleij D, de Jong EC, Schipper K, van Capel TM, Zaat BA, Yazdanbakhsh M, Wierenga EA, van Kooyk Y, Kapsenberg ML (2005) Selective probiotic bacteria induce IL-10-producing regulatory T cells in vitro by modulating dendritic cell function through dendritic cell-specific intercellular adhesion molecule 3-grabbing nonintegrin. J Allergy Clin Immunol 115:1260–1267

Spahn TW, Weiner HL, Rennert PD, Lugering N, Fontana A, Domschke W, Kucharzik T (2002) Mesenteric lymph nodes are critical for the induction of high-dose oral tolerance in the absence of Peyer's patches. Eur J Immunol 32:1109–1113

Steidler L, Robinson K, Chamberlain L, Schofield KM, Remaut E, Le Page RW, Wells JM (1998) Mucosal delivery of murine interleukin-2 (IL-2) and IL-6 by recombinant strains of Lactococcus lactis coexpressing antigen and cytokine. Infect Immun 66(7):3183–3189

Steidler L, Hans W, Schotte L, Neirynck S, Obermeier F, Falk W, Fiers W, Remaut E (2000) Treatment of murine colitis by Lactococcus lactis secreting interleukin-10. Science 289:1352–1355

Sun CM, Hall JA, Blank RB, Bouladoux N, Oukka M, Mora JR, Belkaid Y (2007) Small intestine lamina propria dendritic cells promote de novo generation of Foxp3 T reg cells via retinoic acid. J Exp Med 204:1775–1785

Suzuki K, Fagarasan S (2008) How host-bacterial interactions lead to IgA synthesis in the gut. Trends Immunol 29:523–531

Suzuki H, Sekine S, Kataoka K, Pascual DW, Maddaloni M, Kobayashi R, Fujihashi K, Kozono H, McGhee JR (2008) Ovalbumin-protein sigma 1 M-cell targeting facilitates oral tolerance with reduction of antigen-specific CD4+ T cells. Gastroenterology 135:917–925

Tlaskalová-Hogenová H, Stepánková R, Hudcovic T, Tucková L, Cukrowska B, Lodinová-Zádníková R, Kozáková H, Rossmann P, Bártová J, Sokol D, Funda DP, Borovská D, Reháková Z, Sinkora J, Hofman J, Drastich P, Kokesová A (2004) Comensal bacteria (normal microflora), mucosal immunity and chronic inflammatory and autoimmune diseases. Immunol Lett 93:97–108

Uhlig HH, Coombes J, Mottet C, Izcue A, Thompson C, Fanger A, Tannapfel A, Fontenot JD, Ramsdell F, Powrie F (2006) Characterization of Foxp3+CD4+CD25+ and IL-10-secreting CD4+CD25+ T cells during cure of colitis. J Immunol 177:5852–5860

Vandenbroucke K, Huyck L, Demetter P, Dreier T, Beirnaert E, Lauwereys M, Cuvelier C, Remaut E, Rottiers P (2006) Treatment of Murine Colitis by *Lactococcus lactis* Secreting anti-tumor necrosis factor nanobodiesTM. 4th annual broad medical research program investigator meeting for inflammatory bowel disease: abstract

Wing K, Onishi Y, Prieto-Martin P, Yamaguchi T, Miyara M, Fehervari Z, Nomura T, Sakaguchi S (2008) CTLA-4 control over Foxp3+ regulatory T cell function. Science 322:271–275

Worbs T, Bode U, Yan S, Hoffmann MW, Hintzen G, Bernhardt G, Forster R, Pabst O (2006) Oral tolerance originates in the intestinal immune system and relies on antigen carriage by dendritic cells. J Exp Med 203:519–527

Yasui H, Shida K, Matsuzaki T, Yokokura T (1999) Immunomodulatory function of lactic acid bacteria. Antonie Van Leeuwenhoek 76(1–4):383–389

Young SL, Simon MA, Baird MA, Tannock GW, Bibiloni R, Spencely K, Lane JM, Fitzharris P, Crane J, Town I, Addo-Yobo E, Murria CS, Woodcock A (2004) Bifidobacterial species differentially affect expression of cell surface markers and cytokines of dendritic cells harvested from cord blood. Clin Diagn Lab Immunol 11:686–690

Zhou L, Lopes JE, Chong MM, Ivanov II, Min R, Victora GD, Shen Y, Du J, Rubtsov YP, Rudensky AY, Ziegler SF, Littman DR (2008) TGF-β-induced Foxp3 inhibits T(H)17 cell differentiation by antagonizing RORγ T function. Nature 453:236–240

Chapter 11
The Protective Role of Probiotics in Disturbed Enteric Microbiota

Denis Roy and Véronique Delcenserie

11.1 Introduction

The human intestine is inhabited by a large number of microorganisms, primarily bacteria, which form a microbial ecosystem. This ecosystem is mainly confined to the ileum and colon. The colon is the most heavily colonised area of the gastrointestinal tract, with bacteria concentrations ranging from 10^{11} to 10^{12} cfu/mL of contents (Salminen and Isolauri 2006; Tannock 2003). These microorganisms form a microbiota that performs a number of biochemical and physiological functions, including helping to break down food particles, acting as a protective barrier against pathogenic bacteria and stimulating the development of the immune system.

Various disease states are associated with an imbalance of the intestinal microbiota. These can include susceptibility to pathogens such as *Clostridium difficile* as well as chronic diseases such as Crohn's disease and ulcerative colitis. The enteric microbiota has also been found altered in other conditions, such as irritable bowel syndrome, Type 1 and 2 diabetes, acute gastroenteritis, food hypersensitivity and allergies, colon cancer and obesity (Rochet et al. 2006; Gibson and Vallance 2008). In addition, antibiotic use is often associated with adverse gastrointestinal effects on the microbial population. The use of probiotics has been proposed to minimise these disturbances by stabilising the intestinal bacterial communities and minimising potential alterations in microbial community structures. The enteric microbiota is therefore considered the main target for probiotics, and this chapter focuses on the role and impact of probiotics in disturbed enteric microbiota. The discussion is limited to the relationship between probiotics and disturbed enteric microbiota, and the significance of this relationship in immune system modulation.

D. Roy (✉)
Department of Food science and Nutrition, Institute of Nutraceutical and Functional Foods, Université Laval, 2440 Boul. Hochelaga, G1 V 0A6 QC, Canada
Tel.: +1-418-656-2131
Fax: +1-418-656-5877
e-mail: denis roy@inaf.ulaval.ca

J. J. Malago et al. (eds.), *Probiotic Bacteria and Enteric Infections*,
DOI 10.1007/978-94-007-0386-5_11, © Springer Science+Business Media B.V. 2011

11.2 The Enteric Microbiota

The enteric microbiota can be described in general terms only, because each human harbours a unique microbial ecosystem (Zoetendal et al. 1998). Hundreds and even thousands of bacterial species are likely to reside within it (Frank and Pace 2008). The phylogenetic core of the healthy adult's faecal microbiota consists of a set of five phyla: *Firmicutes, Bacteroidetes, Actinobacteria, Proteobacteria*, and *Verrucomicrobia* (Tap et al. 2009). Strict anaerobes are more numerous than facultative anaerobes. The predominant genera in the enteric microbiota, identified by molecular methods, include *Faecalibacterium, Ruminococcus, Eubacterium, Dorea, Bacteroides, Alistipes* and *Bifidobacterium* (Tap et al. 2009). Facultative anaerobes include lactobacilli,enterococci, streptococci, and *Enterobacteriaceae* (Rastall 2004). The carbon and energy needs of the billions of bacteria in the colon are met by two sources: complex carbohydrates, proteins and fats that have escaped digestion in the small intestine; and the components of host secretions (mucins) and sloughed epithelial cells (Lebeer et al. 2008).

One of the most prevalent species is *Faecalibacterium prausnitzii*, which was described as dominant in healthy individuals (Suau et al. 2001) and under-represented in patients with inflammatory bowel disease (Manichanh et al. 2006). This species is believed to possess anti-inflammatory properties (Sokol et al. 2008), and might be crucial to protect homeostasis, since lower *F. prausnitzii* counts are consistently associated with reduced protection of the gut mucosa (Sokol et al. 2009). Of the other genera considered beneficial (Dommels et al. 2009), *Bifidobacterium* is the one of the most common species in the human enteric microbiota (Lay et al. 2005), and *Bifidobacterium longum* subsp. *longum* is one of the 10 most prevalent species found (Tap et al. 2009), whereas the genus *Lactobacillus* belongs to a subdominant group in the faecal microbial community (Sghir et al. 2000).

It has been increasingly recognised that certain bacterial species of the enteric microbiota play a clinical and biological role in improving the health of the host. Bifidobacteria and lactobacilli are thought to improve resistance to intestinal infections by preventing the growth of potentially harmful bacteria, improving the immune response and producing vitamins (Tuohy et al. 2007). The enteric microbiota is a metabolically active microbial environment. The activity of this ecosystem therefore plays a major role in the nutrition and health of the host. Thus, it performs vital protective and metabolic functions, such as growth and development of the epithelial barrier, stimulation of intestinal angiogenesis, regulation of nutrition, metabolic functions,and education of naïve neonatal innate immunity (Dommels et al. 2009). Intestinal microbial colonisation occurs in stages, beginning at birth and continuing throughout the lifetime to form a unique intestinal microbiota. After birth, the host environment and its attributes allow the establishment of a stable intestinal microbiota that favours long-term symbiosis (Gueimonde et al. 2006; Salminen and Isolauri 2006). Dysregulation (dysbiosis) of this process during the early postnatal period can lay the groundwork for later inflammatory, immune and allergic disorders (Gueimonde et al. 2006).

Homeostasis of the enteric microbiota is essential for maintaining human health and well-being (Firmesse et al. 2008). The adult enteric microbiota shows remarkable stability in terms of microbial community composition over time (Zoetendal et al. 1998). This phenomenon is referred to as colonisation resistance or competitive exclusion. It acts as an integral defense against colonisation by orally ingested microorganisms (Plant et al. 2003). The microbiota live in intestinal regions that are relatively static, allowing bacteria to multiply in the digesta at growth rates that compensate for the constant loss of material (faeces). Hence, the number of bacteria never decreases (Tannock 2003). The bacterial strains that form stable populations of a characteristic size in a particular niche of the gut are true residents, or autochthonous. Autochthonous strains are highly compatible with the intestinal environments of their hosts, and are tolerated by the immune system. Other bacteria that transit through the gut are allochthonous (Tannock 2003; Lebeer et al. 2008). They do not persist within the ecosystem, and are detectable for only a limited time. Allochthonous strains may have been dislodged from another gut niche, or they may come from ingested food and water. Probiotic bacterial strains should therefore be considered as allochthonous (Tannock et al. 2000). In summary, the enteric microbiota consists of a bacterial consortium that adapts to the host, becomes stable, and therefore resists change. The administration of probiotics is considered a beneficial practice because it fosters the maintenance of this homeostasis and its resistance to change in healthy individuals. The use of lactobacilli and bifidobacteria in particular is considered appropriate to assess the probiotic effect, since their biological functions in the intestinal ecosystem are well identified (Servin 2004).

11.3 The Probiotic Effect

Probiotics are live microorganisms that, when consumed in sufficient amounts, play a major role in maintaining the equilibrium and stability of the intestinal microbiota, primarily by preventing infection, to the benefit of the health of the host. Probiotics may be eaten as food or taken as a dietary supplement to either reinforce or modify the enteric microbiota. Few studies have rigorously compared the effects of live versus dead bacteria as probiotics (Xiao et al. 2003; Rachmilewitz et al. 2004). Given that certain effects produced by live bacteria may be due to their metabolites, DNA or cell wall components, some microorganisms may produce beneficial health effects without having to survive transit through the digestive mucosa. The death of bacteria in the gastrointestinal tract does not necessarily imply the absence of effect, nor does survival necessarily imply a positive effect. Other studies have reported the greater efficacy of a live strain than the same heat-killed strain (Kaila et al. 1995; Kecskes et al. 2003). In any case, the beneficial effects of dead organisms have not yet been fully established, and for the purposes of this chapter, the current definition of probiotics is used: a monoculture or mixed culture of live microorganisms that, when administered in adequate amounts, confers a health benefit on the host (De Champs et al. 2003). Probiotics may protect the host against intestinal disorders by producing inhibitory substances, blocking adhesion sites, competing for nutrients and stimulating immune systems. Nevertheless, before being able to confer these benefits, probiotics need to survive in the gastrointestinal tract and persist in the intestine at high levels (De Champs et al. 2003).

The quality of a probiotic product depends not only on the type of probiotic used but also its concentration, which must be sufficient, and its metabolic activity. When assessing probiotics, its capacity to resist bile salts and pancreatic secretions must be determined. The concentrations of live probiotics that reach the intestine are influenced by survival percentage and ingested quantity. Some probiotics are destroyed during their passage through the stomach, whereas others transit through the small bowel and even the colon in high concentrations. Many probiotic bacteria can survive in gastric conditions (1.5–2.0 pH during fasting and 4.0–5.0 pH post-meal) and colonise the intestine (at least temporarily) by adhering to the intestinal mucosa with surface adhesins (proteins, polysaccharides and cell wall components). For instance, the *Lactobacillus rhamnosus* GG strain has mucus-binding pili on its surface that enable it to interact with a host. This is one illustration of how a probiotic strain can persist in the host and potentially compete with pathogens for residence sites in the human intestinal tract (Kankainen et al. 2009). These bacteria offer new dietetic alternatives to stabilise the intestinal microbial flora. However, the most widely studied and best-known probiotic strains do not necessarily offer the best pharmacokinetic properties, and strains currently understudied should receive more attention.

11.3.1 Pharmacokinetic Considerations

Many clinical pharmacokinetics studies posit that the probiotic culture must be continuously ingested to achieve an exogenous probiotic effect. The quantity of bacteria recuperated from faeces provides a good marker of comparison between different strains, provided that comparable ingestion doses and enumeration methods are used. Faecal recuperation before and after ingestion also enables an assessment of the increase in specific bacteria in the faeces following consumption. Table 11.1 presents some pharmacokinetics studies of different probiotic strains. Certain strains have received more attention than others, namely *Bifidobacterium lactis* B94, *Lactobacillus rhamnosus* GG, *Lactobacillus salivarius* UCC118, *Lactobacillus plantarum* 299v and NCIMB 8826, *Lactobacillus johnsonii* La1, and *Lactobacillus casei* Shirota or DN-114001. Other strains have been investigated less often. Nevertheless, they do not appear any less promising in terms of results. We may note from Table 11.1 that *B. lactis, Bifidobacterium longum* subsp. *longum* and *Lactobacillus paracasei* are found in good proportions in the faeces, which should make them the organisms of choice for probiotic purposes. However, faecal recuperation of *L. rhamnosus, Lactobacillus fermentum* KDL and *Lactobacillus gasseri* ADH is extremely low. Moreover, lactic acid bacteria (*Streptococcus thermophilus, Lactobacillus delbrueckii* subsp. *bulgaricus* and *Lactococcus lactis*) and *Entero-*

Table 11.1 Summary of the pharmacokinetics of potential probiotic bacterial strains

Strain	Dose	Faecal recuperation	Persistence (day)	References
B. lactis B94	1×10^{11} cells×7 days	1.8×10^{9} cfu/g	28	Su et al. (2005)
B. lactis B94	1×10^{10} cells	2.5×10^{6} cfu/mL	ND	Pochart et al. (1992)
B. lactis Bb12	1×10^{11} cfu	8×10^{7} cfu/g	14	Larsen et al. (2006)
Bifidobacterium sp.	1×10^{10} cells	1×10^{9} cell total	ND	Pochart et al. (1992)
B. longum SB T2928	7×10^{11} cfu×7 days	1×10^{9} cfu/g	>30	Fujiwara et al. (2001b)
L. rhamnosus GG	6×10^{10} cfu×12 days	4×10^{4} cfu	14	Alander et al. (1999a)
L. rhamnosus GG	1×10^{11} cfu	2.5×10^{6}–5.0×10^{7} cfu/g	7	Goldin et al. (1992)
L. rhamnosus GG	1×10^{10} cfu	1×10^{6} cfu/g	ND	Saxelin et al. (1995)
L. rhamnosus DR20	1.6×10^{9} cells×182 days	6.3×10^{5} cfu/g	60	Tannock et al. (2000)
L. casei Shirota	$3,1\times10^{11}$ cfu	1×10^{7} cfu/g	ND	Spanhaak et al. (1998)
L. casei Shirota	1×10^{10} cells×3 days	1×10^{7} cell/g	ND	Yuki et al. (1999)
L. casei DN-114001	1×10^{10} cells×7 days	4×10^{7} cell/g	ND	Oozeer et al. (2006)
L. paracasei LTH 2579	1×10^{9} cfu×3 days	1.3×10^{7}–1.6×10^{8} cfu/g	ND	Bunte et al. (2000)
L. salivarus UCC118	1.6×10^{10} cfu	2×10^{6} cfu/mL	ND	Collins et al. (2002)
L. salivarus UCC118	$1\times10^{10}\times21$ days	1×10^{3}–1×10^{7} cfu/g	>21	Collins et al. (2002)
L. plantarum 299v	$2\times10^{10}\times21$ days	1×10^{7} cfu/g	>8	Johansson et al. (1998)
L. plantarum 299v	2×10^{11} cfu×14 days	1.3×10^{7} cfu/g	ND	Goossens et al. (2006)
L. plantarum NCIMB 8826	1.6×10^{10} cfu	1×10^{8} cfu/mL	0.4	Vesa et al. (2000)
L. plantarum NCIMB 8826	5×10^{10} cfu×7 days	1×10^{8} cfu/g	14	Vesa et al. (2000)
L. helveticus GCL1001	1.6×10^{10} cfu	3×10^{4}–6×10^{7} cells/g	5	Saito et al. (2004)
L. helveticus CP53	1×10^{11} cfu×5 days	4×10^{6} cfu/g	ND	Shinoda et al. (2001)
L. fermentum KLD	1.5×10^{9} cfu	3.2×10^{4} cfu/mL	0.2	Vesa et al. (2000)
L. acidophilus	1×10^{10} cells	1×10^{6} cfu/mL	ND	Marteau et al. (1992)
L. johnsonii La1	1×10^{9} cfu×21 days	8×10^{6} cfu/g	4	Yamano et al. (2006)
L. johnsonii La1	5×10^{10} cfu×7 days	1×10^{6} cfu/g	17	Garrido et al. (2005)

Table 11.1 (continued)

Strain	Dose	Faecal recuperation	Persistence (day)	References
L. gasseri SBT2055	1×10^{11} cfu×7 days	1×10^{7} cfu/g	>31	Fujiwara et al. (2001a)
L. gasseri ADH	4×10^{10} cells×12 days	3×10^{2} cfu/mL	ND	Pedrosa et al. (1995)
L. bulgaricus	4.3×10^{10} cells	2.5×10^{5} cfu/mL	ND	Pochart et al. (1989)
L. bulgaricus	1×10^{9} cfu/g	10^{5}–10^{6} cfu/g	ND	Lindwall and Fonden (1984)
S. thermophilus	4.3×10^{10} cells	3.2×10^{6} cfu/mL	ND	Pochart et al. (1989)
S. thermophilus	2.5×10^{8}–2.2×10^{11} cfu	4×10^{5} cfu/g	9	Brigidi et al. (2003)
S. thermophilus	1.2×10^{12} cfu	5×10^{6} cfu/g	6	Brigidi et al. (2003)
Lc lactis MG 1363	1.6×10^{10} cfu	1.6×10^{5} cfu/mL	ND	Vesa et al. (2000)
Lc lactis MG 1363	1×10^{11} cfu×4 days	1×10^{4} cfu/g	3	Klijn et al. (1995)
Enterococcus faecium	1×10^{9} cfu×10 days	1.3×10^{3}–4×10^{6} cfu/g	ND	Lund et al. (2002)

coccus faecium show poor resistance to the intestinal transit. The required dose to obtain adequate faecal recuperation appears to be at least 10^{10} bacteria.

11.3.1.1 Probiotic Survival

In summary, the research on the pharmacokinetics of probiotics has focused more on comparing bacterial strains before and after ingestion than on obtaining precise data on bacterial survival rates. Gastric acid is the key defense mechanism that probiotics must confront. Bile salts and pancreatic secretions are the second line of defense. Motility can also act as a barrier, and a decrease in peristalsis fosters bacterial colonisation in the small intestine. The ability of the *Bifidobacterium* sp. strain to survive gastrointestinal transit when ingested in fermented milk was determined by Pochart et al. (1992), who studied 6 volunteer subjects using *in vivo* ileal perfusion. The average number of *Bifidobacterium* sp. recuperated in the terminal ileum during the 8 hours after milk consumption was 10^9 bacteria, or 23.5±10.4% of the ingested amount. Bouhnik et al. (1992) obtained a nearly 30% faecal recuperation of *Bifidobacterium* sp. following administration in fermented milk. Survival of the *Bifidobacterium* probiotic strain contained in the fermented product Ofilus® was determined by Marteau et al. (1992), with an estimated 37.5% survival rate in the ileum. In another study, skim milk containing about 10^{10} cfu of an unidentified exogenous bifidobacterial strain was consumed daily by 6 subjects for 8 days (Kullen et al. 1997). The consumed strain was undetectable in the faeces prior to consumption, but rapidly became the predominant bifidobacterial strain, attaining 57.8 ±9.2% of the total bifidobacteria after 4 days of consumption and 67.2±8.5% after 8 days of consumption. Four days after cessation of consumption, the proportion of the strain dropped to 15.8±12.2%, and was undetectable 8 days after milk ingestion ceased. Collins et al. (2002) measured the pharmacokinetics of the *L. salivarius* UCC118 strain in the intestine and its impact on intestinal microbiota and the mucosal immune system. Survival in the ileum in 6 volunteers who consumed 10^8 cfu was estimated at around 11.8%. Survival of the *L. plantarum* NCIMB 8826 strain in the ileum was investigated in a study in 6 volunteers who consumed 10^8/gr cfu in fermented milk (Vesa et al. 2000). Survival in the ileum was tested by intubation. *L. plantarum* NCIMB 8826 showed a high capacity to survive in this environment, rising to a population peak of 10^8 cfu/mL after 2 hours. Survival rate was estimated at 7% in the ileum, and bacterial passage in this section occurred simultaneously with the transit of the marker, suggesting passive transit with no colonisation. Cell counts dropped to zero at 8–10 hours after ingestion. After a daily consumption of 10^8 cfu for 7 days, faecal concentration was 10^8 cfu/g at day 7, for an estimated survival rate of 25±29%. *L. plantarum* NCIMB 8826 was undetectable in the faeces 2 weeks after the end of the administration period. Using a dynamic *in vitro* model, Marteau et al. (1993) showed that 26% of ingested *L. bulgaricus* survived passage through the stomach. Pochart et al. (1989) collected intestinal fluid after intubation of 10 volunteers who consumed 430 g of yogurt containing 10^8 bacteria/g each of *L. bulgaricus* and *S. thermophilus*. The tube was placed in the third portion of the duodenum and duodenal fluid was recuperated by aspiration. Fifteen minutes after yogurt ingestion, the concentration of live bacteria reaching the duodenum was $10^{5\ 4}$ cfu/mL for *L. bulgaricus* and $10^{6\ 5}$ cfu/mL for *S. thermophilus*. This represents only 0.2% of all the *L. bulgaricus* and 0.6% of all the *S. thermophilus* ingested.

11.3.1.2 Persistence

The persistence of a probiotic depends on its interaction with the enteric microbiota, which competes for the nutritional substrate, adhesion sites and environmental changes due to the production of bacteriocins and other bacterial products. Because the immune system helps control the microbiota, it probably influences the persistence of probiotics as well. Faecal recuperation is not a sufficient indicator of probiotic survival in the upper intestine, where the probiotic organisms are better positioned to wield their biological effects. For this reason, intubation at specific intestinal sites where the bacteria are more likely to colonise, proliferate and produce active metabolites would yield better physiological results. A biopsy of this portion of the digestive tract to confirm bacterial colonisation would be even more informative. Table 11.2 presents some of the clinical studies that have estimated the intestinal survival and persistence of probiotic strains.

The persistence of *L. rhamnosus* GG in the faeces and colonic mucosa was monitored by Alander et al. (1999b). They confirmed that this strain adheres *in vivo* to the colonic mucosa and persists for lengthy periods after consumption cessation. Volunteers consumed the strain twice a day for 12 days at a dose of about 6×10^{10} cfu/day. They then underwent a biopsy either immediately or one or two weeks after ceasing consumption. The amount of recuperated *L. rhamnosus* GG in the biopsy immediately after the 12-day consumption period was around 6×10^{1}–4×10^{4} cfu per biopsy sample. The quantity of *L. rhamnosus* GG found in the faeces decreased progressively after consumption ceased, and the bacteria was undetectable 14 days later. However, the bacteria persisted in the mucosa up to 21 days before disappearing at 28 days. The persistence of *L. rhamnosus* GG in the mucosa can be explained by its capacity to attach itself to the colonic surface and multiply at a pace that partially offsets faecal excretion. Moreover, a human intervention study found that the GG strain persisted at higher levels than *L. rhamnosus* LC705 throughout the study and for 7 days longer in the digestive tract of healthy volunteers, suggesting that a protein-mediated strain-specific adhesion mechanism such as mucus-binding pili was responsible for the strong binding properties of the GG strain (Kankainen et al. 2009). Tannock et al. (2000) administered *L. rhamnosus* DR20 bacteria in powdered milk to 10 volunteers every day for 6 months at a concentration of $10^{9\ 2}$ lactobacilli per day. The mean quantity of *L. rhamnosus* DR20 found in faeces following ingestion of the powdered milk was around $10^{5\ 8}$ cfu/g. A predominance of the *L. rhamnosus* DR20 strain in the microbiota lactobacilli population was observed in several subjects (about 40%) who had a pre-existing, stable *Lactobacillus* population residing in

Table 11.2 Intestinal survival and persistence of probiotics: clinical studies

Organisms	Daily dose	Results	References
Lactobacillus GG	6×10^{10} cfu	GG strain detected in biopsy specimens and final faecal samples of all volunteers (6) after the 12-day GG administration period; in 7 of 8 biopsy samples 1 week after stopping GG administration; None of the 7 subjects 2 weeks after stopping GG administration	Alander et al. (1999b)
L. rhamnosus Lcr35	10^8 or 10^{10} or 10^{12} cfu	Increases in Lcr35-like bacteria in the faeces during the intake periods, whatever the dose (10^8 or 10^{10} or 10^{12} cfu)	De Champs et al. (2003)
L. rhamnosus R11 *L. acidophilus R52*	4 capsules daily containing 2×10^9 *L. rhamnosus* R11 and 1×10^8 *L. acidophilus R52*	*L. rhamnosus* R11 detectable in 5 of 14 volunteers after 1 week of the wash-out period; two weeks after the end of the consumption period, no *L. rhamnosus* R11 detected; *L. acidophilus* R52 not found at the end of the first week of the wash-out period	Firmesse et al. (2008)
L. reuteri DSM 17938 *L. rhamnosus* GG	1×10^9 cfu of *L. reuteri* DSM 17938 5×10^9 cfu of *L. rhamnosus* GG	Detectable levels of *L. reuteri* in 11 out of 13 volunteers after a 3-week intervention with the *L. reuteri* DSM 17938; detectable levels of *L. rhamnosus* GG in 15 out of the 16 volunteers after a 3-week intervention	Dommels et al. (2009)
L. casei DN-114 001	Ileal 300 mL once; Faecal 3×100 mL of fermented milk containing about 10^8 cfu/mL of *L. casei* DN-114 001	The total recovery of *L. casei* DN-114 001^{Rif} in the ileum over the entire 8-hour period was estimated at 9.2 ± 0.5 $\log_{10}$ cfu, (3.6% of the total ingested quantity); *L. casei* DN-114 001^{Rif} survival was approximately 28.4% in the faeces	Oozeer et al. (2006)
L. casei Shirota	2×65 mL of 4×10^8 cfu/mL	*L. casei* Shirota recovered at a population level of 7.1 log cfu/g faeces 7 days after commencement of feeding; maintained at this level until day 21; persisted in 6 of 10 volunteers until day 28 at 5 log cfu/g	Tuohy et al. (2007)
L. paracasei strain B21060 and strain B21070; *L. gasseri* strain B21090	5×10^9 of both *L. paracasei* strain B21060 and strain B21070 and 0.5×10^9 of *L. gasseri* strain B21090	Strains seem to persist in the colon for at least 3 days after discontinuation of oral intake	Morelli et al. (2003)

Table 11.2 (continued)

Organisms	Daily dose	Results	References
B. animalis subsp. *lactis* Bb-12, *L. acidophilus* NCFB 1748 and *Lactobacillus paracasei* subsp. *paracasei* LMG P-17806 (Lactobacillus F19)	250 mL twice daily for 14 days containing 10^8 cfu/mL of the strains *L. acidophilus* NCFB 1748, *B. lactis* Bb12 and *Lactobacillus paracasei* subsp. *paracasei* F19	*B. lactis* was recovered from 4 samples of 3 subjects (range $\log_{10}$ 3.0–6.6 cfu/g faeces)	Sullivan et al. (2003)
B. lactis Bb-12	5×10^9 cfu *B. lactis* Bb-12.oat-based cereal bar	At the end of the intervention period (day 7), *B. lactis* Bb-12 detected in the faeces of 4 subjects. During follow-up, 4 days after consumption (day 11), detected in 5 subjects. One week after consumption ceased (day 14), 5 subjects were still excreting *B. lactis* Bb-12 in the faeces	Ouwehand et al. (2004)
B. animalis subsp. *lactis* Bb-12, *L. acidophilus* NCFB 1748 and *Lactobacillus paracasei* subsp. *paracasei* LMG P-17806 (Lactobacillus F19)	5×10^8 cfu/g of Bb12 1×10^8 cfu/g of two lactobacilli	*Lactobacillus* F19 and *B. animalis* subsp. *lactis* Bb-12 showed good survival in the GI-tract (detected in 14 and 11 of the 14 volunteers, respectively). *Lactobacillus* F19 was not detected in any of the biopsy samples, whereas *L. acidophilus* NCFB 1748 and *B. animalis* subsp. *lactis* Bb-12 were detected in biopsy samples of 1 and 3 subjects, respectively	Matto et al. (2006)

their gut. The strain completely disappeared 2 months following cessation of consumption. The impact of probiotics consumption on the presence of *Lactobacillus* in the intestinal tract cannot be predicted, because it depends on factors such as the pre-existing composition of the enteric microbiota and the degree of competitive exclusion that it produces (Tannock 2003). The individual compositions of autochthonous *Lactobacilli* considerably influence the implantation of probiotic lactobacilli. The increase in the number of *L. rhamnosus* Lcr35-like bacteria was highest in subjects harbouring the lowest initial colonisation levels (<4 log cfu/g). In contrast, the number of Lcr53-lactobacilli decreased in individuals harbouring the highest colonisation levels in the pre-treatment period (>5 log cfu/g). No major variation was observed in subjects with initial colonisation levels between 4 and 5 log cfu/g. Finally, no relationship was found between the average number of cfu in the faeces and the doses ingested by the subjects (10^8 or 10^{10} or 10^{12} cfu) (De Champs et al. 2003).

In contrast, Firmesse et al. (2008) suggest that the dose of the ingested probiotic is an important factor to obtain high concentrations in the various parts of the gastrointestinal tract. After cessation of capsule consumption, *L. rhamnosus* R11 was rapidly eliminated from faecal samples, with no lasting colonisation. *L. acidophilus* R52 was not recovered from faecal samples after consumption of capsules, suggesting that either this strain was destroyed during transit or it was consumed in insufficient quantity (4×10^8 cfu) to be detected in the faeces. *L. reuteri* DSM 17938 and *L. rhamnosus* GG were found to survive passage through the human GI tract after consumption of a low-fat probiotic spread. Faecal viable plate counts of *L. reuteri* were 5.6 log cfu/g of faeces after 3 weeks of consumption of the *L. reuteri* DSM 17938 spread, which contained 5.7×10^9–1.0×10^{10} cfu/20 g of spread on average. Faecal viable plate counts of *L. rhamnosus* GG were on the order of 6.6 log cfu/g of faeces after 3 weeks of consumption of the *L. rhamnosus* GG spread containing 3.3×10^{10}–5.6×10^{10} cfu/20 g of spread on average (Dommels et al. 2009).

Many studies (Table 11.2) indicate that some strains can be isolated from faecal samples (or endoscopy samples) for only a few days after consumption has ceased. These bacteria can persist or remain metabolically active only during passage through the gut (Tuohy et al. 2007). For the purposes of stabilising disturbed enteric microbiota, therefore, bacteria that persist in the faeces would be preferable to bacteria that survive for only a short time.

11.3.1.3 Stabilisation

Before addressing the capacity of probiotic strains to stabilise the enteric microbiota or to modulate the immune response, it would be important to ascertain whether they can reinforce populations of autochtonous bifidobacteria or lactobacilli without altering the balanced microbial ecosystem of healthy individuals. Only a few well-designed human clinical trials have studied these aspects. Yamano et al. (2006) examined the survival ability of *L. johnsonii* La1 in the human gastrointestinal tract of children and adults. Increases in the number and ratio of *Bifidobacterium* in the faeces were observed after the ingestion of fermented milk containing *L. johnsonii* La1. However, no changes were observed in *Bifidobacterium* or *Lactobacillus* populations with the fermented milk placebo. No major alterations in the bacteriology or biochemistry of the faecal microbiota in the 12 healthy human subjects were observed during a short-term consumption of *L. casei* DN-114001. This strain did not affect the communities of obligate anaerobes, which are the numerically dominant members of the faecal microbiota. The average level of *L. paracasei* equivalents observed at the end of the supplementation ($10^{9\,89}$ cfu/g faeces) was not accompanied by an increase in the *Lactobacillus–Enterococcus* group in the total faecal microbiota (Rochet et al. 2006). However, transient high levels of DNA from the probiotic were measured in the upper part of the intestine, compatible with interaction with the immune system (Rochet et al. 2008).

In contrast, Garrido et al. (2005) noted that regular ingestion of an La1-containing product affected the faecal bacterial populations of healthy human volunteers. *Bifidobacterium, Lactobacillus* and *Clostridium histolyticum* cluster populations increased, whereas the cluster corresponding to *F. prausnitzii* decreased, and those for *Bacteroides, Atopobium* and *Eubacterium rectale* were not affected by La1 ingestion. In a randomised placebo-controlled study, Marzotto et al. (2006) observed that *L. paracasei* A survived in faecal samples in most of the infants examined (92%), particularly after 1 week of oral consumption of fermented milk, demonstrating the strain's ability to pass the gastrointestinal barrier and temporarily dominate the intestinal *Lactobacillus* community of different subjects with diverse intestinal environments. Increased faecal lactobacilli was the main impact following the strain A consumption. Consumption of *L. paracasei* A did not affect the counts of bifidobacteria, enterococci, total anaerobes or *Bacteroides*, and decreased clostridia count was measured in infant faeces after 4 weeks of treatment. These results suggest that the intestinal microbiota of young infants, which is not yet at stable equilibrium, is more susceptible than the faecal microbiota of healthy adults to the introduction of an exogenous strain (Rochet et al. 2008). Interestingly, a 2-week supplementation of probiotic bacteria *B. longum* and a yogurt enriched with *B. animalis* modified the amount of bacteria and increased β-galactosidase activity in faeces from lactose-intolerant subjects, although the composition of the predominant bacterial groups in faeces remained unchanged (He et al. 2008).

11.4 Disturbance of the Enteric Microbiota

Very few studies have examined the effects of probiotics on disturbed enteric microbiota. However, many authors have described the impact of probiotics on disease states such as acute diarrhoea (Guandalini 2008; Madsen 2008; Misra et al. 2009; Zanello et al. 2009), antibiotic-associated diarrhoea (Alam and Mushtaq 2009; Rohde et al. 2009), and allergies (Saavedra 2007; Kopp and Salfeld 2009).The body of evidence on these aspects being beyond the scope of this chapter, the focus will be limited to human clinical studies on probiotic interventions to treat disturbed enteric microbiota caused by antibiotics, or in the case of irritable bowel syndrome, on stabilisation of the intestinal microbial population.

11.4.1 Antibiotic Treatment

Antibiotic treatment can cause perturbations of the enteric microbiota, and may also affect the predominant bacterial groups, allowing naturally opportunistic organisms such as yeasts and potentially pathogenic bacteria to proliferate (Sullivan et al. 2003). Antibiotics have also been implicated in the pathogenesis of inflammatory bowel disease. In a revealing study, changes in the intestinal microbiota of three healthy humans before and after treatment with the antibiotic ciprofloxacin were characterised using a metagenomic approach. Ciprofloxacin reduced the diversity of intestinal microbiota, with significant effects on about one-third of the bacterial taxa. Despite this pervasive disturbance, community memberships largely returned to their pre-treatment state within 4 weeks, indicating the presence of factors that promoted community resilience (Dethlefsen et al. 2008).

Exposure to antibiotics in neonates triggers the subsequent development of certain diseases, particularly due to immune response disorders. The host–microbe interaction is crucial in the neonatal period, because the establishment of a normal microbiota increases the stimulatory effect on the maturation of the gut-associated lymphoid tissue in response to antigens. Aberrant microbiota caused by antibiotic treatment during childhood may predispose an individual to both inflammatory gut disease and diarrhoea. The early postnatal period (about the first week of life) is considered critical, because facultative anaerobes are essential for the establishment of reductive conditions favourable for colonisation by strict anaerobes, including bifidobacteria. Bacteria belonging to the genus *Bifidobacterium* are the most predominant and metabolically active organisms in the enteric microbiota of healthy infants. Changes in the quantitative and qualitative composition of specific *Bifidobacterium* species may be useful indicators of deviations from the balanced microbiota (Salminen and Isolauri 2006). Early colonisation by *Bifidobacterium* was greatly attenuated by early antibiotic treatment, whereas overgrowth of *Enterococcus* and *Enterobacteriaceae* occurred in infants treated with antibiotics (Tanaka et al. 2009). Prenatal administration of probiotics may have a strong impact on microbiota development, with significant effects on neonates. It was observed that the mother's consumption of the probiotic *L. rhamnosus* GG affected faecal *Bifidobacterium* transfer and composition during early infancy. The probiotic treatment appeared to reinforce the development of a more complex and diverse *Bifidobacterium* microbiota (Gueimonde et al. 2006). Clostridia numbers at 6 months of age were higher in infants receiving probiotics, whereas the inverse was observed at 24 months, with higher numbers of clostridia found in infants receiving a placebo than in infants receiving probiotics. Hence, bifidobacteria and clostridia species may prove to be markers of healthy versus deviant microbiota succession (Rinne et al. 2006). The acute effects of antibiotic treatment on the enteric microbiota range from diarrhoea to pseudomembranous colitis. The long-term consequences of these disturbances are not known, but attempts to minimise these side effects have included the use of probiotics with the aim of stabilising intestinal bacterial communities and minimising potential alterations in the microbial community structure (Engelbrektson et al. 2009).

In adults, the impact of *L. acidophilus, B.lactis* and *Lactobacillus* F19 on ecological disturbances in the intestinal microbiota was examined in connection with clindamycin administration. For the placeboyoghurt and probiotic yoghurt, the numbers of enterococci increased, whereas other Gram-positive microorganisms decreased. In both treatments, the numbers of *Escherichia coli* decreased, particularly in the placebo group, whereas other Gram-negative bacilli, mainly *Klebsiella* spp., increased. In the placebo group, the numbers of lactobacilli, bifidobacteria, eubacteria, veillonella and bacteroides decreased during treatment and increased at the end of the study period. In contrast, in subjects receiving probiotic yoghurt, the numbers of lactobacilli and bacteroides remained stable throughout the study period, whereas the numbers of bifidobacteria and veillonella decreased. Unfortunately, the probiotic treatment did not prevent colonisation with *C. difficile*, and one subject developed *C. difficile*-associated disease (CDAD). The main finding was that the probiotic strains prevented ecological disturbances in the numbers of intestinal *Bacteroides fragilis* group species during clindamycin administration.These authors suggest that mutually beneficial interactions between lactobacilli and bacteroides might have occurred in complement to create a more favourable environment, or as the nutrients degraded (Sullivan et al. 2003). The capacity of a probiotic mixture of *B. lactis* Bl-04, *B. lactis* Bi-07, *L. acidophilus* NCFM, *L. paracasei* Lpc-37 and *Bifidobacterium bifidum* Bb-02 to minimise disruption of the intestinal microbiota was investigated in individuals during and after antibiotic therapy (amoxicillin and clavulanic acid). Antibiotic-induced alterations in enteric microbiota were identified and the probiotic mixture was found to have minimised antibiotic disturbances. The baseline microbiota was restored more rapidly in the probiotic-treated group, with a particularly pronounced effect on concentrations of *Enterobactereaceae*, *Bifidobacterium* and *Bacteroides*. For these authors, part of the benefit of probiotics is that they increase the numbers of *Bifidobacteria*, which may limit antibiotic disturbance of gut microbiota, thereby stabilizing concentrations of *Enterobactereaceae* and *Bacteroides* in particular, which is consistent with the previously mentioned study. These studies propose a potential mechanism whereby probiotics limit gastrointestinal-adverse events associated with antibiotics.

The role of triple therapy in eradicating *H. pylori* has been well established. A clinical trial examined the impact of probiotic supplements (two strains of *L. acidophilus*: CUL60 and CUL21, and two strains of *Bifidobacterium* spp.) on the regrowth of intestinal microbiota following antibiotic therapy. In the placebo group, the numbers of facultative anaerobes and enterobacteria increased during the study, and at day 35 the numbers were significantly higher in the placebo than the active group. When the microbiota of the two groups were compared, the total numbers of facultative anaerobes and *Enterobacteriaceae* in the active group were significantly lower than in the placebo group. This suggests that, despite the probiotic supplement, the microbiota of the placebo and active groups were susceptible to the effects of the antibiotics administered to eradicate *H. pylori*.

11.4.2 Irritable Bowel Syndrome

Irritable bowel syndrome (IBS) is a functional gastrointestinal disorder that affects up to 20% of the population, with a higher prevalence in women. It is characterised by abdominal pain or discomfort and intestinal dysfunction. Enteric infection and intestinal inflammation appear to play a role in the development of IBS (Barbara et al. 2009; Collins et al. 2009). An altered enteric microbiota may also play an important role in the development and continuation of IBS. Quantitative differences in the enteric microbiota were found between healthy controls (higher numbers of *Lactobacillus* and *Collinsella*) and IBS patients (higher numbers of *Bacteroides* and *Allisonella* in IBS patients with mixed bowel habits, and ruminococci and streptococci in constipation-predominant IBS). Total counts of the *Clostridium coccoides* group and *B. catenulatum* group were significantly lower in IBS patients than in healthy subjects (Malinen et al. 2005). In addition, faecal microbiota differed between the different types of IBS (diarrhoea, constipation, pain). These results support earlier observations that IBS was associated with altered enteric microbiota (Madden and Hunter 2002). In contrast, no difference in the culturable faecal numbers of bacteroides, bifidobacteria, spore-forming bacteria, lactobacilli, enterococci or yeasts were observed between IBS and control groups (Matto et al. 2005), although the numbers of coliforms as well as the proportions of aerobically growing bacteria were higher in the IBS group than the control group. The putative consequences of changes in the enteric microbiota are increased fermentation of food with gas production, bile acid malabsorption, changes in intestinal motor and sensory function, mucosal immune activation and minimal inflammation (Barbara et al. 2008).

Few randomised controlled trials on IBS and probiotics have described the impact on the enteric microbiota. The effects of *L. plantarum* 299V on colonic fermentation were evaluated in a four-week trial in 12 previously untreated patients with IBS. No benefit was observed following the administration of *L. plantarum* 299V, although this strain reduced hydrogen production in the colon, but not sufficiently to be clinically effective (Sen et al. 2002). In another study, *L. plantarum* DSM 9843 was found in faecal samples from 84% of the 30 participants in the probiotic group, and was also found in rectal biopsies from 32% of these patients. No major changes were found in *Enterobacteriaceae* in either group, either before or after the administration period, except that *Enterococusi* increased in the placebo group and remained unchanged in the test group (Nobaek et al. 2000). Seventy-seven subjects with IBS were randomised to receive either *L. salivarius* UCC4331 or *Bifidobacterium infantis* 35624. *Lactobacillus* and *Bifidobacterium* were detected in faecal samples. *B. infantis* 35624 alleviates symptoms in IBS, suggesting an immune-modulating role of this organism in this disorder. In a short-term study, *L. paracasei* subsp. *paracasei F19* was administered twice daily (24×10^9) for 14 days to evaluate its efficacy and tolerability in IBS patients presenting with diarrhoea or with constipation (Lombardo et al. 2009). At the end of the treatment period, pain was no longer present in 49/52 IBS patients with diarrhoea (94.2%) or in 42/48 of the IBS patients with constipation (87.5%). In a series of 20 consecutive patients, *L. paracasei* subsp. *paracasei F19* content was evaluated in faeces, before and after treatment. Microbiological analysis performed on faeces revealed a low microbial *L. paracasei* subsp. *paracasei F19* load prior to treatment and a marked increase in load following treatment.

A multispecies probiotic significantly alleviates IBS symptoms. Over a 6-month study period, each subject (a total of 55) daily received either a probiotic capsule containing *L. rhamnosus* GG, *L. rhamnosus* Lc705, *Propionibacterium freudenreichii* subsp. *shermanii* JS and *Bifidobacterium breve* Bb99, or a placebo capsule. LGG and Bb99 were detected in most of the subjects in the probiotic group at 6 months. The analysed bacterial groups of the enteric microbiota remained relatively stable during the supplementation, with the exception of *Bifidobacterium* spp., which increased in the placebo group and decreased in the probiotic group. The decrease in bifidobacteria due to probiotics was significant only in subjects who had above-median counts at baseline, whereas in subjects with below-median counts, no significant difference between groups was detected. In addition, the probiotic tended to decrease the β-glucuronidase activity (67% of subjects in the probiotic group versus 38% in the placebo group), although no significant change in β-glucosidase activity was observed during the intervention (Kajander et al. 2007). A second clinical trial was conducted to clarify the mechanisms by which the multispecies probiotic alleviates IBS symptoms. The effect of *L. rhamnosus* GG, *L. rhamnosus* Lc705, *P. freudenreichii* subsp. *shermanii* JS and *B. animalis* subsp. *lactis* Bb12 on enteric microbiota stabilisation was estimated with a high-throughput microarray, enabling the simultaneous analysis of all currently known intestinal bacterial species (Kajander et al. 2008). Microbiota stabilisation was observed, as the microbiota similarity index increased with the probiotic supplementation, whereas it decreased with the placebo. This multispecies probiotic could therefore be appropriate for alleviating IBS symptoms and stabilising the intestinal microbiota.

11.5 Immunomodulation

Intestinal microbiota constitutes the largest source of microbial stimulus that exerts both harmful and beneficial effects on human health. It therefore acts as a critical factor in the development and maturation of the intestinal immune system. Indigenous microbiota plays a major role in the stimulation and development of both systemic and mucosal immunity in newborn infants. Enteric microbiota alterations may also be involved in the inflammatory responses in allergic and inflammatory bowel diseases. If so, probiotics might prevent these inflammatory processes by stabilising the intestinal microbial environment and intestinal barrier permeability, promoting enteric antigen degradation, and altering immunogenicity. Immunostimulation and immunomodulation are the most compelling explanations for the action mechanisms of probiotics against bacterial pathogens.

11.5.1 Innate Immune System and Interaction with Probiotics

The cells that participate in innate immunity react rapidly to challenges by infectious agents, providing early protection of the host. The ensuing inflammatory reaction triggers a cascade of events in an attempt to eliminate the invading agent. This first line of defense is not specific in its ability to recognise its target. Key players in the innate immune response include the phagocytic cells (neutrophils, monocytes and macrophages) and natural killer cells (NK).

Phagocytic cells are attracted to the infection site by chemotaxis, and macrophages can produce cytokines that recruit other inflammatory cells, such as neutrophils. Natural killer cells rapidly react to the presence of virus-infected cells in the early stages of infection by killing the infected target cells. Other cells, called dendritic cells (DC), along with macrophages and monocytes, provide an interface between the innate and adaptive immune systems, as they can act as antigen-presenting cells (APC). This "bridging" role is crucial for initiating the adaptive immune response. Cells of the innate immune system also influence the adaptive immune response through the production of cytokines. Cytokines are proteins that enable communication between cells. They are essential for regulating the outcomes of an immune response at both the innate and adaptive levels. The innate immune system must distinguish between self and non-self, based on the emission of particular signs by pathogens. These signs are common to pathogens, but are different from the signal of human body cells.

Many probiotic strains can influence innate defense mechanisms, such as phagocytosis and NK activity, as shown in several studies (Table 11.3). Furthermore, the positive effects of probiotics on phagocytosis and NK cell function have been observed not just in clinical situations; immune system stimulation has also been observed in healthy people, especially elderly subjects (Gill et al. 2001a, b). It is also noteworthy that an immunostimulatory effect was observed in healthy subjects, whereas a downregulation response was observed in milk-hypersensitive subjects. In healthy subjects, probiotics may stimulate the non-specific immune response, whereas in atopic or allergic subjects, they may lead an improved reaction to hypersensitivity (Roessler et al. 2008).

11.5.2 The Mucosal Immune System and Interaction with Probiotics

The mucosal surfaces of the gastrointestinal tract, which cover an area of more than 300 m^2, continuously encounter foreign antigens and infectious agents (Delves and

Table 11.3 Some probiotics and their effects on innate immune system: clinical studies

Immune system effect	Organism	References
Macrophage and phagocytic activity		
Increased CXCL1, CXCL2, CCL20 production (attract macrophages)	*L. casei* DN-114 001	Tien et al. (2006)
Increased phagocytic activity in a clinical study (elderly subjects with total enteral nutrition by nasogastric tube or gastrostomy)	*L. johnsonii* La1 (NCC533) *S. thermophilus*	Fukushima et al. (2007)
Increased phagocytic activity of granulocytes and monocytes in healthy adults	*B. bifidum* Bb12 *Lactobacillus* strain La1	Schiffrin et al. (1995, 1997)
Significantly increased Polymorphonuclear (PMN) cell phagocytic capacity in healthy elderly subjects	*Bifidobacterium lactis* HN019	Arunachalam et al. (2000)
Increased activity of neutrophils in healthy subjects and downregulated neutrophils and monocyte activity in milk sensitive subjects	*Lactobacillus* GG (ATCC 53103)	Pelto et al. (1998)
Significantly increased monocyte and granulocyte activity in healthy subjects and non-significantly decreased in patients suffering from atopic dermatitis	*L. paracasei* Lpc-37 *L. acidophilus* 74-2 *B. lactis* 420	Roessler et al. (2008)
Increased monocyte and granulocyte activity in healthy subjects	*L. paracasei* LTH 2579	Jahreis et al. (2002)
Increased monocyte and granulocyte activity in healthy subjects	*B. animalis* subsp. *lactis* DGCC 420 *L. acidophilus*	Klein et al. (2008)
Natural killer (NK) cells		
IL-15 stimulation (NK cell activation)	*L. casei* subsp. *casei*	Ogawa et al. (2006)
Increased NK cell activity	Daily yoghurt intake	Meyer et al. (2006)
Suppressed lymphocyte proliferation Induced apoptosis	*L. rhamnosus* GG *B. lactis* *L. acidophilus* *L. delbrueckii* subsp. *bulgaricus* *S. thermophilus* *L. paracasei* *E. coli* Nissle 1917	Pessi et al. (1999); von der Weid et al. (2001); Sturm et al. (2005); Carol et al. (2006)
Phagocytic activity and NK cell activity		
Increased PMN and NK activity in a clinical trial (adults presenting milk intolerance)	*B. lactis* HN019 + galactooligosaccharides	Chiang et al. (2000)
Increased PMN and NK activity in a clinical trial (healthy elderly living independently)	*B. lactis* HN019	Gill et al. (2001a)
Significantly enhanced systemic phagocytic activity of leukocytes and cell lysis by NK cells	*L. rhamnosus* NH001	Sheih et al. (2001)

Roitt 2000). When probiotics are taken orally, they remain in the gastrointestinal tract and are not absorbed. They are therefore ideally suited to influence the immune response at the mucosal "frontier" of the gastrointestinal tract (Tlaskalová-Hogenová et al. 2004).

The immune response of the intestinal mucosa has certain distinctive properties. It acts as the first immune organ, because it contains the Gut-associated lymphoid tissue (GALT). The GALT is composed of lymphoid aggregates, including Peyer's patches (located mainly in the small intestinal distal ileum) and mesenteric lymphoid nodes. In addition, there are large amounts of immune-competent cells in the lamina propria and the mucosal epithelium. The intestine also protects us because its epithelium is covered by mucus, which prevents direct contact with microorganisms.

The intestinal immune system must encounter all the antigens in order to determine which ones require an immune response and which ones can be safely tolerated. Intestinal antigens are acquired through various pathways. First, the enterocytes transport antigens from the intestinal lumen to the lamina propria. Enterocytes can even act as APC. Antigen sampling also occurs in Peyer's patches, where a specialised epithelium, called the follicular-associated epithelium (FAE), covers one or more lymphoid follicle composed of B and T cells, DC and macrophages. The FAE is made up of enterocytes and M cells. It has been shown that DC, using their dendrites, also act as guard cells in the intestinal lumen without disturbing the integrity of their tight surface junctions. Finally, the intestinal epithelium also contains intraepithelial lymphocytes (IEL) located between the enterocytes. The intestinal immune system is governed by complex regulation processes that enable pathogenic micro-organisms to be eliminated while maintaining a tolerance for food antigens and endogenous flora.

The interaction between probiotic strains and enterocytes is critical for the controlled production of cytokines and chemokines that are secreted by epithelial cells. Indeed, some probiotic organisms have been shown to modulate the *in vitro* expression of pro- and anti-inflammatory molecules in a strain-dependent manner (Table 11.4).

The interaction between probiotics and Peyer's patch M cells has been established, and the importance of these cells in antigen transport across the intestinal epithelium has received considerable attention (Muscettola et al. 1994; Claassen et al. 1995). It has been shown that probiotics, or at least their products, can access the intestinal mucosal immune system, persist for a certain amount of time and trigger a specific immune response.

Some probiotics can regulate IL8 production by attaching to enterocytes (Table 11.4). The quality and dose of probiotic preparations may impact this IL-8 production. IL-8 appears to be a major cytokine produced by enterocytes following an encounter with a probiotic organism. The IL-8 cytokine functions primarily as a neutrophil chemoattractant. Probiotic strains differ in their capacity to augment IL-8 expression, however, and some strains even appear to decrease epithelial-cell production of IL-8. Because all these findings are based on the use of cell lines as experimental models, they do not necessarily represent the actual *in vivo* situa-

Table 11.4 Cytokines produced following the interaction of probiotics and enterocytes

Organism	Cytokines	Cells	References
Lactobacillus sakei	IL-1b, IL-8, TNF-a (pro-inflammatory, apoptosis inducer)	Caco-2	Haller et al. (2000)
E. coli Nissle 1917	IL-8 (Pro-inflammatory, neutrophiles stimulator; bactericid activity by oxygen activation)	T-84, HT-29	Lammers et al. (2002); Otte and Podolsky (2004)
B. lactis Bb12	IL-6 (Pro-inflammatory, growth factor of B cells; supports blood plate production)	Primary intestinal epithelial cells, Mode-k	Ruiz et al. (2005)
L. rhamnosus GG	IL-8	Caco-2	Zhang et al. (2005)
L. casei CRL 431	IL-6	Primary intestinal epithelial cells	Vinderola et al. (2005)
L. helveticus R389	IL-6	Primary intestinal epithelial cells	Vinderola et al. (2005)
L. rhamnosus GG	IL-6		Viljanen et al. (2005b)[a]

[a] Clinical study on infants presenting atopic eczema/dermatitis syndrome

tion. Aside from IL-8, enterocytes can excrete other cytokines, such as IL-6 in the presence of probiotic organisms, as demonstrated with more physiological models (Table 11.4).

Taken together, these studies suggest that the interaction between probiotic bacteria and the intestinal epithelium is a key determinant for cytokine production by enterocytes, and is probably the initiating event in probiotic immunomodulatory activity, as it occurs prior to the encounter with the immune system cells.

11.5.3 Adaptive Immune System and Interaction with Probiotics

Lymphocytes (B and T) are essential players in the adaptive immune response. Adaptive immune response takes longer to develop than innate immune response. Specificity and memory are the distinguishing characteristics of adaptive immune response. The adaptive immune system can provide more effective protection against pathogens through its ability to recognise and remember an impressive number of antigens. Armed with this immune memory, it provides a strong defense against secondary infections. Lymphocytes have specific antigen receptors. Thus, each naive lymphocyte has an antigen receptor with a unique specificity. Together, they build a repertoire of polyclonal lymphocytes that can respond to a multitude of antigens.

B cells contribute to the immune response by secreting antibodies (humoural immunity), whereas T cells act primarily to provide cell-mediated immunity. T cells can be subdivided into T helper cells (CD4+, also called Th) and T cytotoxic cells (CD8+). B cells recognise their antigens via their B cell receptor (BCR).

T cells cannot recognise the antigen without some assistance, however. The antigenic determinant must be presented by an appropriate major histocompatibility complex (MHC) molecule. Dendritic cells are the major antigen-presenting cells, and they play a critical role in triggering the adaptive immune response. Macrophages and B lymphocytes can also act as specialised APCs. Once activated, naive T lymphocytes proliferate and differentiate into effector cells. CD8+ T cells become cytotoxic (CTL), at which point they can target infected cells. CD4+ T helper (Th) cells control the immune response by activating and regulating other cells, such as macrophages and B cells.

11.5.4 Stimulation of IgA Production by Probiotics

Many probiotic strains appear to effectively stimulate the production of IgA by B cells, which helps maintain intestinal humoural immunity by binding to antigens, thereby limiting their access to the epithelium (Table 11.5). It has been demonstrated that consumption of some probiotic strains following vaccination against *Salmonella typhi* (Link-Amster et al. 1994), rotavirus (Isolauri et al. 1995) or polio vaccine (Fukushima et al. 1998) result in an increase in the total amount of IgA in faeces, and more particularly the specific IgA against that particular disease. Moreover, probiotics can increase IgA seroconversion during the remission phase of an intestinal disease (Kaila et al. 1995; Majamaa et al. 1995).

Park et al. (2002) showed that *B. bifidum* significantly induced IgA production in Peyer's patches and mesenteric lymph nodes. Surprisingly, rather than inducing a specific humoural immune response, *B. bifidum* apparently had a more systemic immune effect.

11.5.5 Stimulation of Dendritic Cells by Probiotics

Antigen-presenting cells, and more particularly DC, are key players in both the determination of the Th1/Th2 balance and the development of tolerance. DCs can instruct naive CD4+ T cells to differentiate into Th1, Th2 and even Th3 cells. Given the key role of DCs in the orchestration of the immune response, it has been hypothesised that probiotic organisms modulate the immune response by influencing DC maturation.

Several studies have shown that probiotics can induce DC maturation. Taken together, these studies indicate that many probiotic organisms act as anti-inflammatory agents by influencing DCs to induce a non-response state, more particularly by encouraging the development of T cells with immunoregulatory properties (Table 11.5). Meanwhile, other studies have suggested that some probiotic strains

Table 11.5 Effects of probiotics on the adaptive immune system

Organism	Effect	Reference
IgA		
B. bifidum *L. johnsonii* La1 *L. rhamnosus* GG *B. lactis* Bb12	Stimulated IgA production	Link-Amster et al. (1994); Isolauri et al. (1995); Kaila et al. (1995); Majamaa et al. (1995); Park et al. (2002); Ibnou-Zekri et al. (2003)
L. rhamnosus GG (LGG)	Increased specific IgA response to rotavirus in children presenting acute diarrhoea	Kaila et al. (1992)
B. lactis Bb12	Increased IgA response to the attenuated S. typhi Ty21a oral Vivotif vaccine capsule	Link-Amster et al. (1994)
L. rhamnosus GG and *B. lactis* Bb-12	Increased cow's milk-specific IgA antibodies at the time of introduction of cow's milk into the infant's diet	Rautava et al. (2006)
L. fermentum CECT5716	Increased specific anti-influenza-IgA and significantly increased total IgM following an intramuscular vaccination for influenza	Olivares et al. (2007)
Anti-inflammatory		
L. rhamnosus	↓ T cell proliferation and activation	Braat et al. (2004)
L. reuteri	↓ IL-12, IL-6, TNF-α induced regulatory T cell differentiation	Christensen et al. (2002); Smits et al. (2005)
VSL #3	↑ DC maturation, ↓ lymphocyte proliferation, ↓ IL-12, ↑ IL-10, ↓ Th1	Drakes et al. (2004); Hart et al. (2004)
L. casei	Induced regulatory T cell differentiation	Smits et al. (2005)
L. paracasei NCC2461	↑ IL-10	von der Weid et al. (2001)
Pro-inflammatory		
L. casei subsp. *alactus*	↑ IL-12, IL-6, TNF-α	Christensen et al. (2002)
L. gasseri, *L. johnsonii* and *L. reuteri*	↑ IL-12 and IL-18, but not IL-10	Mohamadzadeh et al. (2005)
L. casei	↑ IL-12 via macrophages stimulation	Shida et al. (2006)
Anti- and pro-inflammatory		
B. longum	↑ IL-10, ↑ IL-12	Rigby et al. (2005)

induce DC to regulate T cell responses toward a pro-inflammatory pathway by stimulating the secretion of high levels of IL-12, but not IL-10 (Table 11.5).

These results show that, although the cytokine profiles secreted by DC are more often directed toward an anti-inflammatory response, they remain strain-dependent. Regulation of DC cytokines by probiotics may contribute to the development of more effective molecules for treating intestinal diseases.

11.5.6 Th1/Th2 Balance and Probiotics

Helper T cells are found in two distinct cell types: Th1 and Th2 cells. They are distinguished by the cytokines they produce and respond to and by the immune responses they are involved in. Th1 cells produce pro-inflammatory cytokines such as interferon gamma (IFNγ), tumor necrosis α (TNFα) and interleukin 2 (IL-2), while Th2 cells produce the cytokines IL-4, IL-5, IL-6 and IL-13. The cytokines produced by Th1 cells stimulate the phagocytosis and destruction of microbial pathogens, while Th2 cytokines such as IL-4 and IL-5 generally stimulate the immune response toward large extracellular parasites. The differentiation of naïve T cells into either Th1 or Th2 cells generally depends on environmental conditions (e.g., DCs, cytokines in the milieu, nature and dose of antigen). The process by which commitment develops is called polarisation. It now appears that Th1 differentiation is reliant on IFN-γ and IL-12. The presence of IL-12 induces a signalling cascade that coordinates the expression towards Th1 differentiation, implying a T-bet transcription factor. The T-bet factor also prevents differentiation towards Th2 by suppressing the expression of the factors required for the Th2 subset differentiation process. Th2 differentiation implies the presence of IL-4 in the extracellular milieu, which induces a reaction cascade that leads to the expression of GATA-3, a master regulator towards Th2 commitment. GATA-3 also suppresses the critical elements for the Th1 differentiation process. Th1 and Th2 cells can antagonise each other's action, either by blocking polarised maturation of the opposite cell type or by blocking its receptor functions. This regulatory loop allows full differentiation towards one subset or the other once the decisional process has been triggered.

The balance between Th1 and Th2 cytokine production can determine the direction and outcome of an immune response. A true balance between Th1 and Th2 profiles can be difficult to maintain, as Th1 and Th2 cells inhibit each other. The theory is that the subclasses Th1 and Th2 direct the immune response towards intracellular pathogen eradication (Th1) or towards parasitic and extracellular infections (Th2). An exaggeratedly deviated response towards Th1 is associated with chronic inflammatory diseases, whereas an abnormal Th2 type response is characteristic of an allergic reaction.

The mucosal immune system normally maintains itself in a state that favours tolerance and IgA production, with a slight deviation towards the Th2 response over the Th1 response. Nevertheless, this deviation is not absolute, as some chronic inflammatory bowel diseases such as ulcerative colitis are somewhat Th2-driven, whereas others, such as Crohn's disease, show a predominantly Th1-mediated cytokine profile (Strober et al. 2002; Korzenik and Podolsky 2006). The cytokine profile therefore plays an important role in the maintenance of intestinal immune homeostasis.

Probiotics can exert an immunomodulatory effect by influencing the cytokine production of a wide variety of immune cells in the intestine, especially enterocytes. The effect appears to be strain-dependent, as shown previously for DCs. Depending on the strain, probiotics can induce a pro- or anti-inflammatory effect by upsetting the Th1/Th2 equilibrium, which disrupts the T CD4+ and helper cell differentiation process. Many studies have shown that each probiotic appears to influence the immune system in a unique fashion. Studies also confirm that immunomodulation properties are bacteria-specific. Moreover, it must be emphasised that the studies mentioned in this section have examined cytokine production in response to probiotic organisms under different conditions: *in vitro, in vivo*, physiological and clinical conditions. Results obtained when stimulating peripheral blood mononuclear cells (PBMC) with probiotics therefore differ from those obtained when measuring local cytokine production in the intestine following oral consumption of probiotics. When probiotics are consumed, the microbial agents initially interact with immune cells at an intestinal induction site rather than cells found in the blood.

In vitro studies examine isolated cells that are directly exposed to probiotic cells in culture. *In vitro* system studies are typically highly unphysiological in nature, because they investigate cells in isolation from other components, with which they would normally interact. That is why models that mimic the mucosa should be preferred over blood cell models whenever possible to assess the probiotic immune function. Although an *in vitro* investigation is a more practical way to assess the stimulation of the blood cell immune response by probiotics, it does not represent the physiological reality, and an examination of the intestinal cell immune response would be the best option. Extrapolations from *in vitro* studies to the whole body should be made with caution, and probiotic effects observed *in vitro* should be backed up by properly conducted clinical studies (i.e., randomised double-blind). The use of transgenic or knockout mice and other animal models allows a better understanding of these mechanisms. Interest in the impact of probiotics on the Th1/Th2 balance has led to many clinical trials to assess the effects of probiotics on the treatment of a number of chronic inflammatory diseases and allergies.

The following tables (Tables 11.6 and 11.7) summarise the *in vitro, in vivo* and clinical effects of probiotics against a Th1/Th2 alteration. Table 11.6 covers the effects of probiotics in the case of Th1/Th2 alteration toward Th1. This is the case of Crohn's disease, a chronic inflammatory disease similar to arthritis, and the autoimmune model of encephalomyelitis.

Table 11.7 covers the effects of probiotics in the case of Th1/Th2 alteration toward Th2. This is the case of allergies, atopic dermatitis and some chronic inflammatory bowel diseases such as ulcerative colitis.

These studies suggest that probiotics can be useful to treat and prevent allergies when administered in infancy, but are less useful for adults. For instance, no protective effect was observed for other atopic symptoms in adults such as food allergies or asthma. Further clinical studies are needed to determine the efficacy of probiotics in preventing atopic diseases.

Although a particularly significant clinical effect of the use of probiotics remains to be clearly established, with the exception of the treatment of pouchitis, bacteriotherapy and the use of probiotics remains a major research focus in the treatment of intestinal disorders aiming to re-establish homeostasis of the intestinal microbiota (Borody et al. 2004).

Table 11.6 Effect of probiotics on Th1 deviation conditions and diseases

Th1 deviation diseases				
Disease or conditions tested	Tested microorganisms	Effects on Th1/Th2 balance toward Th2	Results	References
In vitro studies				
Mucosa samples isolated from patients with Crohn's disease	*L. casei* DN114001 or *L. bulgaricus* LB10	Reduced TNF-α Decreased the proportion of CD4+ T cells and TNF-α producing cells by lymphocytes of the mucosa		Borruel et al. (2002)
Mucosa samples isolated from patients with Crohn's disease	*L. casei* DN 114 001	Reduced IL-6 and TNF-α Inhibiting inflammatory T cell activation through its ability to sensitise T cells to apoptose		Carol et al. (2006)
In vivo studies (animal models of chronic inflammatory bowel disease)				
Colitis in IL-10 deficient mice	*L. reuteri*			Madsen et al. (1999)
IL-10 knockout mice	*L. salivarius* and *B. infantis*	Decreased IL-12, IFN-γ and TNF-α Increased TGF-β		O'Mahony et al. (2001); McCarthy et al. (2003)
Intestinal inflammation in IL-10 knockout mice	*L. salivarius* 118	Reduced the production of pro-inflammatory Th1 cytokines, e.g., IL-12 and TNF-α		Sheil et al. (2004)
Intestinal inflammation caused by *Helicobacter hepaticus* in IL-10-deficient mice	*L. reuteri* and *L. paracasei*	Reduced the expression of TNF-α and IL-12 Did not reduce *Helicobacter hepaticus* counts		Pena et al. (2005)
In vivo studies (other animal models)				
Murine rheumatoid arthritis	*L. casei* Shirota	Reduced development of arthritis by reducing the Th1 response		Kato et al. (1998)
Murine collagen-induced arthritis model	*L. salivarius* 118 (Sheil et al. 2004)	Reduced symptoms, probably by reducing IL-12 and TNF-α and increasing TGF-β, as shown in the colitis model		Sheil et al. (2004)

Table 11.6 (continued)

Th1 deviation diseases

Disease or conditions tested	Tested microorganisms	Effects on Th1/Th2 balance toward Th2	Results	References
Arthritis model in rats	*L. rhamnosus* GG	Decreased arthritis inflammation		Baharav et al. (2004)
Experimental autoimmune encephalomyelitis	*L. casei* and *Lactobacillus murines*	Beneficial to inducted mice		Maassen et al. (1998)
Clinical studies				
Patients with acute dermatitis	*L. rhamnosus GG*	Reduced intestinal TNF-α		Majamaa and Isolauri (1997)
Patients with arthritis	*L. rhamnosus* GG		No clinical effect against arthritis	Hatakka et al. (2003)
Patients suffering from Crohn's disease	*L. rhamnosus* GG		No clinical effect.	Prantera et al. (2002); Schultz et al. (2004)
Patients suffering from Crohn's disease	*Saccharomyces boulardii* + mezalamine		Reduced the number of relapses	Guslandi et al. (2000)
Healthy subjects	*L. paracasei* LTH 2579	Significantly decreased CD54/ICAM-1 (an indicator of intestinal inflammation)		Jahreis et al. (2002)

Table 11.7 Effect of probiotics on Th2 deviation conditions and diseases

Disease or conditions tested	Tested microorganisms	Effects on Th1/Th2 balance toward Th1	Results	References
Th2 deviation diseases				
In vitro studies				
Stimulation of blood PBMCs with specific allergens	*L. plantarum* *L. lactis* *L. casei* *L. rhamnosus* GG	Reduced IL-4 and IL-5 production by peripheral blood mononuclear cells (PBMC) prior to stimulation with allergens		Pochard et al. (2002)
Stimulation of blood PBMCs	*L. brevis* subsp. *coagulans* (Labre) *B. lactis* HN019	Produced IFN-α	Contributed to inflammatory response	Kishi et al. (1996); Arunachalam et al. (2000)
Stimulation of blood PBMCs	*L. rhamnosus* E509 *L. rhamnosus* GG *L. bulgaricus* E585	Produced IL-1, IL-6, IL-18 and TNF-α Moderately increased IL-12 and IL-10		Miettinen et al. (1998)
In vivo studies				
Food allergy model in mice	Peritoneal injection of heat-killed *L. casei* Shirota	Increased serum IL-12 and IFN-γ Decreased IL-4 and IL-5 Decreased IgE secretion	Prevented systemic anaphylactic reaction	Shida et al. (2002)
	Orally administered: *L. reuteri* ML1 *L. brevis* ML12	Increased TNF-α, IL-2 and IL-1b		Maassen et al. (2000)
	Orally administered: *L. casei*, *L. delbrueckii* subsp. *bulgaricus* and *L. acidophilus*	Increased TNF-α and IFN-γ (three strains). Increased IL-2 and IL-12 (acidophilus only). Increased IL-4 and IL-10 (delbrueckii and casei only)		Perdigon et al. (2002)

Table 11.7 (continued)

Th2 deviation diseases

Disease or conditions tested	Tested microorganisms	Effects on Th1/Th2 balance toward Th1	Results	References
Clinical studies				
Infants allergic to cow's milk or presenting IgE associated dermatitis	*L. rhamnosus* GG	Increased IFN-γ production in PBMC		Pohjavuori et al. (2004)
Infants with atopic eczema during exclusive breast-feeding	*B. lactis* Bb-12 *Lactobacillus* strain GG (ATCC 53103)	Reduced sCD4 in serum and EPX (eosinophilic protein X) in urine	Significantly alleviated allergic inflammation	Isolauri et al. (2000)
Infant with atopic eczema and presenting or not intolerance to hydrolysed whey formula (EHF)	*B. lactis* Bb-12	Reduced IgE directly correlated with reduced *E. coli* in faeces	*B. lactis* Bb-12 supplementation appears to modify the gut microbiota and may alleviate allergic inflammation	Kirjavainen et al. (2002)
Infants presenting atopic dermatitis (AD) and suspected of CMA (Cow milk allergy)	*Lactobacillus rhamnosus* *Lactobacillus* GG	No significant decrease in IL-4, IL-5 or IFN-γ No significant decrease in EPX (eosinophilic protein X) in urine	No clinical or immunological effects of either probiotic bacteria (*L. rhamnosus* NT-Lrh or GG) used in infants with AD	Brouwer et al. (2006)
Infants presenting moderate to severe AD	*L. rhamnosus* GG	No significant decrease in serum IgE levels	Results could not confirm *L. rhamnosus* GG as an effective treatment of AD in infancy	Folster-Holst et al. (2006)
Infants presenting AD and symptoms suggestive of milk allergy=AEDS (Atopic eczma/dermatitis syndrome)	*Lactobacillus rhamnosus* GG	Increased IL-6 secretion. Decreased TNF-α. Increased IgA	Treatment with Lactobacillus GG may alleviate AEDS symptoms in IgE-sensitised infants, induce systemically detectable low-grade inflammation and alleviate intestinal inflammation	Viljanen et al. (2005a, b, c)

Table 11.7 (continued)

Th2 deviation diseases				
Disease or conditions tested	Tested microorganisms	Effects on Th1/Th2 balance toward Th1	Results	References
Infants whose mothers were allergic or atopic	*L. acidophilus* LAVRI-AI (Probiomics, Australia)	Decreased interleukins TGF-β and TNF-α) Increased colonisation by *L. acidophilus* No change in IgA levels	No significant protection against allergies	Taylor et al. (2007)
Infants presenting moderate to severe atopic dermatitis	*Lactobacillus fermentum* VRI 003 PCC™ (Protract Probiomics Eveleigh, Australia)	Increased INF-γ	Improved atopic dermatitis symptoms linked to the increased Th1/INF-γ response	Prescott et al. (2005)
Pregnant women whose unborn children were at high risk for being atopic	*L. rhamnosus* GG		Reduced the atopic eczema prevalence in children to up to four years of age	Kalliomaki et al. (2001, 2003)
Pregnant women receiving probiotics two to four weeks before delivery	*L. rhamnosus* GG *L. rhamnosus* LC705 *B. breve* Bb99 *Propionibacterium freudenreichii* ssp. *shermanii* JS	Inverse association between atopic diseases and colonisation of the gut by probiotics	No clinical effects on allergy overall Reduced the prevalence of eczema and atopic dermatitis	Kukkonen et al. (2007)
Healthy adults	*L. fermentum* CECT5716 + an intramuscular vaccination for influenza	Increased the T-helper type 1 (Th1) response. Increased TNF-α and specific anti-influenza-IgA	Provided enhanced systemic protection from infection	Olivares et al. (2007)
Adults presenting allergy to birch pollen	*L. rhamnosus* GG		No clinical effect	Helin et al. (2002)

Table 11.7 (continued)

Disease or conditions tested	Tested microorganisms	Effects on Th1/Th2 balance toward Th1	Results	References
Th2 deviation diseases				
Patients with ulcerative colitis (UC) having a restorative ileal pouch-anal anastomosis	VSL#3	Increased the percentage of mucosal lamina propria CD4+CD25+ and CD4+ LAP-positive mononuclear cells	Slight but significant reduction in the PDAI score (Pouchitis disease activity index)	Pronio et al. (2008)
Patients suffering from ulcerative colitis	VSL#3		Improvements in patients with mild to moderate symptoms	Bibiloni et al. (2005)
Patients suffering from ulcerative colitis	VSL#3		Prevented pouchitis following antibiotic-induced remission.	Gionchetti et al. (2000, 2003); Mimura et al. (2004)
Patients suffering from ulcerative colitis	*L. rhamnosus* GG		Reduced pouchitis episodes from about 30% to 7% over a three-year period	Gosselink et al. (2004)
Patients suffering from ulcerative colitis	*E. coli* Nissle 1917		Clinical effect similar to mezalamine	Kruis et al. (1997); Rembacken et al. (1999); Kruis et al. (2004)

11.6 Conclusion

The persistence of probiotic organisms in the digestive tract is an important factor for their beneficial effects on the host, and is related to the administration method, dosage and strain. The enteric microbiota is a highly competitive environment that limits the ability of probiotic strains to persist in the gastrointestinal tract. Although the evidence on using probiotic bacteria for stabilisation in IBS is controversial, the alteration in the enteric microbiota of IBS subjects suggests that probiotics could be used to prevent damage to the intestinal microbial ecosystem following gastroenteritis or the administration of antibiotics.

The intestinal microbiota acts as a primary agent in the development of the postnatal immune system, including oral tolerance and immunity. The interaction between probiotics and enterocytes is crucial to triggering immunomodulation. Probiotics act on a wide variety of cells in the intestine to modulate the immune system towards pro- or anti-inflammatory action, depending on the strain, the setting and the immunological parameters measured, as well as the type of cells acted upon.

Models that mimic the mucosa should be preferred whenever possible when evaluating probiotic immune function. Albeit more practical to implement, assessment of the stimulation of the blood cell immune response by probiotics does not represent the physiological reality, and an examination of the intestinal cell immune response would be the better option. For each new probiotic strain, profiles of the cytokines secreted by lymphocytes, enterocytes and/or DCs that come into contact with the strain should be established. This would enable the pro- and anti-inflammatory properties of the strain to be determined, along with its specific clinical uses. Probiotic products purporting to be immune boosters should be supported by accurate data to substantiate the claim of each proposed strain. Probiotic effects observed *in vitro* should be backed up by properly conducted clinical studies (such as randomised double-blind trials).

Nevertheless, the findings so far do not explain how these organisms induce their effects, nor do they explain which bacterial molecules or cellular receptors are responsible for them. Further in-depth studies are needed to determine the precise mechanisms of probiotics in developing both mucosal and systemic immunity. The use of transgenic or knockout mice and other animal models would allow a better understanding of these mechanisms.

References

Alam S, Mushtaq M (2009) Antibiotic associated diarrhoea in children. Indian Pediatr 46:491–496

Alander M, De Smet I, Nollet L, Verstraete W, von Wright A, Mattila-Sandholm T (1999a) The effect of probiotic strains on the microbiota of the simulator of the human intestinal microbial ecosystem (SHIME). Int J Food Microbiol 46:71–79

Alander M, Satokari R, Korpela R, Saxelin M, Vilpponen-Salmela T, Mattila-Sandholm T, von Wright A (1999b) Persistence of colonization of human colonic mucosa by a probiotic strain, *Lactobacillus rhamnosus* GG, after oral consumption. Appl Environ Microbiol 65:351–354

Arunachalam K, Gill HS, Chandra RK (2000) Enhancement of natural immune function by dietary consumption of *Bifidobacterium lactis* (HN019). Eur J Clin Nutr 54:263–267

Baharav E, Mor F, Halpern M, Weinberger A (2004) *Lactobacillus* GG bacteria ameliorate arthritis in Lewis rats. J Nutr 134:1964–1969

Barbara G, Stanghellini V, Cremon C, De Giorgio R, Gargano L, Cogliandro R, Pallotti F, Corinaldesi R (2008) Probiotics and irritable bowel syndrome: rationale and clinical evidence for their use. J Clin Gastroenterol 42:S214–S217

Barbara G, Cremon C, Pallotti F, De Giorgio R, Stanghellini V, Corinaldesi R (2009) Postinfectious irritable bowel syndrome. J Pediatr Gastroenterol Nutr 48:S959–S957

Bibiloni R, Fedorak RN, Tannock GW, Madsen KL, Gionchetti P, Campieri M, De Simone C, Sartor RB (2005) VSL#3 probiotic-mixture induces remission in patients with active ulcerative colitis. Am J Gastroenterol 100:1539–1546

Borody TJ, Warren EF, Leis SM, Surace R, Ashman O, Siarakas S (2004) Bacteriotherapy using faecal flora: toying with human motions. J Clin Gastroenterol 38:475–483

Borruel N, Carol M, Casellas F, Antolin M, de Lara F, Espin E, Naval J, Guarner F, Malagelada JR (2002) Increased mucosal tumour necrosis factor alpha production in Crohn's disease can be downregulated ex vivo by probiotic bacteria. Gut 51:659–664

Bouhnik Y, Pochart P, Marteau P, Arlet G, Goderel I, Rambaud JC (1992) Faecal recovery in humans of viable *Bifidobacterium* sp. ingested in fermented milk. Gastroenterol 102:875–878

Braat H, van den Brande J, van Tol E, Hommes D, Peppelenbosch M, van Deventer S (2004) *Lactobacillus rhamnosus* induces peripheral hyporesponsiveness in stimulated CD4+ T cells via modulation of dendritic cell function. Am J Clin Nutr 80:1618–1625

Brigidi P, Swennen E, Vitali B, Rossi M, Matteuzzi D (2003) PCR detection of *Bifidobacterium* strains and *Streptococcus thermophilus* in faeces of human subjects after oral bacteriotherapy and yogurt consumption. Int J Food Microbiol 81:203–209

Brouwer ML, Wolt-Plompen SA, Dubois AE, van der Heide S, Jansen DF, Hoijer MA, Kauffman HF, Duiverman EJ (2006) No effects of probiotics on atopic dermatitis in infancy: a randomized placebo-controlled trial. Clin Exp Allergy 36:899–906

Bunte C, Hertei C, Hammes WP (2000) Monitoring and survival of *Lactobacillus paracasei* LTH 2579 in food and the human intestinal tract. Syst Appl Microbiol 23:260–266

Carol M, Borruel N, Antolin M, Llopis M, Casellas F, Guarner F, Malagelada JR (2006) Modulation of apoptosis in intestinal lymphocytes by a probiotic bacteria in Crohn's disease. J Leukoc Biol 79:917–922

Chiang BL, Sheih YH, Wang LH, Liao CK, Gill HS (2000) Enhancing immunity by dietary consumption of a probiotic lactic acid bacterium (*Bifidobacterium lactis* HN019): optimization and definition of cellular immune responses. Eur J Clin Nutr 54:849–855

Christensen HR, Frokiaer H, Pestka JJ (2002) Lactobacilli differentially modulate expression of cytokines and maturation surface markers in murine dendritic cells. J Immunol 168:171–178

Claassen E, Van Winsen R, Posno M, Boersma WJ (1995) New and safe "oral" live vaccines based on lactobacillus. Adv Exp Med Biol 371B:1553–1558

Collins J, Dunne C, Murphy L, Morrisey D, O'Mahony L, O'Sullivan E, Fitzgerald G, Kiely B, O'Sullivan G, Daly C, Marteau P, Shanahan F (2002) A randomised controlled trial of a probiotic *Lactobacillus* strain in healthy adults: assessment of its delivery, transit and influence on microbial flora and enteric immunity. Microbial Ecol Health Dis 14:81–89

Collins SM, Denou E, Verdu EF, Bercik P (2009) The putative role of the intestinal microbiota in the irritable bowel syndrome. Dig Liver Dis 41:850–853

De Champs C, Maroncle N, Balestrino D, Rich C, Forestier C (2003) Persistence of colonization of intestinal mucosa by a probiotic strain, *Lactobacillus casei* subsp. *rhamnosus* Lcr35, after oral consumption. J Clin Microbiol 41:1270–1273

Delves PJ, Roitt IM (2000) The immune system. First of two parts. N Engl J Med 343:37–49

Dethlefsen L, Huse S, Sogin ML, Relman DA (2008) The pervasive effects of an antibiotic on the human gut microbiota, as revealed by deep 16S rRNA sequencing. PLoS Biol 6:e280

Dommels YEM, Kemperman RA, Zebregs YEMP, Draaisma RB, Jol A, Wolvers DAW, Vaughan EE, Albers R (2009) Survival of *Lactobacillus reuteri* DSM 17938 and *Lactobacillus rham-*

nosus GG in the human gastrointestinal tract with daily consumption of a low-fat probiotic spread. Appl Environ Microbiol 75:6198–6204
Drakes M, Blanchard T, Czinn S (2004) Bacterial probiotic modulation of dendritic cells. Infect Immun 72:3299–3309
Engelbrektson A, Korzenik JR, Pittler A, Sanders ME, Klaenhammer TR, Leyer G, Kitts CL (2009) Probiotics to minimize the disruption of faecal microbiota in healthy subjects undergoing antibiotic therapy. J Med Microbiol 58:663–670
Firmesse O, Mogenet A, Bresson JL, Corthier G, Furet JP (2008) *Lactobacillus rhamnosus* R11 consumed in a food supplement survived human digestive transit without modifying microbiota equilibrium as assessed by real-time polymerase chain reaction. J Mol Microbiol Biotechnol 14:90–99
Folster-Holst R, Muller F, Schnopp N, Abeck D, Kreiselmaier I, Lenz T, von Ruden U, Schrezenmeir J, Christophers E, Weichenthal M (2006) Prospective, randomized controlled trial on *Lactobacillus rhamnosus* in infants with moderate to severe atopic dermatitis. Br J Dermatol 155:1256–1261
Frank DN, Pace NR (2008) Gastrointestinal microbiology enters the metagenomics era. Curr Opin Gastroenterol 24:4–10
Fujiwara S, Seto Y, Kimura A, Hashiba H (2001a) Establishment of orally-administered *Lactobacillus gasseri* SBT2055SR in the gastrointestinal tract of humans and its influence on intestinal microflora and metabolism. J Appl Microbiol 90:343–352
Fujiwara S, Seto Y, Kimura A, Hashiba H (2001b) Intestinal transit of an orally administered streptomycin-rifampicin-resistant variant of *Bifidobacterium longum* SBT2928: its long-term survival and effect on the intestinal microflora and metabolism. J Appl Microbiol 90:43–52
Fukushima Y, Kawata Y, Hara H, Terada A, Mitsuoka T (1998) Effect of a probiotic formula on intestinal immunoglobulin A production in healthy children. Int J Food Microbiol 42:39–44
Fukushima Y, Miyaguchi S, Yamano T, Kaburagi T, Iino H, Ushida K, Sato K (2007) Improvement of nutritional status and incidence of infection in hospitalised, enterally fed elderly by feeding of fermented milk containing probiotic *Lactobacillus johnsonii* La1 (NCC533). Br J Nutr 98:969–977
Garrido D, Suau A, Pochart P, Cruchet S, Gotteland M (2005) Modulation of the faecal microbiota by the intake of a *Lactobacillus johnsonii* La1-containing product in human volunteers. FEMS Microbiol Lett 248:249–256
Gibson DL, Vallance BA (2008) Intestinal microbiota are transiently altered during *Salmonella*-induced gastroenteritis. Expert Rev Gastroenterol Hepatol 2:525–529
Gill HS, Rutherfurd KJ, Cross ML (2001a) Dietary probiotic supplementation enhances natural killer cell activity in the elderly: an investigation of age-related immunological changes. J Clin Immunol 21:264–271
Gill HS, Rutherfurd KJ, Cross ML, Gopal PK (2001b) Enhancement of immunity in the elderly by dietary supplementation with the probiotic *Bifidobacterium lactis* HN019. Am J Clin Nutr 74:833–839
Gionchetti P, Rizzello F, Venturi A, Brigidi P, Matteuzzi D, Bazzocchi G, Poggioli G, Miglioli M, Campieri M (2000) Oral bacteriotherapy as maintenance treatment in patients with chronic pouchitis: a double-blind, placebo-controlled trial. Gastroenterol 119:305–309
Gionchetti P, Rizzello F, Helwig U, Venturi A, Lammers KM, Brigidi P, Vitali B, Poggioli G, Miglioli M, Campieri M (2003) Prophylaxis of pouchitis onset with probiotic therapy: a double-blind, placebo-controlled trial. Gastroenterol 124:1202–1209
Goldin BR, Gorbach SL, Saxelin M, Barakat S, Gualtieri L, Salminen S (1992) Survival of Lactobacillus species (strain GG) in human gastrointestinal tract. Dig Dis Sci 37:121–128
Goossens DA, Jonkers DM, Russel MG, Stobberingh EE, Stockbrugger RW (2006) The effect of a probiotic drink with *Lactobacillus plantarum* 299v on the bacterial composition in faeces and mucosal biopsies of rectum and ascending colon. Aliment Pharmacol Ther 23:255–263
Gosselink MP, Schouten WR, van Lieshout LM, Hop WC, Laman JD, Ruseler-van Embden JG (2004) Delay of the first onset of pouchitis by oral intake of the probiotic strain *Lactobacillus rhamnosus* GG. Dis Colon Rectum 47:876–884

Guandalini S (2008) Probiotics for children with diarrhoea: an update. J Clin Gastroenterol 42:S53–S57

Gueimonde M, Sakata S, Kalliomaki M, Isolauri E, Benno Y, Salminen S (2006) Effect of maternal consumption of *Lactobacillus* GG on transfer and establishment of faecal bifidobacterial microbiota in neonates. J Pediatr Gastroenterol Nutr 42:166–170

Guslandi M, Mezzi G, Sorghi M, Testoni PA (2000) *Saccharomyces boulardii* in maintenance treatment of Crohn's disease. Dig Dis Sci 45:1462–1464

Haller D, Bode C, Hammes WP, Pfeifer AM, Schiffrin EJ, Blum S (2000) Non-pathogenic bacteria elicit a differential cytokine response by intestinal epithelial cell/leucocyte co-cultures. Gut 47:79–87

Hart AL, Lammers K, Brigidi P, Vitali B, Rizzello F, Gionchetti P, Campieri M, Kamm MA, Knight SC, Stagg AJ (2004) Modulation of human dendritic cell phenotype and function by probiotic bacteria. Gut 53:1602–1609

Hatakka K, Martio J, Korpela M, Herranen M, Poussa T, Laasanen T, Saxelin M, Vapaatalo H, Moilanen E, Korpela R (2003) Effects of probiotic therapy on the activity and activation of mild rheumatoid arthritis—a pilot study. Scand J Rheumatol 32:211–215

He T, Priebe MG, Zhong Y, Huang C, Harmsen HJ, Raangs GC, Antoine JM, Welling GW, Vonk RJ (2008) Effects of yogurt and bifidobacteria supplementation on the colonic microbiota in lactose-intolerant subjects. J Appl Microbiol 104:595–604

Helin T, Haahtela S, Haahtela T (2002) No effect of oral treatment with an intestinal bacterial strain, *Lactobacillus rhamnosus* (ATCC 53103), on birch-pollen allergy: a placebo-controlled double-blind study. Allergy 57:243–246

Ibnou-Zekri N, Blum S, Schiffrin EJ, von der Weid T (2003) Divergent patterns of colonization and immune response elicited from two intestinal *Lactobacillus* strains that display similar properties *in vitro*. Infect Immun 71:428–436

Isolauri E, Joensuu J, Suomalainen H, Luomala M, Vesikari T (1995) Improved immunogenicity of oral D × RRV reassortant rotavirus vaccine by *Lactobacillus casei* GG. Vaccine 13:310–312

Isolauri E, Arvola T, SÜtas Y, Moilanen E, Salminen S (2000) Probiotics in the management of atopic eczema. Clin Exp Allergy 30:1605–1610

Jahreis G, Vogelsang H, Kiessling G, Schubert R, Bunte C, Hammes WP (2002) Influence of probiotic sausage (*Lactobacillus paracasei*) on blood lipids and immunological parameters of healthy volunteers. Food Res Int 35:133–138

Johansson ML, Nobaek S, Berggren A, Nyman M, Bjorck I, Ahrne S, Jeppsson B, Molin G (1998) Survival *of Lactobacillus plantarum* DSM 9843 (299v), and effect on the short-chain fatty acid content of faeces after ingestion of a rose-hip drink with fermented oats. Int J Food Microbiol 42:29–38

Kaila M, Isolauri E, Soppi E, Virtanen E, Laine S, Arvilommi H (1992) Enhancement of the circulating antibody secreting cell response in human diarrhoea by a human Lactobacillus strain. Pediatr Res 32:141–144

Kaila M, Isolauri E, Saxelin M, Arvilommi H, Vesikari T (1995) Viable versus inactivated *lactobacillus* strain GG in acute rotavirus diarrhoea. Arch Dis Child 72:51–53

Kajander K, Krogius-Kurikka L, Rintillä T, Karjalainen H, Palva A, Korpela R (2007) Effects of multispecies probiotic supplementation on intestinal microbiota in irritable bowel syndrome. Aliment Pharmacol Ther 26:463–473

Kajander K, Myllyluoma E, Rajilic-Stojanovic M, Kyronpalo S, Rasmussen M, Jarvenpaa S, Zoetendal EG, De Vos WM, Vapaatalo H, Korpela R (2008) Clinical trial: multispecies probiotic supplementation alleviates the symptoms of irritable bowel syndrome and stabilizes intestinal microbiota. Aliment Pharmacol Ther 27:48–57

Kalliomaki M, Salminen S, Arvilommi H, Kero P, Koskinen P, Isolauri E (2001) Probiotics in primary prevention of atopic disease: a randomised placebo-controlled trial. Lancet 357:1076–1079

Kalliomaki M, Salminen S, Poussa T, Arvilommi H, Isolauri E (2003) Probiotics and prevention of atopic disease: 4-year follow-up of a randomised placebo-controlled trial. Lancet 361:1869–1871

Kankainen M, Paulin L, Tynkkynen S, von Ossowski I, Reunanen J, Partanen P, Satokari R, Vesterlund S, Hendrickx APA, Lebeer S, De Keersmaecker SCJ, Vanderleyden J, Hämäläinen T, Laukkanen S, Salovuori N, Ritari J, Alatalo E, Korpela R, Mattila-Sandholm T, Lassig A,

Hatakka K, Kinnunen KT, Karjalainen H, Saxelin M, Laakso K, Surakka A, Palva A, Salusjärvi T, Auvinen P, de Vos WM (2009) Comparative genomic analysis of *Lactobacillus rhamnosus* GG reveals pili containing a human- mucus binding protein. Proc Natl Acad Sci 106:17193–17198

Kato I, Endo-Tanaka K, Yokokura T (1998) Suppressive effects of the oral administration of *Lactobacillus casei* on type II collagen-induced arthritis in DBA/1 mice. Life Sci 63:635–644

Kecskes G, Belagyi T, Olah A (2003) Early jejunal nutrition with combined pre- and probiotics in acute pancreatitis—prospective, randomized, double-blind investigations. Magy Seb 56:3–8

Kirjavainen PV, Arvola T, Salminen SJ, Isolauri E (2002) Aberrant composition of gut microbiota of allergic infants: a target of bifidobacterial therapy at weaning? Gut 51:51–55

Kishi A, Uno K, Matsubara Y, Okuda C, Kishida T (1996) Effect of the oral administration of *Lactobacillus brevis* subsp. *coagulans* on interferon-alpha producing capacity in humans. J Am Coll Nutr 15:408–412

Klein A, Friedrich U, Vogelsang H, Jahreis G (2008) *Lactobacillus acidophilus* 74-2 and *Bifidobacterium animalis* subsp. *lactis* DGCC 420 modulate unspecific cellular immune response in healthy adults. Eur J Clin Nutr 62:584–593

Klijn N, Weerkamp AH, de Vos WM (1995) Genetic marking of *Lactococcus lactis* shows its survival in the human gastrointestinal tract. Appl Environ Microbiol 61:2771–2774

Kopp MV, Salfeld P (2009) Probiotics and prevention of allergic disease. Curr Opin Clin Nutr Metab Care 12:298–303

Korzenik JR, Podolsky DK (2006) Evolving knowledge and therapy of inflammatory bowel disease. Nat Rev Drug Discov 5:197–209

Kruis W, Schutz E, Fric P, Fixa B, Judmaier G, Stolte M (1997) Double-blind comparison of an oral *Escherichia coli* preparation and mesalazine in maintaining remission of ulcerative colitis. Aliment Pharmacol Ther 11:853–858

Kruis W, Fric P, Pokrotnieks J, Lukas M, Fixa B, Kascak M, Kamm MA, Weismueller J, Beglinger C, Stolte M, Wolff C, Schulze J (2004) Maintaining remission of ulcerative colitis with the probiotic *Escherichia coli* Nissle 1917 is as effective as with standard mesalazine. Gut 53:1617–1623

Kukkonen K, Savilahti E, Haahtela T, Juntunen-Backman K, Korpela R, Poussa T, Tuure T, Kuitunen M (2007) Probiotics and prebiotic galacto-oligosaccharides in the prevention of allergic diseases: a randomized, double-blind, placebo-controlled trial. J Allergy Clin Immunol 119:192–198

Kullen MJ, Amann MM, O'Shaughnessy MJ, O'Sullivan DJ, Busta FF, Brady LJ (1997) Differentiation of ingested and endogenous bifidobacteria by DNA fingerprinting demonstrates the survival of an unmodified strain in the gastrointestinal tract of humans. J Nutr 127:89–94

Lammers KM, Helwig U, Swennen E, Rizzello F, Venturi A, Caramelli E, Kamm MA, Brigidi P, Gionchetti P, Campieri M (2002) Effect of probiotic strains on interleukin 8 production by HT29/19A cells. Am J Gastroenterol 97:1182–1186

Larsen CN, Nielsen S, Kaestel P, Brockmann E, Bennedsen M, Christensen HR, Eskesen DC, Jacobsen BL, Michaelsen KF (2006) Dose-response study of probiotic bacteria *Bifidobacterium animalis* subsp. *lactis* BB-12 and *Lactobacillus paracasei* subsp. *paracasei* CRL-341 in healthy young adults. Eur J Clin Nutr 60:1284–1293

Lay C, Rigottier-Gois L, Holmstrom K, Rajilic M, Vaughan EE, de Vos WM, Collins MD, Thiel R, Namsolleck P, Blaut M, Dore J (2005) Colonic microbiota signatures across five northern European countries. Appl Environ Microbiol 71:4153–4155

Lebeer S, Vanderleyden J, De Keersmaecker SCJ (2008) Genes and molecules of Lactobacilli supporting probiotic action. Microbiol Mol Biol Rev 72:728–764

Lindwall S, Fonden R (1984) Passage and survival of *L. acidophilus* in the human gastrointestinal tract. Int Dairy Fed Bull 21:179

Link-Amster H, Rochat F, Saudan KY, Mignot O, Aeschlimann JM (1994) Modulation of a specific humoral immune response and changes in intestinal flora mediated through fermented milk intake. FEMS Immunol Med Microbiol 10:55–63

Lombardo L, Vernetto A, Blanco I (2009) Clinical evaluation of *Lactobacillus paracasei* subsp. *paracasei* F19 with gluco-oligosaccharides in the short-term treatment of irritable bowel syndrome. Microbial Ecol Health Dis 21:28–32
Lund B, Adamsson I, Edlund C (2002) Gastrointestinal transit survival of an *Enterococcus faecium* probiotic strain administered with or without vancomycin. Int J Food Microbiol 77:109–115
Maassen CB, van Holten JC, Balk F, Heijne den Bak-Glashouwer MJ, Leer R, Laman JD, Boersma WJ, Claassen E (1998) Orally administered *Lactobacillus* strains differentially affect the direction and efficacy of the immune response. Vet Q 20:S81–S83
Maassen CBM, van Holten-Neelen C, Balk F, den Bak-Glashouwer JH, Leer RJ, Laman JD, Boersma WJA, Claassen E (2000) Strain-dependent induction of cytokine profiles in the gut by orally administered *Lactobacillus* strains. Vaccine 18:2613–2623
Madden JAJ, Hunter JO (2002) A review of the role of the gut microflora in irritable bowel syndrome and the effects of probiotics. Br J Nutr 88:S67–S72
Madsen K (2008) Probiotics in critically ill patients. J Clin Gastroenterol 42:S116–S118
Madsen KL, Doyle JS, Jewell LD, Tavernini MM, Fedorak RN (1999) *Lactobacillus* species prevents colitis in interleukin 10 gene-deficient mice. Gastroenterol 116:1107–1114
Majamaa H, Isolauri E (1997) Probiotics: a novel approach in the management of food allergy. J Allergy Clin Immunol 99:179–185
Majamaa H, Isolauri E, Saxelin M, Vesikari T (1995) Lactic acid bacteria in the treatment of acute rotavirus gastroenteritis. J Pediatr Gastroenterol Nutr 20:333–338
Malinen E, Rinttila T, Kajander K, Matto J, Kassinen A, Krogius L, Saarela M, Korpela R, Palva A (2005) Analysis of the faecal microbiota of irritable bowel syndrome patients and healthy controls with real-time PCR. Am J Gastroenterol 100:373–382
Manichanh C, Rigottier-Gois L, Bonnaud E, Gloux K, Pelletier E, Frangeul L, Nalin R, Jarrin C, Chardon P, Marteau P, Roca J, Dore J (2006) Reduced diversity of faecal microbiota in Crohn's disease revealed by a metagenomic approach. Gut 55:205–211
Marteau P, Pochart P, Bouhnik Y, Zidi S, Goderel I, Rambaud JC (1992) Survival of *Lactobacillus acidophilus* and *Bifidobacterium* sp. in the small intestine following ingestion in fermented milk. A rational basis for the use of probiotics in man. Gastroenterol Clin Biol 16:25–28
Marteau P, Pochart P, Bouhnik Y, Rambaud JC (1993) The fate and effects of transiting, nonpathogenic microorganisms in the human intestine. World Rev Nutr Diet 74:1–21
Marzotto M, Maffeis C, Paternoster T, Ferrario R, Rizzotti L, Pellegrino M, Dellaglio F, Torriani S (2006) *Lactobacillus paracasei* A survives gastrointestinal passage and affects the faecal microbiota of healthy infants. Res Microbiol 157:857–866
Matto J, Maunuksela L, Kajander K, Palva A, Korpela R, Kassinen A, Saarela M (2005) Composition and temporal stability of gastrointestinal microbiota in irritable bowel syndrome–a longitudinal study in IBS and control subjects. FEMS Immunol Med Microbiol 43:213–222
Matto J, Fonden R, Tolvanen T, von Wright A, Vilpponen-Salmela T, Satokari R, Saarela M (2006) Intestinal survival and persistence of probiotic *Lactobacillus* and *Bifidobacterium* strains administered in triple-strain yoghurt. Int Dairy J 16:1174–1180
McCarthy J, O'Mahony L, O'Callaghan L, Sheil B, Vaughan EE, Fitzsimons N, Fitzgibbon J, O'Sullivan GC, Kiely B, Collins JK, Shanahan F (2003) Double blind, placebo controlled trial of two probiotic strains in interleukin 10 knockout mice and mechanistic link with cytokine balance. Gut 52:975–980
Meyer AL, Micksche M, Herbacek I, Elmadfa I (2006) Daily intake of probiotic as well as conventional yogurt has a stimulating effect on cellular immunity in young healthy women. Ann Nutr Metab 50:282–289
Miettinen M, Matikainen S, Vuopio-Varkila J, Pirhonen J, Varkila K, Kurimoto M, Julkunen I (1998) Lactobacilli and streptococci induce interleukin-12 (IL-12), IL-18, and gamma interferon production in human peripheral blood mononuclear cells. Infect Immun 66:6058–6062
Mimura T, Rizzello F, Helwig U, Poggioli G, Schreiber S, Talbot IC, Nicholls RJ, Gionchetti P, Campieri M, Kamm MA (2004) Once daily high dose probiotic therapy (VSL#3) for maintaining remission in recurrent or refractory pouchitis. Gut 53:108–114

Misra S, Sabui TK, Pal NK (2009) A randomized controlled trial to evaluate the efficacy of *Lactobacillus* GG in infantile diarrhoea. J Pediatr 155:129–132

Mohamadzadeh M, Olson S, Kalina WV, Ruthel G, Demmin GL, Warfield KL, Bavari S, Klaenhammer TR (2005) Lactobacilli activate human dendritic cells that skew T cells toward T helper 1 polarization. Proc Natl Acad Sci 102:2880–2885

Morelli L, Zonenschain D, Callegari M, Grossi E, Maisano F, Fusillo M (2003) Assessment of a new synbiotic preparation in healthy volunteers: survival, persistence of probiotic strains and its effect on the indigenous flora. Nutri J 2:11

Muscettola M, Massai L, Tanganelli C, Grasso G (1994) Effects of lactobacilli on interferon production in young and aged mice. Ann N Y Acad Sci 717:226–232

Nobaek S, Johansson M-L, Molin G, Ahrne S, Jeppsson B (2000) Alteration of intestinal microflora is associated with reduction in abdominal bloating and pain in patients with irritable bowel syndrome. Am J Gastroenterol 95:1231–1238

O'Mahony L, Feeney M, O'Halloran S, Murphy L, Kiely B, Fitzgibbon J, Lee G, O'Sullivan G, Shanahan F, Collins JK (2001) Probiotic impact on microbial flora, inflammation and tumour development in IL-10 knockout mice. Aliment Pharmacol Ther 15:1219–1225

Ogawa T, Asai Y, Tamai R, Makimura Y, Sakamoto H, Hashikawa S, Yasuda K (2006) Natural killer cell activities of synbiotic *Lactobacillus casei* ssp. *casei* in conjunction with dextran. Clin Exp Immunol 143:103–109

Olivares M, Diaz-Ropero MP, Sierra S, Lara-Villoslada F, Fonolla J, Navas M, Rodriguez JM, Xaus J (2007) Oral intake of *Lactobacillus fermentum* CECT5716 enhances the effects of influenza vaccination. Nutrition 23:254–260

Oozeer R, Leplingard A, Mater DDG, Mogenet A, Michelin R, Seksek I, Marteau P, Dore J, Bresson J-L, Corthier G (2006) Survival of *Lactobacillus casei* in the human digestive tract after consumption of fermented Milk. Appl Environ Microbiol 72:5615–5617

Otte JM, Podolsky DK (2004) Functional modulation of enterocytes by gram-positive and gram-negative microorganisms. Am J Physiol Gastrointest Liver Physiol 286:G613–G626

Ouwehand AC, Kurvinen T, Rissanen P (2004) Use of a probiotic *Bifidobacterium* in a dry food matrix, an *in vivo* study. Int J Food Microbiol 95:103–106

Park JH, Um JI, Lee BJ, Goh JS, Park SY, Kim WS, Kim PH (2002) Encapsulated *Bifidobacterium bifidum* potentiates intestinal IgA production. Cell Immunol 219:22–27

Pedrosa MC, Golner BB, Goldin BR, Barakat S, Dallal GE, Russell RM (1995) Survival of yogurt-containing organisms and *Lactobacillus gasseri* (ADH) and their effect on bacterial enzyme activity in the gastrointestinal tract of healthy and hypochlorhydric elderly subjects. Am J Clin Nutr 61:353–359

Pelto L, Isolauri E, Lilius EM, Nuutila J, Salminen S (1998) Probiotic bacteria down-regulate the milk-induced inflammatory response in milk-hypersensitive subjects but have an immunostimulatory effect in healthy subjects. Clin Exp Allergy 28:1474–1479

Pena JA, Rogers AB, Ge Z, Ng V, Li SY, Fox JG, Versalovic J (2005) Probiotic Lactobacillus spp. diminish Helicobacter hepaticus-induced inflammatory bowel disease in interleukin-10-deficient mice. Infect Immun 73:912–920

Perdigon G, Maldonado Galdeano C, Valdez JC, Medici M (2002) Interaction of lactic acid bacteria with the gut immune system. Eur J Clin Nutr 56:S21–S26

Pessi T, Sutas Y, Saxelin M, Kallioinen H, Isolauri E (1999) Antiproliferative effects of homogenates derived from five strains of candidate probiotic bacteria. Appl Environ Microbiol 65:4725–4478

Plant L, Lam C, Conway PL, O'Riordan K (2003) Gastrointestinal microbial community shifts observed following oral administration of a *Lactobacillus fermentum* strain to mice. FEMS Microbiol Ecol 43:133–140

Pochard P, Gosset P, Grangette C, Andre C, Tonnel AB, Pestel J, Mercenier A (2002) Lactic acid bacteria inhibit T(H)2 cytokine production by mononuclear cells from allergic patients. J Allergy Clin Immunol 110:617–623

Pochart P, Dewit O, Desjeux JF, Bourlioux P (1989) Viable starter culture, beta-galactosidase activity, and lactose in duodenum after yogurt ingestion in lactase-deficient humans. Am J Clin Nutr 49:828–831

Pochart P, Marteau P, Bouhnik Y, Goderel I, Bourlioux P, Rambaud JC (1992) Survival of bifidobacteria ingested via fermented milk during their passage through the human small intestine: an *in vivo* study using intestinal perfusion. Am J Clin Nutr 55:78–80

Pohjavuori E, Viljanen M, Korpela R, Kuitunen M, Tiittanen M, Vaarala O, Savilahti E (2004) *Lactobacillus* GG effect in increasing IFN-gamma production in infants with cow's milk allergy. J Allergy Clin Immunol 114:131–136

Prantera C, Scribano ML, Falasco G, Andreoli A, Luzi C (2002) Ineffectiveness of probiotics in preventing recurrence after curative resection for Crohn's disease: a randomised controlled trial with *Lactobacillus* GG. Gut 51:405–409

Prescott SL, Dunstan JA, Hale J, Breckler L, Lehmann H, Weston S, Richmond P (2005) Clinical effects of probiotics are associated with increased interferon-gamma responses in very young children with atopic dermatitis. Clin Exp Allergy 35:1557–1564

Pronio A, Montesani C, Butteroni C, Vecchione S, Mumolo G, Vestri A, Vitolo D, Boirivant M (2008) Probiotic administration in patients with ileal pouch-anal anastomosis for ulcerative colitis is associated with expansion of mucosal regulatory cells. Inflamm Bowel Dis 14:662–668

Rachmilewitz D, Katakura K, Karmeli F, Hayashi T, Reinus C, Rudensky B, Akira S, Takeda K, Lee J, Takabayashi K, Raz E (2004) Toll-like receptor 9 signaling mediates the anti-inflammatory effects of probiotics in murine experimental colitis. Gastroenterology 126:520–528

Rastall RA (2004) Bacteria in the gut: friends and foes and how to alter the balance. J Nutr 134:2022S–2026S

Rautava S, Arvilommi H, Isolauri E (2006) Specific probiotics in enhancing maturation of IgA responses in formula-fed infants. Ped Res 60:221–224

Rembacken BJ, Snelling AM, Hawkey PM, Chalmers DM, Axon AT (1999) Non-pathogenic *Escherichia* coli versus mesalazine for the treatment of ulcerative colitis: a randomised trial. Lancet 354:635–639

Rigby RJ, Knight SC, Kamm MA, Stagg AJ (2005) Production of interleukin (IL)-10 and IL-12 by murine colonic dendritic cells in response to microbial stimuli. Clin Exp Immunol 139:245–256

Rinne M, Kalhomaki M, Salminen S, Isolauri E (2006) Probiotic intervention in the first months of life: short-term effects on gastrointestinal symptoms and long-term effects on gut microbiota. J Pediatr Gastroenterol Nutr 43:200–205

Rochet V, Rigottier-Gois L, Sutren M, Krementscki MN, Andrieux C, Furet JP, Tailliez P, Levenez F, Mogenet A, Bresson JL, Meance S, Cayuela C, Leplingard A, Dore J (2006) Effects of orally administered *Lactobacillus casei* DN-114 001 on the composition or activities of the dominant faecal microbiota in healthy humans. Br J Nutr 95:421–429

Rochet V, Rigottier-Gois L, Levenez F, Cadiou J, Marteau P, Bresson JL, Goupil-Feillerat N, Dore J (2008) Modulation of *Lactobacillus casei* in ileal and faecal samples from healthy volunteers after consumption of a fermented milk containing *Lactobacillus casei* DN-114 001(Rif). Can J Microbiol 54:660–667

Roessler A, Friedrich U, Vogelsang H, Bauer A, Kaatz M, Hipler UC, Schmidt I, Jahreis G (2008) The immune system in healthy adults and patients with atopic dermatitis seems to be affected differently by a probiotic intervention. Clin Exp Allergy 38:93–102

Rohde CL, Bartolini V, Jones N (2009) The use of probiotics in the prevention and treatment of antibiotic-associated diarrhoea with special interest in *Clostridium difficile*-associated diarrhoea. Nutr Clin Pract 24:33–40

Ruiz PA, Hoffmann M, Szcesny S, Blaut M, Haller D (2005) Innate mechanisms for *Bifidobacterium lactis* to activate transient pro-inflammatory host responses in intestinal epithelial cells after the colonization of germ-free rats. Immunology 115:441–450

Saavedra JM (2007) Use of probiotics in pediatrics: rationale, mechanisms of action, and practicall aspects. Nutr Clin Pract 22:351–365

Saito Y, Sakamoto M, Takizawa S, Benno Y (2004) Monitoring the cell number and viability of *Lactobacillus helveticus* GCL1001 in human faeces by PCR methods. FEMS Microbiol Lett 231:125–130

Salminen S, Isolauri E (2006) Intestinal colonization, microbiota, and probiotics. J Pediatr 149:S115–S120

Saxelin M, Pessi T, Salminen S (1995) Faecal recovery following oral administration of Lactobacillus strain GG (ATCC 53103) in gelatine capsules to healthy volunteers. Int J Food Microbiol 25:199–203

Schiffrin EJ, Rochat F, Link-Amster H, Aeschlimann JM, Donnet-Hughes A (1995) Immunomodulation of human blood cells following the ingestion of lactic acid bacteria. J Dairy Sci 78:491–497

Schiffrin EJ, Brassart D, Servin AL, Rochat F, Donnet-Hughes A (1997) Immune modulation of blood leukocytes in humans by lactic acid bacteria: criteria for strain selection. Am J Clin Nutr 66:515S–520S

Schultz M, Timmer A, Herfarth HH, Sartor RB, Vanderhoof JA, Rath HC (2004) Lactobacillus GG in inducing and maintaining remission of Crohn's disease. BMC Gastroenterol 4:5

Sen S, Mullan MM, Parker TJ, Woolner JT, Tarry SA, Hunter JO (2002) Effect of *Lactobacillus plantarum* 299v on colonic fermentation and symptoms of irritable bowel syndrome. Dig Dis Sci 47:2615–2620

Servin AL (2004) Antagonistic activities of lactobacilli and bifidobacteria against microbial pathogens. FEMS Microbiol Rev 28:405–440

Sghir A, Gramet G, Suau A, Rochet V, Pochart P, Dore J (2000) Quantification of bacterial groups within human faecal flora by oligonucleotide probe hybridization. Appl Environ Microbiol 66:2263–2266

Sheih YH, Chiang BL, Wang LH, Liao CK, Gill HS (2001) Systemic immunity-enhancing effects in healthy subjects following dietary consumption of the lactic acid bacterium *Lactobacillus rhamnosus* HN001. J Am Coll Nutr 20:149–156

Sheil B, McCarthy J, O'Mahony L, Bennett MW, Ryan P, Fitzgibbon JJ, Kiely B, Collins JK, Shanahan F (2004) Is the mucosal route of administration essential for probiotic function? Subcutaneous administration is associated with attenuation of murine colitis and arthritis. Gut 53:694–700

Shida K, Takahashi R, Iwadate E, Takamizawa K, Yasui H, Sato T, Habu S, Hachimura S, Kaminogawa S (2002) *Lactobacillus casei* strain Shirota suppresses serum immunoglobulin E and immunoglobulin G1 responses and systemic anaphylaxis in a food allergy model. Clin Exp Allergy 32:563–570

Shida K, Kiyoshima-Shibata J, Nagaoka M, Watanabe K, Nanno M (2006) Induction of interleukin-12 by lactobacillus strains having a rigid cell wall resistant to intracellular digestion. J Dairy Sci 89:3306–3317

Shinoda T, Kusuda D, Ishida Y, Ikeda N, Kaneko K, Masuda O, Yamamoto N (2001) Survival of *Lactobacillus helveticus* strain CP53 in the human gastrointestinal tract. Lett Appl Microbiol 32:108–113

Smits HH, Engering A, van der Kleij D, de Jong EC, Schipper K, van Capel TM, Zaat BA, Yazdanbakhsh M, Wierenga EA, van Kooyk Y, Kapsenberg ML (2005) Selective probiotic bacteria induce IL-10-producing regulatory T cells *in vitro* by modulating dendritic cell function through dendritic cell-specific intercellular adhesion molecule 3-grabbing nonintegrin. J Allergy Clin Immunol 115:1260–1267

Sokol H, Pigneur B, Watterlot L, Lakhdari O, Bermudez-Humaran LG, Gratadoux JJ, Blugeon S, Bridonneau C, Furet JP, Corthier G, Grangette C, Vasquez N, Pochart P, Trugnan G, Thomas G, Blottiere HM, Dore J, Marteau P, Seksik P, Langella P (2008) *Faecalibacterium prausnitzii* is an anti-inflammatory commensal bacterium identified by gut microbiota analysis of Crohn disease patients. Proc Natl Acad Sci 105:16731–16736

Sokol H, Seksik P, Furet JP, Firmesse O, Nion-Larmurier L, Beaugerie L, Cosnes J, Corthier G, Marteau P, Dore J (2009) Low counts of *Faecalibacterium prausnitzii* in colitis microbiota. Inflamm Bowel Dis 15:1183–1189

Spanhaak S, Havenaar R, Schaafsma G (1998) The effect of consumption of milk fermented by *Lactobacillus casei* strain Shirota on the intestinal microflora and immune parameters in humans. Eur J Clin Nutr 52:899–907

Strober W, Fuss IJ, Blumberg RS (2002) The immunology of mucosal models of inflammation. Ann Rev Immunol 20:495–549

Sturm A, Rilling K, Baumgart DC, Gargas K, Abou-Ghazale T, Raupach B, Eckert J, Schumann RR, Enders C, Sonnenborn U, Wiedenmann B, Dignass AU (2005) *Escherichia coli* Nissle 1917 distinctively modulates T-cell cycling and expansion via toll-like receptor 2 signaling. Infect Immun 73:1452–1465

Su P, Henriksson A, Tandianus JE, Park JH, Foong F, Dunn NW (2005) Detection and quantification of *Bifidobacterium lactis* LAFTI B94 in human faecal samples from a consumption trial. FEMS Microbiol Lett 244:99–103

Suau A, Rochet V, Sghir A, Gramet G, Brewaeys S, Sutren M, Rigottier-Gois L, Dore J (2001) *Fusobacterium prausnitzii* and related species represent a dominant group within the human faecal flora. Syst Appl Microbiol 24:139–145

Sullivan A, Barkholt L, Nord CE (2003) *Lactobacillus acidophilus*, *Bifidobacterium lactis* and *Lactobacillus* F19 prevent antibiotic-associated ecological disturbances of Bacteroides fragilis in the intestine. J Antimicrob Chemother 52:308–311

Tanaka S, Kobayashi T, Songjinda P, Tateyama A, Tsubouchi M, Kiyohara C, Shirakawa T, Sonomoto K, Nakayama J (2009) Influence of antibiotic exposure in the early postnatal period on the development of intestinal microbiota. FEMS Immunol Med Microbiol 56:80–87

Tannock GW (2003) Probiotics: time for a dose of realism. Curr Issues Intest Microbiol 4:33–42

Tannock GW, Munro K, Harmsen HJM, Welling GW, Smart J, Gopal PK (2000) Analysis of the faecal microflora of human subjects consuming a probiotic product containing *Lactobacillus rhamnosus* DR20. Appl Environ Microbiol 66:2578–2588

Tap J, Mondot S, Levenez F, Pelletier E, Caron C, Furet JP, Ugarte E, Munoz-Tamayo R, Paslier DLE, Nalin R, Dore J, Leclerc M (2009) Towards the human intestinal microbiota phylogenetic core. Environ Microbiol 11:2574–2584

Taylor AL, Dunstan JA, Prescott SL (2007) Probiotic supplementation for the first 6 months of life fails to reduce the risk of atopic dermatitis and increases the risk of allergen sensitization in high-risk children: a randomized controlled trial. J Allergy Clin Immunol 119:184–191

Tien MT, Girardin SE, Regnault B, Le Bourhis L, Dillies MA, Coppee JY, Bourdet-Sicard R, Sansonetti PJ, Pedron T (2006) Anti-inflammatory effect of *Lactobacillus casei* on *Shigella*-infected human intestinal epithelial cells. J Immunol 176:1228–1237

Tlaskalová-Hogenová H, Stepánková R, Hudcovic T, Tucková L, Cukrowska B, Lodinová-Zádníková R, Kozáková H, Rossmann P, Bártová J, Sokol D, Funda DP, Borovská D, Reháková Z, Sinkora J, Hofman J, Drastich P, Kokesová A (2004) Commensal bacteria (normal microflora), mucosal immunity and chronic inflammatory and autoimmune diseases. Immunol Lett 93:97–108

Tuohy KM, Pinart-Gilberga M, Jones M, Hoyles L, McCartney AL, Gibson GR (2007) Survivability of a probiotic *Lactobacillus casei* in the gastrointestinal tract of healthy human volunteers and its impact on the faecal microflora. J Appl Microbiol 102:1026–1032

Vesa T, Pochart P, Marteau P (2000) Pharmacokinetics of *Lactobacillus plantarum* NCIMB 8826, *Lactobacillus fermentum* KLD, and *Lactococcus lactis* MG 1363 in the human gastrointestinal tract. Aliment Pharmacol Ther 14:823–828

Viljanen M, Kuitunen M, Haahtela T, Juntunen-Backman K, Korpela R, Savilahti E (2005a) Probiotic effects on faecal inflammatory markers and on faecal IgA in food allergic atopic eczema/dermatitis syndrome infants. Pediatr Allergy Immunol 16:65–71

Viljanen M, Pohjavuori E, Haahtela T, Korpela R, Kuitunen M, Sarnesto A, Vaarala O, Savilahti E (2005b) Induction of inflammation as a possible mechanism of probiotic effect in atopic eczema-dermatitis syndrome. J Allergy Clin Immunol 115:1254–1259

Viljanen M, Savilahti E, Haahtela T, Juntunen-Backman K, Korpela R, Poussa T, Tuure T, Kuitunen M (2005c) Probiotics in the treatment of atopic eczema/dermatitis syndrome in infants: a double-blind placebo-controlled trial. Allergy 60:494–500

Vinderola G, Matar C, Perdigon G (2005) Role of intestinal epithelial cells in immune effects mediated by gram-positive probiotic bacteria: involvement of toll-like receptors. Clin Diagn Lab Immunol 12:1075–1084 von der Weid T, Bulliard C, Schiffrin EJ (2001) Induction by a lactic acid bacterium of a population of CD4+ T cells with low proliferative capacity that produce transforming growth factor β and Interleukin-10. Clin Diagn Lab Immunol 8:695–701

Xiao SD, Zhang de Z, Lu H, Jiang SH, Liu HY, Wang GS, Xu GM, Zhang ZB, Lin GJ, Wang GL (2003) Multicenter, randomized, controlled trial of heat-killed *Lactobacillus acidophilus* LB in patients with chronic diarrhoea. Adv Ther 20:253–260

Yamano T, Iino H, Takada M, Blum S, Rochat F, Fukushima Y (2006) Improvement of the human intestinal flora by ingestion of the probiotic strain *Lactobacillus johnsonii* La1. Br J Nutr 95:303–312

Yuki N, Watanabe K, Mike A, Tagami Y, Tanaka R, Ohwaki M, Morotomi M (1999) Survival of a probiotic, *Lactobacillus casei* strain Shirota, in the gastrointestinal tract: selective isolation from faeces and identification using monoclonal antibodies. Int J Food Microbiol 48:51–57

Zanello G, Meurens F, Berri M, Salmon H (2009) *Saccharomyces boulardii* effects on gastrointestinal diseases. Curr Issues Mol Biol 11:47–58

Zhang L, Li N, Caicedo R, Neu J (2005) Alive and dead *Lactobacillus rhamnosus* GG decrease tumor necrosis factor-alpha-induced interleukin-8 production in Caco-2 cells. J Nutr 135:1752–1756

Zoetendal EG, Akkermans ADL, De Vos WM (1998) Temperature gradient gel electrophoresis analysis of 16S rRNA from human faecal samples reveals stable and host-specific communities of active bacteria. Appl Environ Microbiol 64:3854–3859

Chapter 12
Modulation of Immune System by Probiotics to Protect Against Enteric Disorders

Joshua J. Malago and Jos F. J. G. Koninkx

12.1 Introduction

Immunity, defined as resistant to diseases, is achieved naturally (natural immunity) or following exposure to disease causing agents or immunogens such as vaccines (acquired immunity). The function of the immune system is to defend the body against invaders. Microbes (germs or microorganisms), cancer cells and transplanted tissues or organs are all recognised by the immune system as non-self against which the body must be defended. Thus the immune system recognises the enemy, mobilises forces and attacks. Failure or excessive reaction of the immune system in response to the non-self agents leads to immune disorders that could either be immunodeficiency or hypersensitivity, respectively. On the other hand challenges of the gut by pathogens may lead to breakdown of the immune barriers and subsequent development of infectious diseases. Whether the disorder is fundamentally immune mediated or infectious in origin, the intestinal immune system has means of fighting against them.

The intestinal mucosal immunity is a function of its epithelial physical barrier that excludes most of the antigens before they stimulate the immune system. The barrier also functions through production of mucus and other secretions, regulation of paracellular permeability, and synthesis of antimicrobial peptides. In addition, intestinal immunity is maintained by a complex of immune responses from the epithelium and the gut-associated lymphoid tissue (GALT). GALT composes of intra-epithelial lymphocytes, lamina propria, mesenteric lymph nodes, Peyer's patches, isolated lymphoid follicles, and cryptopatches. It is associated with production of cytokines, chemokines and effector and regulatory T cells. Intestinal microbiota is another component of the intestinal immune system. Its contribution to the intestinal mucosal immunity has been a focus of interest in recent years. Mounting evidence shows that it influences the development and regulation of host's immune (i.e. the GALT) and non-immune defences, regulate mucin gene expression by goblet cells, modify glycosylation of mu-

J. J. Malago (✉)
Department of Veterinary Pathology, Faculty of Veterinary Medicine,
Sokoine University of Agriculture, P.O. Box 3203, Morogoro, Tanzania
e-mail: malagojj@yahoo.com

J. J. Malago et al. (eds.), *Probiotic Bacteria and Enteric Infections*,
DOI 10.1007/978-94-007-0386-5_12,

cus to interfere with bacterial adhesion, colonization and invasion (Freitas et al. 2005; Caballero-Franco et al. 2007), induce secretion of antimicrobial peptides by intestinal Paneth cells (Vaishnava et al. 2008), regulate alterations of intestinal permeability caused by infection, stress, and inflammation (Lutgendorff et al. 2008), and influence development of mucosal and systemic immunity (Tlaskalová-Hogenová et al. 2004).

The immunological importance of the intestinal microbiota is exemplified in antibody production, development and persistence of oral tolerance to food antigens and the formation of germinal centres within lymphoid follicles (Gaboriau-Routhiau and Moreau 1996; Backhed et al. 2005). *Bacteroides fragilis* and *Bifidobacterium* species are important in this respect, as infants harbouring these organisms had more circulating immunoglobulin (Ig) A-secreting and IgM-secreting cells. These immunological roles are critical during infancy as the microbiota forms the newborn infant's first major microbial challenge and throughout life, the microbiota becomes the greatest microbial exposure in an individual. This persistent exposure has made the host to develop tolerance against the microbiota, which in return, contributes significantly to the host immunity through modulation of both cellular and humoral immune system. Imbalances to the microbiota could therefore account for infectious diseases as well as immune-mediated disorders like allergy, autoimmunity, or immunodeficiency.

According to Tlaskalová-Hogenová et al. (2004), absence of microbiota affects the systemic immunity by decreasing serum IgA and the size of the spleen. These defects disappear following intestinal colonization by microbiota or administration of bacterial components. Based on this principle, some probiotic bacteria are used to stimulate the immune system enough to enhance defences against human enteric infections and benefit patients with certain chronic inflammatory bowel conditions. It is therefore important to explore the role of probiotics in a broken immunity. This chapter describes the ability of probiotics to modulate the immune system and benefit patients with immunologically mediated disease of affluence which are at an increasing incidence in a developed world. Such diseases include allergies and autoimmune disorders like inflammatory bowel disease (IBD), celiac disease and type 1 diabetes (T1D). Studies have suggested that these diseases are consequences of disruption of factors that could induce and maintain tolerance to allergens and autoantigens. The underlying mechanism is proposed to be a disturbed immune regulation of the link between T helper (Th)1-dependent autoimmune disease and Th2-linked atopic allergy involving T regulatory cells. Changes in the microbiota composition and colonisation of the gut as a result of improved lifestyle are proposed to be an important determinant. Modifying this imbalance through probiotic bacteria may rectify the illness, partly through modulation of the immune system.

12.2 Disruption of Intestinal Immunity

12.2.1 Impaired Intestinal Permeability

Disruption of intestinal immunity, in particular its physical barrier is implicated in the pathogenesis of acute illnesses such as bacterial translocation that leads to sepsis and multiple organ system failure (Macintire and Bellhorn 2002). In addition, many immune mediated disorders, both allergy and autoimmune, are characterized by impaired intestinal permeability, partly due to disrupted intestinal barrier, a vital process in disease development. Breaching of the intestinal epithelium and the subsequent leaking of luminal antigens into the submucosa lead to pathologic stimulation of the highly immunoreactive subepithelium. Moreover, the impaired intestinal permeability allows antigens to circulate throughout the blood stream, stimulate the immune system, and in response, immune system cells react to these antigens as they would to any foreign protein. In so doing, the immune cells initiate an inflammatory reaction that often leads to allergy or autoantibody production which results in development of autoimmune disease. This theory is evidenced by occurrence of systemic allergic reactions like eczema, food allergies, celiac enteropathy, T1D, and asthma, as well as presence of gastrointestinal tissue damage in patients with various autoimmune diseases like Crohn's disease, irritable bowel syndrome (IBS), ankylosing spondylitis, rheumatoid arthritis, multiple sclerosis, vasculitis, thryroiditis, and dermatitis herpetiformis. In many of these conditions, a reduction of digestive inflammation correlates with a reduction or remission of symptoms.

12.2.2 Causes of Impaired Permeability

12.2.2.1 Disorders of Digestion

The term 'leaky gut syndrome' is used to describe an inability of the gastrointestinal tract to digest and absorb protein molecules. The 'leaky gut syndrome' can develop due to improper or incomplete digestion and impaired nutrient absorption that leads to formation of toxic by-products of digestion. These by-products and other toxins, as well as allergens cause intestinal permeability which initiates enteric inflammation. The resulting inflammation allows for large molecules to pass across the intestinal barrier, including molecules from proteins. Because human tissues have protein antigens similar in structure to these other proteins, the scene is set for autoimmune disease development. Further, extraintestinal allergic symptoms occur when allergens are the underlying antigens.

12.2.2.2 Disruption of Tight Junctions

Intestinal epithelial cells adhere to each other through three major lateral junctions: tight junctions (also called zonula occludens), adherens junctions and desmosomes. The tight junctions represent the major barrier within the paracellular pathway between neighbouring intestinal epithelial cells. They are made up of multiple sealing transmembrane proteins: occluding, claudin, and junctional adhesion protein (Gonzalez-Mariscal et al. 2003). Occludin and claudin interact with cytoplasmic plaques that consist of different types of cytosolic proteins, such as zonula-occludin proteins ZO-1, ZO-2, and ZO-3. The zonula-occludin proteins function as adaptors between the tight junction proteins and actin and myosin contractile elements within the cell during opening and closing of the paracellular junctions. While the opening and closing of the paracellular junctions is tightly regulated under normal conditions, dysregulation and subsequent loss of cell junction integrity have been indicated to contribute to disease states (Di Pierro 2001). Food antigens and pathogenic microbes are best known factors that affect proper functioning of the junctions leading to increased paracellular permeability.

Factors Affecting Tight Junctions

Food Antigens

A good example of food antigens is the gluten component, gliadin, which is implicated in the pathogenesis of gluten-induced enteropathy also known as celiac disease. In this disease, the breakdown of intestinal barrier function plays a key role in its pathogenesis. Gliadin disrupts intercellular junctions and triggers several other autoimmune diseases such as T1D (Clemente et al. 2003). On the contrary, some anti-inflammatory agents such as fatty acid eicosapentaenoic acid and gamma linolenic acid up-regulate tight junctions by increasing occluding (Jiang et al. 1998). We have also shown that butyrate, a major carbohydrate fermentation product in the colon strengthens intestinal epithelial cell tight junctions and protects against enteric pathogens (Malago et al. 2003). These findings allude that probiotic bacteria capable of inducing production of high levels of butyrate are potential in protecting against enteric pathologies.

Infectious Agents

It is evident that microbial pathogens target the intercellular tight junctions and disrupt them through various virulent factors (Hecht 2001). Some examples include enteropathogenic *Escherichia coli* that alters occludin distribution from the tight junctions into the cytosol (Simonovic et al. 2000), *Clostridium difficile* toxins A and B that dissociates occludin, ZO-1, and ZO-2 from the lateral tight junctions membrane (Nusrat et al. 2001), *Vibrio cholera* that disassembles intercellular tight junctions via interaction with cell membrane receptors (Wang et al. 2000) and rotavirus that causes paracellular leak and F-actin alteration of epithelial cells (Tafazoli et al. 2001).

12.3 Disruption of the Microbiota

12.3.1 The Hygiene Hypothesis

While there is no definitive evidence for the reason behind the rise in immune mediated diseases like allergy and autoimmunity over the past decade or so, there is evidence that declines in rates of endemic infectious diseases correlate within the same population with increases in conditions of immune dysregulation, including asthma, multiple sclerosis, IBD, and other autoimmune diseases (Bach 2002). Less-industrialized nations have consistently lower incidences of such immune-mediated conditions. One of the most popular explanations for the pathophysiologic mechanisms for these disorders is the *hygiene hypothesis.* The main premise of this hypothesis is that some characteristics of modern Western lifestyles (e.g. increased household cleanliness and socioeconomic status) have led to a decrease in exposure to germs and other disease-causing substances and quantitative or qualitative deficiencies in the normal microbiota. This disturbance occurs in early life during development of the immune system and has therefore affected the human immune system's opportunity to develop standard immune responses. Because of this lack of opportunity, the immune system becomes dysregulated and prone to respond by reacting to otherwise harmless substances or self proteins leading to development of allergies or autoimmune disorders, respectively (Wills-Karp et al. 2001; Noverr and Huffnagle 2004).

Several studies have indicated that reduced exposure to infectious organisms like helminths and in particular hepatitis A is strongly associated with allergic sensitization (Sporik et al. 1990; Lynch et al. 1993; Matricardi et al. 1997) while increased exposure to wild type measles infection correlates with reduced allergic sensitization (Shaheen et al. 1996). Other studies have shown that both allergic sensitization and allergic rhinitis are uncommon in individuals that spent their childhood on a farm (Braun-Fahrlander et al. 1999; Riedler et al. 2001). This could reflect increased exposure to organisms specific to the farms or to less specific bacterial products like the lipopolysaccharide which is present at higher levels in the homes of farm children than other children (von Mutius et al. 2000). Several studies have suggested that the environment in early life strongly affects the individual's risk of becoming allergic (Strachan 1989; Ball et al. 2000; Riedler et al. 2001).

The mechanism behind the *hygiene hypothesis* is that early exposure to infectious agents helps direct the immune system toward a Th1 cell–predominant response that, in turn, inhibits the production of Th2 cells (Fig. 12.1). Allergic reactions are associated with the generation of Th2- associated cytokines (IL-4, IL-5, and IL-13), which promote IgE production and eosinophilia. These cytokines may be balanced by cytokines secreted by Th1, Th3, and T regulatory cells, partially as a result of stimulation by the gut microbiota (Rautava et al. 2005).

12.4 Disorders of the Immune System

12.4.1 Allergy

12.4.1.1 Microbiota and Allergy

It has been observed that children who later developed allergic sensitization to common allergens and develop allergy have lower numbers of *Bifidobacteria, Lacto-*

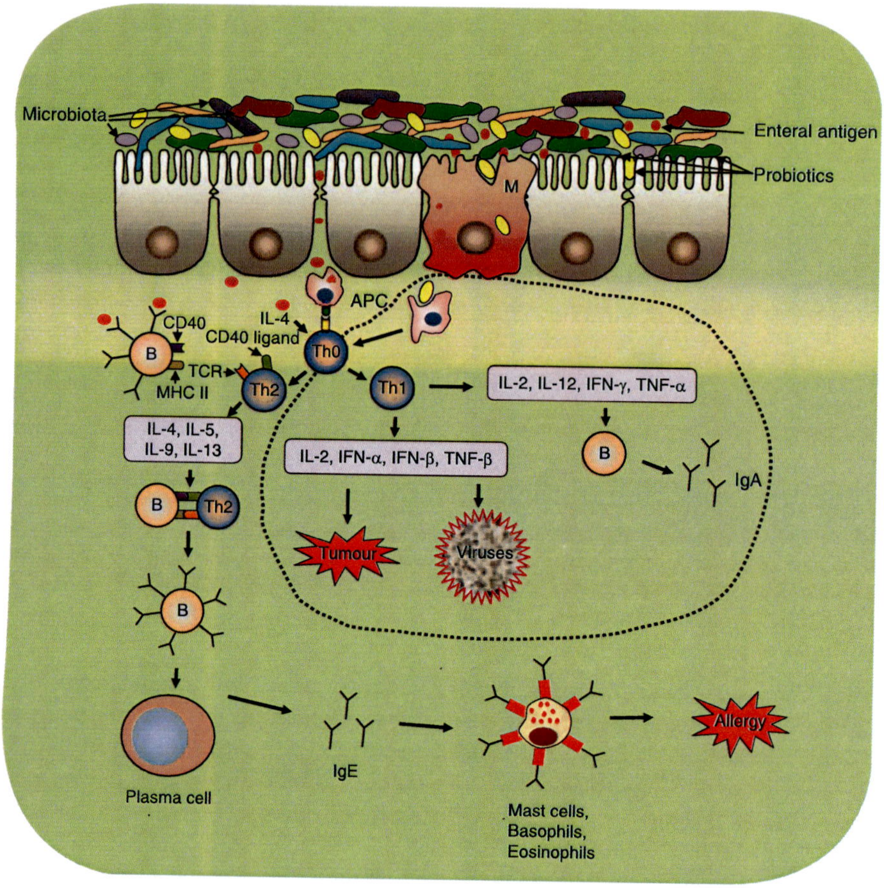

Fig. 12.1 Pathogenesis of allergy and the immunomodulatory role of probiotic bacteria. Microbiota in the intestinal lumen contribute to intestinal mucosal immunity. Increased intestinal paracellular permeability due to disrupted tight junctions allows translocation of enteral antigen into subepithelial location where it comes in contact with cells of the immune system. The antigen is processed by APC, presented to Th0 cell, then, in the presence of IL-4, the Th0 cell develops into Th2 cell. Antigen is also processed by B cell and presented to Th2 cell via MHC II. The CD40 on B cell binds to CD40 ligand on Th2 cell surface. The Th2 cell produces cytokines which stimulate B cell to produce antigen-specific IgE. B cell differentiates into plasma cells which produce more IgE. The IgE then bind to immune cells via high-affinity IgE receptor and can result into allergic reaction following re-exposure of an individual to the antigen. Probiotic bacteria strengthen intestinal barrier integrity to prevent translocation of enteral antigens. They also favour Th1 response associated with protection. M, M cell; APC, antigen presenting cell; Th0, naïve T cell; Th1 and Th2, T helper cells; TCR, T cell receptor; MHC II, major histocompatibility class II

bacilli and *Enterococci* and increased numbers of *Clostridia* and *Staphylococcus aureus,* and altered bacterial fatty acid profiles in faeces from the first weeks of life (Kalliomaki et al. 2001; Furrie 2005). The lower levels of faecal *Bifidobacteria* species correlate with both presence and severity of eczema in children of 2–12 years of age (Watanabe et al. 2003). Additionally, these children have alterations in the faecal composition of *Bifidobacteria*, with *B. adolescentis* (a strain more commonly found in adults) being predominant, whereas their age-match healthy infants are predominated with *B. bifidum* (Ouwehand et al. 2001). Furthermore, a comparison of formula-fed versus breast-fed infants has shown a lower incidence of allergies in the breast-fed infants that correlates with higher levels of *Bifidobacteria,* as opposes to formula-fed infants that are known to have higher levels of *Clostridia* (Kalliomaki et al. 2001; Watanabe et al. 2003). *Bifidobacteria* species have also been shown to alter the cytokine production pattern in allergic infants. A study by He et al. (2002) demonstrated that *Bifidobacteria* from allergic infants induce less production of anti-inflammatory cytokines like IL-10 and more proinflammatory cytokines than those from non-allergic infants. All these findings attest that microbiota modulate the immune system and could be central to prevention of several allergic reactions if their profile is not disrupted or is properly changed in early life.

12.4.1.2 Mechanisms of Allergy

The route for the first allergic responses in an individual's life usually arises from the gastrointestinal tract with food allergy being a common problem in infants with atopic eczema. In these infants, disrupted intestinal and skin barrier functions lead to greater antigen transfer across the mucosal barrier and the routes of transport are altered. The translocation of antigens evokes aberrant immune responses and release of proinflammatory cytokines with further impairment of the barrier functions which exacerbates further the increased intestinal permeability. The net result is a vicious circle of increasing allergenic responses, and a more permanent dysregulation of the immune responses to ubiquitous antigens in genetically susceptible individuals (Isolauri and Salminen 2008).

Allergic reactions are mediated by IgE following stimulation of Th1, Th2 and B cells by an allergen. Regulatory T cells may also play a role (Xystrakis et al. 2006). Following exposure to an antigen, antigen presenting cells (APC) such as macrophages and dendritic cells are activated to phagocytise and process the allergen (antigen). The APC then migrate to lymph nodes where they present the antigen to naive Th (Th0) cells that in response, release both Th1 and Th2 cytokines and can develop into either cell type. In case the antigen is an allergen, the Th0 cells are thought to be exposed to IL-4 (from as yet unidentified sources, but including germinal-centre B cells) and possibly to histamine-primed dendritic cells, both of which favour them to develop into Th2 cells. The Th2 cells in return, produce more IL-4 and IL-13, which then act on B cells to promote the production of antigen-specific IgE. For this to occur, B cells must also bind to the allergen via allergen-specific receptors, internalize the antigen, process it, and present it to the Th2 cells. During presentation, the B cell must also bind to the Th2 cell by binding the CD40 expressed on its surface to the CD40 ligand on the surface of the Th2 cell. The produced IL-4 and IL-13 by the Th2 cells act on the B cell to promote antibody class switching from IgM to antigen-specific IgE production. The produced IgE bind to high-affinity receptors located on the surfaces of cells of the immune system (particularly mast cells, basophils and eosinophils) and circulate in the body awaiting for re-exposure. Exposure for the second time to the antigen may cause antigenic binding of two adjacent IgE antibodies causing cross-linking (bridging) of the bound IgE on the mast cells, basophils or eosinophils. This is followed by degranulation and release of preformed mediators as well as newly synthesized mediators, and cytokines that account for multiple allergic clinical symptoms (Fig. 12.1).

In allergic reactions, Th2 cells play a central role, and patients who are atopic are thought to have a higher Th2 to Th1 cell ratio. The functions of Th1 cells in these individuals are production of interferon (IFN)-γ, IL-2, and tumour necrosis factor (TNF)-β, and promotion of a cell-mediated immune response. The Th1 produced IFN-γ diminishes the production of Th2 cells to suppress allergic reactions. Additionally, regulatory T cells actively inhibit Th2 responses to allergens (Xystrakis et al. 2006).

12.4.1.3 The Role of Probiotics in Allergic Reactions

Stemming on the pathogenesis of allergy, probiotic bacteria may be useful in preventing and treating allergy through different ways including degradation or structural modification of enteral antigens, normalization of the properties of aberrant indigenous microbiota and of gut barrier functions, regulation of the secretion of inflammatory mediators, and promoting development of the immune system, prevents food allergy by promoting endogenous barrier mechanisms and alleviating intestinal inflammation, stimulating immune response and reduction of serum IgE levels, and reduction of Th2 cytokine response.

12.4.1.4 Maintenance of Barrier Integrity

Under normal conditions, food-borne antigens can traverse the epithelial barrier of the gastrointestinal tract through two mechanisms. The first method is IFN-γ-mediated transcellular movement of antigen through the epithelial cell by endocytosis (Buning et al. 2005). This allows effective presentation of food antigens to the immune response and induction of oral tolerance. Reduction in the IFN-γ activity in infants with cow's milk allergy results in loss of tolerance and development of allergy (Pohjavuori et al. 2004). The second mechanism is by a paracellular route through tight junctions that allow passive but selective movement of molecules through epithelial barriers. Both the microbiota and the probiotic bacteria strengthen the paracellular transport of molecules, rendering the gut 'less leaky' and thereby reversing the increased intestinal permeability observed in cow's milk allergy (Baumgart and Dignass 2002; Qin et al. 2005).

12.4.1.5 Degradation/Structural Modification of Enteral Antigens

In order for the symptoms of food allergy to develop, the antigen must exit the gut to reach and stimulate the immune response in the immune environment. The antigen must be delivered to the immune environment inappropriately, since in a normal situation there is oral tolerance associated with hyporesponsiveness to the food antigens (Mowat et al. 2004). The beneficial effects of probiotics in this case is via their ability to subtly alter the transport of antigen across the epithelium while increasing Th1-type cytokines, therefore re-establishing oral tolerance and removing the allergic stimuli (Furrie 2005). Moreover, probiotics have been shown to exert distinct effects on antigen transport, depending on the food matrix, such as the quality of protein in the usual diet. Pessi et al. (1998) demonstrated in a rodent model, that mucosal transport of degraded macromolecules is stimulated when *Lactobacillus* GG is administered together with unhydrolyzed protein, but reduced when administered with hydrolyzed protein. Such protein may thus stimulate the humoral immunity in the gut, but also affect the induction of oral tolerance, as antigen degradation is an indispensable component in the acquisition of mucosal tolerance (Rautava et al. 2005).

12.4.1.6 Modulation of IgE Production

Several studies have shown that the elevated IgE in allergic patients can be suppressed by probiotic bacteria resulting in reduced clinical illness. Adult eczema patients with gastrointestinal symptoms have been found to have high serum IgE levels and increased counts of intestinal immunoreactive IgE cells (Kalimo et al. 1988). In these patients, a positive association between eczema severity and serum IgE levels occurs. Administration of probiotic bacteria *L. rhamnosus* and *L. reuteri* reduces the clinical severity of eczema with a more pronounced effect in patients with elevated IgE (Rosenfeldt et al. 2003). Moreover, intake of these strains by pregnant mothers and then by their newborn babies for 6–12 months leads to a modest reduction of IgE-mediated eczema in infants.

As described in preceding sections, allergic diseases are initiated and maintained by disruption of the Th1/Th2-cytokine balance, which is characterised by a relative predominance of Th2-cytokines (IL-4, IL-5, IL-9, and IL-13) and insufficient secretion of Th1-cytokines (IFN-γ, IL-12, TNF-α) (van der Velden et al. 2001). The Th2-associated cytokines activate the production of IgE and the activity of mast cells and eosinophils whereas Th1-associated cytokines suppresses IgE synthesis, as well as IgE-mediated allergies but stimulate secretion of IgA (van der Velden et al. 2001; Behera et al. 2002; O'Byrne 2006; Ghadimi et al. 2008). Probiotic bacteria influence this regulation in favour of a shift from Th2- to Th1-associated cytokines and the subsequent antiallergic effects. Studies have demonstrated that oral administration of *Bifidobacterium breve* and *Lactobacillus casei* strain Shirota activates the humoral immune system leading to increased Th1 cells, augmented production of IgA and IgG and inhibited IgE production (Yasui et al. 1999).

12.4.2 Autoimmune Diseases

A disease in which cytotoxic cells are directed against self-antigens in the body's tissues is considered autoimmune in nature. Autoimmune diseases often cause prolonged inflammation and subsequent tissue destruction. The gastrointestinal tract and the underlying mucosal immune system are important mediators of various autoimmune diseases. Genes, the environment, and the gut are intimately linked in establishing disease (Fasano and Shea-Donohue 2005; Vaarala et al. 2008). In this work, the common autoimmune disorders affecting the gastrointestinal tract and the role of probiotic bacteria in preventing or treating them is described.

12.4.2.1 Celiac Disease

Celiac disease is caused by ingestion of gluten-containing cereals in genetically susceptible individuals (Catassi and Fasano 2008). It is associated with disruption of barrier function marked by increased paracellular permeability. Gliadin peptides contained in gluten undergo deamidation by tissue transglutaminase 2 to produce epitopes that are presented to T cells by HLA-DQ2 or HLA-DQ8 molecules (Molberg et al. 1998; van de Wal et al. 1998). The activated T cells then mount a Th1-like immune response toward gliadin essentially characterized by IFN-γ and TNF-α productions both in the systemic (especially spleen) and intestinal compartments, with undetectable Th2-associated cytokines IL-10 and IL-4 (D'Arienzo et al. 2009). An inflammatory response then develops in the intestine, resulting in villous atrophy, increased lymphocytic infiltration of the lamina propria and epithelium, and the development of autoantibodies to tissue transglutaminase 2 (Jabri and Sollid 2006). Intake of probiotic bacteria improves the balance of Th1/Th2-associated pro- and anti-inflammatory cytokines and stabilises intestinal barrier integrity, and thus prevents inflammatory reactions in celiac diseases (Salvatore et al. 2007; Mennigen and Bruewer 2009).

12.4.2.2 Type 1 Diabetes

Another autoimmune condition is T1D. In this disorder, the autoimmune effector T cells destroy the insulin-producing pancreatic β-cells in the islets of Langerhans. The disorder is due to dietary antigens, mostly gluten, that shift gut homeostasis towards an inflammatory autoimmune reaction by inducing disruption of intestinal barrier and alteration of mucosal immune system (Lefebvre et al. 2006). According to Mordes et al. (2004), diabetes-prone rats fed wheat-based diets develop spontaneous T1D characterized by a Th1- biased lymphocytic infiltration into the islets (insulitis) followed by T cell-mediated destruction of insulin-producing β-cells and clinical symptoms resembling those observed in human T1D. In addition, a celiac-like enteropathy, characterized by small intestinal villous shortening and infiltration of lamina propria by immune cells develops prior to T1D (Graham et al. 2004). This enteropathy is a reflection of increased intestinal permeability that precedes insulitis and is associated with increased tight junction modulator zonulin, the blockage of which prevents T1D (Watts et al. 2005). Feeding the rats a protective hydrolysed casein or gluten-free formula diet decreases the gut permeability and the severity of the disease (Courtois et al. 2005). Moreover, in the course of the disease, there is an increase in the percentage of IFN-γ- and TNF-α-producing Th1 lymphocytes and a decrease in Th2 cytokine IL-5 and IL-10 in the mesenteric lymph nodes draining the gut (Caicedo et al. 2008). IL-17A and IL-6 can also be produced as proinflammatory mediators. In paediatric patients with T1D there is an enhanced immune activation characterised by increased expression of HLA-DR, HLA-DP, and intracellular adhesion molecule-1, and increased density of IL-1α+ and IL-4+ intestinal lamina propria cells (Westerholm-Ormio et al. 2003). In addition, there is an increased recruitment of regulatory T cells to the inflammatory sites during overt disease. The ability of probiotic bacteria to reverse the increased intestinal paracellular permeability and modulate the immune responses accounts for their beneficial effects to T1D.

12.4.2.3 Inflammatory Bowel Disease

Inflammatory bowel disease (IBD) patients, which include patients with Crohn's disease, ulcerative colitis, and pouchitis, have aggressive T cell responses to discrete antigens from the complex luminal microbiota (Duchmann et al. 1999). The overly T cell response triggers chronic intestinal inflammation. It appears that the pathogenesis involves activation of genetic polymorphisms encoding defective immunoregulation, microbial killing, and mucosal barrier function leading to aggressive immune responses to commensal bacteria. The first IBD susceptibility locus was identified as NOD2/CARD15 (Hugot et al. 2001; Ogura et al. 2001), a highly expressed cytoplasmic protein in monocytes/macrophages that binds to bacterial muramyl dipeptide (MDP) and activates nuclear factor kappa B (NF-κB) signaling. Polymorphisms in the ligand-binding domain of NOD2 lead to increased Toll-like receptor 2 (TLR2) ligand-stimulated activation of NF-κB in response to MDP, resulting in the production of higher amounts of IFN-γ, IL-1β, IL-2, and IL-12/IL-23 p40, defective killing and clearance of invasive bacteria, and decreased production of antimicrobial α-defensins by Paneth cells (Hisamatsu et al. 2003; Maeda et al. 2005; Wehkamp et al. 2005). NOD2 mutations also diminish the secretion of IL-10 in peripheral blood mononuclear cells stimulated with TLR2 ligands (Netea et al. 2004). Additional studies in rodents indicate that genetically susceptible hosts develop pathogenic Th1/Th17 or Th2 responses to their commensal microbiota that lead to progressive chronic intestinal inflammation, while in normal hosts, the microbiota induce homeostatic epithelial responses and tolerogenic T-cell responses to enteric bacterial antigens (Sartor 2007), indicating a loss of tolerance to microbiota by the immune system.

The defective immune response to commensal bacteria also leads to dysbiosis of the microbiota characterized by decrease in *Clostridium* species and increase in *Escherichia coli* and other proteobacteria in the intestine of IBD patients. This disturbance could decrease production of protective SCFA and result in colonization with bacteria that attach to and invade epithelial cells and produce antigens that activate effector T-cell responses. Functional changes to the commensal bacteria have also been observed to follow the defective immune response to microbiota.

Darfeuille-Michaud et al. (2004) observed that adherent/invasive *E. coli* (AIEC) adhere to and invade epithelial cells and persist and even replicate within macrophages. AIEC adhesion to epithelial cells is stimulated by IFN-γ and TNF-α which could be produced by the overly reactive Th1 cells. Further, disruption of the microbiota leads to increased paracellular permeability and exposure of the reactive subepithelial tissue. This intestinal hyperpermeability is essential to the pathogenesis of IBD and occurs even in the non-inflamed ileal mucosa of IBD patients (Söderholm et al. 1999; Hollander 2002). Persistent production of proinflammatory cytokines subsequent to increased permeability results in prolonged inflammatory reaction. Probiotic bacteria could benefit against both increased permeability and cytokine secretion. Indeed, therapeutic intervention for IBD aims at suppressing the cytokine production (Ogata and Hibi 2003).

12.4.2.4 Irritable Bowel Syndrome

The etiopathogenesis of irritable bowel syndrome (IBS) is not fully understood but it is suggested to be multifactorial, attributed to alterations in gastrointestinal motility, visceral hypersensitivity, dysfunction of the brain-gut axis or certain psychosocial factors. Disturbances to the immune system emanate from disruption of microbiota and inflammatory responses. The disruption of microbiota is due to alterations of intestinal neural plexuses and its associated hypomotility that leads to bacterial overgrowth. It is also associated with lack of *Lactobacilli* and *Bifidobacteria* but increase of *Clostridia* and *Enterobacteriaceae.* Intestinal hypersensitivity reaction in IBS is due to inflammatory disorders characterized by increase in T lymphocytes and mastocytes in the lamina propria, and local and systemic increase in pro-inflammatory cytokines, or imbalance between anti-inflammatory (IL-10) and pro-inflammatory (IL-12) cytokines (Bixquert Jiménez 2009).

Current treatment for IBD focuses on an etiopathogenesis- or pathophysiology-based therapeutic approach, attempting to influence the possible existence of intestinal dysbiosis, altered intestinal fermentation, excess production or alteration in the management of intestinal gas, and mucosal inflammation. This approach earmarks probiotic bacteria as an emerging therapeutic agent of choice. Several studies have shown that the use of probiotic bacteria *L. plantarum, B. infantis, E. coli Nisle* and *S. faecalis* by IBS patients improves the pain and the overall illness of the disease (Nobaek et al. 2000; Talley 2008). This protection could be due to attempt to restore the disrupted microbiota as well as prevention of excessive immune responses by modulating the Th1/Th2-associated cytokines favouring the anti-inflammatory responses.

12.4.2.5 Autism

Autism is a developmental disability typically appearing during the first 3 years of life. Children with this disorder seem to have a widespread gastrointestinal pathology that plays an important role in the symptomatology. A leaky mucosa and subsequent induction of dietary antigenic (gluten and cow milk) responses have been observed to be the underlying features in the development of the disease (D'Eufemia et al. 1996). The antigens result from incomplete digestion of dietary gluten and casein in the lumen of the small intestine by the action of peptidases. These antigens are able to enter the blood through the leaky mucosa and induce antigenic responses. They are structurally similar to endorphins and are called exorphins. Exorphins are able to enter the mammalian central nervous system and evoke the disorders (Hemmings 1978). The beneficial role of probiotics in autism could be mediated through improvement of barrier integrity to prevent a leaky intestinal mucosa.

12.5 Modulation of Immune System by Probiotic Bacteria

It is now evident that probiotic bacteria modulate the innate and adaptive immunity in animal models and humans (Fig. 12.2). Elucidation of the mechanisms that mediate probiotic-driven immunomodulation may facilitate their therapeutic

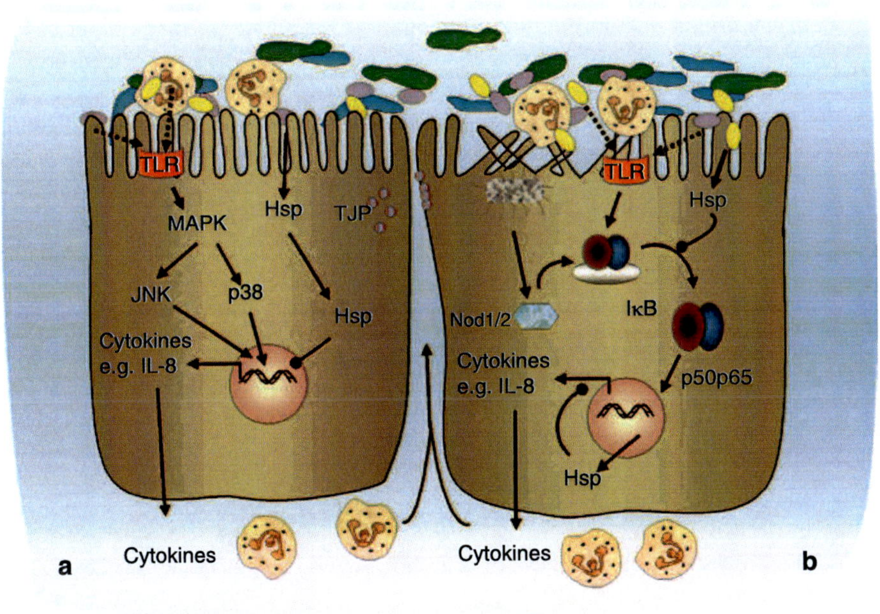

Fig. 12.2 **a** Enteric pathogens stimulate TLR to induce production of inflammatory cytokine such as IL-8 through activation of JNK and p38 pathways of MAPK. IL-8 in turn attracts subepithelial neutrophils, which migrate to the site of infection in the intestinal lumen via disrupted intercellular tight junctions. **b** Intracellular parasites such as Salmonella stimulate Nod1/2 to cause activation of NF-κB that leads to activation of inflammatory cytokine production. Probiotics can inhibit this activation by inducing production of Hsps. TLR, toll-like receptors; MAPK, mitogen activated protein kinase; NF-κB, nuclear factor kappa B; IL, interleukin; c-Jun N kinase

application for specific immune-mediated diseases or for prophylaxis. Recently, D'Arienzo et al. (2009) explored the effect of several *Lactobacillus* species (*L. plantarum, L. paracasei, L. casei,* and *L. fermentum*) and *Bifidobacterium lactis* in transgenic mice expressing the human DQ8 heterodimer, a HLA molecule linked to celiac disease. In their setting, the probiotic bacteria were co-administered in mice mucosally immunized with the gluten component gliadin. They found that all strains up-regulated, with different intensity, surface B7-2 (CD86) on immature bone marrow-derived dendritic cells, indicative of dendritic cell maturation. In addition, *L. paracasei* and *L. fermentum* elicited TNF-α expression and increased the antigen-specific TNF-α secretion *in vivo*. No strain induced appreciable levels of IL-10 or IL-12 in the immature dendritic cells. Their results suggest that probiotics are capable of exerting strain-specific inductive rather than suppressive effects both on the innate and adaptive immunity in a mouse model of food antigen sensitivity.

12.5.1 Cell Regulation

Various studies have considered the immune competent cells, particularly dendritic cells that play a pivotal role in immunological responses, as main target of the modulatory activity of probiotics. Dendritic cells are involved in antigen uptake and processing and promotion of the development of unprimed naive T cells toward Th1, Th2, or unpolarized T cell responses (Pochard et al. 2002; Furrie 2005; Mohamadzadeh et al. 2005). Since dendritic cells are present at different sites in the gastrointestinal tract and since the gut is colonized by commensal bacteria, interactions between these cells and the gut microbes, including probiotics, are expected. Some *Lactobacillus* species and *B. lactis* do alter cell surface antigen expression and cytokine production in dendritic cells. Several studies have shown up-regulation of CD80 and CD86 on dendritic cells in mice models given probiotic bacteria like *Lactobacillus* species (Christensen et al. 2002; Drakes et al. 2004; D'Arienzo et al. 2009).

Further to modulation of antigen presenting cell responses, regulation of T cell responses by probiotic bacteria is evident. Ghadimi et al. (2008) assessed the effect of different probiotics, *Lactobacillus rhamnosus, Lactobacillus gasseri, Bifidobacterium bifidum,* and *Bifidobacterium longum,* on the Th1 and Th2 responses of peripheral blood mononuclear cells (PBMCs) from healthy subjects and from patients with allergy against house dust mite following pretreatment with superantigen *Staphylococcus* enterotoxin A (SEA) and specific allergen *Dermatophagoides pteronyssinus* (Dpt). They found that the probiotic bacteria and their genomic DNA inhibited SEA- and Dpt-stimulated secretion of Th2-cytokines (IL-4 and IL-5) and enhanced the stimulation of IFN-γ in both healthy individuals and allergic patients. The DNA seemed to contribute to more than 50% of the effect exerted by living bacteria. Consistently, Pochard et al. (2002) and Flinterman et al. (2007) observed that probiotics such as *Lacobacillus plantarum, L. casei, L. rhamnosus,* or *Lactococcus lactis* promote the Th1 response accompanied by enhanced secretion of

IFN-γ and TNF-α by PBMCs following exposure to probiotics. Moreover, a study by Hessle et al. (2000) found that Gram-negative bacteria stimulate the secretion of the Th2-associated cytokine IL-10 in blood monocytes from healthy subjects, whereas Gram-positive bacteria, including *L. plantarum, Bifidobacterium adolescentis,* and *Enterococcus faecalis* are potent inducers of IL-12 which activates Th1 cells to secrete IFN-γ. A correlating *in vivo* study showed that administration of Gram-positive Lactobacillus rhamnosus GG to children with cow's milk allergy and atopic dermatitis stimulates IFN-γ secretion in their PBMCs (Pohjavuori et al. 2004).

The modulation of Th1/Th2 shift by probiotic bacterial DNA could partly be mediated by CpG motifs via TLR9 (Winkler et al. 2007). CpG ODNs are oligodeoxynucleotides containing one or more unmethylated CpG dinucleotides in specific base sequences. They are known to trigger Th1 responses by enhancing secretion of IFN-γ (Hussain et al. 2002) and to inhibit established Th2 responses (Broide et al. 2001). This effect is suggested to be mediated either through IFN-γ and IL-10 (Kitagaki et al. 2002) or independently through increased expression of Th1-associated cytokines, IFN-γ and IL-12 (Kline et al. 1999).

These studies and several others strongly support the fact that probiotic bacteria modulate the immune response in an attempt to prevent or alleviate disorders of the immune system.

12.5.2 *Cytokine Regulation*

Probiotic bacteria, especially lactobacilli and bifidobacteria, have been shown to modulate cytokine production by epithelial and professional immune cells to confer protection against enteric disorders. The modulation could be due to the ability of probiotic bacteria like *Lactobacillus* species (Pochard et al. 2002) to inhibit the activities of Th2 cells and prevent allergy, or other different mechanisms. Probiotic *L. rhamnosus* GG and *L. plantarum* induce secretion of IL-8 by epithelial cells in response to *S. Typhimurium* to enhance their defence against the pathogen (Grangette et al. 2005). Similarly, *L. reuteri* administered to mice induces a transient gene expression of pro-inflammatory cytokines and chemokines, including IL-1α, IL-6, IFN-γ, and macrophage inflammatory protein 2 (Hoffmann et al. 2008). It appears that the probiotic-induced pro-inflammatory cytokine production by epithelial cells is transient since high levels and persistent production of these cytokines lead to chronic inflammation. And to avoid this, probiotic bacteria inhibit persistent production of pro-inflammatory cytokines. In a series of studies done in our laboratory, we have consistently observed suppression of *Salmonella enteritidis* 857-induced IL-8 production by lactic acid bacteria (Nemeth et al. 2006; Malago et al. 2010). Probiotic bacteria also prevent persistent production of inflammatory cytokines and avoid excessive inflammation by inducing regulatory and anti-inflammatory cytokine production, such as IL-10 and TGF-β. IL-10 production by the epithelial cells triggers different anti-inflammatory mechanisms through JAK1/STAT3, p38 MAPK, and NF-κB pathways to confer protection against chronic intestinal inflammation (Malago et al. 2002; Werner and Haller 2007). TGF-β is produced following intake of some probiotic strain such as *B. longum* and *B. lactis* in human peripheral blood (Ouwehand et al. 2008). This cytokine inhibits the TLR-induced NF-κB-dependent pro-inflammatory gene expression by inducing TLR2 degradation via Smad signalling (Werner and Haller 2007).

The anti-inflammatory role of probiotic bacteria is also mediated through induction of heat shock response. We have shown the potency of different probiotic bacteria *L. casei* Shirota, *L. plantarum B. infantis,* and *Lactococcus lactis* in inducing expression of heat shock protein (Hsp)-70 and subsequently suppress the pro-inflammatory cytokine IL-8 in intestinal epithelial cells (Malago et al. 2010). The produced Hsp70 interferes with MAPK and NF-κB transcriptional activities leading to suppression of pro-inflammatory cytokines and protection of the intestine (Malago et al. 2002; Petrof et al. 2004).

Microbial colonization is another way through which probiotic bacteria modulate cytokine synthesis and secretion. Colonisation induces expression of antimicrobial peptides and down-regulation of the expression of pro-inflammatory type I IFN related genes in the large intestine. This modulation plays a role in preventing excessive inflammation caused by continuous microbial exposure (Munakata et al. 2008).

Probiotic bacteria can also mediate their immunomodulatory role by stimulating cells of the innate immune (Schiffrin et al. 1997). Administration of the probiotic *L. fermentum, L. plantarum, L. paracasei, L. reuteri* or *L. brevis* elicits Th1 cells to produce various cytokines including IL-1β, IL-2, IL-12, IFN-γ and TNF-α but not Th2 cytokines such as IL-10 and IL-4 (Maassen et al. 2000; D'Arienzo et al. 2009). It is apparent that IL-12 is a critical factor in switching naïve or memory T cells in Th1 responses, disfavouring Th2, that lead to vigorous immunity against infection and other diseases (Mohamadzadeh et al. 2005).

12.5.3 Influence on Toll Like Receptors

Immunomodulation by probiotics on inflammation is mainly via TLR and their subsequent release of cytokines. TLR are pattern-recognition receptors expressed by epithelial cells and antigen-presenting cells (dendritic cells and macrophages) responsible for the initial recognition of specific pathogen-associated molecular patterns and the discrimination between pathogens and harmless microbes and the development of appropriate innate and acquired immune responses (Michelsen and Arditi 2007). They can be located at the cell surface (e.g. TLR1, 2, 4, 5) or intracellularly (e.g. TLR3, 7, 8, 9). Activation of TLR signalling through recognition of pathogen-associated molecular patterns leads to the transcriptional activation of genes encoding for pro-inflammatory cytokines, chemokines and co-stimulatory molecules, which subsequently control the activation of antigen-specific adaptive immune response. TLRs have been pursued as potential therapeutic targets for various inflammatory diseases and cancer. As therapeutics, probiotics are known to influence TLR expression and, thereby, modulate their mediated immune responses.

Several recent studies have reported that Gram-negative and Gram-positive bacteria, as well as commensal and pathogenic bacteria influence TLR to mediate different gastrointestinal immune responses. In mice, Gram-positive probiotic bacteria (*L. acidophilus, L. delbrueckii* subsp. *bulgaricus*, and *L. casei*) increase the number of TLR2 positive cells, while Gram-negative *E. coli* increases that of TLR4 positive cells parallel to different cytokine induction (IL-10 versus IL-12) (Dogi et al. 2008). In human cells, *L. plantarum* and *L. rhamnosus* GG up-regulate TLR2 and TLR9 transcription levels as well as protein levels of TLR2 and TLR5 (Vizoso et al. 2009). The signalling pattern for Gram-negative bacteria slightly differs from that of Gram-positive but with seemingly the same outcome: anti-inflammatory role. For instance, the Gram-negative probiotic *E. coli* Nissle increases TLR2 and TLR4 protein expression and NF-κB activity of human T cells. Additionally, *E. coli* Nissle decreases pro-inflammatory cytokine secretion and ameliorates colitis via TLR2- and TLR4-dependent pathways (Grabig et al. 2006). In most cases however, Gram-positive probiotic bacteria induce TLR2 expression via interactions with lipoteichoic acids but not TLR4 expression. A recent study by Voltan et al. (2007) for instance, demonstrated that the Gram-positive *L. crispatus* increases TLR2 and reduces TLR4 mRNA and protein levels in colonic mucosa of mice, and inhibits the LPS-induced IL-6 production while enhancing TLR2-mediated IL-10 secretion in colonic epithelial cells. The probiotic *Bifidobacteria* are also capable of influencing the TLR signalling and their responses. For instance, administration of mice with *B. animalis* dose dependently influences TLR2 gene expression in the lymph nodes and the bacterial DNA negatively correlates with the TNF-α gene expression (Trevisi et al. 2008). Needless to say, all these findings and several others, describe that probiotic bacteria influence the TLR signalling favouring anti-inflammatory responses as opposes to pathogenic bacteria that favour pro-inflammatory signalling. Although probiotic and pathogenic bacteria may share TLR signalling, their downstream substrates and outcome could be different. For example both *L. rhamnosus* GG and *Streptococcus pyogenes* enhance TLR2 expression and the subsequent NF-κB activation depends on this expression. However, in addition to enhanced TLR2 gene expression, *S. pyogenes* also up-regulates TLR3 and TLR7 expression resulting to a different response (Miettinen et al. 2008).

Probiotic bacteria are also capable of influencing intracellular TLR via their DNAs to exhibit different responses compared from pathogenic bacteria. For instance, while *B. breve* DNA does not elicit any change in TLR9 mRNA levels or protein localization, pathogenic *Salmonella* and *E. coli* strains increase TLR9 mRNA expression. Additionally, *Salmonella* strains increase surface TLR9 protein, activate NF-κB and MAPK to increase IL-8 secretion, a chemokine pivotal to intestinal inflammation (Malago et al. 2002; Ewaschuk et al. 2007). On the contrary, probiotic bacterial DNA from probiotic mixture VSL#3 delays NF-κB activation by stabilising the levels of I-κB and inhibiting ubiquitin dependent proteasome degradation, suppress the expression of inflammatory p38 pathway of MAPK, and inhibit IL-8 secretion (Jijon et al. 2004). Thus targeting TLR9 to affect these anti-inflammatory responses is one of key sites where probiotic bacteria act to prevent chronic intestinal inflammation. In addition, probiotic bacteria are capable of regulating the different inflammatory components of TLR signalling pathways like NF-κB, and MAPK at different steps. For example, the probiotic *B. breve* and its released soluble product inhibit secretion of the chemokine CXCL8 by epithelial cells driven by both AP-1 of MAPK and NF-κB pathways. This inhibition decreases colitis score and inflammatory cytokine expression in mice (Heuvelin et al. 2009). Another study showed that the probiotic *L. reuteri* inhibits activation of MAPK-regulated c-Jun and the transcription factor AP-1 leading to suppression of TNF production by LPS-activated monocytes and primary monocyte-derived macrophages. This effect benefits children with Crohn's disease (Lin et al. 2008).

12.6 Conclusion

Available literature from research and knowledge and experience from medical practice are clear that probiotic bacteria are beneficial in prevention and treatment of human enteric diseases caused by disruption of microbiota. Probiotic bacteria benefit patients by attempting to restore the disrupted intestinal microbiota, modifying enteral antigens, normalising the properties of aberrant intestinal barrier functions, influencing inflammatory pathways such as NF-κB and MAPK, thereby regulating secretion of inflammatory mediators, influence T cell responses and their cytokine production by suppressing Th2-associated cytokines but favouring Th1-associated cytokines, and by reducing serum IgE levels to prevent allergic reactions. Their functions involve regulation of the antigen presenting cells (macrophages and dendritic cells) to influence their immune responses and cytokine production and alterations of TLR signalling responses. Through these and yet undiscovered mechanisms, probiotic bacteria reduce and prevent occurrence of autoimmune diseases like IBD, T1D, and celiac disease as well as allergic reactions like eczema. Exploring further the mechanisms by which probiotic bacteria modulate the immune system could provide insight in choosing the right probiotic bacterial strains or combination of different bacterial species for a particular disorder.

References

Bach JF (2002) The effect of infections on susceptibility to autoimmune and allergic diseases. N Engl J Med 347:911–920

Backhed F, Ley RE, Sonnenburg JL, Peterson DA, Gordon JI (2005) Host-bacterial mutualism in the human intestine. Science 307:1915–1920

Ball TM, Castro-Rodriguez JA, Griffith KA, Holberg CJ, Martinez FD, Wright AL (2000) Siblings, day-care attendance, and the risk of asthma and wheezing during childhood. N Engl J Med 343:538–543

Baumgart DC, Dignass AU (2002) Intestinal barrier function. Curr Opin Clin Nutr Met Care 5:685–694

Behera AK, Kumar M, Lockey RF, Mohapatra SS (2002) Adenovirus-mediated interferon g gene therapy for allergic asthma: involvement of interleukin 12 and STAT4 signaling. Hum Gene Ther 13:1697–1709

Bixquert Jiménez M (2009) Treatment of irritable bowel syndrome with probiotics. An etiopathogenic approach at last? Rev Esp Enferm Dig 101:553–564

Braun-Fahrlander C, Gassner M, Grize L, Neu U, Sennhauser FH, Varonier HS, Vuille JC, Wüthrich B (1999) Prevalence of hay fever and allergic sensitization in farmer's children and their peers living in the same rural community. SCARPOL team. Swiss study on childhood allergy and respiratory symptoms with respect to air pollution. Clin Exp Allergy 29:28–34

Broide DH, Stachnick G, Castaneda D, Nayar J, Miller M, Cho J, Rodriquez M, Roman M, Raz E (2001) Immunostimulatory DNA mediates inhibition of eosinophilic inflammation and airway hyperreactivity independent of natural killer cells in vivo. J Allergy Clin Immunol 108:759–763

Buning J, Schmitz M, Repenning B, Ludwig D, Schmidt MA, Strobel S, Zimmer K-Ph (2005) Interferon γ mediates antigen trafficking to MHC class II positive late endosomes of enterocytes. E J Immunol 35:831–842

Caballero-Franco C, Keller K, De Simona C, Chadee K (2007) The VSL#3 probiotic formula induces mucin gene expression and secretion in colonic epithelial cells. Am J Physiol Gastrointest Liver Physiol 292:G315–G322

Caicedo RA, Li N, Des Robert C, Scumpia PO, Hubsher CP, Wasserfall CH, Schatz DA, Atkinson MA, Neu J (2008) Neonatal formula feeding leads to immunological alterations in an animal model of type 1 diabetes. Pediatr Res 63:303–307

Catassi C, Fasano A (2008) Celiac disease. Curr Opin Gastroenterol 24:687–691

Christensen HR, Frokiaer H, Pestka JJ (2002) Lactobacilli differentially modulate expression of cytokines and maturation surface markers in murine dendritic cells. J Immunol 168:171–178

Clemente MG, De Virgiliis S, Kang JS, Macatagney R, Musu MP, Di Pierro MR, Drago S, Congia M, Fasano A (2003) Early effects of gliadin on enterocyte intracellular signalling involved in intestinal barrier function. Gut 52:218–223

Courtois P, Nsimba G, Jijakli H, Sener A, Scott FW, Malaisse WJ (2005) Gut permeability and intestinal mucins, invertase, and peroxidase in control and diabetes-prone BB rats fed either a protective or a diabetogenic diet. Dig Dis Sci 50:266–275

D'Arienzo R, Maurano F, Lavermicocca P, Ricca E, Rossi M (2009) Modulation of the immune response by probiotic strains in a mouse model of gluten sensitivity. Cytokine 48:254–259

Darfeuille-Michaud A, Boudeau J, Bulois P, Neut C, Glasser AL, Barnich N, Bringer MA, Swidsinski A, Beaugerie L, Colombel JF (2004) High prevalence of adherent-invasive Escherichia coli associated with ileal mucosa in Crohn's disease. Gastroenterology 127:412–421

D'Eufemia P, Celli M, Finocchiaro R, Pacifico L, Viozzi L, Zaccagnini M, Cardi E, Giardini O (1996) Abnormal intestinal permeability in children with autism. Acta Paediatr 85:1076–1079

Di Pierro M, Lu R, Uzzau S, Wang W, Margaretten K, Pazzani C, Maimone F, Fasano A (2001) Zonula occludens toxin structure-function analysis. Identification of the fragment biologically active on tight junctions and of the zonulin receptor binding domain. J Biol Chem 276:19160–19165

Dogi CA, Galdeano CM, Perdigón G (2008) Gut immune stimulation by non pathogenic Gram(+) and Gram(−) bacteria. Comparison with a probiotic strain. Cytokine 41:223–231

Drakes M, Blanchard T, Czinn S (2004) Bacterial probiotic modulation of dendritic cells. Infect Immun 72:3299–3309

Duchmann R, May E, Heike M (1999) T cell specificity and cross reactivity towards Enterobacteria, Bacteroides, Bifidobacterium, and antigens from resident intestinal flora in humans. Gut 44:812–818

Ewaschuk JB, Backer JL, Churchill TA, Obermeir F, Krause DO, Madsen KL (2007) Surface expression of Toll-like receptor 9 is upregulated on intestinal epithelial cells in response to pathogenic bacterial DNA. Infect Immun 75:2572–2579

Fasano A, Shea-Donohue T (2005) Mechanisms of disease: the role of intestinal barrier function in the pathogenesis of gastrointestinal autoimmune diseases. Nat Clin Pract Gastroenterol Hepatol 2:416–422

Flinterman AE, Knol EF, van Ieperen-van Dijk AG, Timmerman HM, Knulst AC, Bruijnzeel-Koomen CA, Pasmans SG, van Hoffen E (2007) Probiotics have a different immunomodulatory potential in vitro versus ex vivo upon oral administration in children with food allergy. Int Arch Allergy Immunol 143:237–244

Freitas M, Axelsson LG, Cayuela E (2005) Indigenous microbes and their soluble factors differentially modulate intestinal glycosylation steps in vivo. Use of a "lectin assay" to survey in vivo glycosylation changes. Histochem Cell Biol 124:423–433

Furrie E (2005) Probiotics and allergy. Proc Nutr Soc 64:465–469

Gaboriau-Routhiau V, Moreau MC (1996) Gut flora allows recovery of oral tolerance to ovalbumin in mice after transient breakdown mediated by cholera toxin or Escherichia coli heat-labile enterotoxin. Pediatr Res 39:625–629

Ghadimi D, Fölster-Holst R, de Vrese M, Winkler P, Heller KJ, Schrezenmeir J (2008) Effects of probiotic bacteria and their genomic DNA on TH1/TH2-cytokine production by peripheral blood mononuclear cells (PBMCs) of healthy and allergic subjects. Immunology 213:6677–6692

Gonzalez-Mariscal L, Betanzos A, Nava P, Jaramillo BE (2003) Tight junction proteins. Prog Biophys Mol Biol 81:1–44

Grabig A, Paclik D, Guzy C, Dankof A, Baumgart DC, Erckenbrecht J, Raupach B, Sonnenborn U, Eckert J, Schumann RR, Wiedenmann B, Dignass AU, Sturm A (2006) Escherichia coli strain Nissle 1917 ameliorates experimental colitis via toll-like receptor 2- and toll-like receptor 4-dependent pathways. Infect Immun 74:4075–4082

Graham S, Courtois P, Malaisse WJ, Rozing J, Scott FW, Mowat AM (2004) Enteropathy precedes type 1 diabetes in the BB rat. Gut 53:1437–1444

Grangette C, Nutten S, Palumbo E, Morath S, Hermann C, Dewulf J, Pot B, Hartung T, Hols P, Mercenier A (2005) Enhanced antiinflammatory capacity of a *Lactobacillus plantarum* mutant synthesizing modified teichoic acids. Proc Natl Acad Sci U S A 102:10321–10326

He F, Morita H, Hashimoto H, Hosoda M, Kurisaki J, Ouwehand AC, Isolauri E, Benno Y, Salminen S (2002) Intestinal Bifidobacterium species induce varying cytokine production. J Allergy Clin Immunol 109:1035–1036

Hecht G (2001) Microbial pathogens that affect tight junctions in tight junctions, 2nd ed. CRC Press, Boca Raton, pp 493–516

Hemmings WA (1978) The entry into the brain of large molecules derived from dietary protein. Proc R Soc Lond B Biol Sci 200:175–192

Hessle C, Andersson B, Wold AE (2000) Gram-positive bacteria are potent inducers of monocytic interleukin-12 (IL-12) while Gram-negative bacteria preferentially stimulate IL-10 production. Infect Immun 68:3581–3586

Heuvelin E, Lebreton C, Grangette C, Pot B, Cerf-Bensussan H, Heyman M (2009) Mechanisms involved in alleviation of intestinal inflammation by *Bifidobacterium breve* soluble factors. PLoS One 4:e5184

Hisamatsu T, Suzuki M, Reinecker HC, Nadeau WJ, McCormick BA, Podolsky DK (2003) CARD15/NOD2 functions as an anti-bacterial factor in human intestinal epithelial cells. Gastroenterology 124:993–1000

Hoffmann M, Rath E, Hölzlwimmer G, Quintanilla-Martinez L, Loach D, Tannock G, Haller D (2008) *Lactobacillus reuteri* 100-23 transiently activates intestinal epithelial cells of mice that have a complex microbiota during early stages of colonization. J Nutr 138:1684–1691

Hollander D (2002) Crohn's disease, TNF-alpha, and the leaky gut. The chicken or the egg? Am J Gastroenterol 97:1867–1868

Hugot JP, Chamaillard M, Zouali H, Lesage S, Cézard JP, Belaiche J, Almer S, Tysk C, O'Morain CA, Gassull M, Binder V, Finkel Y, Cortot A, Modigliani R, Laurent-Puig P, Gower-Rousseau C, Macry J, Colombel JF, Sahbatou M, Thomas G (2001) Association of NOD2 leucine-rich repeat variants with susceptibility to Crohn's disease. Nature 411:599–603

Hussain I, Jain VV, Kitagaki K, Businga TR, O'Shaughnessy P, Kline JN (2002) Modulation of murine allergic rhinosinusitis by CpG oligodeoxynucleotides. Laryngoscope 112:1819–1826
Isolauri E, Salminen S (2008) Probiotics: use in allergic disorders. J Clin Gastroenterol 42:S91–S96
Jabri B, Sollid LM (2006) Mechanisms of disease: immunopathogenesis of celiac disease. Nat Clin Pract Gastroenterol Hepatol 3:516–525
Jiang WG, Bryce RP, Horrobin DF, Mansel RE (1998) Regulation of tight junction permeability and occludin expression by polyunsaturated fatty acids. Biochem Biophys Res Commun 244:414–420
Jijon H, Backer J, Diaz H, Yeung H, Thiel D, McKaigney C, De Simone C, Madsen K (2004) DNA from probiotic bacteria modulates murine and human epithelial and immune function. Gastroenterology 126:1358–1373
Kalimo K, Lammintausta K, Klemi P (1988) Mast cells and IgE in intestinal mucosa in adult atopic dermatitis patients. Br J Dermatol 86:101–104
Kalliomaki M, Kirjavainen P, Eerola E, Kero P, Salminen S, Isolauri E (2001) Distinct patterns of neonatal gut microflora in infants in whom atopy was and was not developing. J Allergy Clin Immunol 107:129–134
Kitagaki K, Jain VV, Businga TR, Hussain I, Kline JN (2002) Immunomodulatory effects of CpG oligodeoxynucleotides on established Th2 responses. Clin Diagn Lab Immunol 9:1260–1269
Kline JN, Krieg AM, Waldschmidt TJ, Ballas ZK, Jain V, Busing TR (1999) CpG oligodeoxynucleotides do not require TH1 cytokines to prevent eosinophilic airway inflammation in a murine model of asthma. J Allergy Clin Immunol 104:1258–1264
Lin YP, Thibodeaux CH, Peña JA, Ferry GD, Versalovic J (2008) Probiotic Lactobacillus reuteri suppress proinflammatory cytokines via c-Jun. Inflamm Bowel Dis 14:1068–1083
Lefebvre DE, Powell KL, Strom A, Scott FW (2006) Dietary proteins as environmental modifiers of type 1 diabetes mellitus. Ann Rev Nutr 26:175–202
Lutgendorff F, Akkermans LM, Söderholm JD (2008) The role of microbiota and probiotics in stress-induced gastro-intestinal damage. Curr Mol Med 8:282–298
Lynch NR, Hagel I, Perez M, Di Prisco MC, Lopez R, Alvarez N (1993) Effect of anthelmintic treatment on the allergic reactivity of children in a tropical slum. J Allergy Clin Immunol 92:404–411
Maassen CB, van Holten-Neelen C, Balk F, den Bak-Glashouwer MJ, Leer RJ, Laman JD, Boersma WJ, Claassen E (2000) Strain-dependent induction of cytokine profiles in the gut by orally administered Lactobacillus strains. Vaccine 18:2615–2623
Macintire DK, Bellhorn TL (2002) Bacterial translocation: clinical implications and prevention. Vet Clin North Am Small Anim Pract 32:1165–1178
Maeda S, Hsu LC, Liu H, Bankston LA, Iimura M, Kagnoff MF, Eckmann L, Karin M (2005) Nod2 mutation in Crohn's disease potentiates NF-kappaB activity and IL-1beta processing. Science 307:734–738
Malago JJ, Koninkx JFJG, van Dijk JE (2002) The heat shock response and cytoprotection of the intestinal epithelium. Cell Stress Chaperones 7:191–199
Malago JJ, Koninkx JFJG, Douma PM, Dirkzwager A, Veldman A, Hendriks HGCJM, van Dijk JE (2003) Differential modulation of enterocyte-like Caco-2 cells after exposure to short chain fatty acids. Food Addit Contam 20:427–437
Malago JJ, Tooten PCJ, Koninkx JFJG (2010) Anti-inflammatory properties of probiotic bacteria on *Salmonella*-induced IL-8 synthesis in enterocyte-like Caco-2 cells. Benef Microbe 1(2):121–130
Matricardi PM, Rosmini F, Ferrigno L, Pannetta V, Bonini S (1997) Cross sectional retrospective study of prevalence of atopy among Italian military students with antibodies against hepatitis A virus. Br Med J 314:999–1003
Mennigen R, Bruewer M (2009) Effect of probiotics on intestinal barrier function. Ann N Y Acad Sci 1165:183–189
Michelsen KS, Arditi M (2007) Toll-like receptors and innate immunity in gut homeostasis and pathology. Curr Opin Hematol 14:48–54

Miettinen M, Veckman V, Latvala S, Sareneva T, Matikainen S, Julkunen I (2008) Live *Lactobacillus rhamnosus* and *Streptococcus pyogenes* differentially regulate Toll-like receptor (TLR) gene expression in human primary macrophages. J Leukoc Biol 84:1092–1100
Mohamadzadeh M, Olson S, Kalina WV, Ruthel G, Demmin GL, Warfield KL, Bavari S, Klaenhammer TR (2005) Lactobacilli activate human dendritic cells that skew T cells toward T helper 1 polarization. Proc Natl Acad Sci U S A 102:2880–2885
Molberg O, McAdam SN, Korner R, Quarsten H, Kristiansen C, Madsen L, Fugger L, Scott H, Norén O, Roepstorff P, Lundin KE, Sjöström H, Sollid LM (1998) Tissue transglutaminase selectively modifies gliadin peptides that are recognized by gut-derived T cells in celiac disease. Nat Med 4:713–717
Mordes JP, Bortell R, Blankenhorn EP, Rossini AA, Greiner DL (2004) Rat models of type 1 diabetes: genetics, environment, and autoimmunity. ILAR J 45:278–291
Mowat AM, Parker LA, Beacock-Sharp H, Millington OR, Chirdo F (2004) Oral tolerance: overview and historical perspectives. Ann N Y Acad Sci 1029:1–8
Munakata K, Yamamoto M, Anjiki N (2008) Importance of the interferon-alpha system in murine large intestine indicated by microarray analysis of commensal bacteria-induced immunological changes. BMC Genomics 9:192
Nemeth E, Fajdiga S, Malago JJ, Koninkx JFJG, Tooten PCJ, van Dijk JE (2006) Inhibition of *Salmonella*-induced IL-8 synthesis and expression of Hsp70 in enterocyte-like Caco-2 cells after exposure to non-starter lactobacilli. Int J Food Microbiol 112:266–274
Netea MG, Kullberg BJ, de Jong DJ, Franke B, Sprong T, Naber TH, Drenth JP, Van der Meer JW (2004) NOD2 mediates anti-inflammatory signals induced by TLR2 ligands: implications for Crohn's disease. Eur J Immunol 34:2052–2059
Nobaek S, Johansson ML, Molin G, Ahrne S, Jeppsson B (2000) Alteration of intestinal microflora is associated with reduction in abdominal bloating and pain in patients with irritable bowel syndrome. Am J Gastroenterol 95:1231–1238
Noverr MC, Huffnagle GB (2004) Does the microbiota regulate immune responses outside the gut? Trends Microbiol 12:562–568
Nusrat A, von Eichel-Streiber C, Turner JR, Verkade P, Madara JL, Parkos CA (2001) Clostridium difficile toxins disrupt epithelial barrier function by altering membrane microdomain localization of tight junction proteins. Infect Immun 69:1329–1336
O'Byrne PM (2006) Cytokines or their antagonists for the treatment of asthma. Chest 130:244–250
Ogata H, Hibi T (2003) Cytokine and anti-cytokine therapies for inflammatory bowel disease. Curr Pharm Des 9:1107–1113
Ogura Y, Bonen DK, Inohara N, Nicolae DL, Chen FF, Ramos R, Britton H, Moran T, Karaliuskas R, Duerr RH, Achkar JP, Brant SR, Bayless TM, Kirschner BS, Hanauer SB, Nuñez G, Cho JH (2001) A frameshift mutation in NOD2 associated with susceptibility to Crohn's disease. Nature 411:603–606
Ouwehand AC, Isolauri E, He F, Hashimoto H, Benno Y, Salminen S (2001) Differences in Bifidobacterium flora composition in allergic and healthy infants. J Allergy Clin Immunol 108:144–145
Ouwehand AC, Bergsma N, Parhiala R, Lahtinen S, Gueimonde M, Finne-Soveri H, Strandberg T, Ptikälä K, Salminen S (2008) *Bifidobacterium* microbiota and parameters of immune function in elderly subjects. FEMS Immunol Med Microbiol 53:18–25
Pessi T, Sütas Y, Marttinen A, Isolauri E (1998) Probiotics reinforce mucosal degradation of antigens in rats: implications for therapeutic use of probiotics. J Nutr 128:2313–2318
Petrof EO, Kojima K, Ropeleski MJ, Musch MW, Tao Y, De Simone C, Chang EB (2004) Probiotics inhibit nuclear factor-kappaB and induce heat shock proteins in colonic epithelial cells through proteasome inhibition. Gastroenterology 127:1474–1487
Pochard P, Gosset P, Grangette C, Andre C, Tonnel AB, Pestel J, Mercenier A (2002) Lactic acid bacteria inhibit TH2 cytokine production by mononuclear cells from allergic patients. J Allergy Clin Immunol 110:617–623
Pohjavuori E, Viljanen M, Korpela R, Kuitunen M, Tiittanen M, Vaarala O, Savilahti E (2004) Lactobacillus GG effect in increasing IFN-g production in infants with cow's milk allergy J Allergy Clin Immunol 114:131–136

Qin HL, Shen TY, Gao ZG, Fan XB, Hang XM, Jiang YQ, Zhang HZ (2005) Effect of lactobacillus on the gut microflora and barrier function of the rats with abdominal infection. World J Gastroenterol 11:2591–2596

Rautava S, Kalliomäki M, Isolauri E (2005) New therapeutic strategy for combating the increasing burden of allergic disease: probiotics. J Allergy Clin Immunol 116:31–37

Riedler J, Braun-Fahrlander C, Eder W, Schreuer M, Waser M, Maisch S, Carr D, Schierl R, Nowak D, von Mutius E (2001) Exposure to farming in early life and development of asthma and allergy: a crosssectional survey. Lancet 358:1129–1133

Rosenfeldt V, Benfeldt E, Nielsen SD, Michaelsen KF, Jeppesen DL, Valerius NH, Pærregaard A (2003) Effect of probiotic Lactobacillus strains in children with atopic dermatitis. J Allergy Clin Immunol 11:389–395

Salvatore S, Hauser B, Vandenplas Y (2007) Chronic enteropathy and feeding. Nestle Nutr Workshop Ser Pediatr Program 59:115–126

Sartor RB (2007) Bacteria in Crohn's disease: mechanisms of inflammation and therapeutic implications. J Clin Gastroenterol 41:S37–S43

Schiffrin EJ, Brassart D, Servin AL, Rochat F, Donnet-Hughes A (1997) Immune modulation of blood leukocytes in humans by lactic acid bacteria: criteria for strain selection. Am J Clin Nutr 66:515S–20S

Shaheen SO, Aaby P, Hall AJ, Barker DJ, Heyes CB, Shiell AW, Goudiaby A (1996) Measles and atopy in Guinea-Bissau. Lancet 347:1792–1796

Simonovic I, Rosenberg J, Koutsouris A, Hecht G (2000) Enteropathogenic Escherichia coli dephosphorylates and dissociates occludin from intestinal epithelial tight junctions. Cell Microbiol 2:305–315

Söderholm JD, Peterson KH, Olaison G, Franzén LE, Westrom B, Magnusson KE, Sjödahl R (1999) Epithelial permeability to proteins in the noninflamed ileum of Crohn's disease? Gastroenterology 117:65–72

Sporik R, Holgate ST, Platts-Mills TA, Cogswell JJ (1990) Exposure to house-dust mite allergen (Der p I) and the development of asthma in childhood. A prospective study. N Engl J Med 323:502–507

Strachan DP (1989) Hay fever, hygiene, and household size. Br Med J 299:1259–1260

Tafazoli F, Zeng CQ, Estes MK, Magnusson KE, Svensson L (2001) NSP4 enterotoxin of rotavirus induces paracellular leakage in polarized epithelial cells. J Virol 75:1540–1546

Talley NJ (2008) Functional gastrointestinal disorders as a public health problem. Neurogastroenterol Motil 20(Suppl 1):121–129

Tlaskalová-Hogenová H, Stepánková R, Hudcovic T, Tucková L, Cukrowska B, Lodinová-Zádníková R, Kozáková H, Rossmann P, Bártová J, Sokol D, Funda DP, Borovská D, Reháková Z, Sinkora J, Hofman J, Drastich P, Kokesová A (2004) Commensal bacteria (normalmicroflora), mucosal immunity, and chronic inflammatory and autoimmune diseases. Immunol Lett 93:97–108

Trevisi P, De Filippi S, Minieri L, Mazzoni M, Modesto M, Biavati B, Bosi P (2008) Effect of fructo-oligosaccharides and different doses of *Bifidobacterium animalis* in a weaning diet on bacterial translocation and Toll-like receptor gene expression in pigs. Nutrition 24:1023–1029

Vaarala O, Atkinson MA, Neu J (2008) The "perfect storm" for type 1 diabetes: the complex interplay between intestinal microbiota, gut permeability, and mucosal immunity. Diabetes 57:2555–2562

Vaishnava S, Behrendt CL, Ismail AS, Eckmann L, Hooper LV (2008) Paneth cells directly sense gut commensals and maintain homeostasis at the intestinal host-microbial interface. Proc Natl Acad Sci U S A 105:20858–20863

van de Wal Y, Kooy Y, van Veelen P, Peña S, Mearin L, Papadopoulos G, Koning F (1998) Selective deamidation by tissue transglutaminase strongly enhances gliadin-specific T cell reactivity. J Immunol 161:1585–1588

van der Velden VH, Laan MP, Baert MR, de Waal Malefyt R, Neijens HJ, Savelkoul HF (2001) Selective development of a strong Th2 cytokine profile in high-risk children who develop atopy: risk factors and regulatory role of IFN-g, IL-4, IL-10. Clin Exp Allergy 31:997–1006

Vizoso Pinto MG, Rodriguez Gómez M, Seifert S, Watzl B, Holzapfel WH, Franz CM (2009) Lactobacilli stimulate the innate immune response and modulate the TLR expression of HT29 intestinal epithelial cells in vitro. Int J Food Microbiol 133:86–93

Voltan S, Castagliuolo I, Elli M, Longo S, Brun P, D'Incà R, Porzionato A, Macchi V, Palù G, Sturniolo GC, Morelli L, Martines D (2007) Aggregating phenotype in *Lactobacillus crispatus* determines intestinal colonization and TLR2 and TLR4 modulation in murine colonic mucosa. Clin Vaccine Immunol 14:1138–1148

von Mutius E, Braun-Fahrlander C, Schierl R, Riedler J, Ehlermann S, Maisch S, Waser M, Nowak D (2000) Exposure to endotoxin or other bacterial components might protect against the development of atopy. Clin Exp Allergy 30:1230–1234

Wang WL, Lu RL, DiPierro M, Fasano A (2000) Zonula occluding toxin, a microtubule binding protein. World J Gastroenterol 6:330–334

Watanabe S, Narisawa Y, Arase S, Okamatsu H, Ikenaga T, Tajiri Y, Kumemura M (2003) Differences in faecal microflora between patients with atopic dermatitis and healthy control subjects. J Allergy Clin Immunol 111:587–591

Watts T, Berti I, Sapone A, Gerarduzzi T, Not T, Zielke R, Fasano A (2005) Role of the intestinal tight junction modulator zonulin in the pathogenesis of type I diabetes in BB diabetic-prone rats. Proc Natl Acad Sci U S A 102:2916–2921

Wehkamp J, Salzman NH, Porter E, Nuding S, Weichenthal M, Petras RE, Shen B, Schaefferler E, Schwab M, Linzmeier R, Feathers RW, Chu H, Lima H Jr, Fellermann K, Ganz T, Stange EF, Bevins CL (2005) Reduced Paneth cell alpha-defensins in ileal Crohn's disease. Proc Natl Acad Sci U S A 102:18129–18134

Werner T, Haller D (2007) Intestinal epithelial cell signalling and chronic inflammation: from the proteome to specific molecular mechanisms. Mutat Res 622:42–57

Westerholm-Ormio M, Vaarala O, Pihkala P, Ilonen J, Savilahti E (2003) Immunologic activity in the small intestinal mucosa of pediatric patients with type 1 diabetes. Diabetes 52:2287–2295

Wills-Karp M, Santeliz J, Karp CL (2001) The germless theory of allergic disease: revisiting the hygiene hypothesis. Nat Rev Immunol 1:69–75

Winkler P, Ghadimi D, Schrezenmeir J, Kraehenbuhl JP (2007) Molecular and cellular basis of microflora-host interactions. J Nutr 137:756S–772S

Xystrakis E, Boswell SE, Hawrylowicz CM (2006) T regulatory cells and the control of allergic disease. Expert Opin Biol Ther 6:121–133

Yasui H, Shida K, Matsuzaki T, Yokokura T (1999) Immunomodulatory function of lactic acid bacteria. Antonie Van Leeuwenhoek 76:383–389

Part VI
Probiotics for Enteric Therapy

Chapter 13
Probiotic-Pathogen Interactions and Enteric Cytoprotection

Joshua J. Malago and Jos F. J. G. Koninkx

13.1 Introduction

The intestinal epithelium forms a physicochemical barrier that impedes enteric pathogens from invading the epithelium and cause disease. In order for a particular pathogen to colonise the intestinal mucosa, it needs to break and cross this barrier. The barrier consists of a low pH area mainly resulting from carbohydrate fermentation, a mucus layer along the epithelial surface, an epithelial mechanical barrier maintained by intercellular tight junctions, an apical actin cytoskeleton, and the presence of stable microbiota. In addition, this barrier produces inflammatory mediators, mainly cytokines.

The microbiota govern most of other components of intestinal mucosa immunity. For instance, they ferment carbohydrates to produce organic acids that lower the luminal pH and enhance the barrier integrity by regulating tight junction functions (Cani et al. 2009). The microbiota also compete with pathogenic bacteria for binding sites and nutrition, produce antimicrobial substances, promote production of mucus, enhance development of innate immunity, and down-regulate pro-inflammatory cytokine production induced by pathogens. Particular members of the microbiota, namely the *Lactobacillus* and *Bifidobacterium* species, have been found to mediate most of the beneficial effects of the microbiota. As such, they are commercially incorporated in diets like yoghourt and cheese as well as in medications against various intestinal disorders. Consumption of adequate amounts of these microorganisms and other live microorganisms (mainly members of microbiota) which confer a health benefit on the host has been termed 'probiotics' (FAO/WHO 2002).

From the last decade, the use of probiotic bacteria for health benefits has been growing significantly with an immense interest. Probiotic bacteria are used to treat a variety of gastrointestinal diseases, including inflammatory bowel disease (IBD), irritable bowel syndrome, pouchitis, and rotavirus and antibiotic-associated diarrhoea. Being mainly members of the microbiota, the potential mechanisms by which

J. J. Malago (✉)
Department of Veterinary Pathology, Faculty of Veterinary Medicine,
Sokoine University of Agriculture, P.O. Box 3203, Chuo Kikuu, Morogoro, Tanzania
e-mail: malagojj@yahoo.com

J. J. Malago et al. (eds.), *Probiotic Bacteria and Enteric Infections*,
DOI 10.1007/978-94-007-0386-5_13,

probiotic bacteria fight against infectious enteric diseases do not differ much from those employed by microbiota. They mediate their mechanisms through protective, trophic, and anti-inflammatory effects on bowel mucosa. Some of the proposed mechanisms include inhibition of the growth of pathogenic bacteria through production of ammonia, hydrogen peroxide (Ocana et al. 1999; Kullisaar et al. 2002), bacteriocins, and other unidentified substances observed in their culture products, competition with pathogenic bacteria for adhesion sites on intestinal epithelia and nutrition, and an adjuvant-like stimulation of the immune system against pathogenic organisms (Maassen et al. 2000).

To execute their beneficial effects, probiotic bacteria interact with the enteric pathogens to oppose the detrimental effects of the pathogens. The interaction can be a direct relationship between the bacteria and/or their secreted products or indirect through the host epithelial cells or the intestinal luminal contents such as carbohydrate substrates. In all these interactions, probiotic bacteria prevent intestinal colonisation of pathogenic bacteria and subsequently protect the gut against invasion, infection and occurrence of disease. In this chapter, the interaction of probiotic and pathogenic bacteria is discussed. The chapter starts by highlighting the barriers of immune system which the pathogens have to cross, and then it discusses the potential mechanisms employed by enteric pathogens to cause disease. Finally, the beneficial effect of probiotic bacteria achieved through interaction with pathogenic bacteria is discussed.

13.2 Barriers for Pathogens to Cross

13.2.1 Gastric Acidity

The first barrier to cross following oral ingestion of pathogens is the stomach acidity. This acidity has little bactericidal effect on some pathogens like *Escherichia coli* K-12 strain or *Shigella dysentariae*, but significantly reduces the survival of more acid sensitive bacteria like *Vibrio cholerae*. For instance, while it needs 10^8 *V. cholerae* organisms as an infective dose, only 10 *Shigella dysentariae* are needed for infection. The relatively high infective dose in *V. cholerae* is due to poor acid tolerance in the stomach, which is the major physical barrier to *V. cholerae* infection. Indeed, successful colonization of the more acid sensitive *V. cholerae* depends on neutralization of stomach acid (Olivier et al. 2009). In the intestine, the microbiota ferment carbohydrate leading to production of organic acids, mainly short chain fatty acids (SCFA) like butyric acid, formic acid, and propionic acid. These acids lower the intestinal pH, creating another acidic barrier for enteric pathogens.

13.2.2 The Mucous Layer

The mucous layer overlying the intestinal epithelium acts as a non-specific defence mechanism that traps pathogens to clear them from the intestine via propulsive motility. For the pathogens to colonise the intestine, they have to cross the mucosal barrier before they reach the serosal side. This barrier consists of the mucus layer, the glycocalyx, the intestinal cells and the gut innate immune system. For pathogens to reach the glycocalyx of the epithelial cell they need to penetrate the mucus layer. The thickness of the mucus layer as well as the composition of the mucins (Koninkx et al. 1988a, b) produced by the goblet cells in the layer determines the speed by which the mucus layer is penetrated by bacteria. Penetration of the layer occurs through production of enzymes such as mucinase, glycosidase or protease mucous glycoproteins that break down the mucus. *V. cholerae* and *S. flexneri* are good examples of such pathogens. After penetration, the pathogens can attach to carbohydrate moieties on the epithelial cells that serve as binding sites. *E. coli* uses its mannose-sensitive type I fimbriae to attach to the epithelium. Depending upon the type of pathogen, attachment can be followed by elaboration of toxins or invasion.

13.2.3 Intestinal Motility

Motility of the gastrointestinal tract presents a hostile milieu to pathogens. As a result of motility, there is a continuous production of enzymes and secretions that aids digestion, but is harmful to pathogens. In addition, due to motility, enterocytes on which bacteria may adhere are constantly shed and replaced and the mucus barrier is swept away and replenished. Under these conditions, pathogens will colonize the gut with difficulties and most of them fail to do so. Colonization is considerably more successful after reduced motility of the gut (Ratnaike 1999).

13.2.4 Competition of the Microbiota

When the enteric pathogens enter the intestinal lumen, they encounter a large number of microorganisms that make the microbiota. Since the microbiota has established a balanced niche in colonizing the gut, the pathogenic bacteria must compete for binding sites and nutrition. Usually the microbiota is successful in avoiding pathogenic colonization. In most cases therefore, infectious enteric diseases develop following disruption of microbiota and impaired competition that favours pathogenic colonization.

13.2.5 Phagocytosis by Cells of the Immune System

After successful colonization along the gastrointestinal mucosa, pathogenic bacteria need to translocate into the subepithelial surrounding to cause disease. It is in this location where they encounter cells of the immune system (the antigen presenting cells) that engulf and phagocytise them. Some pathogens such as *Salmonella* have developed mechanisms to evade this action. In order to survive and multiply within macrophages, *Salmonella* blocks vesicle trafficking and endosome fusion (Uchiya 1999) leading to failure of fusion of those phagosomes containing the parasite with the lysosomes. As a result, the phagocyte is unable to deliver microbicidal compounds to *Salmonella* causing the bacteria to survive, replicate and cause disease.

13.3 Mechanisms of Enteric Diseases by Bacteria

13.3.1 Enterotoxigenic Organisms

Enterotoxigenic pathogens do not cause considerable acute intestinal inflammation or mucosal destruction. Instead they alter the absorptive and/or secretory processes of the enterocyte after adhering to the small intestinal mucosa and/or elaboration of toxins. Examples include *V. cholerae, Vibrio parahaemolyticus, Enterotoxigenic E. coli* (*ETEC*), *Clostridium perfringens, Bacillus cereus,* and *Staphylococcus aureus.* Adherence of these bacteria to intestinal epithelium is a paramount factor for their virulence. For the sake of space, only *V. cholerae*, ETEC and *C. perfringens* are described here.

Adherence to epithelium by *V. cholerae* is mainly mediated by a fimbrial colonization factor, known as the toxin-coregulated pilus (Sack et al. 2004). However, other pilus structures like the fucose-binding and mannose-binding hemagglutinins occur in different *V. cholerae* biotypes (Shogomori and Futerman 2001).

Detailed studies on ETEC infection have shown that ETEC organisms are positioned 50–100 nm from the microvilli of plasma membrane. They colonize the epithelial cell glycocalyx, a carbohydrate rich coat under the mucous layer that covers the surface of the epithelium, rather than the plasma membrane itself. A ligand-receptor interaction via protein antigens on the surface of fimbriae mediates the adherence and attachment to the mucosa (Evans et al. 1975).

Infection of the gut with *C. perfringens* leads to two different food borne diseases namely type A diarrhoea and type C human necrotic enteritis (Granum 1990). Type A infection follows ingestion of at least 10^7 *C. perfringens* spores from contaminated heat-treated food. The heat-treated food kills competing intestinal microbioata while the *C. perfringens* spores survive, vegetate and multiply to dominate the gut flora. The bacterium then produces the enterotoxin type A that causes the illness (Skelkvåle and Uemura 1977; Sarker et al. 1999). The illness can also occur following uptake of antibiotics that kill off most members of the gut microbiota to pave way to multiplication of *C. perfringens.*

C. perfringens type C disease is a very serious but rare condition (Granum 1990). In this disease, the ingested spores from contaminated food develop into vegetative form that produce β-toxin (major disease agent) as well as δ- and q- toxins (Jolivet-Reynaud et al. 1986; Granum 1990).

13.3.2 Enteropathogenic Organisms

Enteropathogenic organisms target the large bowel, mainly the distal ileum and the colon. They cause considerable acute non-invasive inflammatory reactions and mucosal destruction by secreting toxins (Giannella 2006; Navaneethan and Giannella 2008). Examples include *enteropathogenic E. coli (EPEC)*, *enterohaemorrhagic E. coli* (EHEC), entero-aggregative *E. coli* (EAEC), and *Clostridium difficile*.

EPEC causes attaching and effacing lesions on the intestinal cells. It uses a shiga-like toxin to attach. After attachment, EPEC organisms are found in a close association with the intestinal mucosa on apical surface of the epithelial cells with less than 20 nm distance between the plasma membrane and the bacterium (Donnenberg et al. 1997). The apical plasma membrane partly surrounds the attached organisms in a cup-like or pedestal-like extension on which the pathogens reside to ultimately cause severe disruption of the brush border.

EHEC illness is due to the bacterial shiga toxin produced by hemorrhagic *E. coli* 0157:H7 and O26:H11 organisms (Griffin 2002). The ingested EHEC attaches to the intestinal mucosa via a pilus. The attachment is followed by activation of protein kinase and intracellular release of calcium leading to flattening and dissolution of villi (termed attaching–effacing lesions) (Griffin 2002). In addition, EHEC secrete two verocytotoxic shiga-like toxins, STX cytotoxins I and II (Su and Brandt 1995) that cause inflammation of the small intestine.

EAEC have fimbriae that aggregate tissue culture cells. They diffusely adhere to the epithelial brush border and alter cell function of jejunal, ileal and colonic epithelium (Nataro and Kaper 1998). Adherence is mediated by the aggregative adherence fimbriae, particularly fimbria II (Okeke et al. 2000), while dispersal of EAEC and development of new foci of infection to cause diffuse adherence is via a protein dispersin (Sheikh et al. 2002). EAEC adherence leads to production of hemolysin and a heat stable enterotoxin (EAST1) similar to that of ETEC. Both toxins are linked with the development of acute inflammatory response and the associated enterocyte damage, cytokine release and intestinal secretion. The bacteria also trigger the host to produce IL-8 which attracts inflammatory cells and exacerbates further the epithelial cell destruction and fluid secretion (Huang et al. 2004).

Clostridium difficile causes pseudomembranous colitis which develops due to alteration of microbiota following uptake of broad-spectrum antibiotics such as clindamycin, cephalosporins and chinolonics. The antibiotics kill off competing microbiota in the intestine leaving behind bacteria with less competition for space and nutrients. The net effect is to permit much more extensive growth than normal of *C. difficile.* The highly proliferating *C. difficile* produces UDP-glucose hydrolases and glucosyltransferases toxins A and B, which cause intense inflammation of the colonic mucosa with fluid and electrolyte secretion (Warny et al. 2005). The toxins also regulate cytoskeletal organization and gene expression leading to disruption of protein synthesis and cell death (Warny and Kelly 2003), to activation of transcription factor nuclear factor kappa B (NF-κB) as well as mitogen activated protein (MAP) kinases and their subsequent inflammatory cytokine release (such as IL-1β, tumour necrosis factor, and IL-8). The inflammatory cytokines contribute to the marked intestinal inflammation and secretion following *C. difficile* infection (Giannella 2006).

13.3.3 Enteroinvasive Organisms

Invasive organisms to the intestinal epithelium include *Shigella, Salmonella, Yersinia enterocolitica, Campylobacter jejuni,* and *Enteroinvasive E. coli* (EIEC).

Shigella strains invade epithelial cells preferentially through M cells, which are specialized epithelial cells overlying lymphocyte-rich Peyer's patches. The organisms secrete and inject IpaA, IpaB and IpaC proteins into the host cell membranes to activate Cdc42, Rac (Van Nhieu et al. 1999) and Rho leading to depolymerization of actin filament and formation of membrane ruffles (Blocker et al. 1999). Through membrane ruffling and macropinocytosis, *Shigella* organisms enter the colonic mucosa (Adam et al. 1995; Clerc and Sansonetti 1987). The pathogen then rapidly lyses the surrounding vacuole and is released into the cytosol where it grows, divides, and moves through the cytoplasm to invade neighbouring cells without contacting the extracellular milieu (Palmer et al. 1998). *Shigella* organisms may also spread to underlying macrophages and induce apoptosis and release infective bacteria and large quantities of the proinflammatory cytokines interleukin (IL)-1 and IL-8. The produced IL-1 and IL-8 attract neutrophils into the intestinal lumen to cause inflammation and epithelial cell destruction (Fleckenstein and Kopecko 2001). In addition, the neutrophils loosen the basolateral intercellular junctions, thereby increasing the cellular invasion by the organisms (Sansonetti et al. 1999).

Similar to *Shigella*, EIEC causes a superficial mucosal invasion. The EIEC organisms are closely related to *Shigella* spp. In fact, they both (*Shigella* spp. and EIEC strains) belong to the species *E. coli* (Lan and Reeves 2002). Similar to *Shigella*, EIEC possess the virulence plasmid of about 220 kB and have ability to invade epithelial cells and disseminate from cell to cell. The organisms produce shiga-like toxin that enables penetration of the bacteria into the enterocytes to cause colitis via a mechanism similar to that of *Shigella* spp. Briefly, EIEC attach to M cells overlying lymphoid follicles, induce membrane ruffling, and enter the cells while within vacuole (Van Nhieu and Guignot 2009). After entry, the bacteria rapidly lyse the vacuolar membrane to gain access to host cell cytoplasm, multiply within the cytoplasm, induce actin polymerization at one pole, and move through the cytoplasm to infect adjacent cells.

Campylobacter jejuni invades the intestinal epithelium through the M cells underlying the Peyer's patches and via a microtubule-based entry system (Oelschlaeger et al. 1993). The invasion is followed by spread of the bacteria to adjacent cells by interacting with host invasion receptors located on the basolateral aspects of the enterocytes (Hu and Kopecko 2000).

Salmonella species exploit the host cytoskeleton during invasion resulting in dramatic morphologic changes to the cell. The changes are mediated by the *Salmonella* pathogenicity island 1 (SPI1) that directs the uptake of the bacteria by non-phagocytic cells. Once *Salmonella* is in close contact with the epithelium, the bacterial SPI1 system secretes effector protein SopE that is micro-injected into the host cell membrane. The SopE then activates host Rac and Cdc42 resulting in extensive membrane ruffling and the subsequent bacterial entry (Donnenberg 2000). The entire process occurs within minutes and when completed, *Salmonella* resides within membrane-bound vesicles, and the cytoskeleton returns to its normal distribution (Francis et al. 1993). The latter is mediated by another bacterial protein, the SptP which acts directly on Rac and Cdc42 to antagonize the effects of SopE, and thereby returning the host cell to its normal state (Fu and Galán 1999). While within the enterocytes, *Salmonella* can activate inflammatory cytokine pathways, NF-κB and MAP kinase to induce production of inflammatory IL-8.

13.4 Mechanisms of Enteric Diseases by Viruses

Several viruses infect the gut. In this chapter, we cite only norovirus and rotavirus which are frequent causative agents of gastroenteritis. Norovirus causes broadening and blunting of intestinal villi and exerts a direct action on the activity of enzymes of the brush border by diminishing the activity of intestinal disaccharidases (Musher and Musher 2004). These events lead to malabsorption and diarrhoea without inflammatory reactions. On the contrary, rotavirus induces inflammatory reactions to the gut. The virus infects mature enterocytes of the villous tip of the small intestine and produces an enterotoxin, the nonstructural rotavirus protein (NSP4) that accounts for inflammatory changes and diarrhoea.

13.5 Parasites

Several parasites are known to cause intestinal disorders. They include *Entamoeba histolytica, Giardia lamblia, Cryptosporidium*, and *Cyclospora cayetanensis.*

Ingested *Entamoeba histolytica* trophozoites adhere to the colonic mucins and epithelial cells via their galactose–*N*-acetyl-[D]-galactosamine-inhibitable surface lectin (Ravdin and Guerrant 1981). After adhesion, channel-forming peptides called amebophores are formed that render the colonic cells immobile. Subsequently, these cells are going to loose their cytoplasmic granules, structures, and eventually their nuclei resulting in cytolysis (Ravdin et al. 1980; Berninghausen and Leippe 1997). The trophozoites also disrupt the tight junction proteins causing a decrease in transepithelial resistance that leads to increased intestinal permeability. The parasite then invades the epithelial mucosa into the submucosal tissues to cause amoebic colitis. In addition, *E. histolytica* trophozoites secrete cysteine proteinases encoded by six different genes (EhCP1 to EhCP6) that digest extracellular matrix proteins to facilitate trophozoite invasion by burrowing through colonic mucus and matrix proteins into and within the submucosal tissues (Que and Reed 2000).

Following ingestion of *G. lamblia* cysts, the cysts excyst to release trophozoites that adhere to the epithelium of the upper small intestine (mainly the duodenum) and damage the mucosal brush border without invasion (Hill 1990). The exact mechanism by which *G. lamblia* causes enteritis remains obscure.

Another potential intestinal parasite is *Cryptosporidium*. Ingested oocysts of this parasite excysts in the small intestine and become located in the brush border of the small intestinal epithelial cells, mainly in the jejunum. It then attaches to an 85 kDa surface protein on intestinal epithelial cells via its apical glycoprotein CSL complex (Langer et al. 2001). When the organisms touch the epithelial membrane, the CSL complex envelops the parasite to make it intracellular but extracytoplasmic (Ryan et al. 2004). The parasite then damages the microvilli where it attaches, increases local levels of prostaglandins and the subsequent cAMP-mediated chloride secretion, inhibition of electroneutral sodium chloride and decreased water absorption by the villous cells leading to diarrhoea (Simon et al. 1994; Guarino et al. 1995; Winn et al. 2006).

Last but not least, is the intestinal parasite *Cyclospora cayetanensis* which is transmitted through ingestion of water or food contaminated with sporulated oocysts. In the gut these oocysts release their sporozoites which then proceed to invade the epithelial cells of the small intestine (Ortega et al. 1997).

13.6 Probiotic-Pathogenic Bacteria Interaction

The described mechanisms of enteric pathogens could be summarised as mediated by the ability of pathogens to overcome the intestinal acidity and the mucus layer, to adhere and attach to the epithelial cell, to secrete toxins, to disrupt the cell cytoskeleton and to cause membrane ruffling that enables pathogens to invade the enterocytes. Subsequent disruption of the lateral inter-epithelial tight junctions enables translocation of pathogens into the subepithelial tissues. Once inside the cell the bacteria trigger the production of pro-inflammatory cytokines by immune as well as epithelial cells that attract inflammatory cells to the site of infection and cause inflammatory reactions; and to inhibit intracellular phagocytosis. Most of these mechanisms are counteracted by probiotic bacteria as they interact with the pathogens. This is achieved through various ways including competitive exclusion, stabilization of cellular cytoskeleton, maintenance of barrier integrity, impairment of flagella motility, production of antimicrobial compounds, production of organic acids, co-aggregation, modulation of the immune system characterized by inhibition of pro-inflammatory cytokines that avoids excessive inflammation and tissue damage, and improvement of anti-oxidative status of enteric mucosa to inhibit translocation of intracellular pathogens.

13.6.1 Competitive Exclusion

For enteropathogens, attachment to intestinal epithelial cells represents an essential step in establishing infection. It is subsequently followed by establishment of the parasite and inflammation. Preventing this attachment could be an important approach to avoid enteric infections. One of the mechanisms by which probiotic bacteria protect the gut against pathogenic bacteria is through competition for adhesion sites and nutrition. In doing so, probiotic bacteria prevent intestinal colonization and subsequent infection by pathogenic bacteria. Different strains of probiotic bacteria vary in their effectiveness in blocking the adhesion sites for pathogens.

Specific adhesin–receptor interactions and nonspecific hydrophobic group interactions have been suggested as the major mechanisms for the adhesion of bacteria to gastrointestinal surfaces. The stereo-specific adhesin–receptor interaction involves carbohydrate moieties on the intestinal surface and carbohydrate-binding adhesins on the bacterial cell surface (Ofek and Doyle 1994). The carbohydrate moieties vary between and within individuals as well as between crypt cells and villus cells. Moreover, there might be major impact on these moieties in the brush border membrane of the intestinal cells due to variations in diets (Pusztai et al. 1995; Ovelgönne et al. 2000) which governs the composition of intestinal microbiota. A study by Lee and Puong (2002) clearly indicated a competition for adhesion between probiotic *Lactobacillus rhamnosus* GG or *Lactobacillus casei* Shirota and *E. coli* or *Salmonella* spp. (*Salmonella typhimurium, Salmonella enteritidis* and *Salmonella bellurup*) of high adhesion-receptor interaction for intestinal epithelial cells. The interference of adhesion of these gastrointestinal pathogens by probiotics was suggested to be through steric hindrance, and the degree of inhibition was related to the distribution of the adhesin receptors and hydrophobins on the intestinal cell surface.

Another study by Candela et al. (2008) demonstrated that probiotics *B. longum* Bar33, *B. lactis* Bar30, *L. acidophilus* Bar13 and *L. plantarum* Bar10 compete and effectively inhibit and displace adhesion of *S. cholerasuis* serovar Typhimurium and 40–70% for *E. coli* ETEC. Moreover, all probiotic strains tested showed the capability to adhere to the enterocyte layer previously colonized by the enteropathogenic bacteria. This behaviour attests that probiotic bacteria can be beneficial after pathogenic colonization of the gut since it allows the probiotic bacteria to exert antagonistic activities against enteropathogens, such as the competition for nutrients, reduction of luminal pH, and production of inhibitory compounds (Bernet et al. 1994). Based on this competition, it is reasonable to suggest that probiotic bacteria also compete for nutrients with pathogenic bacteria although vivid evidence is lacking. However, it has been shown that lactobacilli have very limited abilities to utilize complex carbohydrates as opposes to bifidobacteria (Amaretti et al. 2006; Walter 2008). Instead, they thus utilize simple sugars liberated from the microbial degradation of complex carbohydrates that opens specific niches in the gastrointestinal tract that are also occupied by pathogenic enterobacteria. This predisposes probiotic lactobacilli to high nutritional competition with enteric pathogens. In fact, several studies have observed that enterobacteria could use prebiotics as a substrate for their growth (Hartemink et al. 1997; Martín-Peláez et al. 2008). On the other hand, some lactobacilli such as *L. acidophilus* and *L. plantarum* are capable of metabolising complex carbohydrates like fructans that widen their niche for colonisation in the gut (Walter 2008). Bifidobacteria have the ability to utilize a variety of plant dietary fibres due to their ability to produce several cell-bound and extracellular depolymerizing enzymes (Rossi et al. 2005). This diversity could enable the probiotic bacteria to diffusely colonize the gut and inhibit infections by enteric pathogens.

13.6.2 Stabilization of the Cellular Cytoskeleton

During invasion, invasive parasites like *Salmonella, Shigella*, EIEC, *Y. enterocolitica,* and *C. jejuni* exploit the host cytoskeleton resulting in dramatic morphologic changes to the cell. We will cite *Salmonella* as an example on how probiotic bacteria protect the host epithelial cell against disrupted cytoskeleton. As described earlier, for *Salmonella,* the epithelial cytoskeletal changes are mediated by the SPI1 that directs the uptake of the bacteria by non-phagocytic cells. Once *Salmonella* is in close contact with the epithelium, the bacterial SPI1 system secretes effector protein SopE that is microinjected into the host cell membrane. The SopE then activates host Rac and Cdc42 resulting in extensive membrane ruffling at the site of entry on apical epithelial surface and the subsequent bacterial entry (Donnenberg 2000). At these sites, a rearrangement of actin filaments takes place. Both *Salmonella*-induced ruffles and the subsequent entry of bacteria are sensitive to inhibitors of actin filament polymerization (Francis et al. 1993). Actin disruption of polarized intestinal epithelial cells can augment internalization of bacteria (Wells et al. 1998).

Several studies have emphasized the relationship between the intestinal epithelial cytoskeleton and heat shock proteins, (Hsps). For instance, induction of Chinese hamster Hsp27 gene expression in mouse NIH/3T3 cells prevents actin depolymerization during acute exposure to cytochalasin (Lavoie et al. 1993). In human colonic epithelial Caco-2/bbe cells, Hsp72 protects the integrity of the actin cytoskeleton against oxidant-induced injury (Musch et al. 1999). Liang and MacRae (1997) concluded in their review that Hsp60, Hsp70, Hsp90, and Hsp100 have different but cooperative roles in the formation and function of the eukaryotic cell cytoskeleton. Studies on the function of Hsp70 and Hsp90 revealed the actin-binding activity of these proteins (Liang and MacRae 1997), which stabilizes the actin filaments by cross-linking (Koyasu et al. 1986). With respect to Hsp90, this protein was found to bind to at most 10 actin molecules in the polymerized form (Nishida et al. 1986), and its localization in membrane ruffles was revealed by immunofluorescence staining using specific antisera (Koyasu et al. 1986). We have proposed earlier that intestinal epithelial cells could be protected against *S. enteritidis* invasion from previously induced high levels of Hsps (Malago et al. 2003). In several other studies, different probiotic bacteria have been shown to stimulate intestinal epithelial cells to induce production of protective Hsps (Nemeth et al. 2006, Malago et al. 2010). Recently, we have observed the potency of probiotic *Bifidobacterium infantis, Lac-*

tobacillus casei and *Lactococcus lactis* to induce Hsp70 and protect cells against *Salmonella*-induced infection (Malago et al. 2010). In part, the protective effect of probiotic bacteria against *Salmonella* could be via stabilisation of the apical cytoskeleton to prevent membrane ruffling and thus impede bacterial invasion. References have also shown that *L. acidophilus* LB antagonizes cytoskeleton rearrangements by virulent *E. coli* and prevents cellular damage. In one study for instance, Liévin-Le et al. (2002) examined the effect of *L. acidophilus* LB on *E. coli* (Afa/Dr DAEC) C1845 strain-induced cellular injuries in human intestinal Caco-2 cells and observed a protective effect against the brush border lesions promoted by the pathogen. The protection was due to alterations in expression of F-actin, sucrase-isomaltase, dipeptidylpeptidase IV, alkaline phosphatase, and fructose transporter induced by the lactobacillus.

13.6.3 Stabilization of the Barrier Integrity

Several studies have shown that some enteric pathogens including *Salmonella* species, *Escherichia coli, Clostridium difficile, Vibrio cholera*, and rotavirus disrupt the integrity of the epithelial barrier leading to increased paracellular permeability. They target and alter the interepithelial tight junction proteins namely the occludin, claudin, and junctional adhesion protein (Gonzalez-Mariscal et al. 2003). Occludin and claudin interact with cytoplasmic plaques that consist of different types of cytosolic proteins, such as zonula-occludin proteins ZO-1, ZO-2, and ZO-3. Some of the disturbances to these proteins include alterations in occludin, claudin-1, JAM-1, and ZO-1 distribution from the tight junctions into the cytosol following *E. coli* infection (Simonovic et al. 2000), dissociation of occludin, ZO-1, and ZO-2 from the lateral tight junctions membrane caused by *Clostridium difficile* (Nusrat et al. 2001), delocalization of claudin-1 and occludin by *Shigella dysenteria*, disassembling of intercellular tight junctions via interaction with cell membrane receptors in *Vibrio cholera* infections (Wang et al. 2000), and paracellular leak and F-actin alteration induced by rotavirus (Tafazoli et al. 2001). Several studies have shown an inhibitory effect upon these disturbances by probiotic bacteria. *E. coli* caused disturbances are inhibited by *Lactobacillus acidophilus* (Resta-Lenert and Barrett 2003), *L. plantarum* (Qin et al. 2009), and *L. casei* (Parassol et al. 2005), whereas probiotic *L. rhamnosus* and *L. acidophilus* antagonise the *Shigella dysenteria*-mediated tight junction disruption (Moorthy et al. 2009). Probiotic mixture VSL#3 (*L. casei, L. plantarum, L. acidophilus, L. delbrueckii* subsp. *bulgaricus, B. longum, Bifidobacterium infantis, B. breve* and *Streptococcus salivarius* subsp. *thermophilus*) has also been shown to protect the intestinal epithelial barrier against acute colitis by preventing redistribution of tight junction proteins occludin, ZO-1, and claudin-1, 3, 4, and 5 (Mennigen and Bruewer 2009). These studies, together with several others, affirm that probiotic bacteria interact with pathogenic bacteria to prevent disruption of tight junctions induced by the pathogens. In doing so, they prevent increased paracellular permeability, translocation of the pathogens and the resulting pathology.

13.6.4 Impairment of Flagella Motility

Another mechanism utilized by probiotic bacteria to prevent pathogens from intestinal colonization is impairment of flagella motility which results in preclusion of bacterial internalization. Recently, Pai and Kang (2008) demonstrated that *L. johnsonii* La1 and *L. casei* inhibit flagella motility and prevent internalization of *S. enterica* serovar *typhimurium* to the epithelial cell. A similar mechanism could apply for other enteric pathogens that use flagella for pathogenicity.

13.6.5 Production of Antimicrobial Compounds

13.6.5.1 Compounds Inhibiting Pathogen Adhesion

Adherence to intestinal epithelial cells by pathogenic bacteria is pivotal to their colonization and virulence. Probiotic bacteria are known to produce compounds that inhibit adhesion of pathogenic bacteria to the intestinal epithelium and thus prevent their establishment and development of disease. Studies done by Fujiwara et al. (1997, 2001) have characterized a proteinaceous compound named Bacillus inhibitory factor (BIF) from the culture supernatant of *Bacillus longum* BL 2928 to inhibit adhesion of Gram-negative bacteria. The BIF interferes with Enterobacteriaceae binding to the glycolipid binding receptor, gangliotetrasylceramide (GA1), sites on the intestinal brush border membrane. Its inhibitory activity increases with cultivation time reaching a maximum after 72 hours. Specifically, BIF inhibits adhesion of *ETEC* strain Pb176 expressing the colonization factor adhesion II to the GA1 molecule. It is likely that by preventing binding of pathogens to intestinal epithelial surfaces, BIF could work as a physiological component in the intestinal tract to protect individuals against enteric infections (Fujiwara et al. 2001).

Another compound of about 3,500 Da produced by two bifidobacterial strains (CA1 and F9) strongly inhibit adhesion of *S. typhimurium* SL1344 and *E. coli* C1845 (Liévin et al. 2000). The characteristics of this compound resemble those produced by several lactobacilli strains including *L. casei rhamnosus* GG, *L. johnsonii* LA1, and *L. acidophilus* LB which inhibit *S. typhimurium* infection both *in vitro* and *in vivo* (Cheikhyoussef et al. 2008). Several studies have shown the ability of *B. lactis* to inhibit adhesion of enteric pathogens to intestinal epithelial cells. Gopal et al. (2001) reported that *B. lactis* DR10 adheres to intestinal epithelial cells and inhibits colonization by *E. coli* O157:H7 by producing proteinaceous substances. Furthermore, Matsumoto et al. (2002) observed inhibition of *C. perfringens* adhesion by *B. lactis* LKM 512 through adhering to the intestinal mucin. The ability of *B. bifidum* RBL 71 and *B. bifidum* RBL 460 to reduce adhesion of *E. coli* O157:H7 to intestinal epithelial cells has also been reported (Cheikhyoussef et al. 2008). These bacteria and some other probiotic bifidobacteria and lactobacilli strains could mediate their effects against pathogenic bacteria adhesion through secretion of lectin-like bacteriocins and lectin-like complexes following stimulation by cell-surface adhesion factors such as lectin/adhesin proteins of S-layers (Kravtsov et al. 2008).

13.6.5.2 Bacteriocins

Production of bacteriocins by probiotic bacteria is another way through which inhibition of pathogenic colonisation and occurrence of enteric disease occurs. The bacteriocins may limit colonization by altering the intestinal environment or may directly kill the pathogens. *Bifidobacterium bifidum* NCFB produces bifidocin B that is dependent on the bacterial cell number. The bacteriocin inhibits the growth of selected species of the genera *Listeria, Bacillus, Enterococcus, Lactobacillus, Leuconostoc* and *Pediococcus*, but is not active against other Gram-positive and most Gram-negative bacteria (Yildirim and Johnson 1998). Purification studies have shown that bifidocin B does not adsorb to the Gram-negative bacteria, but adsorbs to the Gram-positive bacteria. The difference in adsorption between Gram-negative and Gram-positive bacteria is due to their cell wall/membrane composition, since addition of purified lipoteichoic acid (LTA) to the cells completely blocks the adsorption of bifidocin B, suggesting a possible involvement of the cell wall (LTA), present in Gram-positive bacteria but absent in gram-negative bacteria (Yildirim et al. 1999).

Other studies have shown that *Enterococcus faecium* CCM7420 and *E. faecium* NRRL B-30746 produce bacteriocins that reduce the number of *Staphylococci* spp. colonising the intestine to protect the gut against the pathogen (Simonova et al. 2009). In addition, a bacteriocin E 50-52 produced by *E. faecium* NRRL B-30746 is effective against methicillin-resistant *S. aureus* (Svetoch et al. 2008). The latter effect is also exhibited by bacteriocin OR-7 produced by *Lactobacillus salivarius* NRRL B-30514. Another strain, *L. salivarius* NRRL B-3501 also produces the bacteriocin OR-7 and eliminates *Campylobacter coli* in turkeys and *C. jejuni* in chicken intestines. This pathogenic elimination results from bacteriocin-induced reduction in crypt depth and the number of goblet cells that lead to reduction in mucin production and composition which subsequently limits colonization of the pathogen (Cole et al. 2006; Stern et al. 2006). Another study by Corr et al. (2007) demonstrated that *L. salivarius* UCC118 produces bacteriocin Abp 118 that limits epithelial infection with *Listeria monocytogenes* in mice to protect the host. A similar inhibition of this pathogen is exhibited by pediocin AcH from *L. plantarum* (Bernbom et al. 2006).

Bacteriocins bifilact Bb-12 and bifilong Bb-46 produced by *B. lactis* Bb-12 and *B. longum* Bb-46, respectively, inhibit microbial activities of *Staphylococcus aureus, Salmonella typhimurium, Bacillus cereus* and *E. coli* in a species- and pH-dependent manner. For instance, the minimal inhibition concentration (MIC) of bifilact Bb-12 and bifilong Bb-46 against *E. coli* has been found to be 20 and 16 mg/ml, respectively, whereas that for *S. aureus* is 40 and 20 mg/ml, respectively. Further, both bacteriocins lose their activities at pH 3 and pH 9 but work optimally at a pH between 4 and 7 (Abd El-Salam et al. 2004). Studies by Fayol-Messaoudi et al. (2005) have shown that several bacteriocins from lactobacilli probiotics such as *L. johnsonii, L. rhamnosus, L. casei* Shirota, and *L. rhamnosus* very effectively kill *Salmonella enterica* serovar *typhimurium* thereby preventing intestinal infection and pathology.

In summary, probiotic bacteria use their bacteriocins to effectively interact with pathogenic bacteria by killing them or suppressing their growth, which then results in a limited colonization of the gut by pathogenic bacteria. In doing so, they protect the host against enteric infections.

13.6.5.3 Bacteriocin-Like Compounds

Bacteriocin-like compounds are antagonistic substances that are incompletely defined or do not fit the typical criteria defining bacteriocins and tend to have a broader spectrum of activity. They are identified on the basis that their inhibitory activities are not solely dependent on acid or hydrogen peroxide production, but on other chemical compounds present in culture supernatants. Meghrous et al. (1990) identified one such chemical produced by a strain of *Bifidobacterium.* This chemical was of a protein type, heat-stable and active at pH values ranging from 2 to 10. It displayed inhibitory activity against lactic streptococci strains and other species of bifidobacteria, *Clostridium* and *Lactobacillus,* but showed no activity towards Gram-negative bacteria. *B. infantis* has been shown to produce antimicrobial compound(s) that inhibit the effects of *E. coli* or *Clostridium perfringens* (Gibson and Wang 1994). A study by O'Riordan and Fitzgerald (1998) identified twelve strains of *Bifidobacterium* with a broad spectrum of antagonistic activity against both Gram-positive and Gram-negative pathogens. Several other *Bifidobacterium* species produce bacteriocin-like substances. *B. animalis* 31 and *B. breve* J show inhibitory activities to *Salmonella enteritidis* (Bielecka et al. 1998), *B. longum* and *B. infantis* to *B. cereus* and *E. coli* (Cheikhyoussef et al. 2008), *B. infantis* ATCC 15697 to *L. monocytogenes, B. infantis* and *L. salivarius* to *C. difficile* (Lee et al. 2003), and *B. longum* BB 536 to *E. coli* C1845 (Makras and De Vuyst (2006). Our own studies have shown inhibition of *Salmonella enteritidis* 857 growth by culture supernatants of non-starter lactobacilli *Lactobacillus casei* subsp. *casei* 2756, *Lactobacillus curvatus* 2775, and *Lactobacillus plantarum* 2142 in enterocyte-like Caco-2 cells (Nemeth et al. 2006).

13.6.5.4 Siderophores

Siderophores are organic compounds (e.g. polypeptides) secreted by micro-organisms that inhibit the growth of other microorganisms by depriving them of iron. They do so by reducing the amount of free iron available to micro-organisms. O'Sullivan (2004) isolated a siderophore from *Bifidobacterium* strains that inhibits growth of *L. lactis, C. difficile,* and *C. perfringens.*

13.6.6 Production of Organic Acids

Fermentation of carbohydrates in the colon by the intestinal microbioata and probiotics leads to production of intermediary products (ethanol and formic, succinic and lactic acids) as well as short-chain fatty acids (SCFA) (acetate, propionate, and butyrate) (Malago et al. 2003). The intermediary products are converted to produce CO_2 and H_2 that are rapidly absorbed, whereas the SCFA can be absorbed from the colon to serve as source of energy to the colonocytes. Of these products, butyrate forms the largest proportion and is pivotal to the probiotic-pathogenic bacteria interaction. Butyrate affects the intestinal immunity in various ways. It promotes intestinal immunity through induction of epithelial cell proliferation and differentiation, thereby enhancing barrier integrity which hampers epithelial invasion by enteric pathogens. Additionally, Peng et al. (2009) suggested that butyrate enhances the intestinal barrier by regulating the assembly of tight junction proteins. Butyrate also inhibits histone deacetylase activity, resulting in hyperacetylation of histones, and as a consequence, suppression of NF-κB activation. This antagonizes the pathogen-induced activation of inflammatory cytokine production by epithelial cells that would otherwise cause chronic inflammation and tissue damage.

Some authors have proposed that butyrate reinforces the colonic defence barrier by increasing production of mucins and antimicrobial peptides. The produced mucins form a physicochemical barrier that protects epithelial cells and limits microbial adhesion and subsequent invasion, whereas antimicrobial peptides eliminate pathogenic bacteria in the gut. These effects are exemplified by *Lactobacillus plantarum* 299v and *L. rhamnosus* GG that induce mucin gene (MUC-2) expression to inhibit pathogenic *Escherichia coli* (Mack et al. 1999). Raqib et al. (2006) demonstrated that butyrate induces antimicrobial peptide cathelicidin LL-37 that eliminates *Shigella* spp. in the gut and protects the host against infection. Butyrate also decreases intestinal epithelial permeability by increasing the expression of tight junction proteins, thereby preventing bacterial translocation into subepithelial location to cause inflammation. Therefore, it will be highly likely that butyrate-producing probiotic bacteria like *Butyrivibrio fibrisolvens* (Asanuma et al. 2001), *Butyrivibrio hungatei* and *Pseudobutyrivibrio xylanivorans* (Kopecný et al. 2003) will be very effective in enteric cytoprotection.

Production of acids in the intestine lowers the local pH, rendering the intestinal milieu acidic. This effect eliminates pathogenic bacteria from gaining access to the gut. Carey et al. (2008) observed that co-incubation of Shiga toxigenic *Escherichia coli* with different strains of *Lactobacillus, Pediococcus* and *Bifidobacterium* species resulted in specific down-regulation of Stx gene expression and inhibition of growth of the pathogen. These effects were attributable to production of organic acids by the probiotic species which lowered the local intestinal pH. Several other authors have suggested that production of organic acids by probiotics and the subsequent decrease in the gut pH create more favourable ecological conditions for the resident microbiota and decrease the risk of pathogen colonisation (Fuller 1989; Ballongue 1998; Servin 2004).

13.6.7 Co-aggregating

Probiotic bacteria may also elicit anti-pathogenic effects by co-aggregating with the pathogens (Collado et al. 2007).

13.6.8 Modulation of the Immune System

Another mechanism by which probiotic bacteria interact with intestinal pathogens is through antagonising pathogenic inflammatory responses. This involves interference of inflammatory pathways, particularly NF-κB and MAP kinase. Pathogenic bacteria such as *Salmonella* species activate these pathways to induce production of pro-inflammatory cytokines, including chemokines (e.g. IL-8) that attract inflammatory cells (e.g. neutrophils) to the site of infection resulting in inflammation. Probiotic bacteria suppress these cytokines and avoid excessive inflammation. We have shown a significant suppression of *Salmonella enteritidis* 857 induced IL-8 levels by *Bifidobacterium infantis* W52, *Lactobacillus casei* W56 and *Lactococcus lactis* W58 in both co- and pre-incubation exposures (Malago et al. 2010). Our studies also suggested that the anti-inflammatory role of probiotic bacteria is mediated, at least in part, by Hsps. A significant time-dependent increase in Hsp70 expression was obvious after exposure to *Bifidobacterium infantis* W52, *Lactobacillus casei* W56, and *Lactococcus lactis* W58 (Malago et al. 2010). These Hsps mediate their action by selectively suppressing the pathogen-induced activation of NF-κB and MAPK at various steps (Malago et al. 2002). Another study showed that *Lactococcus casei* attenuates the pro-inflammatory signalling induced by invasion of the intestinal mucosa by *Shigella flexneri* by inhibition of the NF-κB signalling pathway (Tien et al. 2006). There could be mechanisms other than induction of Hsp expression by which probiotic bacteria inhibit the pro-inflammatory cytokine production. The detailed mechanisms of the immunomodulatory role of probiotic bacteria are discussed in a separate chapter.

13.6.9 Improvement of the Enteric Mucosal Anti-oxidative Status

Studies have shown that after intracellular infection phagocytes produce reactive oxygen species (ROS) which is important in killing the pathogen. However, excessive amounts of ROS damage the collateral intestinal epithelial cells leading to inflammation and extra-intestinal translocation of pathogens. Neutralizing these chemical products by superoxide dismutase from probiotic *L. fermentum* has been shown to be the mechanism through which *L. fermentum* prevents granulomatous lesions in the liver (Kullisaar et al. 2002). In another study Truusalu et al. (2008) showed that treatment of mice with *L. fermentum* reduces the translocation and invasion of *Salmonella enterica* serotype *typhimurium* into other body organs as shown by a decrease in the number of viable *S. enterica* serotype *typhimurium* isolated from the ileum, blood and the liver. This effect was slightly more potent than that induced by the antibiotic ofloxacin, a fluoroquinolone drug advocated for treating typhoid fever. Compared to untreated and ofloxacin-treated mice the anti-oxidative status of the gut mucosa indicated by a significant reduction in oxidative stress indices, particularly lipid peroxides and glutathione redox ratio, appeared to be significantly elevated in *L. fermentum*-treated mice.

13.7 Conclusion

In a healthy gut, the microbiota play a great role in preventing establishment of pathogens by effectively interacting with invading pathogen. They do this in various ways ranging from physical interaction among the micro-organisms to produced chemical products. Further, the interaction occurs at both luminal or intracellular (enterocytes) levels as the microbiota antagonise the deleterious effects of pathogenic bacteria. There are enough studies supporting the fact that disruption of the microbiota (e.g. by oral antibiotics) leads to enteric diseases partly due to inefficient interaction between the microbiota and the pathogenic micro-organisms. The disruption could involve reduction in numbers of microbiota or specific group or strain of microbiota. The latter is accounted for by the fact that individual members of the microbiota may elicit different effect following interaction with pathogens. As a result, selection of probiotics to institute the effect of microbiota in treating enteric disorders should be done intelligently and selectively to meet the specific desired effect of the probiotics. This means the species or strain, the number, and proportion of a particular probiotic bacterium in any preparation mixture is critical for optimal protective effect. To achieve this, future studies, both *in vitro* and *in vivo* are needed to explore the optimal balance of probiotic bacteria that will effectively interact with specific pathogens to protect against enteric disorders. *In vivo* studies involving larger cohorts of volunteers and accurate/representative placebo controls could be preferred to truly establish the effectiveness of probiotics in treating and preventing infectious enteric diseases.

References

Abd El-Salam MH, Saleh FA, Kholif AM, El-Sayed EM, Abdou SM, El-Shibiny S (2004) Isolation and characterization of bacteriocins produced by Bifidobacterium lactis BB-12 and Bifidobacterium longum BB-46. 9th Egyptian conference for dairy science and technology, Cairo, Egypt, 9–11 Oct

Adam T, Arpin M, Prevost MC, Gounon P, Sansonetti PJ (1995) Cytoskeletal rearrangement and the functional role of T-plastin during entry of *Shigella flexneri* into HeLa cells. J Cell Biol 129:367–381

Amaretti A, Tamburini E, Bernardi T, Pompei A, Zanoni S, Vaccari G, Matteuzzi D, Rossi M (2006) Substrate preference of *Bifidobacterium adolescentis* MB 239: compared growth on single and mixed carbohydrates. Appl Microbiol Biotechnol 73:654–662

Asanuma N, Kawato M, Hino T (2001) Presence of *Butyrivibrio fibrisolvens* in the digestive tract of dogs and cats, and its contribution to butyrate production. J Gen Appl Microbiol 47:313–319

Ballongue J (1998) Bifidobacteria and probiotic action. In: Salminen S, Von Wright A (eds) Lactic acid bacteria: microbiology and functional aspects. Marcel Dekker, New York, pp 519–587

Bernbom N, Licht TR, Saadbye P, Vogensen FK, Norrung B (2006) *Lactobacillus plantarum* inhibits growth of *Listeria monocytogenes* in an *in vitro* continuous flow gut model, but promotes invasion of *L. monocytogenes* in the gut of gnotobiotic rats. Int J Food Microbiol 108:10–14

Bernet MF, Brassart D, Neeser JR, Servin AL (1994) Lactobacillus acidophilus LA 1 binds to cultured human intestinal cell lines and inhibits cell attachment and cell invasion by enterovirulent bacteria. Gut 35:483–489

Berninghausen O, Leippe M (1997) Necrosis versus apoptosis as the mechanism of target cell death induced by *Entamoeba histolytica.* Infect Immun 65:3615–3621

Bielecka M, Biedrzycka E, Biedrzycka E, Smoragiewicz W, Smieszek M (1998) Interaction of Bifidobacterium and Salmonella during associated growth. Int J Food Microbiol 45:151–155

Blocker A, Gounon P, Larquet E, Niebuhr K, Cabiaux V, Parsot C, Sansonetti P (1999) The tripartite type III secreton of *Shigella flexneri* inserts IpaB and IpaC into host membranes. J Cell Biol 147:683–693

Candela M, Perna F, Carnevali P, Vitali B, Ciati R, Gionchetti P, Rizzello F, Campieri M, Brigidi P (2008) Interaction of probiotic Lactobacillus and Bifidobacterium strains with human intestinal epithelial cells: adhesion properties, competition against enteropathogens and modulation of IL-8 production. Int J Food Microbiol 125:286–292

Cani PD, Possemiers S, Van de Wiele T, Guiot Y, Everard A, Rottier O, Geurts L, Naslain D, Nevrinck A, Lambert DM, Muccioli GG, Delzenne NM (2009) Changes in gut microbiota control inflammation in obese mice through a mechanism involving GLP-2-driven improvement of gut permeability. Gut 58:1091–1103

Carey CM, Kostrzynska M, Ojha S, Thompson S (2008) The effect of probiotics and organic acids on Shiga-toxin 2 gene expression in enterohemorrhagic *Escherichia coli* O157:H7. J Microbiol Methods 73:125–132

Cheikhyoussef A, Pogori N, Chen W, Zhang H (2008) Antimicrobial proteinaceous compounds obtained from bifidobacteria: from production to their application. Int J Food Microbiol 125:215–222

Clerc P, Sansonetti PJ (1987) Entry of *Shigella flexneri* into HeLa cells: evidence for directed phagocytosis involving actin polymerization and myosin accumulation. Infect Immun 55:2681–2688

Cole K, Farnell MB, Donoghue AM, Stern NJ, Svetoch EA, Eruslanov BN, Volodina LI, Kovalev YN, Perelygin VV, Mitsevich EV, Mitsevich IP, Levchuk VP, Pokhilenko VD, Borzenkov VN, Svetoch OE, Kudryavtseva TY, Reyes-Herrera I, Blore PJ, De Los Santos FS, Donoghue DJ (2006) Bacteriocins reduce *Campylobacter* colonization and alter gut morphology in turkey poults. Poultry Sci 85:1570–1575

Collado MC, Meriluoto J, Salminen S (2007) Role of commercial probiotic strains against human pathogen adhesion to intestinal mucus. Lett Appl Microbiol 45:454–460

Corr SC, Li Y, Riedel CU, O'toole PW, Hill C, Gahan CGM (2007) Bacteriocin production as a mechanism for the antiinfective activity of *Lactobacillus salivarius* UCC118. Proc Natl Acad Sci U S A 104:7617–7621

Donnenberg MS (2000) Pathogenic strategies of enteric bacteria. Nature 406:768–774

Donnenberg MS, Kaper JB, Finlay BB (1997) Interactions between enteropathogenic *Escherichia coli* and host epithelial cells. Trends Microbiol 5:109–114

Evans DG, Silver RP, Evans DJ Jr, Chase DG, Gorbach SL (1975) Plasmid-controlled colonization factor associated with virulence in *Escherichia coli* enterotoxigenic for humans. Infect Immun 12:656–667

FAO/WHO (2002) Joint working group report on guidelines for the evaluation of probiotics in food, London, Ontario, Canada, April 30 and May 1. http:www.who.int/foodsafety/publications/fs_management/probiotics2/en/index html
Fayol-Messaoudi D, Berger CN, Coconnier-Polter MH, Liévin-Le Moal V, Servin AL (2005) pH-, lactic acid-, and non-lactic acid-dependent activities of probiotic lactobacilli against *Salmonella enterica* serovar *Typhimurium*. Appl Environ Microbiol 71:6008–6013
Fleckenstein JM, Kopecko DJ (2001) Breaching the mucosal barrier by stealth: an emerging pathogenic mechanism for enteroadherent bacterial pathogens. J Clin Invest 107:27–30
Francis CL, Ryan TA, Jones BD, Smith SJ, Falkow S (1993) Ruffles induced by *Salmonella* and other stimuli direct macropinocytosis of bacteria. Nature 364:639–642
Fu YX, Galán JE (1999) A *Salmonella* protein antagonizes Rac-1 and Cdc42 to mediate host-cell recovery after bacterial invasion. Nature 401:293–297
Fujiwara S, Hashiba H, Hirota T, Forstner JF (1997) Proteinaceous factor(s) in culture supernatant fluids of bifidobacteria which prevents the binding of enterotoxigenic Escherichia coli to gangliotetraosylceramide. Appl Environ Microbiol 63:506–512
Fujiwara S, Hashiba H, Hirota T, Forstner JF (2001) Inhibition of the binding of enterotoxigenic Escherichia coli Pb176 to human intestinal epithelial cell line HCT-8 by an extracellular protein fraction containing BIF of Bifidobacterium longum SBT2928: suggestive evidence of blocking of the binding receptor gangliotetraosylceramide on the cell surface. Int J Food Microbiol 67:97–106
Fuller R (1989) Probiotics in man and animals. J Appl Bacteriol 66:365–378
Giannella RA (2006) Infectious enteritis and proctocolitis and food poisoning. In: Feldman M (ed) Sleisenger & Fordtran's gastrointestinal and liver disease, 8th edn. WB Saunders, Philadelphia, pp 2333–2391
Gibson GR, Wang X (1994) Regulatory effects of bifidobacteria on the growth of othercolonic bacteria. J Appl Bacteriol 77:412–420
González-Mariscal L, Betanzos A, Nava P, Jaramillo BE (2003) Tight junction proteins. Prog Biophys Mol Biol 81:1–44
Gopal PK, Prasad J, Smart J, Gill HS (2001) In vitro adherence properties of Lactobacillusrhamnosus DR20 and Bifidobacterium lactis DR10 strains and theirantagonistic activity against an enterotoxigenic Escherichia coli. Int J Food Microbiol 67:207–216
Granum PE (1990) *Clostridium perfringens* toxins involved in food poisoning. Int J Food Microbiol 10:101–112
Griffin PM (2002) *Escherichia coli* O157:H7 and other enterohemorrhagic *E. coli*. In: Blaser M (ed) Infections of the GI tract, 2nd edn. Lippincott Williams & Wilkins, Philadelphia, pp 627–642
Guarino A, Canani RB, Casola A, Pozio E, Russo R, Bruzzese E, Fontana M, Rubino A (1995) Human intestinal cryptosporidiosis: secretory diarrhoea and enterotoxic activity in Caco-2 cells. J Infect Dis 171:976–983
Hartemink R, Van Laere KM, Rombouts FM (1997) Growth of enterobacteria on fructo-oligosaccharides. J Appl Microbiol 83:367–374
Hill DR (1990) Giardia lamblia. In: Mandell GL et al (eds) Principles and practice of infectious diseases. Churchill Livingstone, Philadelphia, pp 2487–2493
Hu L, Kopecko DJ (2000) Interactions of *Campylobacter* with eukaryotic cells: gut luminal colonization and mucosal invasion mechanisms. In: Nachamkin I, Blaser MJ (eds) Campylobacter, 2nd edn. ASM Press, Washington, pp 191–205
Huang DB, Okhuysen PC, Jiang ZD, DuPont HL (2004) Enteroaggregative *Escherichia coli:* an emerging enteric pathogen. Am J Gastroenterol 99:383–389
Jolivet-Reynaud C, Popoff MR, Vinit MA, Ravisse P, Moreau H, Alouf JE (1986) Enteropathogenicity of Clostridium perfringens b-toxin and other clostridial toxins. Zb Bacteriol Microbiol Hyg Suppl 15:145–151
Koninkx JFJG, Stemerdink AFE, Mirck MH, Egberts HJA, van Dijk JE, Mouwen JVJM (1988a) Histochemical changes in the composition of mucins in goblet cells during methotrexate-induced atrophy in rats. Exp Pathol 34:125–133

Koninkx JFJG, Mirck MH, Hendriks HGCJM, Mouwen JMVM, van Dijk JE (1988b) Nippostrongylus brasiliensis: Histochemical changes in the composition of mucins in goblet cells during infection in rats. Exp Parasitol 65:84–90

Kopecný J, Zorec M, Mrázek J, Kobayashi Y, Marinsek-Logar R (2003) *Butyrivibrio hungatei* sp. nov. and *Pseudobutyrivibrio xylanivorans* sp. nov., butyrate-producing bacteria from the rumen. Int J Syst Evol Microbiol 53:201–209

Koyasu S, Nishida E, Kadowaki T, Matsuzaki F, Iida K, Harada F, Kasuga M, Sakai H, Yahara I (1986) Two mammalian heat shock proteins, HSP90 and HSP100, are actin binding proteins. Proc Natl Acad Sci U S A 83:8054–8058

Kravtsov EG, Yermolavev AV, Anokhina IV, Yashina NV, Chesnokova VL, Dalin MV (2008) Adhesion characteristics of lactobacillus is a criterion of the probiotic choice. Bull Exp Biol Med 145:232–234

Kullisaar T, Zilmer M, Mikelsaar M, Vihalemm T, Annuk H, Kairane C, Kilk A (2002) Two antioxidative lactobacilli strains as promising probiotics. Int J Food Microbiol 72:215–224

Lan RT, Reeves PR (2002) Escherichia coli in disguise: molecular origins of *Shigella*. Microb Infect 4:1125–1132

Langer RC, Schaefer DA, Riggs MW (2001) Characterization of an intestinal epithelial cell receptor recognized by the *Cryptosporidium parvum* sporozoite ligand CSL. Infect Immun 69:1661–1670

Lavoie JN, Gingras-Breton G, Tanguy RM, Landry J (1993) Induction of Chinese hamster HSP27 gene expression in mouse cells confers resistance to heat shock. HSP27 stabilization of the microfilament organization. J Biol Chem 268:3420–3429

Lee YK, Puong KY (2002) Competition for adhesion between probiotics and human gastrointestinal pathogens in the presence of carbohydrate. Br J Nutr 88(Suppl 1):S101–S108

Lee YJ, Yu WK, Heo TR (2003) Identification and screening for antimicrobial activity against Clostridium difficile of Bifidobacterium and Lactobacillus species isolatedfrom healthy infant faeces. Int J Antimicrob Agents 21:340–346

Liang P, MacRae TH (1997) Molecular chaperones and the cytoskeleton. J Cell Sci 110:1431–1440

Liévin V, Peiffer I, Hudault S, Rochat F, Brassart D, Neeser JR, Servin AL (2000) Bifidobacterium strains from resident infant human gastrointestinal microflora exert antimicrobial activity. Gut 47:646–652

Liévin-Le Moal V, Amsellem R, Servin AL, Coconnier MH (2002) Lactobacillus acidophilus (strain LB) from the resident adult human gastrointestinal microflora exerts activity against brush border damage promoted by a diarrhoeagenic *Escherichia coli* in human enterocyte-like cells. Gut 50:803–811

Maassen CB, van Holten-Neelen C, Balk F, den Bak-Glashouwer MJ, Leer RJ, Laman JD, Boersma WJ, Claassen E (2000) Strain-dependent induction of cytokine profiles in the gut by orally administered Lactobacillus strains. Vaccine 18:2613–2623

Mack DR, Michail S, Wei S, McDougall L, Hollingsworth MA (1999) Probiotics inhibit enteropathogenic *E. coli* adherence in vitro by intestinal mucin gene expression. Am J Physiol Gastrointest Liver Physiol 276:941–950

Makras L, De Vuyst L (2006) The in vitro inhibition of Gram-negative pathogenic bacteria by bifidobacteria is caused by the production of organic acids. Int Dairy J 16:1049–1057

Malago JJ, Koninkx JFJG, van Dijk JE (2002) The heat shock response and cytoprotection of the intestinal epithelium. Cell Stress Chaperones 7:191–199

Malago JJ, Koninkx JFJG, Ovelgönne HH, van Asten FJAM, Swennenhuis JF, van Dijk JE (2003) Expression levels of heat shock proteins in enterocyte-like Caco-2 cells after exposure to *Salmonella enteritidis.* Cell Stress Chaperones 8:194–203

Malago JJ, Tooten PCJ, Koninkx JFJG (2010) Anti-inflammatory properties of probiotic bacteria on *Salmonella*-induced IL-8 synthesis in enterocyte-like Caco-2 cells. Benef Microbe 1:121–130

Martín-Peláez S, Gibson GR, Martín-Orúe SM, Klinder A, Rastall RA, La Ragione RM, Woodward MJ, Costabile A (2008) *In vitro* fermentation of carbohydrates by porcine faecal inocula and their influence on *Salmonella* Typhimurium growth in batch culture systems. FEMS Microbiol Ecol 66:608–619

Matsumoto M, Tani H, Ono H, Ohishi H, Benno Y (2002) Adhesive property of Bifidobacterium lactis LKM512 and predominant bacteria of intestinal microflora to human intestinal mucin. Curr Microbiol 44:212–215
Meghrous J, Euloge P, Junelles AM, Ballongue J, Petitidemange H (1990) Screening of Bifidobacterium strains for bacteriocins production. Biotechnol Lett 12:575–580
Mennigen R, Bruewer M (2009) Effect of probiotics on intestinal barrier function. Ann N Y Acad Sci 1165:183–189
Moorthy G, Murali MR, Devaraj SN (2009) Lactobacilli facilitate maintenance of intestinal membrane integrity during Shigella dysenteriae 1 infection in rats. Nutrition 25:350–358
Musch MW, Sugi K, Straus D, Chang EB (1999) Heat shock protein 72 protects against oxidant-induced injury of barrier function of human colonic epithelial Caco-2/bbe cells. Gastroenterology 117:115–122
Musher DM, Musher BL (2004) Contagious acute gastrointestinal infections. N Engl J Med 351:2417–2427
Nataro JP, Kaper JB (1998) Diarrhoeaegenic *E. coli.* Clin Microbiol Rev 11:142–201
Navaneethan U, Giannella RA (2008) Mechanisms of infectious diarrhoea. Nat Clin Pract Gastroenterol Hepatol 5:637–647
Nemeth E, Fajdiga S, Malago JJ, Koninkx JFJG, Tooten PCJ, van Dijk JE (2006) Inhibition of Salmonella-induced IL-8 synthesis and expression of Hsp70 in enterocyte-like Caco-2 cells after exposure to non-starter lactobacilli. Int J Food Microbiol 112:266–274
Nishida E, Koyasu S, Sakai H, Yahara I (1986) Calmodulin-regulated binding of the 90 kDa heat shock protein to actin filaments. J Biol Chem 261:16033–16036
Nusrat A, von Eichel-Streiber C, Turner JR, Verkade P, Madara JL, Parkos CA (2001) Clostridium difficile toxins disrupt epithelial barrier function by altering membrane microdomain localization of tight junction proteins. Infect Immun 69:1329–1336
Ocana VS, Pesce de Ruiz Holgado AA, Nader-Macias ME (1999) Selection of vaginal H_2O_2-generating Lactobacillus species for probiotic use. Curr Microbiol 38:279–284
Oelschlaeger TA, Guerry P, Kopecko DJ (1993) Unusual microtubuledependent endocytosis mechanisms triggered by *Campylobacter jejuni* and *Citrobacter freundii.* Proc Natl Acad Sci U S A 90:6884–6888
Ofek I, Doyle RJ (1994) Bacterial adhesion to cells and tissues. Chapman and Hall, New York
Okeke IN, Lamikanra A, Czeczulin J, Dubovsky F, Kaper JB, Nataro JP (2000) Heterogeneous virulence of enteroaggregative *Escherichia coli* strains isolated from children in Southwest Nigeria. J Infect Dis 181:252–260
Olivier V, Queen J, Satchell KJF (2009) Successful small intestine colonization of adult mice by Vibrio cholerae requires ketamine anesthesia and accessory toxins. PLoS One 4:e7352
O'Riordan K, Fitzgerald GF (1998) Evaluation of Bifidobacteria for the evaluation of antimicrobial compounds and assessment of performance in cottage cheese at refrigeration temperature. J Appl Microbiol 85:103–114
Ortega Y, Nagle R, Gilman RH, Watanabe J, Miyagui J, Kanugusuku P (1997) Pathologic and clinical findings in patients with cyclosporiasis and a description of intracellular parasite life-cycle stages. J Infect Dis 176:1584–1589
O'Sullivan G (2004) Isolated bifidobacteria that produce siderophores which inhibit growth of Lactococcus lactis. US patent 6,746,672 B2
Ovelgönne JH, Koninkx JFJG, Pusztai A, Bardocz S, Kok W, Ewen SW, Hendriks HG, van Dijk JE (2000) Decreased levels of heat shock proteins in gut epithelial cells after exposure to plant lectins. Gut 46:679–687
Pai R, Kang G (2008) Microbes in the gut: a digestable account of host-symbiont interactions. Indian J Med Res 128:587–594
Palmer LE, Hobbie S, Galán JE, Bliska JB (1998) YopJ of *Yersinia pseudotuberculosis* is required for the inhibition of macrophage TNF-a production and downregulation of the MAP kinases p38 and JNK. Mol Microbiol 27:953–965
Parassol N, Freitas M, Thoreux K, Dalmasso G, Bourdet-Sicard R, Rampal P (2005) *Lactobacillus casei* DN-114 001 inhibits the increase in paracellular permeability of enteropathogenic *Escherichia coli*-infected T84 cells. Res Microbiol 156:256–262

Peng L, Li ZR, Green RS, Holzman, IR, Lin J (2009) Butyrate enhances the intestinal barrier by facilitating tight junction assembly via activation of AMP-activated protein kinase in Caco-2 cell monolayers. J Nutr 139:1619–1625

Pusztai A, Ewen SWB, Grant G, Peumans WJ, van Damme EJM, Coates ME, Bardocz S (1995) Lectins and also bacteria modify the glycosylation of gut surface receptors in the rat. Glycoconjugate J 12:22–35

Qin H, Zhang Z, Hang X, Jiang Y (2009) *L. plantarum* prevents enteroinvasive *Escherichia coli*-induced tight junction proteins changes in intestinal epithelial cells. BMC Microbiol 9:63

Que X, Reed SL (2000) Cysteine proteinases and the pathogenesis of amebiasis. Clin Microbiol Rev 13:196–206

Raqib R, Sarker P, Bergman P, Ara G, Lindh M, Sack DA, Nasirul Islam KM, Gudmundsson GH, Andersson J, Agerberth B (2006) Improved outcome in shigellosis associated with butyrate induction of an endogenous peptide antibiotic. Proc Natl Acad Sci U S A 103:9178–9183

Ratnaike RN (1999) Diarrhoea and constipation in geriatric practice. Cambridge University Press, New York

Ravdin JI, Guerrant RL (1981) Role of adherence in cytopathogenic mechanisms of *Entamoeba histolytica:* study with mammalian tissue culture cells and human erythrocytes. J Clin Invest 68:1305–1313

Ravdin JI, Croft BY, Guerrant RL (1980) Cytopathogenic mechanisms of *Entamoeba histolytica.* J Exp Med 152:377–390

Resta-Lenert S, Barrett KE (2003) Live probiotics protect intestinal epithelial cells from the effects of infection with enteroinvasive *Escherichia coli* (EIEC). Gut 52:988–997

Rossi M, Corradini C, Amaretti A, Nicolini M, Pompei A, Zanoni S, Matteuzzi D (2005) Fermentation of fructooligosaccharides and inulin by bifidobacteria: a comparative study in pure and faecal cultures. Appl Environ Microbiol 71:6150–6158

Ryan Kenneth J, George CR (2004) Sherris medical microbiology: an introduction to infectious disease, 4th edn. McGraw-Hill, New York, pp 727–730

Sack DA, Sack RB, Nair GB, Siddique AK (2004) Cholera. Lancet 363:223–233

Sansonetti PJ, Tran VN, Egile C (1999) Rupture of the intestinal epithelial barrier and mucosal invasion by *Shigella flexneri.* Clin Infect Dis 28:466–475

Sarker MR, Carman RJ, McClane BA (1999) Inactivation of the gene (cpe) encoding Clostridium perfringens enterotoxin eliminates the ability of two cpe-positive C. perfringens type A human gastrointestinal disease isolates to affect rabbit ileal loops. Mol Microbiol 33:946–958

Servin AL (2004) Antagonistic activities of lactobacilli and bifidobacteria against microbial pathogens. FEMS Microbiol Rev 28:405–440

Sheikh J, Czeczulin JR, Harrington S, Hicks S, Henderson IR, Le Bouguénec C, Gounon P, Phillips A, Nataro JP (2002) A novel dispersin protein in enteroaggregative *Escherichia coli.* J Clin Invest 110:1329–1337

Shogomori H, Futerman AH (2001) Cholera toxin is found in detergent insoluble rafts/domains at the cell surface of hippocampal neurons but is internalized via a raft-independent mechanism. J Biol Chem 276:9182–9188

Simon D, David J, Brandt LJ (1994) Studies on the pathogenesis of cryptosporidia-incuced diarrhoea in HIV-infected individuals. Am J Gastroenterol 89:2277–2278

Simonova MP, Laukova A, Chrastinova L, Strompfova V, Faix S, Vasilkova Z, Ondruska L, Jurcik R, Rafay J (2009) *Enterococcus faecium* CCM7420, bacteriocin PPB CCM7420 and their effect in the digestive tract of rabbits. Czech J Anim Sci 54:376–386

Simonovic I, Rosenberg J, Koutsouris A, Hecht G (2000) Enteropathogenic Escherichia coli dephosphorylates and dissociates occluding from intestinal epithelial tight junctions. Cell Microbiol 2:305–315

Skelkvåle R, Uemura T (1977) Experimental diarrhoea in human volunteers following oral administration of Clostridium perfringens enterotoxin. J Appl Bacteriol 43:281–286

Stern NJ, Svetoch EA, Eruslanov BV, Perelygin VV, Mitsevich EV, Mitsevich IP, Pokhilenko VD, Levchuk VP, Svetoch OE, Seal BS (2006) Isolation of a *Lactobacillus salivarius* strain and purification of its bacteriocin, which is inhibitory to *Campylobacter jejuni* in the chicken gastrointestinal system. Antimicrob Agents Chemoth 50:3111–3116

Su C, Brandt LJ (1995) *Escherichia coli* O157:H7 infection in humans. Ann Intern Med 123:698–714

Svetoch EA, Levchuk VP, Pokhilenko VD, Eruslanov BV, Mitsevich EV, Mitsevich IP, Perelygin VV, Stepanshin YG, Stern NJ (2008) Inactivating methicillin-resistant *Staphylococcus aureus* and other pathogens by use of bacteriocins OR-7 and E 50–52. J Clin Microbiol 46:3863–3865

Tafazoli F, Zeng CQ, Estes MK, Magnusson KE, Svensson L (2001) NSP4 enterotoxin of rotavirus induces paracellular leakage in polarized epithelial cells. J Virol 75:1540–1546

Tien MT, Girardin SE, Regnault B, Le Bourhis L, Dillies MA, Coppée JY, Bourdet-Sicard R, Sansonetti PJ, Pédron T (2006) Anti-inflammatory effect of *Lactobacillus casei* on *Shigella*-infected human intestinal epithelial cells. J Immunol 176:1228–1237

Truusalu K, Mikelsaar RH, Naaber P, Karki T, Kullisaar T, Zilmer M, Mikelsaar M (2008) Eradication of *Salmonella* Typhimurium infection in a murine model of typhoid fever with the combination of probiotic *Lactobacillus fermentum* ME-3 and ofloxacin. BMC Microbiol 8:132

Uchiya K (1999) A *Salmonella* virulence protein that inhibits cellular trafficking. EMBO J 18:3924–3933

Van Nhieu GT, Caron E, Hall A, Sansonetti PJ (1999) IpaC induces actin polymerization and filopodia formation during *Shigella* entry into epithelial cells. EMBO J 18:3249–3262

Van Nhieu GT, Guignot J (2009) When Shigella tells the cell to hang on. J Mol Cell Biol 1:64–65

Walter J (2008) Ecological role of lactobacilli in the gastrointestinal tract: implications for fundamental and biomedical research. Appl Environ Microbiol 74:4985–4996

Wang W, Uzzau S, Goldblum SE, Fasano A (2000) Human zonulin, a potential modulator of intestinal tight junctions. J Cell Sci 113:4435–4440

Warny M, Kelly CP (2003) Pathogenicity of *Clostridium difficile* toxins. In: Hecht G (ed) Microbial pathogenesis and the intestinal epithelial cell. ASM Press, Washington, p 503

Warny M, Pepin J, Fang A, Killgore G, Thompson A, Brazier J, Frost E, McDonald LC (2005) Toxin production by an emerging strain of Clostridium difficile associated with outbreaks of severe disease in North America and Europe. Lancet 366:1079–1084

Wells CL, van de Westerlo EM, Jechorek RP, Haines HM, Erlandsen SL (1998) Cytochalasin-induced actin disruption of polarized enterocytes can augment internalization of bacteria. Infect Immun 66:2410–2419

Winn W Jr, Allen S, Janda W, Koneman E, Procop G, Schreckenberger P, Woods G (2006) Koneman's color atlas and textbook of diagnostic microbiology, 6th edn. Lippincott Williams & Wilkins, Philadelphia, pp 1267–1270

Yildirim Z, Johnson MG (1998) Characterization and antimicrobial spectrum of bifidocin B, a bacteriocin produced by Bifidobacterium bifidum NCFB 1454. J Food Protect 61:47–51

Yildirim Z, Winters DK, Johnson MG (1999) Purification, amino acid sequence and mode of action of bifidocin B produced by Bifidobacterium bifidum NCFB 1454. J Appl Microbiol 86:45–54

Chapter 14
Bacteriocins of Probiotics and Enteric Cytoprotection

Bojana Bogovič-Matijašić and Irena Rogelj

14.1 Introduction

The production of bacteriocins is widespread among Gram-negative as well as Gram-positive bacteria, including LAB and *Bifidobacteria* (*B.*) representing the most common probiotics (Gillor et al. 2008; Cheikhyoussef et al. 2008; Nissen-Meyer et al. 2009; Bierbaum and Sahl 2009). The use of bacteriocins, namely nisin and pediocin AcH, as antimicrobials has already been applied in food preservation, but the role of bacteriocins in the intestinal tract of humans and animals has been less investigated so far. One of the main reasons for this is that the health-contributing effects of bacteriocin-producing bacteria are very difficult to measure (Flynn et al. 2002). Probiotics possessing the ability to transiently colonize the gastrointestinal tract (GIT), are excreting in the lumen their metabolic products including bacteriocins (Sherman et al. 2009). Those *in situ* produced molecules may interact with gastrointestinal (GI) microbiota and with the host. It is believed that the ability to directly inhibit the growth or virulence of possible pathogenic bacteria in the GIT will not only give this bacterium a competitive advantage within this environment and enable them to become established in the parts of the gut, but will also directly benefit the human host. The probiotic bacteria, especially bacteriocinogenic strains, may also act indirectly on the GI microbiota balance by increasing the concentration of beneficial microbes. Beside this there is an emerging interest in the oral application of bacteriocins in a purified form, as an alternative to the antibiotics commonly used for fighting GI infections in animals and humans (Flynn et al. 2002; Kirkup 2006; Sit and Vederas 2008).

It is however not so easy to distinguish between the probiotic and non-probiotic strains described in literature. Several bacteriocinogenic strains studied so far have not been examined for probiotic properties but are members of the species known

I. Rogelj (✉)
Department of Animal Science, Biotechnical Faculty,
University of Ljubljana, Groblje 3, 1230 Domžale, Slovenia
Tel.: +386-1-7217903
Fax: +386-1-721-79-03
e-mail: irena rogelj@bfro.uni-lj.si

J. J. Malago et al. (eds.), *Probiotic Bacteria and Enteric Infections*,
DOI 10.1007/978-94-007-0386-5_14,

to be indigenous in human and animal gut. The main challenge related to this review was therefore to decide which bacteriocin producers to include—only the ones with well-characterised probiotic properties, the ones described in the publications as potential probiotics, or also other bacteria of GI origin but not examined for probiotic properties so far. An example is bacteria-producing lantibiotics, with nisin as the most studied one, which are interesting because of wide range of activity. Among the producers of lantibiotics there are also representatives of *Lactobacillus plantarum, Lb. sakei* or *Bacillus* species, which are undoubtedly associated with probiotics. The lactococci, typical nisin producers, are already more difficult to classify as potential probiotics, since traditionally they have not been considered natural inhabitants of the human GIT. Since opinion about the probiotic potential of lactococci has been changing recently, it is probably reasonable to include among bacteriocinogenic probiotics also the nisin producers (Kimoto-Nira et al. 2007).

Beside LAB and *Bifidobacterium*, which are by far the most studied probiotics, some bacteriocinogenic Gram-negative bacteria, especially the *E. coli* producing microcins, colicins, should not be overlooked. The producers of these bacteriocins are common inhabitants of the GIT, therefore their bacteriocins are probably naturally present in the gut, although the designation of probiotic bacteria has been reserved to date only for few *E. coli* strains, such as Nissle 1917.

14.2 Classification of Bacteriocins

14.2.1 Bacteriocins of LAB (Gram-Positive Bacteria)

Until recently the classification of bacteriocins of lactic acid bacteria on the basis of structure and mode of activity into the following four groups has been widely accepted: (1) lantibiotics (class I), (2) small heat stable, non-lanthionine containing bacteriocins (class II), (3) large heat-labile (>30 kDa) bacteriocins (class III), and (4) complex bacteriocins composed not only from protein (also lipid or carbohydrate) (class IV) (Klaenhammer 1993). The present classification is basically very similar regarding the first two groups which also contain far most of the bacteriocins of probiotic bacteria characterised so far (Cotter et al. 2005; Nes et al. 2007). According to Cotter et al. (2005), the antimicrobial proteins formerly described as class III bacteriocins should not be named bacteriocins at all, but bacteriolysins. The class IV is presently reserved for cyclic bacteriocins (Nes et al. 2007; Maqueda et al. 2008). Some authors, however, consider the classification of cyclic bacteriocins into class IIc proposed by Cotter et al. (2005) (Nissen-Meyer et al. 2009). There are some discrepancies in the division of class II bacteriocins into subgroups, however, we will follow in this review the one proposed by Cotter et al. (2005), and used also recently in the review of class II bacteriocins (Nissen-Meyer et al. 2009).

Although this division was originally introduced for LAB bacteriocins, being very extensively studied due to their application value in food preservation mainly, it has been adopted also for description of bacteriocins of Gram-positive bacteria in general (Nes et al. 2007; Nissen-Meyer et al. 2009; Maqueda et al. 2008).

14.2.1.1 Lantibiotics

The common characteristic of all lantibiotics is that they contain lantionin, and optionally also dehydroalanine or dehydrobutirine, which are formed by post-translational modification and the introduction of intramolecular cyclic structures. The type-A designation was introduced for the elongated and positively charged lantibiotics, and the type-B designation for globular and non-charged lantibiotics. They are able to inhibit the cell wall biosynthesis by binding to lipid II—the main transporter of peptidoglycan units (Type A), to disturb the membrane by formation of pores (Type A and some of Type B), to interfere with cellular enzymatic activities (some of Type B) or combine these activities (Chatterjee et al. 2005; Brotz et al. 1998). Gram-negative bacteria are not susceptible to lantibiotics, because the outer membrane prevents the access of the lantibiotics to the cytoplasmic membrane.

On the basis of structure and mode of action, they are divided into Type AI, Type AII, Type B and others, including morphogenetic peptides and two-peptide lantibiotics. Two-peptide lantibiotics are composed of two peptides, one of them belonging to type-A, and another to type-B lantibiotics. The designation I or II refers to the way of biosynthesis. While class I lantibiotics are modified by two modification enzymes (LanB and LanC), the class II lantibiotics possess a GG cleavage site in their leader peptide and are modified by LanM enzymes and often exported by a LanT exporter (Willey and van der Donk 2007). Lantibiotics are synthesised as prepeptides consisting of an N-terminal leader sequence of up to 59 amino acids in length, and of the C-terminal propeptide which needs to be modified to be able to exert bacteriocin activity. Some excellent recent reviews dealing with genetic organisation and biosynthesis of lantibiotics, which require several genes often located also on transposons, are available (Bierbaum and Sahl 2009; Lawton et al. 2007; Lubelski et al. 2008; McAuliffe et al. 2001). Lantibiotics are produced by representatives of *Lactococcus (L.), Streptococcus (Str.), Bacillus, Microbiospora, Staphylococcus (St.), Micrococcus, Lactobacillus (Lb.), Butyrivibrio, Ruminococcus, Actinoplanes, Clavibacter, Enterococcus (Ent.), Planomonospora, Pediococcus, Carnobacterium* and *Paenibacillus* genera (Bierbaum and Sahl 2009; Willey and van der Donk 2007; Lawton et al. 2007; Chatterjee et al. 2005). Lantibiotic producers can be found also in *L. lactis, Bacillus subtilis, Str. salivarius, Micrococcus* spp., *Lb. plantarum, Lb. sakei, Butyrivibrio fibrisolvens, Ent. faecalis, Carnobacterium* sp. and *Pediococcus* sp., species which are usually included in probiotic products or at least tested for probiotic application (Cosseau et al. 2008; Leisner et al. 2007; Lim and Kim 2009; Mandal et al. 2009; Robertson et al. 2000; Rychlik and Russell 2002).

Nisin represents a prototype of Type A1 lantibiotics. It's widely used for food preservation, and has been tested for many other applications (Lubelski et al. 2008).

Lactococci, typical but not only producers of nisin may also fulfil the criteria for probiotic organisms, and are included in several commercial probiotic preparations (Kimoto-Nira et al. 2007). Some interesting reports on the effectiveness of lantibiotics such as nisin, lacticin 3147 and mutacins and their effectiveness against intestinal pathogens are presented in the following chapters.

14.2.1.2 Class II LAB (or Gram-Positive) Bacteriocins

Bacteriocins of class II are non-modified heat-stable cationic peptides with molecular mass ranging from 2 to 9 kDa. The division of this heterogeneous group of bacteriocins into subgroups is not completely uniform, however, in all recent classifications groups of pediocin-like (IIa) and two-peptide (IIb) bacteriocins exist (Cotter et al. 2005; Nes et al. 2007; Nissen-Meyer et al. 2009). Another important group comprises cyclic bacteriocins designated as IIc, or classified into a new class (IV) according to Nes et al. (2007).

Several bacteriocins of probiotic LAB or those with application potential in antibiotic treatment of infectious diseases are members of the Class IIa (Drider et al. 2006). The main common characteristic of IIa bacteriocins is a highly conserved hydrophilic and charged N-terminal part and two cystein residues linked with disulfide bridge, and a more variable hydrophobic and/or amphiphilic C-terminal part. The three-dimensional structure analysis of pediocin-like bacteriocins revealed the C-terminal hairpin-like structures, N-terminal beta-sheet-like structures stabilised by a disulfide bridge and alfa-helical C-terminal tail (Nissen-Meyer et al. 2009). Such structure enables them to insert into the membrane and to mediate membrane-leakage, disruption of proton-motive force and consequently the cell death. Those bacteriocins may inhibit a wide spectrum of G+ bacteria belonging to LAB bacteria and also *Clostridium* and *Listeria*, and is strain-specific. Inside this group is found some bacteriocins, for instance enterocin A, pediocin AcH and plantaricin 423 produced by bacteria with established probiotic properties (Millette et al. 2008a, b; Maré et al. 2006; Aymerich et al. 1996). The genetic organisation of IIa bacteriocins has been very well studied (reviewed in Ennahar et al. 2000; Drider et al. 2006). They are synthesised as prepeptides which comprise 13–30 aminoacid residues long N-terminal leader sequence with two glycine residues at the N-terminal end, which is removed from the active part of a molecule during the transport of bacteriocins from the cell, by a dedicated ABC-transporter (Nes et al. 2007; Oppegard et al. 2007a).

Bacteriocins of class IIb are composed of two different peptides, both required for the antibacterial activity which is based on the permeabilization of the bacterial membrane which leads to dissipation of the proton motive force and cell death. This group also comprises bacteriocins produced by bacteria of the species referred to as probiotics, such as *Lb. plantarum, Lb. acidophilus, Lb. gasseri, L. lactis, Lb. salivarius, Enterococcus faecalis* (Pingitore et al. 2009; Zorič Peternel 2007; Oppegard et al. 2007a; Nissen-Meyer et al. 2009; Maldonado-Barragán et al. 2009). For most of these bacteriocins, the individual peptides do not display antimicrobial activity when the second peptide is absent. However, there are some exceptions, like lactacin F and plantaricins E/F and J/K (Anderssen et al. 1998; Allison et al. 1994). Many two-peptide bacteriocins have been genetically characterised. One of their important features is that genes for both peptides of such bacteriocins and a common immunity gene are located next to each other in the same operon, indicating that both are produced simultaneously. The recent structural studies also indicate that they act together on the same site of the target cell membrane. Probably they bind individually to the cell wall, but penetration into the membrane takes place only after the interaction between the two peptides (Oppegard et al. 2007b). The three-dimensional structure of the two-peptide bacteriocin lactococcin G (LcnG), the best-characterised bacteriocin of this group has been determined by nuclear magnetic resonance (NMR) spectroscopy (Rogne et al. 2008). It seems that LcnG-α and LcnG-β have a parallel orientation and interact through helix–helix interactions. Interaction of both peptides with membrane causes membrane-leakage.

14.2.1.3 Circular Bacteriocins

The group of circular bacteriocins produced by Gram-positive bacteria has been extensively studied (reviewed by Maqueda et al. 2008). Among these bacteriocins, in which the N- and C-ends are linked to form a circular backbone, are found several bacteriocins produced by probiotic bacteria or at least by the representatives of species traditionally considered as probiotic, like gassericin A produced by *Lb. gasseri,* acidocin B by *Lb. acidophilus*, reutericin 6 by *Lb. reuteri* and AS-48 by *Ent. faecalis* (Maqueda et al. 2004; Kawai 1998a, b; Toba et al. 1991; Leer et al. 1995). The circularisation of bacteriocin molecules results in the increased resistance to protease digestion and enhanced thermodynamic stability and integrity in the protein structure. These properties consecutively improve their biological activity *in vivo*, stability across considerable pH and temperature ranges, and widen the spectrum of antimicrobial activity. Enterocin AS-48 exhibits bactericidal activity against a wide variety of bacteria, including food-spoilage and pathogenic Gram-positive bacteria such as *Bacillus, Clostridium, Brochothrix thermosphacta, Staphylococcus aureus* and *Listeria monocytogenes* and also against some Gram-negative species (Cobos et al. 2001; Abriouel et al. 2002). The genes of this new family of circular bacteriocins are usually chromosomally encoded, or in more rare cases, as acidocin B and enterocin AS-48, on plasmids. They are organised into polygenic operons, which enable their co-ordinated regulation. The gene cluster involved in the production of and immunity to the circular bacteriocins involves several genes, such as structural genes encoding prebacteriocins, putative biosynthetic and processing genes, immunity genes, ABC transporter genes and regulatory genes. The genetic characteristics of the most-studied representatives of circular bacteriocins have been reviewed by Maqueda et al. (2008). Functional studies of AS-48 have shown that this peptide causes non-selective pore formation in lipid bilayers, thereby allowing for the free diffusion of ions and low-molecular-weight solutes across the membrane. A similar mode of action has been reported for gassericin A and reutericin 6. Due to their secondary structures, organised mainly in α-helices, these bacteriocins are very stable, and maintain their activity after having been heated to 100°C for 60 min.

14.2.2 Bacteriocins of *E. coli*

Colicins of *E. coli* were discovered before the Gram-positive bacteria bacteriocins and are the most extensively studied bacteriocins (Cascales et al. 2007). They have higher molecular weight as bacteriocins from Gram-positive bacteria. The mechanisms of their bacteriocinogenic activity include membrane permeabilisation or nucleic acid degradation. Usually the colicin genes are on type A or type B plasmids respectively, the former being smaller (6–10 kb) and the latter bigger (40 kb), and less frequently also/or on chromosome. In addition to the domain responsible for bactericidal activity, the N-terminal domain responsible for the translocation of colicins, a domain responsible for immunity and a domain for the recognition of specific colicin receptors on the target cells have been identified. Microcins also produced by *E. coli,* are smaller from colicins, thermostable, resistant to some proteases and active in a wide range of pH. Some of microcins are also post-translationally modified by specific enzymes (Pons et al. 2002; Duquesne et al. 2007a, b; Severinov et al. 2007).

14.2.3 Bacteriocins of Probiotic Bacteria

Among the first characterised bacteriocins of human-derived probiotic bacteria were those of *Lactobacillus gasseri* LF221 and K7, two strains of human origin with *in vivo* established probiotic properties (Rogelj et al. 1999; Matijašić and Rogelj 2000; Rogelj et al. 2002; Matijašić et al. 2003; Matijašić et al. 2004, 2006; Rogelj and Matijašić 2006). While acidocins LF221 A and B of *Lb. gasseri* LF221 (formerly *Lb. acidophilus* 221, later reclassified into *Lb. gasseri* species) had already been identified on an amino acid sequences basis in 1998 (Bogovič-Matijašić et al. 1998), and later also genetically (Majhenič et al. 2004; acidocin LF221 A, GenBank AY295874; acidocin LF221 B, GenBank AY297947), the genetic characterisation of bacteriocins of the K7 strain, which is much more interesting from the application point of view due to its technological adequacy and higher expression of bacteriocins, has been completed recently (Zorič Peternel 2007; gassericin K7A, GenBank EF392861; gassericin K7B, GenBank AY307382). Bacteriocins of both strains have a wide range of activity against G+ bacteria, and particularly the potential of *Lb. gasseri* K7 to fight *Clostridium* (*Cl.*) *difficile* and *Cl. perfringens* by production of anti-clostridial bacteriocins has been emphasised by the authors (Matijašić and Rogelj 2000). K7 strain was found in gnotobiotic piglets even to play some protective role against enterotoxigenic *E. coli* infection (Rogelj and Matijašić 2006), however, this might not be directly ascribed to bacteriocin complex, since it does not inhibit *E. coli in vitro.*

Another purified and genetically characterised bacteriocins of probiotic bacteria were lactacin B produced by well known probiotic strain *Lb. acidophilus* NCFM (Sanders and Klaenhammer 2001) and the two-component bacteriocin ABP-118, produced by the probiotic strain *Lb. salivarius* subsp. *salivarius* UCC118, an isolate from the human ileum (Flynn et al. 2002). ABP-118 bacteriocin complex is particularly interesting because of the broad antimicrobial spectrum, including a number of food-borne and medically significant pathogens, belonging to *Bacillus, Listeria, Enterococcus* and *Staphylococcus* species (Dunne et al. 1999).

Several other bacteriocins of well-known probiotic bacteria still remain to be characterised. Often they are referred to as BLIS—bacteriocin-like inhibitory substances. In Table 14.1 are listed well-known and potential probiotic bacteria, producers of bacteriocins or BLIS.

More difficult than *in vitro* studies of bacteriocins, is establishing their effect *in vivo*. In most studies, the authors only speculated about the role of bacteriocins in observed effects on the different pathogenic strains used in animal challenge tests. One of the main reasons for this is that the health-contributing effects of bacteriocins produced by bacteriocin-producing bacteria are very difficult to measure. This problem may be overcome by the use of appropriate controls such as bacteriocin-deficient mutant strains. Such studies are few, and among the most convincing is the study of Corr et al. (2007) who demonstrated in a mice model, that an isogenic mutant of *Lb. salivarius* UCC118, deficient in the gene coding for bacteriocin ABP-118, was less effective at protecting mice against invasive *Listeria monocytogenes* infection than the original strain.

A limited number of reports are found which demonstrate production of antimicrobial compounds from bifidobacterial strains and an overview has been recently done by Cheikhyoussef et al. (2008). Until now, only a few researchers have attempted to purify and characterize the antimicrobial peptides from bifidobacteria and the production of bacteriocins by *Bifidobacterium* (*B.*) are largely unknown. The bifidocin B from *B. bifidum* NCFB 1454 is the unique bacteriocin from *Bifidobacterium* characterized to date. The amino acid sequencing homology search revealed that bifidocin B shared significant homology with other class IIa LAB bacteriocins (Yildirim et al. 1999).

Saleh and El-Sayed (2004) isolated and partially characterised bacterocins Bifilact Bb-12 and Bifilong Bb-46 produced by *B. lactis* Bb-12 and *B. longum* Bb-46. Bifilact Bb-12 and Bifilong Bb-46 were sensitive to pepsin and trypsin, and exhibited inhibition activity against *St. aureus, S. typhimurium, Bacillus ceureus* and *E. coli*. *B. animalis* subsp. *lactis* strain Bb-12, isolated from a healthy adult is one of the most thoroughly studied probiotic *Bifidobacterium* strain currently on the market, effective in preventing traveler's diarrhoea, decreasing the risk of constipation, and in the modulation of the immune response (Ouwehand et al. 2002). Recently it was demonstrated in a double-blind, placebo controlled, randomized clinical study, performed on 69 preterm infants that supplementation of preterm infants with *B. lactis* Bb12 had a beneficial effect on gut microbiota composition (Mohan et al. 2006). The involvement of Bifilact-Bb12 was not mentioned.

Table 14.1 An overview of well known or potential probiotic bacteria which produce characterised or partially characterised bacteriocins

Producer strain	Bacteriocin	Isolated from	Class	Activity	*In vivo* evidence (effects of strain)
Lactobacillus gasseri SBT 2055	Gassericin T	Human infant faeces (Kawai et al. 1997)	Class IIb—two-peptide bacteriocin, chromosomally-encoded (Kawai et al. 2000)	Wide spectral activity, including *Bacillus cereus, Listeria monocytogenes,* and *St. aureus* (Kawai et al. 1997)	Successfully established in the human intestinal tract, resulting in the alteration of the intestinal microbiota and the physical characteristics of faeces (Fujiwara et al. 2001, Takahashi et al. 2006)
Lactobacillus gasseri LA39	Gassericin A	Human faeces, 4 months old infant (Ito et al. 2009; Kawai et al. 2001)	Class IIc—cyclic, 58 amino acid residues, genes on chromosome (Kawai et al. 2009)	Wide spectral activity	Data not found
Lactobacillus reuteri LA6	Reutericin 6, identical to Gassericin A isolated from the same infant	Human faeces, 2 months old infant (Kawai et al. 2001)	Class IIc—cyclic, 58 amino acid residues (Kawai et al. 2001)	Wide spectral activity, lytic activity against *Lb. delbrueckii* subsp. *bulgaricus* JCM 1002 (Kabuki et al. 1997)	Data not found
Lactobacillus gasseri KT7	Gassericin KT7	Faeces of a 45-day old breast-fed infant (Zhu et al. 2000)	Partially characterized, probably class II bacteriocins (Zhu et al. 2000)	Wide spectral activity, including *Clostridium, Listeria* and *Enterococcus*	Data not found
Lactobacillus gasseri LF221 (formerly *Lb. acidophilus* LF221)	Acidocins LF221 A and B	Faeces of breast-fed infant (Matijašić and Rogelj 1999)	Two component Class IIb two-peptide bacteriocins (Majhenič et al. 2004)	Wide spectral activity (Matijašić and Rogelj 1999; Bogovič-Matijašić et al. 1998)	The role of bacteriocins in the observed protection of animals against infections not established so far (Rogelj and Matijašić 2006)

Table 14.1 (continued)

Producer strain	Bacteriocin	Isolated from	Class	Activity	*In vivo* evidence (effects of strain)
Lactobacillus gasseri K7	Gassericin K7 A and B	Faeces of breast-fed infant (Matijašić and Rogelj 1999, 2000)	Two component Class IIb two-peptide bacteriocins; primary structure and nucleotide sequences identical to acidocins LF221 A and B (Zorič Peternel 2007)	Wide spectral activity, including *Clostridium difficile* (Matijašić and Rogelj 1999, 2000)	Bacteriocins probably contributed to the inhibition of clostridia in cheese (Matijašić et al. 2007); the role of bacteriocins in the protection against infection not established so far (Rogelj and Matijašić 2006)
Lactobacillus salivarius UCC118	ABP-118	Human ileal-caecal region (Ryan et al. 2008)	Class IIb two-peptide bacteriocin, Abp118α 45 aa and Abp118β 46 aa, *abp118* gene cluster completely identified (Flynn et al. 2002)	Wide spectral antibacterial activity	Protection of mice against invasive *L. monocytogenes* infection, effect absent with Abp-118 negative mutant (Corr et al. 2007)
Lactobacillus acidophilus NCFM/N2	Lactacin B	Human source (Sanders and Klaenhammer 2001)	Class II, chromosomally-encoded bacteriocin, lactacin B operon identified (Dobson et al. 2007)	Activity against closely related species (*Lactobacillus* strains and *Ent. faecalis*) (Barefoot and Klaenhammer 1983, 1984)	The survival of NCFM in humans and effect on lactobacilli population (Sui et al. 2002); effects on cold and influenza-like symptom incidence and duration in children (Leyer et al. 2009); consumption of strain combined with lactitol may improve microbiota composition and mucosal functions (Ouwehand et al. 2009); inoculation of mice with NCFM strain stimulated the function of dendritic cells (Chen et al. 2009)

Table 14.1 (continued)

Producer strain	Bacteriocin	Isolated from	Class	Activity	*In vivo* evidence (effects of strain)
Lactobacillus acidophilus La-5	Lactacin B	Data not found	Lactacin B encoding region is similar in *L. acidophilus* La-5 and NCFM strains (Tabasco et al. 2009)	Narrow inhibitory spectrum—members of the genus *Lactobacillus* (Tabasco et al. 2009)	Prevention of antibiotic-associated diarrhoea (Wenus et al. 2008); influence on gut barrier function and sepsis in critically ill patients (Jain et al. 2004); effect of molecules secreted by La-5 on *E. coli* O157:H7 colonization in mice (Medellin-Peña and Griffiths 2009)
Lactobacillus acidophilus CH5	Acidocin CH5	Dairy starter culture	Class II, the lactacin B homologous bacteriocin (Chumchalova et al. 2004)	Wide spectral antibacterial activity including *Brevibacterium* sp., *Lactobacillus* sp., *Micrococcus* sp., *Arthrobacter* sp., *Bacillus* sp., *Corynebacterium* sp.	Data not found
Lactococcus lactis MM19	Nisin Z	Human stool	Class I—lantibiotics	Wide spectral antibacterial activity, including clinical isolates of vancomycin resistant *Ent.* (VRE) (Millette et al. 2008a, b)	Modulation of intestinal microbiota of mice, reduction of the intestinal colonization of VRE-infected mice by nisin and by pediocin producing strains; pediocin-negative mutant not effective (Millette et al. 2008a)
Pediococcus acidilactici MM33	Pediocin PA-1/AcH	Human stool	Class IIa		
Pediococcus pentosaceus FBB61	Pediocin A	Isolated from cucumber fermentations (Costilow et al. 1956)	Class IIa, plasmid encoded (Henderson et al. 1992; Marugg et al. 1992)	Wide spectral antibacterial activity	Modulation of swine intestinal microflora metabolism; inhibition of clostridia and lactic acid bacteria (Casadei et al. 2009)

Table 14.1 (continued)

Producer strain	Bacteriocin	Isolated from	Class	Activity	*In vivo* evidence (effects of strain)
Lactobacillus plantarum DDEN 11007	Pediocin PA-1	Data not found	Class IIa	Wide spectral antibacterial activity	No effect on persistence of *Listeria monocytogenes* in the GIT was seen in gnotobiotic rats colonized with either the pediocin AcH producing or the non-bacteriocin producing variant of *Lb. plantarum* (Bernbom et al. 2006b); pediocin producing strain or its supernatant did not affect composition of the intestinal microbiota of human flora-associated rats (Bernbom et al. 2009)
Lactobacillus acidophilus M46	Acidocin B	Isolated from human dental plaque (Ten Brink et al. 1994)	Class IIc—cyclic, plasmid encoded, 59 aa with sec-dependent leader (Leer et al. 1995)	Wide spectral antibacterial activity including *Listeria monocytogenes, Cl. sporogenes* and *Brochotrix thermosphacta,* and some lactobacilli	Data not found
Lactobacillus acidophilus JCM 1132	Acidocin J1132	Data not found	Class IIb two-peptide, lactacin B homologous bacteriocin (Tahara et al. 1996)	Narrow spectral antibacterial activity	Data not found
Lactobacillus delbrueckii subsp. *lactis* UO004	Bacteriocin UO004	Infant faeces (Boris et al. 2001)	Partially purified, hydrophobic, heat-stable polypeptide	Narrow spectral antibacterial activity	Data not found

Table 14.1 (continued)

Producer strain	Bacteriocin	Isolated from	Class	Activity	*In vivo* evidence (effects of strain)
Lactobacillus plantarum 423	Plantaricin 423	Sorghum beer (Van Reenen et al. 1998)	Possibly plasmid-encoded; related to the pediocin PA-1 gene cluster (Van Reenen et al. 1998)	Wide spectral antibacterial activity including *Bacillus cereus, Cl. sporogenes, Ent. faecalis, Listeria innocua, Listeria monocytogenes, St. carnosus*, and *Str. thermophilus*	*Ent. faecalis* inhibited *in vivo* in post weaned piglets, probably by plantaricin 423 (Maré et al. 2006)
Lactobacillus plantarum K11	Bacteriocin (BLIS?) K11	Dongchimi (Lim and Im 2007)	Not characterized	Active against some LAB (*Lactobacillus* sp., *Enterococcus* sp., and *Streptococcus* sp) and some Gram-negative (*Enterobacter aerogenes* and *E. coli* O157)	The inhibition of *E. coli* O157 adhesion to Caco-2 may result from the bacteriocin production (Lim et al. 2009)
Lactobacillus johnsonii VPI 11088 (NCK88)	Lactacin F	Data not found	Class IIb two-peptide (Allison and Klaenhammer 1996)	Narrow spectral antibacterial activity (several strains of lactobacilli and *Enterococcus faecalis*) (Barefoot and Klaenhammer 1983)	Data not found
Lactococcus lactis, Lactococcus lactis strain CHCC5826 (Nis+), transconjugant strain	Nisin A	Data not found	Class I—lantibiotic, 34 aa, produced by several *L. lactis* and some other LAB (Lubelski et al. 2008)	Wide spectral antibacterial activity including Gram+ bacteria and spores	No nisin-mediated disturbance of the commensal microbiota of human flora associated HFA rats resulting from dosing with nisin or a nisin producing strain (Bernbom et al. 2006)

Table 14.1 (continued)

Producer strain	Bacteriocin	Isolated from	Class	Activity	*In vivo* evidence (effects of strain)
Lactobacillus salivarius DPC6005	Salivaricin P	Porcine intestinal isolate	Class IIb—two-peptide antilisterial bacteriocin (Barrett et al. 2007)	Wide spectral antibacterial activity	Data not found
Streptococcus, Lactobacillus salivarius K12	Salivaricin A2 Salivaricin B	Isolated from tongue swabbing (Tagg 2004)	Class I lantibiotic, 25 aa, genetic loci encoding SalA2 and SboB in strain K12 are fully characterized and are localized on megaplasmid (Hyink et al. 2007)	Wide spectral antibacterial activity	Bacterium, widely used as a probiotic for the treatment of halitosis and the maintenance of throat health (Burton et al. 2006a, b)
Lactobacillus salivarius NRRL B-30514	Bacteriocin OR-7	Cecal contents of broiler chickens. (Stern et al. 2006)	Class IIa, 54 aa, similar sequence to acidocin A (Stern et al. 2006)	Inhibits *Campylobacter jejuni* (Stern et al. 2006)	Data not found
E. coli Nissle 1917 (DSM 6601)	Microcin B12	*Isolated* from human intestinal flora	Contains two microcin determinants and produces at least one microcin	Wide spectral antibacterial activity	Significant reduction of neonatal calf diarrhoea (von Buenau et al. 2005); reduction of acute diarrhoea in infants and toddlers (Henker et al. 2007). The isogenic microcin-negative mutant was as effective as the parent strain in inhibition of invasion of *S. enterica* serovar *typhimurium* and of *Y. enterocolitica* in human embryonic intestinal epithelial cells. (Altenhoefer et al. 2004)

Table 14.1 (continued)

Producer strain	Bacteriocin	Isolated from	Class	Activity	*In vivo* evidence (effects of strain)
Bifidobacterium bifidum NCFB 1454	Bifidocin B	Human isolate (Collado et al. 2005)	Significant homology with class IIa LAB bacteriocins, (Yildirim et al. 1999)	Wide spectral activity (Yildirim and Johnson 1998)	Data not found
Bifidobacterium lactis Bb-12	Bifilact Bb-12	Healthy adult	Partly purified, sensitive to pepsin and trypsin (Saleh and El-Sayed 2004)	Wide spectral activity (*Staphylococcus aureus, Salmonella typhimurium, Bacillus cereus* and *E. coli*)	Effective in preventing traveler's diarrhoea, and in the modulation of the immune response (Ouwehand et al. 2002). Beneficial effect on gut microbiota composition of preterm infants (Mohan et al. 2006)
Bifidobacterium longum Bb-46	Bifilong Bb-46	Data not found	Partly purified, sensitive to pepsin and trypsin (Saleh and El-Sayed 2004)	Wide spectral activity (*Staphylococcus aureus, Salmonella typhimurium, Bacillus cereus* and *E. coli*)	Data not found

14.3 Regulation of Bacteriocin Production by Quorum Sensing

Speaking about the role of bacteriocins of LAB probiotics in enteric protection, the first question is whether bacteriocinogenic strain produces the bacteriocin also *in situ* in GIT and if so under what circumstances. At least part of the gut microbiota forms the biofilms consisting of mixed consortia of commensal bacteria attached to gut epithelial cells that represent a formidable barrier against food-borne pathogens (Lee et al. 2000). Researches have shown that bacterial cells within biofilms are physiologically distinct from their corresponding planktonic counterparts and that they function in a co-ordinated manner as cooperative consortia. Moreover, the profiles of gene transcriptions of cells in biofilms are distinct as well (Parsek and Fuqua 2004; Hall-Stoodley et al. 2004). In order to survive and proliferate in such complex consortia, bacteria have developed an outspoken interactivity with their neighbouring micro-organisms, ranging from competition for nutrients to collaborative cross-feeding and protective shielding. This highly dynamic nature of biofilms renders them extremely robust to environmental fluctuations (Moons et al. 2009).

Bacteriocin production does grant the producing strain a competitive advantage but it also presents a metabolic burden to the cell. Some bacteria start the production of such components only when their cell density is high enough. Differences in bacteriocin production by LAB in either solid or liquid media, or even absence of bacteriocin production in liquid media, have been observed in several studies (Barefoot and Klaenhammer 1983; Cintas et al. 1995). However, further investigations showed that most of these bacteriocins are produced also in liquid media when specific growth conditions are achieved and a dedicated three-component regulatory system, involved in a quorum sensing (QS) mechanism, is switched on. Such regulatory mechanism consisting of a small induction peptide, a histidine protein kinase and a response regulator is widespread among LAB bacteriocin producers (Kleerebezem et al. 1997; Nes and Eijsink 1999). Qi et al. (2004) found out that the inactivation of a gene encoding a histidine protein kinase in *Str. mutans* resulted in the reduction of competence development, bacteriocin production, acid tolerance, as well as biofilm formation. These data suggested that bacteriocin regulation might be a component of a larger global regulatory network involved in communal survival under various environmental conditions. Along with their role as antibacterial toxins, bacteriocins were found to act as the signalling and coordinating agents necessary for invading, establishing and competing in natural environments. According to Maldonado-Barragán et al. (2009) who studied constitutive versus regulated bacteriocin production on solid media in two different QS-regulated plantaricin-producing strains, the biofilms present optimal conditions for QS-regulated bacteriocin production in LAB. Bacteriocin production can be triggered by accumulation of small diffusible QS signals, produced autonomously or by co-inhabiting populations or an antibacterial compound may serve itself as a QS signal, like in the case of nisin and some other lantibiotics (Moons et al. 2009; Lubelski et al. 2008). Bacteriocin production in *Lb. salivarius* UCC118 for instance is controlled by an auto-induction mechanism involving the secreted peptide pheromone AbpIP (Flynn et al. 2002).

Another type of induction was observed in the case of lactacin B produced by *Lb. acidophilus* NCFM, a probiotic strain commercially available in the United States since the mid-1970s (Sanders and Klaenhammer 2001). Previous reports have demonstrated that lactacin B was not detectable in broth cultures of *Lb. acidophilus* NCFM and activity was only observed using solid media (Barefoot and Klaenhammer 1984). Later, Barefoot et al. (1994) observed that associative cultivation of the lactacin B producer with the sensitive indicator, *Lb. delbrueckii* subsp. *lactis* ATCC 4797, or other Gram-positive bacteria resulted in early or increased lactacin B production. Recently Tabasco et al. (2009) have demonstrated that *Lb. acidophilus* La-5 a well known probiotic bacteria that has demonstrated benefits in the ability of colonisation and competition in the intestinal tract (Jain et al. 2004; Wenus et al. 2008) produces lactacin B and that the expression of lactacin B is controlled by an auto-induction mechanism involving the secreted peptide IP_1800 that forms part of a three-component regulatory system encoded in the lactacin B operon. Production of bacteriocin by *Lb. acidophilus* La-5 was achieved when it was grown in co-cultures with the yoghurt starter species *Str. thermophilus* and *Lb. delbrueckii* subsp. *bulgaricus*. Bacteriocin structural gene (*lba*B) was transcribed constitutively in uninduced *Lb. acidophilus* La-5 cells, but the levels of the secreted bacteriocin were too low to be detected by the agar diffusion assay. In *Lb. acidophilus* La-5 and *Str. thermophilus* STY-31 co-cultures, a remarkable increase of the *lba*B transcription was observed.

14.4 Bacteriocin Detection in Intestinal and Faecal Samples

The most commonly used analytical tools for the determination of bacteriocin concentration are biological assays in which a selected sensitive indicator strain is used and bacteriocin activity is observed on the basis of growth inhibition of sensitive strain. The main drawbacks of these methods are lack of specificity (inability to differentiate among different antimicrobial substances) and limited sensitivity.

The growth inhibition of sensitive strains can be observed by measurement of optical density in liquid medium, by observing the zones of inhibition on agar plates (agar diffusion tests), or by absorbance measurements of growth media containing indicators such as brom cresol or similar (Parente et al. 1995; Mortvedt and Nes 1990). Such methods are useful for measurements of bacteriocin activity in pure cultures, but not always in complex samples such as faeces, intestinal content or food. In agar diffusion and in critical dilution assays, bacteriocins must be able to diffuse or be free in the solution. Bacteriocins however may adsorb to different macromolecules, to the surface of bacteria or particles of complex samples and consequently cannot be detected.

Alternative methods were also developed for some bacteriocins (nisin, lacticin RM) such as specific immunochemistry-based methods or methods based on luminescence or fluorescence measurements (Daoudi et al. 2001; Immonen and Karp 2007; Keren et al. 2004).

Reunanen and Saris (2003) developed a green fluorescent protein (GFP)-based nisin microplate bioassay by constructing a *L. lactis* strain LAC240, which contained a plasmid with nisR and nisK and the reporter gfp gene encoding Green fluorescent protein (GFPuv) under the control of the nisF promoter. The sensitivity of this method was much improved by Hakovirta et al. (2006), who constructed a new nisin indicator strain *L. lactis* LAC275 containing the gfpuv reporter gene and the nisin regulatory genes nisR and nisK in its chromosome. This strain can be used in microplate bioassay for nisin determination in foods. This bioassay was used recently also for nisin detection ex vivo in jejunal chyme from fistulated dogs (Reunanen and Saris 2009).

Gardiner et al. (2007), who tested the stability of lacticin 3147 in the pigs' intestines, have found that the well-diffusion method was not specific enough for the analysis of stomach, jejunum, and ileum samples. Therefore, they examined the samples for the presence of the lacticin 3147 peptides by mass spectrometry. Neither LtnA1 nor LtnA2 component was detected in the digesta of any of the pigs that had ingested lacticin 3147. In the samples spiked with this bacteriocin, only LtnA2 was detectable by mass spectrometry.

Bernbom et al. (2009) used biological assay, namely agar well diffusion assay with indicator strain *Lactobacillus sakei* NCFB 2714 also for determination of pediocin PA-1 in faecal and intestinal samples of rats fed pediocin producing *Lb. plantarum* or culture supernatant containing pediocin. Pediocin was detected in the samples obtained from duodenum, jejunum and colon of animals which were given pediocin containing culture supernatant. Also one faecal sample of an animal fed *Lb. plantarum* strain was positive. However, the authors took into consideration also the possibility of false negative results for the faecal samples, because of its binding to surfaces of the producer strain, other bacteria, food molecules or intestinal surfaces.

In the study of Bernbom et al. (2006a), nisin was successfully determined in the intestinal and faecal samples of rats fed nisin or nisin-producing *L. lactis,* by biological assay using *Micrococcus flavus* strain NCIB 8166 as indicator strain as well as by competitive ELISA immunoassay. The concentrations of nisin in intestinal samples estimated by ELISA were approximately 10-fold higher than the concentrations estimated by the biological assay, while in faecal samples, the difference was even around 200-fold. According to the authors, this difference was due to the presence of proteolytic degradation fragments of nisin which were detected by competitive ELISA, but not by biological assay.

14.5 Direct Effects of Antibacterial Peptides Including Bacteriocins on the Host

Antimicrobial peptides, i.e. short (12–100 amino acids), positively charged and amphiphilic molecules, are produced by a wide variety of organisms. In addition to direct microbicidal effect on bacteria, viruses, fungi, and some parasites, they have an important role as effectors' molecules of the innate immune system and also inflammatory responses: they may enhance phagocytosis, stimulate prostaglandin release, neutralise the septic effects of LPS, promote recruitment and accumulation of various immune cells at inflammatory sites, promote angiogenesis, and induce wound repair (Jenssen et al. 2006). Antimicrobial peptides of mammalian origin (defensins, cathelicidins etc.) in addition were also demonstrated to affect monocytes and T cells, and influence dendritic cell development. Therefore they are referred to as host defence peptides. Bacteriocins, i.e. antimicrobial peptides produced by bacteria, which are structurally very similar molecules, do not protect the host organism against infection in the classical sense, however, they may contribute to the host defence by killing other (pathogenic) bacteria in the gut.

Usually the antibacterial peptides insert into the bacterial membrane to form transmembrane pores which result in membrane permeabilisation. Beside this, several antimicrobial peptides are able to translocate across the membrane and accumulate intracellularly, where they exert different activities including inhibition of nucleic acid synthesis, protein synthesis, enzymatic activity, and cell wall synthesis. This was first established for mammalian antibacterial peptides, and recently also for bacteriocins, namely lantibiotics such as nisin and mersacidin, shown to bind to lipid II, inhibit its transglycosylation, and consequently to inhibit the synthesis of peptidoglycan.

Several probiotic bacteria are known to produce and excrete different extracellular proteins which are believed to play an important biological role in the GI environment (Sanchez et al. 2008). Some of them are involved in the bacterial adhesion to intestinal surfaces, some can modulate the functions of epithelial and immune cells and thus participate in immunomodulation and cross-talking between probiotics and the host. The p75 (75 kDa) and p40 (40 kDa), two proteins produced and excreted by probiotic bacterium *Lactobacillus rhamnosus* GG (LGG) for instance prevent cytokine-induced apoptosis in human and mouse intestinal epithelial cells by regulating signalling pathways (Yan et al. 2007). Except some studies with nisin, there is however, no clear evidence for bacteriocins to be directly involved as signalling molecules in such interactions.

Nisaplin, a commercial preparation containing nisin, administered orally to mice, caused the changes in several immunological parameters, such as increase of macrophage/monocyte population isolated from peripheral blood (de Pablo et al. 1999). An increase of both CD4 and CD8 T-lymphocyte cell counts and a decrease of B-lymphocyte counts were also observed after short-term administration (30 or 75 days). The counts of T-cells, the macrophage/monocyte fraction isolated from peripheral blood and the phagocytic activity of peritoneal cells after prolonged nisin administration (100 days) were affected in a concentration-dependent way. Additional study on mice dealt with the effect of nisin on the pro-inflammatory cytokine production (IL-6, TNF-α or IFN-γ) (Puertollano et al. 2003). Only some temporal changes in IL-6 and TNF-α productions were detected, such as slightly reduced levels of IL-6 and increased levels of TNF-α, therefore it was concluded that proinflammatory cytokine production and NK cells activity were not significantly affected.

Lin (2009) pointed out that to date it is not clear whether or not the bacteriocins which come orally to the gut could be absorbed. Considering their molecular weight, they may not enter the intestinal epithelium directly intact. But they may directly enhance the innate immune response to infections, such as *Salmonella* infections. There are some *in vivo* indications for such an activity, such as reduction of *Salmonella enteritidis* in the liver and spleen of challenged chickens receiving enterocin E 50-52 (Svetoch et al. 2008a).

The lack of evidence for different effects of bacteriocins on the host is at least partially a consequence of the lack of suitable methods and research tools needed to accomplish such studies. We should not forget that the complexity of the GIT with associated microbiota where several interactions are going on simultaneously, makes such studies extremely difficult to perform.

14.6 Effects of Bacteriocinogenic Probiotic Bacteria on GI Microbiota and on Pathogenic Bacteria

Probiotic bacteria which at least temporarily colonise the gut are supposed to be metabolically active also in the GIT, and to affect the surrounding intestinal microbiota by competition for specific nutrients and/or adhesion sites or by direct inhibition of bacteria in close proximity. Also bacteriocins may be produced and excreted by the producers in GIT (Fig. 14.1). The main mechanisms of competitive exclusion of pathogenic bacteria by probiotic ones in the gut are competition for nutrients, immunomodulation of the host, competition for adhesion sites (receptors), and production of inhibitory metabolites including bacteriocins. In several *in vitro* studies, the effect of LAB bacteriocins against gastrointestinal pathogenic bacteria such as *St. aureus, Cl. difficile, Listeria monocytogenes* or pathogenic *Enterococcus* was documented. Although bacteriocins of Gram-positive bacteria including LAB and bifidobacteria, were traditionally considered to be active against G+ bacteria only, it seems that some can be active also against G-like *Helicobacter pylori, E. coli* or *Salmonella* (*S.*) (Gillor et al. 2008; Morency et al. 2001).

14.6.1 Effects on GI Microbiota

Bernbom et al. (2006a) tested nisin-producing *L. lactis* CHCC5826 and the isogenic non-nisin-producing *L. lactis* CHCC2862 in gnotobiotic rats associated with human faecal microbiota (HMA rats). The numbers of total anaerobes, lactobacilli, bifidobacteria coliforms, enterococci/streptococci, and total aerobes were determined, and DGGE profiles of PCR-amplified 16S rRNA genes from DNA extracted from faecal samples. Since feeding of rats with either of the *L. lactis* strains increased the number of bifidobacteria in the faeces, it was concluded that this effect was not related to the production of nisin.

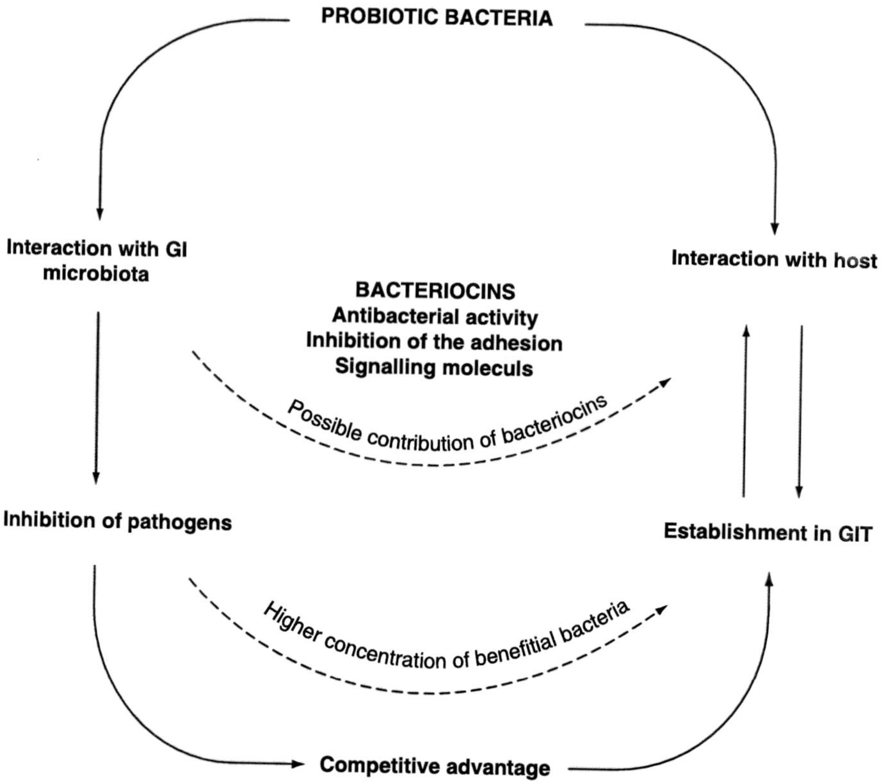

Fig. 14.1 Beneficial effects of probiotic bacteria and their bacteriocins against enteric pathogens

The same model was used to study the effect of pediocin PA-1 producing *Lb. plantarum* DDEN 11007 on the composition of the intestinal microbiota (Bernbom et al. 2009). Animals received pediocin PA-1 producing strain or its bacteriocin-negative derivative. DGGE analysis of total faecal DNA did not reveal any difference between the samples obtained from animals of different groups, indicating that either pediocin PA-1 was not produced or it did not cause any detectable changes of the rat microbiota, indicating the resistance of the majority of common intestinal bacteria species to pediocin PA-1.

In the recent study of Simonova et al. (2009), rabbits were fed *Ent. faecium* CCM7420, bacteriocin-producing strain with probiotic properties. They observed a significant reduction of coagulase-positive staphylococci in the caecum.

The application of *Ent. faecium* EK 13, enterocin A-producing strain to fifty weaned rabbits (Laukova et al. 2006) for four weeks, resulted in reduced *Enterobacteria* and *E. coli* counts but the contribution of enterocin A to this effect remained unknown.

Some studies have shown that bacteriocin production ability may confer a competitive advantage to such strains and consequently better survival in the gut. An example is the study of Walsh et al. (2008) who investigated *in vivo* on weaned piglets, a probiotic preparation composed of five strains. Bacteriocin-producing *Lb. salivarus* DPC6005 dominated over co-administered strains both in the ileum digesta and in mucosa indicating the positive role of bacteriocins.

14.6.2 *Clostridium difficile*

The activity against opportunistic *Cl. difficile* deserves particular interest since these opportunistic bacteria may cause antibiotic-associated diarrhoea or even life-threatening pseudomembraneus colitis. There is little evidence about the role of bacteriocins.

Human isolates *Lb. gasseri* LF221 and *Lb. gasseri* K7 studied by our research group were shown to produce bacteriocins active also against pathogenic human and animal isolates of *Cl. difficile* (Bogovič-Matijašić et al. 1998; Matijašić and Rogelj 2000). Minimal inhibitory concentrations (MIC) of bacteriocins of the *Lactobacillus* strain K7 (later characterised as gassericins K7 A and B) for vegetative cells and spores of the strains *Cl. tyrobutyricum, Cl. difficile* and *Cl. perfringens* were determined as well. The possibility of preventing *Cl. difficile* infections in patients treated with antibiotics with the help of *Lb. gasseri* probiotic strains was recognised. There is a clinical study on children with antibiotic-associated diarrhoea in the course to support this hypothesis.

Trejo et al. (2006) observed *in vitro* the growth inhibition of *Cl. difficile* (ATCC 9689 and ATCC 43593) by the supernatants of bifidobacteria isolated from healthy infants, as well as the inhibition of clostridia adhesion to Caco-2 cells. Although the substances responsible for inhibition were unaffected by proteolytic cleavage (proteinase K and chymotrypsin) we cannot exclude the presence/involvement of bacteriocins. Naaber et al. (2004) reported that the antagonistic activity of 23 (from 50) intestinal *Lactobacillus* spp. strains against 23 pathogenic *Cl. difficile* was strain-specific and seemed to correlate with H_2O_2 and lactic acid production, but the involvement of bacteriocins was not examined. Su et al. (1987) at least proposed that substances other than volatile fatty acids could be involved in colonisation resistance to *Cl. difficile* in gnotobiotic mice.

In the recent review of the mechanisms and efficacy of probiotics in the prevention of *Clostridium difficile*-associated diarrhoea, the possible involvement of bacteriocins is mentioned but there are no publications about this available (Parkes et al. 2009).

14.6.3 *Staphylococcus*

A number of recent studies dealing with the effects of bacteriocins against *St. aureus* infections were focused on the therapeutic potential of LAB bacteriocins for the management of Gram-positive infections, as an alternative for conventional antibiotics. So far, lantibiotics especially were studied, such as MU1140 and mutacin (B-Ny266) which are produced by bacteria not considered as probiotic, or whose potential probiotic properties were not tested (Ghobrial et al. 2009; Kirkup 2006).

One of the first successful demonstrations of *in vivo* efficacy of bacteriocins of G+ bacteria against *St. aureus* was published by Mota-Meira et al. (2005). In their study, mice were injected intraperitoneally with mutacin B-Ny266 a bacteriocin produced by *Streptococcus mutans*. The mortality of the mice was completely reduced compared to the 70 and 100% in control group infected with *St. aureus*. The results support the idea of the applicability of bacteriocins as an alternative for common antibiotics. Mutacins, being widely produced by *Streptococcus mutans*, were classified by Morency et al. (2001) into 24 different groups and 11 clusters of mutacins according to their activity spectra and their resistance to the other mutacinogenic strains. Although *Streptococcus* species are not considered probiotic, there have been other bacteriocins produced by LAB which are similar, and could be potentially used the same way.

Simonova et al. (2009) reported a significant reduction of coagulasa positive staphylococci in the caecum of rabbits ingesting *Enterococcus faecium* CCM7420, bacteriocin-producing strain with probiotic properties and its partially purified bacteriocin PPB CCM7420.

Bacteriocins E 50-52 produced by *Enterococcus faecium* NRRL B-30746 and bacteriocin OR-7 produced by *Lactobacillus salivarius* NRRL B-30514 were *in vitro* effective against methicillin-resistant *St. aureus* (Svetoch et al. 2008b). Although the probiotic properties of these two bacteriocin producers have not been reported, many other probiotic strains belong to these species.

Ent. faecium EK13, producing enterocin A, reduced pathogen concentration in japanese quails (Laukova et al. 2003; Gillor et al. 2008). In rabbits, the colonisation of pathogenic *Staphylococcus* was reduced (Laukova et al. 2006).

14.6.4 *Helicobacter pylori*

Helicobacter pylori (*H. pylori*) infection is the most common cause of gastritis, gastric ulcer and adenocarcinoma. It is usually cured with antibiotics, but strains resistant to antibiotics are often observed.

Some probiotic LAB strains, especially lactobacilli were found to inhibit *H. pylori in vitro,* but most probably not due to bacteriocins production (Ryan et al. 2008; Rokka et al. 2006). *H. pylori* is a Gram-negative species, and bacteriocins of LAB are in general active only against Gram-positive bacteria. There are, however, also reports which are in contradiction with this conviction. Lacticins A164 and BH5 produced by *Lactococcus lactis* subsp. *lactis* A164 and *L. lactis* BH5, for instance are active against Gram-negative *H. pylori* in the strain-dependent mode (Kim et al. 2003). Also pediocin PO2 and leucocin K showed certain anti-*H. pylori* activity, although not against all strains, and higher concentrations were needed. Mutacins A, B, C, D, and nisins A and Z inhibited *H. pylori in vitro* and were proposed by

Morency et al. (2001) as candidates for future antibiotics for treatment of bacterial infections including *H. pylori* infections. Rokka et al. (2006), observed *in vitro* inhibition of *H. pylori* by isolates belonging to *Lb. plantarum* species. The representatives of this species are often found also in GIT, and also can have probiotic properties. In their study, the main anti-*Helicobacter* activity of MLBPL1 strain seemed to be associated with cell wall, but a thermo-stable active component, 3 to 10 kDa in size, which could be bacteriocin, also participated in the observed inhibition.

The beneficial effects of probiotic organisms in the treatment of *H. pylori* infections have been reported by several researchers (Gotteland et al. 2006; Bergonzelli et al. 2006; Kim et al. 2003; Lin et al. 2009; Rokka et al. 2006, 2008; Ryan et al. 2008, 2009; Zou et al. 2009). It is believed that LAB could establish a potential complementary means to the conventional antibiotic treatment of *Helicobacter* infection, to suppress the infection.

It is well known that infection with *H. pylori* is associated with immune response characterised by IL-8 production. While some LAB were able to decrease IL-8 production by acid production and other mechanisms not yet explained, the role of bacteriocins in modulation of the immune response against *H. pylori* has not been demonstrated. Ryan et al. (2009) included in their *in vitro* study by AGS gastric epithelial cells model cells infected with *H. pylori*, and compared the effect of *Lb. salivarius* UCC118 strain able to produce bacteriocin Abp118, with mutant strain without this ability. *H. pylori*-induced IL-8 secretion was down-regulated equally by UCC118 wild-type cells as the bacteriocin-deficient mutant, indicating that bacteriocin was not involved in the immune modulation.

14.6.5 *Enterococcus*

Certain representatives of *Enterococcus* genus possess different virulence factors and may be involved in infections. Since some representatives of this genus are intrinsically resistant to a wide variety of antibiotics used to treat infections in humans, including aminoglycosides and vancomycin and have the ability to pass resistant genes to other bacteria, they have attracted special attention (Kauffman 2003).

Several LAB bacteriocins are capable of enterococci inhibition *in vitro* (Millette et al. 2008a; Miguel et al. 2008; Nassif and Zervos 2005). Millete et al. (2008a) demonstrated *in vivo* for the first time the ability of nisin- and pediocin-producing human isolates to modulate the intestinal microbiota of healthy mice and to reduce the intestinal colonisation of vancomycin resistant enterococci (VRE)-infected mice. *L. lactis* MM19 and *Pediococcus acidilactici* MM33, producers of nisin or pediocin PA-1/AcH, together with MM33A, a mutant derived from *P. acidilactici* MM33 that has lost its ability to produce pediocin through a plasmid-curing procedure, were applied to mice. *P. acidilactici* MM33 only significantly reduced the *Enterobacteriaceae* population in mouse faeces after 18 days of feeding. No modification, however, was observed for *Lactobacillus* spp., *Staphylococcus* spp., and *Enterococcus* spp. populations with either bacteriocin-producing bacterium assayed.

14.6.6 *Campylobacter*

The Gram-negative bacterium *Campylobacter* is one of the most common bacterial causes of human gastroenteritis. Infections with Gram-negative *Campylobacter* present a problem and concern especially for poultry and can also be transferred through the food chain on people. Poultry, particularly chicken, is considered a major source of human campylobacteriosis (Lin 2009).

Recently, several potent anti-*Campylobacter* bacteriocins have been identified. Cole et al. (2006) tested bacteriocin B602 of *Paenibacillus polymyxa* (NRRL B-30509), and bacteriocin OR7 of *Lb. salivarius* (NRRL B-3501) *in vivo* in turkeys challenged with a mixture of different *Campylobacter* (C.) strains: 3 *C. coli* isolates, 2 wild-type turkey isolates, and an American Type Culture Collection isolate 43481. Three days' treatment with bacteriocins eliminated detectable cecal *Campylobacter* concentrations in all turkeys. The authors concluded that reduced duodenum crypt depth and the number of goblet cells observed in bacteriocin-treated animals, may result in reduced mucin production and consequently, in the limited *Campylobacter* colonisation.

Bacteriocin OR-7 was evaluated also in chickens challenged with one of four *C. jejuni* isolates (Stern et al. 2006). The purified bacteriocin was encapsulated in polyvinylpyrrolidone and added to chicken feed. Chickens challenged with each of *Campylobacter* strains consistently reduced colonisation at least for one million fold compared with levels found in the caecal content of untreated groups.

Bacteriocin E 50-52 produced by *Ent. faecium* NRRL B-30746 also seems very promising for the treatment of *Campylobacter* infections. The activity of this enterocin from class IIa bacteriocins was demonstrated in therapeutic broiler trials. The cecal *C. jejuni* count was reduced by more than 100,000-fold when the animals were orally treated with E 50-52 (Svetoch et al. 2008a).

14.6.7 *Salmonella*

Some type A lantibiotics, such as nisin A, mutacins and IIa group bacteriocins E 50-52 and OR7, were found to be active against medically important Gram-negative bacteria including *Campylobacter, Haemophilus, Helicobacter, Salmonella* (*S.*) and *Neisseria* (Morency et al. 2001; Svetoch et al. 2008b). Nisin was reported to be active against *Salmonella* and some other Gram-negative organisms by disrupting the outer membrane (Stevens et al. 1991). Liévin et al. (2000) found two human *Bifidobacterium* strains (CA1 and F9) isolated from infant stools which expressed antagonistic activity against pathogens *in vitro,* inhibited cell entry of *S. typhimurium* SL1344 in Caco-2 cells, and killed intracellular salmonella. Anti-salmonella activity was assigned to lipophilic small antibacterial molecules. Both strains also successfully colonised the intestinal tract of axenic C3/He/Oujco mice, and protected them against *S. typhimurium* C5 infection.

Fayol-Messaoudi et al. (2005) investigated the anti-*Salmonella enterica* serovar *typhimurium* killing activity of *Lactobacillus* and *Bifidobacterium* probiotic strains *Lb. johnsonii* La1, *Lb. rhamnosus* GG, *Lb. casei* Shirota YIT9029 (LcS YT9029), *Lb. casei* DN-114 001, *Lb. rhamnosus* GR1, and *Lb. sakei* CWBI 030202. The inhibition that they observed was obviously multifactorial, including lowering of the pH and the production of lactic acid and of antibacterial compounds, including bacteriocins and non-bacteriocin, non-lactic acid molecules. The non-lactic acid substance(s) responsible for the killing activity found in the cell-free supernatant probably included also bacteriocins.

Efficient reduction of *S. enteritidis* by bacteriocins was observed also in broilers challenged with *S. enteritidis*. The bacteriocin applied orally to the animals was E 50-52 from *Ent. faecium* (NRRL B-30746) (Svetoch et al. 2008a).

14.6.8 Escherichia coli

Although most probiotic bacteria are Gram-positive strains, also non-pathogenic *E. coli* has already been successfully used as probiotics. Most of the work on probiotic *E. coli* was focused on the *Escherichia coli* strain Nissle 1917 (*E. coli* DSM 6601, Mutaflor®). *E. coli* Nissle 1917 was tested *in vivo* in mice, in conventional and gnotobiotic pigs, and in newborn infants (Lodinová-Zádniková and Sonnenborn 1997; Barth et al. 2009; Stritzker et al. 2007). There is still no evidence of the role of bacteriocins microcins produced by this strain, in the observed protective effect of the strain against enteric pathogens such as *Salmonella*. Altenhoefer et al. (2004) observed in gnotobiotic piglets the inhibition of invasion of *S. enterica* serovar *typhimurium* but it could not be attributed to the production of microcins by the Nissle 1917 strain because its isogenic microcin-negative mutant SK22D was as effective as the parent strain.

In a recent study reported by Gillor et al. (2009), who investigated in a mice model whether colicin production improves the persistence and colonisation of *E. coli* in the intestines, the results confirmed the role of colicins in the *E. coli* colonisation.

Escherichia coli O157:H7 is a food-borne pathogen that has been frequently linked to outbreaks attributed to the consumption of meat products, and water. The addition of a mixture of eight probiotic colicin E7-producing *E. coli* strains to feed reduced the faecal shedding of *E. coli* serotype O157:H7 in calves, but the role of bacteriocins in this inhibition is unknown (Schamberger et al. 2004).

14.6.9 Listeria monocytogenes

Corr et al. (2007) were successful in demonstration of bacteriocin production as a mechanism for the anti-infective activity of *Lb. salivarius* UCC118, well-studied probiotic strain of human origin. The protection role of its bacteriocin Abp 118 against infection with the invasive *Listeria monocytogenes* EGDe and LO28 was demonstrated by the help of mutant of *Lb. salivarius* UCC118 unable to produce Abp which failed to protect mice against infection. Additional confirmation of the bacteriocin involvement in the direct antagonism was provided using a strain of *Listeria monocytogenes* expressing the cognate Abp118 immunity protein AbpIM and therefore resistant against Abp 118.

The effect of pediocin AcH producing *Lb. plantarum* on *Listeria monocytogenes* EP2 was investigated *in vitro* in a gastrointestinal tract model, and *in vivo* in gnotobiotic rats by Bernbom et al. (2006b). The observation that AcH producing strain inhibited *Listeria monocytogenes in vitro* more successfully from its non-bacteriocin-producing variant, clearly indicated a direct effect of bacteriocin. No effect on persistence of *Listeria monocytogenes* in the GI tract was observed in gnotobiotic rats.

14.7 Effects of Purified Bacteriocins on GI Microbiota

It is assumed that orally ingested bacteriocins are degraded by proteolytic enzymes of the stomach and small intestines (Cleveland et al. 2001; De Vuyst and Leroy 2007). This assumption has been confirmed not only *in vitro* but also *in vivo* with pediocin PA-1, nisin and lacticin 3147 (Bernbom et al. 2009; Gardiner et al. 2007; Reunanen and Saris 2009). Despite this the results of some *in vivo* studies indicate the positive effects of feeding purified bacteriocins to animals. Reunanen and Saris (2009), who studied sensitivity of nisin to proteolytical breakdown in an *ex vivo* model using jejunal chyme from fistulated dogs, found nisin quite resistant to degradation by the components of the jejunal chyme, since it retained 66% of activity after 30 min incubation in it.

From LAB bacteriocins the *in vivo* studies accomplished to date have been limited to nisin, pediocins, enterocins and some other bacteriocins listed in Table 14.2.

The first study where not only bacteriocin-producing strains but also bacteriocins alone were applied to the animals was the one by Bernbom et al. (2006a), which used a gnotobiotic rat model associated with human faecal microbiota (HMA rats). No significant changes in the rat faecal microbiota were observed after dosage with nisin. The detection of nisin in the faeces and small intestines by competitive enzyme-linked immunosorbent assay showed that nisin was probably degraded or inactivated in the gastrointestinal tract. Also pediocin PA-1 tested in the same model did not mediate changes of the rat microbiota. Like for nisin, a biological assay indicated that the bacteriocin was degraded or inactivated during passage through the intestine.

In rabbits fed partially purified bacteriocin PPB CCM7420 produced by *Ent. faecium* CCM7420, Simonova et al. (2009) observed a significant reduction of coagulase-positive staphylococci in the caecum, and non-significant reduction of *St. aureus* and *Clostridium* in the faecal samples.

Table 14.2 An overview of bacteriocins already tested *in vivo* for the effects against enteric pathogens or effects on GI microbiota

Bacteriocin	Producer strain	Class	Activity	Effects of supernatant/bacteriocin Possible mechanism
Bacteriocin B602	*Paenibacillus polymyxa* NRRL B-30509	Amino acid sequences consistent with class IIa bacteriocins (Stern et al. 2005)	Inhibitory to *Campylobacter jejuni*	Reduction of intestinal levels and frequency of chicken *C. jejuni* colonization by bacteriocin B602 (Stern et al. 2005); reduction of *Campylobacter* concentrations in turkeys, reduction in crypt depth and goblet cell density by B602 and OR7 (Cole et al. 2006)
Bacteriocin OR7	*Lactobacillus salivarius* NRRL B-30514	Class IIa, 54 aa, similar sequence to acidocin A (Stern et al. 2006)		
Mutacin B-Ny266	*Streptococcus mutans* Ny266	Class I—lantibiotic, 21 aa (Morency et al. 2001)	Wide spectral antibacterial activity against Gram+ and Gram− bacteria (Morency et al. 2001)	The mortality of mice infected with *S. aureus* was reduced (Mota-Meira et al. 2005)
Bacteriocin PPB CCM7420	*Enterococcus faecium* CCM7420	A thermostable substance of proteinaceous character (Simonova and Laukova 2007)	Data not found	Reduction of coagulasa positive staphylococci in the caecum of rabbits (Simonova et al. 2009)
Enterocin A	*Enterococcus faecium* EK13	4.83 kDa, N-terminal amino acid sequence identical to enterocin A (Class IIa) (Aymerich et al. 1996; Marekova et al. 2003)	Wide spectral activity icluding *Enterococcus, Leuconostoc, Lactobacillus, Streptococcus, Staphylococcus, Bacillus* and *Listeria* (Marekova et al. 2003)	Reduced pathogen concentration in japanese quails (Laukova et al. 2004); reduced colonisation of pathogenic *Staphylococcus* in rabbits (Laukova 2006); lower damage to the intestinal epithelium of gnotobiotic Japanese quails infected with the toxigenic *Salmonella* (Cigankova et al. 2004)

Table 14.2 (continued)

Bacteriocin	Producer strain	Class	Activity	Effects of supernatant/bacteriocin Possible mechanism
E-760	*Enterococcus* sp. NRRL B-30745	62 amino acids, (Line et al. 2008)	Wide spectral antibacterial activity against Gram+ and Gram− bacteria (Line et al. 2008)	Administration of bacteriocin E-760-treated feed significantly (P 0.05) reduced the colonization of young broiler chicks experimentally challenged with two strains of C. jejuni (Line et al. 2008)
Bacteriocin E 50-52	*Enterococcus faecium* NRRL B-30746	Class IIa, 39 aa, (Svetoch et al. 2008a)	Wide spectral antibacterial activity against Gram+ and Gram− bacteria. In vitro effective against methillicin resistant *St. aureus* (Svetoch et al. 2008a)	In therapeutic broiler trials, oral treatment with E 50-52 reduced both *C. jejuni* and *Salmonella enteritidis* by more than 100,000-fold in the ceca, and systemic *S. enteritidis* was reduced in the liver and spleen. (Svetoch et al. 2008a)
Lacticin 3147	*Lactococcus lactis* DPC3147 Isolated from an Irish kefir grain (Ryan et al. 1996), transconjugant, *Lactococcus lactis* subsp. *cremoris* DPC4275	Two-peptide lantibiotic, genetic determinants on a 60-kb plasmid (Dougherty et al. 1998)	Wide spectral antibacterial activity, including methicillin-resistant *St. aureus*—MRSA), enterococci (including VRE), *Cl. botulinum*, *Cl. difficile* and *Propionibacterium acnes* (McAuliffe et al. 1998; Rea et al. 2007)	Orally applied lacticin 3147 did not affect *Lactobacillus,* coliform, or *Enterococcus* counts in either the porcine stomach or jejunum of pigs (Gardiner et al. 2007)

The application of *Ent. faecium* EK 13, enterocin A-producing strain to fifty weaned rabbits (Laukova et al. 2006) for four weeks, resulted in reduced *Enterobacteria* and *E. coli* counts but the contribution of enterocin A to this effect remained unknown. Another experiment with the same bacteriocin was conducted on three-day-old gnotobiotic japanese quails infected with the toxigenic *Salmonella* Dusseldorf SA31. To observe the preventive effect of enterocin A, the bacteriocin was administered to the CG1 quails before infecting them with the SA31. The therapeutic effect of enterocin A was observed in japanese quails from the group EG2 to which enterocin A crude extract was administered 8 h after SA31 infection. Lower damage to the intestinal epithelium occurred in both groups treated with enterocin compared to the untreated control group (Cigankova et al. 2004).

14.8 Applications of Bacteriocins as Gastrointestinal Antibiotics

There is an increase in interest for clinical application of bacteriocins as alternatives for conventional therapeutic antibiotics, as a consequence of the reduced efficiency of several antibacterial substances, such as glycopeptides, due to the increased resistance of pathogenic bacteria. Basically the main limitation of bacteriocin use as oral antibiotics is their degradation in GI. This limitation may be overcome by new delivery systems which enable the therapeutic agents to survive the passage through the stomach and small intestines, and reach the colon without being digested (Wilding 2000).

Svetoch et al. (2008b) compared MIC of bacteriocin E 50-52 from *Ent. faecium* NRRL B-30746, OR-7 from *Lb. salivarius* NRRL B-30514 and nisin, with MIC of some common antibiotics. Different human isolates involved in diverse infections, highly resistant to the large array of antibiotics, were included in the study. MICs of nisin for *St. aureus* were generally lower than those of the antibiotics. Bacteriocins OR-7 and E 50-52 had a wide range of bactericidal activity (*Citrobacter, Acinetobacter, Klebsiella, Proteus, E. coli, Pseudomonas, Staphylococcus*), indicating that they could be a promising alternative for antibiotics.

Among the most promising bacteriocins to fight antibiotic resistant pathogens is also the two-peptide lantibiotic lacticin 3147 found to be active *in vitro* against *St. aureus* (including methicillin-resistant *St. aureus*—MRSA), enterococci (including VRE), streptococci (*Str. pneumoniae, Str. pyogenes, Str. agalactiae, Str. dysgalactiae, Str. uberis, Str. mutans*), *Cl. botulinum*, and *Propionibacterium acnes*. The promising results in treating infections caused by MRSA have been obtained also *in vivo*.

Bacteriocins in general are non-toxic to eukaryotes and are effective at very low concentrations, thus they could be suitable for therapeutic applications in humans or animals. In the publication of Kirkup (2006) the potential of different bacteriocins, especially those of G- bacteria such as microcins and colicins, for medical application has been extensively reviewed.

Because of the sensitivity of bacteriocins to proteolytic enzymes new strategies were introduced for direct application of bacteriocins. Nisin for instance was successfully protected from degradation in the small intestines when formulated into tablets coated with methacrylic acid copolymer USP/NF type C (Eudragit) (Habib and Sakr 1999). Purified bacteriocin OR-7 produced by *Lb. salivarius* NRRL B-30514 was delivered to chickens in a microencapsulated form (polyvinylpyrrolidone—bacteriocin) (Stern et al. 2006). The bacteriocin E 50-52 was also prepared with poly-vinyl-pyrrolidone (PVP) powder in order to be mixed with ground maize feed for poultry (Svetoch et al. 2008a). Cole et al. (2006) used microencapsulated bacteriocins produced by *Bacillus circulans* and *Paenibacillus polymyxa*, in turkeys.

Kirkup (2006) already underlined the problem of resistance to bacteriocins which has not been addressed enough yet.

The low-level resistance of *Listeria* to the anti-listerial LAB bacteriocins is ascribed to alterations in membrane lipid composition. The high-level resistance is linked with the inactivation of the mptACD operon, which encodes the EIItMan, mannose permease of the phosphotransferase system (PTS). A 54 factor and the ManR activator were found to be involved. Calvez et al. (2007), however, identified three genes (rpoN, glpQ and pde) associated with intermediate resistance of *Enterococcus faecalis* to divercin V41, a pediocin-like bacteriocin belonging to the class IIa. Similarly, the genes responsible for the resistance against bacteriocins DvnV41 and MesY105 were identified also in *Listeria monocytogenes* EGDe (lmo0052 and lmo1292 genes) (Calvez et al. 2008). In general, the emergence of resistance against lantibiotics is very rare. It is, however, possible that the producers of different bacteriocins, which are always immune to their own bacteriocins, usually due to the production of specific immunity proteins, may be resistant also to other closely related bacteriocins. Draper et al. (2009) for instance established the cross-immunity of staphylococcin C55 and lacticin 3147 producers. Since such strains may represent a source of lantibiotic (lacticin 3147) immunity genes for other staphylococci as well, the authors concluded that a certain amount of caution is therefore needed regarding the clinical application of bacteriocins. Another phenomenon which they observed is so-called immune mimicry. It means that the possession of immunity gene homologues assure as well the resistance against certain bacteriocins.

14.9 Inhibition of Adhesion of Pathogens to the Intestinal Enterocytes

The inhibitory effect of selected probiotic strains on the adherence and invasion of enteropathogenic bacteria such as enteropathogenic *E. coli*, enterotoxigenic *E. coli*, *K. pneumonia* and *Salmonella* is well known and may prevent colonisation of the GIT by pathogens. Usually viable cells are required for such activity, however, it seems that metabolites of LAB including bacteriocins can also contribute to inhibition of adhesion. This may occur through stearic hindrance of pathogen adhesion sites or biochemical hindrance through probiotic production of bacteriocin and non-bacteriocin antibacterial substances.

In the recent study of Lim et al. (2009), partially purified bacteriocins of *Lb. plantarum* K11 inhibited adhesion of *E. coli* O157 to Caco-2 cells. The adhesion inhibitory effects of the cell free culture supernatant (CFCS) and bacteriocins were dose-dependent. The inhibitory activity of adhesion by the heated bacteriocin for 30 min at 100°C was similar to the activity of non-treated bacteriocin, but the activity disappeared after treatment with protease. The inhibitory substance still remains to be identified.

Coconnier et al. (1992) studied adhesion of *Lb. acidophilus* BG2FO4 on Caco-2 and HT29-MTX cells. The adhesion was increased by adhesion-promoting factor present in the supernatant which was proteinaceous, since trypsin treatment dramatically decreased the adhesion of the *Lb. acidophilus* BG2FO4 strain. No further data are available about that substance.

Trejo et al. (2006) observed *in vitro* the growth inhibition of *Cl. difficile* (ATCC 9689 and ATCC 43593) by the supernatants of bifidobacteria isolated from healthy infants, as well as the inhibition of clostridia adhesion to Caco-2 cells. Although the substances responsible for inhibition were unaffected by proteolytic cleavage (proteinase K and chymotrypsin), the involvement of bacteriocins could not be excluded.

14.10 Conclusions

Currently the designation probiotic is still more or less confined to lactic acid bacteria and bifidobacteria, mostly due to the tradition of consuming probiotics in the form of dairy products. For similar reasons, application of bacteriocins in foods is limited exclusively to those produced by LAB. Since the use of probiotics in a form of dietary supplements and drugs is increasing, there are, however, no convincing reasons for why not to widen the spectrum of probiotic strains to the other genera common in GIT. This way, many interesting new bacteriocins produced by probiotic bacteria may also be identified.

Advances in genetic engineering have already contributed much in the identification of new bacteriocins as well as in discovering their mode of activity, and regulatory mechanisms. New strategies used in functional genomics, like targeted mutagenesis and others, offer new possibilities for studying the function of bacteriocin genes. There are also great potentials in the field of heterologous production of bacteriocins, where probiotic bacteria for instance may serve as host organisms for particular heterologous bacteriocin genes. Such recombinant probiotics producing natural or modified bacteriocins might target specific gastrointestinal disorders. Recombinant probiotics are being increasingly constructed and tested also *in vivo*. Probably this kind of therapeutic applications might be reserved for probiotic food supplements or drugs, but not for functional foods.

The construction and analysis of genetically modified bacteriocins, the structural analysis by circular dichroism spectroscopy (CD) and nuclear magnetic resonance (NMR), and increasing number of the whole genome sequences of probiotic bacteria will enable to better identify the relations between structure and function of bacteriocins. Significant progress has been made and may be expected in the field of protein-engineering used to synthesise modified bacteriocins with improved properties, used to modify the activity and inhibitory spectrum of bacteriocins. These new antimicrobials are expected to be used as an alternative for treating the infections caused by multi drug-resistant pathogens, however, such applications are out of the scope of this review, dealing with bacteriocins of probiotics. Namely, by definition probiotics have to reach the target site alive, and exert probiotic activity including bacteriocin production *in situ* in GIT.

Most of the *in vivo* studies on direct inhibition of pathogens in the gut by bacteriocins performed to date lacked appropriate controls such as mutants of probiotic strains without ability to produce bacteriocins, making conclusions about the role of bacteriocins in the observed inhibition difficult. One of the phenomena which also needs to be further examined is the occurrence of resistance against bacteriocins which was already observed in the *Listeria monocytogenes*. To date there is also no data about the possibility of the transfer of bacteriocin immunity genes to pathogenic bacteria.

Beside *in vitro* and *in vivo* established inhibitory effect of bacteriocins against bacteria including pathogenic ones, bacteriocins can also function as signalling molecules. This is now clear for instance for lantibiotics biosynthesis, where bacteriocin itself serves as a signalling molecule for its biosynthesis. The signalling role of bacteriocins in the interactions with the other bacteria in the GI biofilms or in the interactions with the host immune system, however, remain one of the main challenges and it is expected that the results will enable further development of therapeutic applications of bacteriocins of probiotic bacteria.

Considering the increasing interest in bacteriocins, the progress in genomics and in human and mammalian microbiome research, we can expect that in the future applications such as treatment of GI bacterial infections will also become available.

References

Abriouel H, Maqueda M, Galvez A, Martinez-Bueno M, Valdivia E (2002) Inhibition of bacterial growth, enterotoxin production, and spore outgrowth in strains of *Bacillus cereus* by bacteriocin AS-48. Appl Environ Microbiol 68:1473–1477

Allison GE, Fremaux C, Klaenhammer TR (1994) Expansion of bacteriocin activity and host-range upon complementation of 2 peptides encoded within the lactacin-F operon. J Bacteriol 176:2235–2241

Allison GE, Klaenhammer TR (1996) Functional analysis of the gene encoding immunity to lactacin F, lafI, and its use as a *Lactobacillus*-specific, food-grade genetic marker. Appl Environ Microbiol 62:4450–4460

Altenhoefer A, Oswald S, Sonnenborn U, Enders C, Schulze J, Hacker J, Oelschlaeger TA (2004) The probiotic *Escherichia coli* strain Nissle 1917 interferes with invasion of human intestinal epithelial cells by different enteroinvasive bacterial pathogens. Fems Immun Med Microbiol 40:223–229

Anderssen EL, Diep DB, Nes IF, Eijsink VGH, Nissen-Meyer J (1998) Antagonistic activity of *Lactobacillus plantarum* C11: two new two-peptide bacteriocins, plantaricins EF and JK, and the induction factor plantaricin A. Appl Environ Microbiol 64:2269–2272

Aymerich T, Holo H, Havarstein LS, Hugas M, Garriga M, Nes IF (1996) Biochemical and genetic characterization of enterocin A from *Enterococcus faecium*, a new antilisterial bacteriocin in the pediocin family of bacteriocins. Appl Environ Microbiol 62:1676–1682

Barefoot SF, Chen YR, Hughes TA, Bodine AB, Shearer MY, Hughes MD (1994) identification and purification of a protein that induces production of the Lactobacillus acidophilus bacteriocin lactacin B. Appl Environ Microbiol 60:3522–3528

Barefoot SF, Klaenhammer TR (1983) Detection and activity of lactacin-B, a bacteriocin produced by *Lactobacillus acidophilus*. Appl Environ Microbiol 45:1808–1815

Barefoot SF, Klaenhammer TR (1984) Purification and characterization of the *Lactobacillus acidophilus* bacteriocin lactacin-B. Antimicrob Agents Chemother 26:328–334

Barrett E, Hayes M, O'connor P, Gardiner G, Fitzgerald GF, Stanton C, Ross RP, Hill C (2007) Salivaricin P, one of a family of two-component antilisterial bacteriocins produced by intestinal isolates of *Lactobacillus salivarius*. Appl Environ Microbiol 73:3719–3723

Barth S, Duncker S, Hempe J, Breves G, Baljer G, Bauerfeind R (2009) *Escherichia coli* Nissle 1917 for probiotic use in piglets: evidence for intestinal colonization. J Appl Microbiol 107:1697–1710

Bergonzelli GE, Granato D, Pridmore RD, Marvin-Guy LF, Donnicola D, Corthesy-Theulaz IE (2006) GroEL of *Lactobacillus johnsonii* La1 (NCC 533) is cell surface associated: potential role in interactions with the host and the gastric pathogen *Helicobacter pylori*. *Infect Immun* 74:425–434

Bernbom N, Licht TR, Brogren C-H, Jelle B, Johansen AH, Badiola I, Vogensen FK, Norrung B (2006a) Effects of *Lactococcus lactis* on composition of intestinal microbiota: role of Nisin. Appl Environ Microbiol 72:239–244

Bernbom N, Licht TR, Saadbye P, Vogensen FK, Norrung B (2006b) *Lactobacillus plantarum* inhibits growth of *Listeria monocytogenes* in an *in vitro* continuous flow gut model, but promotes invasion of *L. monocytogenes* in the gut of gnotobiotic rats. Int J Food Microbiol 108:10–14

Bernbom N, Jelle B, Brogren CH, Vogensen FK, Norrung B, Licht TR (2009) Pediocin PA-1 and a pediocin producing *Lactobacillus plantarum* strain do not change the HMA rat microbiota. Int J Food Microbiol 130:251–257

Bierbaum G, Sahl HG (2009) Lantibiotics: mode of action, biosynthesis and bioengineering. Curr Pharm Biotechnol 10:2–18

Bogovič-Matijašić B, Rogelj I (1998) Bacteriocin complex of *Lactobacillus acidophilus* LF221—production studies in MRS media at different pH values and effect against *Lactobacillus helveticus* ATCC 15009. Process Biochem 33:345–352

Bogovič-Matijašić B, Rogelj I, Nes IF, Holo H (1998) Isolation and characterization of two bacteriocins of *Lactobacillus acidophilus* LF221. Appl Microbiol Biotechnol 49:606–612

Boris S, Jimenez-Diaz R, Caso JL, Barbes C (2001) Partial characterization of a bacteriocin produced by *Lactobacillus delbrueckii* subsp *lactis* UO004, an intestinal isolate with probiotic potential. J Appl Microbiol 91:328–333

Brotz H, Josten M, Wiedemann I, Schneider U, Gotz F, Bierbaum G, Sahl HG (1998) Role of lipid-bound peptidoglycan precursors in the formation of pores by nisin, epidermin and other lantibiotics. Mol Microbiol 30:317–327

Burton JP, Chilcott CN, Moore CJ, Speiser G, Tagg JR (2006a) A preliminary study of the effect of probiotic *Streptococcus salivarius* K12 on oral malodour parameters. J Appl Microbiol 100:754–764

Burton JP, Wescombe PA, Moore CJ, Chilcott CN, Tagg JR (2006b) Safety assessment of the oral cavity probiotic *Streptococcus salivarius* K12. Appl Environ Microbiol 72:3050–3053

Calvez S, Rince A, Auffray Y, Prevost H, Drider D (2007) Identification of new genes associated with intermediate resistance of *Enterococcus faecalis* to divercin V41, a pediocin-like bacteriocin. Microbiology 153:1609–1618

Calvez S, Prevost H, Drider D (2008) Identification of a new molecular target of class IIa bacteriocins in *Listeria monocytogenes* EGDe. Folia Microbiol 53:417–422

Casadei G, Grilli E, Piva A (2009) Pediocin A modulates intestinal microflora metabolism in swine in vitro intestinal fermentations. J Anim Sci 87:2020–2028

Cascales E, Buchanan SK, Duche D, Kleanthous C, Lloubes R, Postle K, Riley M, Slatin S, Cavard D (2007) Colicin biology. Microbiol Mol Biol Rev 71:158–229

Chatterjee C, Paul M, Xie LL, Van der Donk WA (2005) Biosynthesis and mode of action of lantibiotics. Chem Rev 105:633–683

Cheikhyoussef A, Pogori N, Chen W, Zhang H (2008) Antimicrobial proteinaceous compounds obtained from bifidobacteria: From production to their application. Int J Food Microbiol 125:215–222

Chen CC, Chiu CH, Lin TY, Shi HN, Walker WA (2009) Effect of probiotics *Lactobacillus acidophilus* on *Citrobacter rodentium* colitis: the role of dendritic cells. Pediatr Res 65:169–175

Chumchalova J, Stiles J, Josephsen J, Plockova M (2004) Characterization and purification of acidocin CH5, a bacteriocin produced by *Lactobacillus acidophilus* CH5. J Appl Microbiol 96:1082–1089

Cigankova V, Laukova A, Guba P, Nemcova R (2004) Effect of enterocin A on the intestinal epithelium of Japanese quails infected by *Salmonella duesseldorf.* Bull Vet Inst Pulawy 48:25–27

Cintas LM, Rodriguez JM, Fernandez MF, Sletten K, Nes IF, Hernandez PE, Holo H (1995) Isolation and characterization of pediocin I50, a new bacteriocin from *Pediococcus acidilactici* with a broad inhibitory spectrum. Appl Environ Microbiol 61:2643–2648

Cleveland J, Montville TJ, Nes IF, Chikindas ML (2001) Bacteriocins: safe, natural antimicrobials for food preservation. Int J Food Microbiol 71:1–20

Cobos ES, Filimonov VV, Galvez A, Maqueda N, Valdivia E, Martinez JC, Mateo PL (2001) AS-48: a circular protein with an extremely stable globular structure. FEBS Lett 505:379–382

Coconnier MH, Klaenhammer TR, Kernéis S, Bernet MF, Servin AL (1992) Protein-mediated adhesion of Lactobacillus acidophilus BG2FO4 on human enterocyte and mucus-secreting cell lines in culture. Appl Environ Microbiol 58:2034–2039

Cole K, Farnell MB, Donoghue AM, Stern NJ, Svetoch EA, Eruslanov BN, Volodina LI, Kovalev YN, Perelygin VV, Mitsevich EV, Mitsevich IP, Levchuk VP, Pokhilenko VD, Borzenkov VN, Svetoch OE, Kudryavtseva TY, Reyes-Herrera I, Blore PJ, De Los Santos FS, Donoghue DJ (2006) Bacteriocins reduce *Campylobacter* colonization and alter gut morphology in turkey poults. Poultry Sci 85:1570–1575

Collado MC, Hernández M, Sanz Y (2005) Production of bacteriocin-like compounds by human faecal *Bifidobacterium* strains. J Food Protect 68:1034–1040

Corr SC, Li Y, Riedel CU, O'toole PW, Hill C, Gahan CGM (2007) Bacteriocin production as a mechanism for the antiinfective activity of *Lactobacillus salivarius* UCC118. Proc Natl Acad Sci U S A 104:7617–7621

Cosseau C, Devine DA, Dullaghan E, Gardy JL, Chikatamarla A, Gellatly S, Yu LL, Pistolic J, Falsafi R, Tagg J, Hancock REW (2008) The commensal *Streptococcus salivarius* K12 downregulates the innate immune responses of human epithelial cells and promotes host-microbe homeostasis. Infect Immun 76:4163–4175

Costilow RN, Coughlin FM, Robach DL, Ragheb HS (1956) A study of the acid-forming bcteria from cucumber fermentations in Michigan. J Food Sci 21:27–33

Cotter PD, Hill C, Ross RP (2005) Bacterial lantibiotics: strategies to improve therapeutic potential. Curr Protein Peptide Sci 6:61–75

Daoudi L, Turcotte C, Lacroix C, Fliss I (2001) Production and characterization of anti-nisin Z monoclonal antibodies: suitability for distinguishing active from inactive forms through a competitive enzyme immunoassay. Appl Microbiol Biotechnol 56:114–119

de Pablo MA, Gaforio JJ, Gallego AM, Ortega E, Galvez AM, Lopez GAD (1999) Evaluation of immunomodulatory effects of nisin-containing diets on mice. Fems Immunol Med Microbiol 24:35–42

De Vuyst L, Leroy F (2007) Bacteriocins from lactic acid bacteria: production, purification, and food applications. J Mol Microbiol Biotechnol 13:194–199
Dobson AE, Sanozky-Dawes RB, Klaenhammer TR (2007) Identification of an operon and inducing peptide involved in the production of lactacin B by *Lactobacillus acidophilus*. J Appl Microbiol 103:1766–1778
Dougherty BA, Hill C, Weidman JF, Richardson DR, Venter JC, Ross RP (1998) Sequence and analysis of the 60 kb conjugative, bacteriocin-producing plasmid pMRC01 from *Lactococcus lactis* DPC3147. Mol Microbiol 29:1029–1038
Draper LA, Grainger K, Deegan LH, Cotter PD, Hill C, Ross RP (2009) Cross-immunity and immune mimicry as mechanisms of resistance to the lantibiotic lacticin 3147. Mol Microbiol 71:1043–1054
Drider D, Fimland G, Hechard Y, Mcmullen LM, Prevost H (2006) The continuing story of class IIa bacteriocins. Microbiol Mol Biol Rev 70:564–582
Dunne C, Murphy L, Flynn S, O'Mahony L, O'Halloran S, Feeney M, Morrissey D, Thornton G, Fitzgerald G, Daly C, Kiely B, Quigley EM, O'Sullivan GC, Shanahan F, Collins JK (1999) Probiotics: from myth to reality. Demonstration of functionality in animal models of disease and in human clinical trials. Antonie Van Leeuwenhoek 76:279–292
Duquesne S, Destoumieux-Garzon D, Peduzzi J, Rebuffat S (2007a) Microcins, gene-encoded antibacterial peptides from enterobacteria. Nat Product Rep 24:708–734
Duquesne S, Petit V, Peduzzi J, Rebuffat S (2007b) Structural and functional diversity of microcins, gene-encoded antibacterial peptides from enterobacteria. J Mol Microbiol Biotechnol 13:200–209
Ennahar S, Sashihara T, Sonomoto K, Ishizaki A (2000) Class IIa bacteriocins: biosynthesis, structure and activity. Fems Microbiol Rev 24:85–106
Fayol-Messaoudi D, Berger CN, Coconnier-Polter M-H, Liévin-Le Moal V, Servin AL (2005) pH-, lactic acid-, and non-lactic acid-dependent activities of probiotic lactobacilli against *Salmonella enterica* Serovar *Typhimurium*. Appl Environ Microbiol 71:6008–6013
Flynn S, Van Sinderen D, Thornton GM, Holo H, Nes IF, Collins JK (2002) Characterization of the genetic locus responsible for the production of ABP-118, a novel bacteriocin produced by the probiotic bacterium *Lactobacillus salivarius* subsp *salivarius* UCC118. Microbiology 148:973–984
Fujiwara S, Seto Y, Kimura A, Hashiba H (2001) Establishment of orally-administered *Lactobacillus gasseri* SBT2055SR in the gastrointestinal tract of humans and its influence on intestinal microflora and metabolism. J Appl Microbiol 90:343–352
Gardiner GE, Rea MC, O'riordan B, O'connor P, Morgan SM, Lawlor PG, Lynch PB, Cronin M, Ross RP, Hill C (2007) Fate of the two-component lantibiotic lacticin 3147 in the gastrointestinal tract. Appl Environ Microbiol 73:7103–7109
Ghobrial OG, Derendorf H, Hillman JD (2009) Pharmacodynamic activity of the lantibiotic MU1140. Int J Antimicrob Agents 33:70–74
Gillor O, Etzion A, Riley MA (2008) The dual role of bacteriocins as anti- and probiotics. Appl Microbiol Biotechnol 81:591–606
Gillor O, Giladi I, Riley MA (2009) Persistence of colicinogenic *Escherichia coli* in the mouse gastrointestinal tract. BMC Microbiol 9:7–13
Gotteland M, Brunser O, Cruchet S (2006) Systematic review: are probiotics useful in controlling gastric colonization by *Helicobacter pylori*? Aliment Pharm Ther 23:1077–1086
Habib WA, Sakr A (1999) Development and human *in vivo* evaluation of a colonic drug delivery system. Pharm Ind 61:1145–1149
Hakovirta J, Reunanen J, Saris PEJ (2006) Bioassay for nisin in milk, processed cheese, salad dressings, canned tomatoes, and liquid egg products. Appl Environ Microbiol 72:1001–1005
Hall-Stoodley L, Costerton JW, Stoodley P (2004) Bacterial biofilms: from the natural environment to infectious diseases. Nat Rev Microbiol 2:95–108
Henderson JT, Chopko AL, Vanwassenaar PD (1992) Purification and primary structure of pediocin PA-1 produced by *Pediococcus acidilactici* PAC-10. Arch Biochem Biophys 295:5–12

Henker J, Laass M, Blokhin BM, Bolbot YK, Maydannik VG, Elze M, Wolff C, Schulze J (2007) The probiotic *Escherichia coli* strain Nissle 1917 (EcN) stops acute diarrhoea in infants and toddlers. Eur J Pediatr 166:311–318

Hyink O, Wescombe PA, Upton M, Ragland N, Burton JP, Tagg JR (2007) Salivaricin A2 and the novel lantibiotic salivaricin B are encoded at adjacent loci on a 190-kilobase transmissible megaplasmid in the oral probiotic strain *Streptococcus salivarius* K12. Appl Environ Microbiol 73:1107–1113

Immonen N, Karp M (2007) Bioluminescence-based bioassays for rapid detection of nisin in food. Biosens Bioelectron 22:1982–1987

Ito Y, Kawai Y, Arakawa K, Honme Y, Sasaki T, Saito T (2009) Conjugative plasmid from *Lactobacillus gasseri* LA39 that carries genes for production of and immunity to the circular bacteriocin gassericin A. Appl Environ Microbiol 75:6340–6351

Jain PK, Mcnaught CE, Anderson ADG, Macfie J, Mitchell CJ (2004) Influence of synbiotic containing *Lactobacillus acidophilus* La5, *Bifidobacterium lactis* Bb 12, *Streptococcus thermophilus, Lactobacillus bulgaricus* and oligofructose on gut barrier function and sepsis in critically ill patients: a randomised controlled trial. Clin Nutr 23:467–475

Jenssen H, Hamill P, Hancock REW (2006) Peptide antimicrobial agents. Clin Microbiol Rev 19:491–511

Kabuki T, Saito T, Kawai Y, Uemura J, Itoh T (1997) Production, purification and characterization of reutericin 6, a bacteriocin with lytic activity produced by *Lactobacillus reuteri* LA6. Int J Food Microbiol 34:145–156

Kauffman CA (2003) Therapeutic and preventative options for the management of vancomycin-resistant enterococcal infections. J Antimicrob Chemother 51:23–30

Kawai Y, Saito T, Uemura J, Itoh T (1997) Rapid detection method for bacteriocin and distribution of bacteriocin-producing strains in *Lactobacillus acidophilus* group lactic acid bacteria isolated from human faeces. Biosci Biotechnol Biochem 61:179–182

Kawai Y, Saito T, Kitazawa H, Itoh T (1998a) Gassericin A; an uncommon cyclic bacteriocin produced by *Lactobacillus gasseri* LA39 linked at N- and C-terminal ends. Biosci Biotechnol Biochem 62:2438–2440

Kawai Y, Saito T, Suzuki M, Itoh T (1998b) Sequence analysis by cloning of the structural gene of gassericin A, a hydrophobic bacteriocin produced by *Lactobacillus gasseri* LA39. Biosci Biotechnol Biochem 62:887–892

Kawai Y, Saitoh B, Takahashi O, Kitazawa H, Saito T, Nakajima H, Itoh T (2000) Primary amino acid and DNA sequences of gassericin T, a lactacin F-family bacteriocin produced by *Lactobacillus gasseri* SBT2055. Biosci Biotechnol Biochem 64:2201–2208

Kawai Y, Ishii Y, Uemura K, Kitazawa H, Saito T, Itoh T (2001) *Lactobacillus reuteri* LA6 and *Lactobacillus gasseri* LA39 isolated from faeces of the same human infant produce identical cyclic bacteriocin. Food Microbiol 18:407–415

Kawai Y, Kusnadi J, Kemperman R, Kok J, Ito Y, Endo M, Arakawa K, Uchida H, Nishimura J, Kitazawa H, Saito T (2009) DNA sequencing and homologous expression of a small peptide conferring immunity to gassericin A, a circular bacteriocin produced by *Lactobacillus gasseri* LA39. Appl Environ Microbiol 75:1324–1330

Keren T, Yarmus M, Halevy G, Shapira R (2004) Immunodetection of the bacteriocin lacticin RM: analysis of the influence of temperature and tween 80 on its expression and activity. Appl Environ Microbiol 70:2098–2104

Kim TS, Hur JW, Yu MA, Cheigh CI, Kim KN, Hwang JK, Pyun YR (2003) Antagonism of *Helicobacter pylori* by bacteriocins of lactic acid bacteria. J Food Protect 66:3–12

Kimoto-Nira H, Mizumachi K, Nomura M, Kobayashi M, Fujita Y, Okamoto T, Suzuki I, Tsuji NM, Kurisakv JI, Ohmomo S (2007) *Lactococcus* sp as potential probiotic lactic acid bacteria. Jarq-Jpn Agric Res Q 41:181–189

Kirkup BC (2006) Bacteriocins as oral and gastrointestinal antibiotics: theoretical considerations, applied research, and practical applications. Curr Med Chem 13:3335–3350

Klaenhammer TR (1993) Genetics of bacteriocins produced by lactic-acid bacteria. FEMS Microbiol Rev 12:39–86

Kleerebezem M, Beerthuyzen MM, Vaughan EE, Devos WM, Kuipers OP (1997) Controlled gene expression systems for lactic acid bacteria: transferable nisin-inducible expression cassettes for *Lactococcus, Leuconostoc*, and *Lactobacillus* spp. Appl Environ Microbiol 63:4581–4584
Laukova A, Guba P, Nemcova R, Vasilkova Z (2003) Reduction of *Salmonella* in gnotobiotic Japanese quails caused by the enterocin A-producing EK13 strain of *Enterococcus faecium*. Vet Res Commun 27:275–280
Laukova A, Guba P, Nemcova R, Marekova M (2004) Inhibition of *Salmonella enterica* serovar *Dusseldorf* by enterocin A in gnotobiotic Japanese quails. Vet Med 49:47–51
Laukova A, Strompfova V, Skrivanova V, Volek Z, Jindrichova E, Marounek M (2006) Bacteriocin-producing strain of *Enterococcus faecium* EK 13 with probiotic character and its application in the digestive tract of rabbits. Biol Bratisl 61:779–782
Lawton EM, Ross RP, Hill C, Cotter PD (2007) Two-peptide lantibiotics: a medical perspective. Mini-Rev Med Chem 7:1236–1247
Lee YK, Lim CY, Teng WL, Ouwehand AC, Tuomola EM, Salminen S (2000) Quantitative approach in the study of adhesion of lactic acid bacteria to intestinal cells and their competition with enterobacteria. Appl Environ Microbiol 66:3692–3697
Leer RJ, van der Vossen J, Vangiezen M, Vannoort JM, Pouwels PH (1995) Genetic-analysis of acidocin-B, a novel bacteriocin produced by *Lactobacillus acidophilus*. Microbiol U K 141;1629–1635
Leisner JJ, Laursen BG, Prevost H, Drider D, Dalgaard P (2007) *Carnobacterium*: positive and negative effects in the environment and in foods. FEMS Microbiol Rev 31:592–613
Leyer GJ, Li SG, Mubasher ME, Reifer C, Ouwehand AC (2009) Probiotic effects on cold and influenza-like symptom incidence and duration in children. Pediatrics 124:E172–E179
Liévin V, Peiffer I, Hudault S, Rochat F, Brassart D, Neeser JR, Servin AL (2000) Bifidobacterium strains from resident infant human gastrointestinal microflora exert antimicrobial activity. Gut 47:646–652
Lim SM, Im DS (2007) Bactericidal effect of bacteriocin of *Lactobacillus plantarum* K11 isolated from dongchimi on *Escherichia coli* O157. J Food Hyg Safety 22:151–158
Lim JH, Kim SD (2009) Selection and characterization of the bacteriocin-producing bacterium, *Bacillus subtilis* BP6 isolated from chicken gut against *Salmonella gallinarum* causing fowl-typhus. J Korean Soc Appl Biol Chem 52:80–87
Lim SM, Ahn DH, Im DS (2009) Inhibitory effect of *Lactobacillus plantarum* K11 on the adhesion of *Escherichia coli* O157 to Caco-2 cells. Food Sci Biotechnol 18:343–349
Lin J (2009) Novel approaches for *Campylobacter* control in poultry. Foodborne Pathog and Dis 6:755–765
Lin WH, Lin CK, Sheu SJ, Hwang CF, Ye WT, Hwang WZ, Tsen HY (2009) Antagonistic activity of spent culture supernatants of lactic acid bacteria against *Helicobacter Pylori* growth and infection in human gastric epithelial AGS cells. J Food Sci 74:M225–M230
Line JE, Svetoch EA, Eruslanov BV, Perelygin VV, Mitsevich EV, Mitsevich IP, Levchuk VP, Svetoch OE, Seal BS, Siragusa GR, Stern NJ (2008) Isolation and purification of enterocin E-760 with broad antimicrobial activity against gram-positive and gram-negative bacteria. Antimicrob Agents Chemother 52:1094–1100
Lodinová-Zádniková R, Sonnenborn U (1997) Effect of preventive administration of a nonpathogenic *Escherichia coli* strain on the colonization of the intestine with microbial pathogens in newborn infants. Biol Neonate 71:224–232
Lubelski J, Rink R, Khusainov R, Moll GN, Kuipers OP (2008) Biosynthesis, immunity, regulation, mode of action and engineering of the model lantibiotic nisin. Cell Mol Life Sci 65:455–476
Majhenič AC, Venema K, Allison GE, Matijašić BB, Rogelj I, Klaenhammer TR (2004) DNA analysis of the genes encoding acidocin LF221 A and acidocin LF221 B, two bacteriocins produced by *Lactobacillus gasseri* LF221. Appl Microbiol Biotechnol 63:705–714
Maldonado-Barragán A, Caballero-Guerrero B, Jimenez E, Jimenez-Diaz R, Ruiz-Barba JL, Rodriguez JM (2009) Enterocin C, a class IIb bacteriocin produced by *E. faecalis* C901, a strain isolated from human colostrum. Int J Food Microbiol 133:105–112

Mandal V, Sen SK, Mandal NC (2009) Effect of prebiotics on bacteriocin production and cholesterol lowering activity of *Pediococcus acidilactici* LAB 5. World J Microbiol Biotechnol 25:1837–1847

Maqueda M, Galvez A, Bueno MM, Sanchez-Barrena MJ, Gonzalez C, Albert A, Rico M, Valdivia E (2004) Peptide AS-48: prototype of a new class of cyclic bacteriocins. Curr Protein Peptide Sci 5:399–416

Maqueda M, Sanchez-Hidalgo M, Fernandez M, Montalban-Lopez M, Valdivia E, Martinez-Bueno M (2008) Genetic features of circular bacteriocins produced by Gram-positive bacteria. Fems Microbiol Rev 32:2–22

Maré L, Wolfaardt GM, Dicks LMT (2006) Adhesion of *Lactobacillus plantarum* 423 and *Lactobacillus salivarius* 241 to the intestinal tract of piglets, as recorded with fluorescent in situ hybridization (FISH), and production of plantaricin 423 by cells colonized to the ileum. J Appl Microbiol 100:838–845

Marekova M, Laukova A, Devuyst L, Skaugen M, Nes IF (2003) Partial characterization of bacteriocins produced by environmental strain *Enterococcus faecium* EK13. J Appl Microbiol 94:523–530

Marugg JD, Gonzalez CF, Kunka BS, Ledeboer AM, Pucci MJ, Toonen MY, Walker SA, Zoetmulder LCM, Vandenbergh PA (1992) Cloning, expression, and nucleotide-sequence of genes involved in production of pediocin PA-1, a bacteriocin from *Pediococcus acidilactici* PAC1.0. Appl Environ Microbiol 58:2360–2367

Matijašić BB, Rogelj I (1999) Bacteriocinogenic activity of lactobacilli isolated from cheese and baby faeces. Food Technol Biotechnol 37:93–100

Matijašić BB, Rogelj I (2000) *Lactobacillus* K7—a new candidate for a probiotic strain. Food Technol Biotechnol 38:113–119

Matijašić BB, Narat M, Zoric M (2003) Adhesion of two *Lactobacillus gasseri* probiotic strains on Caco-2 cells. Food Technol Biotechnol 41:83–88

Matijašić BB, Stojković S, Salobir J, Malovrh S, Rogelj I (2004) Evaluation of the *Lactobacillus gasseri* K7 and LF221 strains in weaned piglets for their possible probiotic use and their detection in the faeces. Anim Res 53:35–44

Matijašić BB, Narat M, Peternel MZ, Rogelj I (2006) Ability of *Lactobacillus gasseri* K7 to inhibit *Escherichia coli* adhesion in vitro on Caco-2 cells and ex vivo on pigs' jejunal tissue. Int J Food Microbiol 107:92–96

Matijašić BB, Rajsp MK, Perko B, Rogelj I (2007) Inhibition of *Clostridium tyrobutyricum* in cheese by *Lactobacillus gasseri.* Int Dairy J 17:157–166

Mcauliffe O, Ryan MP, Ross RP, Hill C, Breeuwer P, Abee T (1998) Lacticin 3147, a broad-spectrum bacteriocin which selectively dissipates the membrane potential. Appl Environ Microbiol 64:439–445

Mcauliffe O, Ross RP, Hill C (2001) Lantibiotics: structure, biosynthesis and mode of action. Fems Microbiol Rev 25:285–308

Medellin-Peña MJ, Griffiths MW (2009) Effect of molecules secreted by *Lactobacillus acidophilus* Strain La-5 on *Escherichia coli* O157:H7 colonization. Appl Environ Microbiol 75:1165–1172

Miguel MAL, De Castro NCD, Leite SFG (2008) Inhibition of vancomycin and high-level aminoglycoside-resistant enterococci strains and *Listeria monocytogenes* by bacteriocin-like substance produced by *Enterococcus faecium* E86. Curr Microbiol 57:429–436

Millette M, Cornut G, Dupont C, Shareck F, Archambault D, Lacroix M (2008a) Capacity of human nisin- and pediocin-producing lactic acid bacteria to reduce intestinal colonization by vancomycin-resistant enterococci. Appl Environ Microbiol 74:1997–2003

Millette M, Dupont C, Shareck F, Ruiz MT, Archambault D, Lacroix M (2008b) Purification and identification of the pediocin produced by *Pediococcus acidilactici* MM33, a new human intestinal strain. J Appl Microbiol 104:269–275

Mohan R, Koebnick C, Schildt J, Schmidt S, Mueller M, Possner M, Radke M, Blauti M (2006) Effects of *Bifidobacterium lactis* Bb12 supplementation on intestinal microbiota of preterm infants: a double-blind, placebo-controlled, randomized study. J Clin Microbiol 44:4025–4031

Moons P, Michiels CW, Aertsen A (2009) Bacterial interactions in biofilms. Crit Rev Microbiol 35:157–168
Morency H, Mota-Meira M, Lapointe G, Lacroix C, Lavoie MC (2001) Comparison of the activity spectra against pathogens of bacterial strains producing a mutacin or a lantibiotic. Can J Microbiol 47:322–331
Mortvedt CI, Nes IF (1990) Plasmid-associated bacteriocin production by a *Lactobacillus sake* strain. J Gen Microbiol 136:1601–1607
Mota-Meira M, Morency H, Lavoie MC (2005) *In vivo* activity of mutacin B-Ny266. J Antimicrob Chemother 56:869–871
Naaber P, Smidt I, Stsepetova J, Brilene T, Annuk H, Mikelsaar M (2004) Inhibition of *Clostridium difficile* strains by intestinal *Lactobacillus* species. J Med Microbiol 53:551–554
Nassif H, Zervos MJ (2005) Vancomycin-resistant enterococcal infections: epidemiology, clinical manifestations, and management. Infect Med 22:311–316
Nes IF, Eijsink VGH (1999) Regulation of group II peptide bacteriocin synthesis by quorum-sensing mechanisms. In: Dunny GM, Winans SC (eds) Cell-cell signaling in bacteria. ASM Press, Washington, pp 175–192
Nes IF, Yoon SS, Diep WB (2007) Ribosomally synthesiszed antimicrobial peptides (Bacteriocins) in lactic acid bacteria: a review. Food Sci Biotechnol 16:675–690
Nissen-Meyer J, Rogne P, Oppegard C, Haugen HS, Kristiansen PE (2009) Structure-function relationships of the non-lanthionine-containing peptide (class II) bacteriocins produced by gram-positive bacteria. Curr Pharm Biotechnol 10:19–37
Oppegard C, Fimland G, Thorbaek L, Nissen-Meyer J (2007a) Analysis of the two-peptide bacteriocins lactococcin G and enterocin 1071 by site-directed mutagenesis. Appl Environ Microbiol 73:2931–2938
Oppegard C, Rogne P, Emanuelsen L, Kristiansen PE, Fimland G, Nissen-Meyer J (2007b) The two-peptide class II bacteriocins: Structure, production, and mode of action. J Mol Microbiol Biotechnol 13:210–219
Ouwehand AC, Salminen S, Isolauri E (2002) Probiotics: an overview of beneficial effects. Antonie Van Leeuwenhoek 82:279–89
Ouwehand AC, Tiihonen K, Saarinen M, Putaala H, Rautonen N (2009) Influence of a combination of *Lactobacillus acidophilus* NCFM and lactitol on healthy elderly: intestinal and immune parameters. Br J Nutr 101:367–375
Parente E, Brienza C, Moles M, Ricciardi A (1995) A comparison of methods for the measurement of bacteriocin activity. J Microbiol Method 22:95–108
Parkes GC, Sanderson JD, Whelan K (2009) The mechanisms and efficacy of probiotics in the prevention of *Clostridium difficile*-associated diarrhoea. Lancet Infect Dis 9:237–244
Parsek MR, Fuqua C (2004) Biofilms 2003: emerging themes and challenges in studies of surface-associated microbial life. J Bacteriol 186:4427–4440
Pingitore EV, Hebert EM, Nader-Macias ME, Sesma F (2009) Characterization of salivaricin CRL 1328, a two-peptide bacteriocin produced by *Lactobacillus salivarius* CRL 1328 isolated from the human vagina. Res Microbiol 160:401–408
Pons AM, Zorn N, Vignon D, Delalande F, Van Dorsselaer A, Cottenceau G (2002) Microcin E492 is an unmodified peptide related in structure to colicin V. Antimicrob Agents Chemother 46:229–230
Puertollano MA, Gaforio JJ, Galvez A, De Pablo MA, De Cienfuegos GA (2003) Analysis of pro-inflammatory cytokine production in mouse spleen cells in response to the lantibiotic nisin. Int J Antimicrob Agents 21:601–603
Qi FX, Merritt J, Lux R, Shi WY (2004) Inactivation of the ciaH gene in Streptococcus mutans diminishes mutacin production and competence development, alters sucrose-dependent biofilm formation, and reduces stress tolerance. Infect Immun 72:4895–4899
Rea MC, Clayton E, O'connor PM, Shanahan F, Kiely B, Ross RP, Hill C (2007) Antimicrobial activity of lacticin 3147 against clinical *Clostridium difficile* strains. J Med Microbiol 56:940–946

Reunanen J, Saris PEJ (2009) Survival of nisin activity in intestinal environment. Biotechnol Lett 31:1229–1232

Reunanen J, Saris PEJ (2003) Microplate bioassay for nisin in foods, based on nisin-induced green fluorescent protein fluorescence. Appl Environ Microbiol 69:4214–4218

Robertson PAW, O'dowd C, Burrells C, Williams P, Austin B (2000) Use of *Carnobacterium* sp as a probiotic for Atlantic salmon (*Salmo salar L.*) and rainbow trout (*Oncorhynchus mykiss, Walbaum*). Aquaculture 185:235–243

Rogelj I, Matijašić BB (2006) *Lactobacillus gasseri* LF221 and K7—from isolation to application. Biol Bratisl 61:761–769

Rogelj I, Narat M, Hocevar I (1999) The immune response in mice immunized with *Lactobacillus acidophilus* LF221—A potential probiotic strain. Food Technol Biotechnol 37:153–158

Rogelj I, Matijašić BB, Majhenič AC, Stojković S (2002) The survival and persistence of *Lactobacillus acidophilus* LF221 in different ecosystems. Int J Food Microbiol 76:83–91

Rogne P, Fimland G, Nissen-Meyer J, Kristiansen PE (2008) Three-dimensional structure of the two peptides that constitute the two-peptide bacteriocin lactococcin G. Biochim Biophys Acta-Proteins Proteomics 1784:543–554

Rokka S, Pihlanto A, Korhonen H, Joutsjoki V (2006) In vitro growth inhibition of *Helicobacter pylori* by lactobacilli belonging to the *Lactobacillus plantarum* group. Lett Appl Microbiol 43:508–513

Rokka S, Myllykangas S, Joutsjoki V (2008) Effect of specific colostral antibodies and selected lactobacilli on the adhesion of *Helicobacter pylori* on AGS cells and the *Helicobacter*-induced IL-8 production. Scand J Immunol 68:280–286

Ryan KA, Daly P, Li Y, Hooton C, O'toole PW (2008) Strain-specific inhibition of *Helicobacter pylori* by *Lactobacillus salivarius* and other lactobacilli. J Antimicrob Chemother 61:831–834

Ryan KA, O'hara AM, Van Pijkeren JP, Douillard FP, O'toole PW (2009) Lactobacillus salivarius modulates cytokine induction and virulence factor gene expression in *Helicobacter pylori*. J Med Microbiol 58:996–1005

Ryan MP, Rea MC, Hill C, Ross RP (1996) An application in cheddar cheese manufacture for a strain of *Lactococcus lactis* producing a novel broad-spectrum bacteriocin, lacticin 3147. Appl Environ Microbiol 62:612–619

Rychlik JL, Russell JB (2002) Bacteriocin-like activity of *Butyrivibrio fibrisolvens* JL5 and its effect on other ruminal bacteria and ammonia production. Appl Environ Microbiol 68:1040–1046

Saleh FA, El-Sayed EM (2004) Isolation and characterization of bacteriocins produced by *Bifidobacterium lactis* BB-12 and *Bifidobacterium longum* BB-46. In: Abd El-Salam MH, Kholif AM, Abdou SM, El-Shibiny S (eds) 9th Egyptian conference for dairy science and technology. International Agriculture Centre, Cairo, Egypt, 9–11 Oct 2004 (Abstract)

Sanchez B, Bressollier P, Urdaci MC (2008) Exported proteins in probiotic bacteria: adhesion to intestinal surfaces, host immunomodulation and molecular cross-talking with the host. FEMS Immunol Med Microbiol 54:1–17

Sanders ME, Klaenhammer TR (2001) The scientific basis of *Lactobacillus acidophilus* NCFM functionality as a probiotic. J Dairy Sci 84:319–331

Schamberger GP, Phillips RL, Jacobs JL, Diez-Gonzalez F (2004) Reduction of *Escherichia coli* O157 : H7 populations in cattle by addition of colicin E7-producing *E. coli* to feed. Appl Environ Microbiol 70:6053–6060

Severinov K, Semenova E, Kazakov A, Kazakov T, Gelfand MS (2007) Low-molecular-weight post-translationally modified microcins. Mol Microbiol 65:1380–1394

Sherman PM, Ossa JC, Johnson-Henry K (2009) Unraveling mechanisms of action of probiotics. Nutr Clin Pract 24:10–14

Simonova M, Laukova A (2007) Bacteriocin activity of enterococci from rabbits. Vet Res Commun 31:143–152

Simonova MP, Laukova A, Chrastinova L, Strompfova V, Faix S, Vasilkova Z, Ondruska L, Jurcik R, Rafay J (2009) *Enterococcus faecium* CCM7420, bacteriocin PPB CCM7420 and their effect in the digestive tract of rabbits. Czech J Anim Sci 54:376–386

Sit CS, Vederas JC (2008) Approaches to the discovery of new antibacterial agents based on bacteriocins. Biochem Cell Biol 86:116–123

Stern NJ, Svetoch EA, Eruslanov BV, Kovalev YN, Volodina LI, Perelygin VV, Mitsevich EV, Mitsevich IP, Levchuk VP (2005) Paenibacillus polymyxa purified bacteriocin to control *Campylobacter jejuni* in chickens. J Food Protect 68:1450–1453

Stern NJ, Svetoch EA, Eruslanov BV, Perelygin VV, Mitsevich EV, Mitsevich IP, Pokhilenko VD, Levchuk VP, Svetoch OE, Seal BS (2006) Isolation of a *Lactobacillus salivarius* strain and purification of its bacteriocin, which is inhibitory to *Campylobacter jejuni* in the chicken gastrointestinal system. Antimicrob Agents Chemother 50:3111–3116

Stevens KA, Sheldon BW, Klapes NA, Klaenhammer TR (1991) Nisin treatment for inactivation of *Salmonella* species and other gram-negative bacteria. Appl Environ Microbiol 57:3613–3615

Stritzker J, Weibel S, Hill PJ, Oelschlaeger TA, Goebel W, Szalay AA (2007) Tumor-specific colonization, tissue distribution, and gene induction by probiotic *Escherichia coli* Nissle 1917 in live mice. Int J Med Microbiol 297:151–162

Su WJ, Waechter MJ, Bourlioux P, Dolegeal M, Fourniat J, Mahuzier G (1987) Role of volatile fatty acids in colonization resistance to Clostridium difficile in gnotobiotic mice. Infect Immun 55:1686–1691

Sui J, Leighton S, Busta F, Brady L (2002) 16S ribosomal DNA analysis of the faecal lactobacilli composition of human subjects consuming a probiotic strain *Lactobacillus acidophilus* NCFM (R). J Appl Microbiol 93:907–912

Svetoch EA, Eruslanov BV, Perelygin VV, Mitsevich EV, Mitsevich IP, Borzenkov VN, Levchuk VP, Svetoch OE, Kovalev YN, Stepanshin YG, Siragusa GR, Seal BS, Stern NJ (2008a) Diverse antlmicrobial killing by *Enterococcus faecium* E 50-52 bacteriocin. J Agric Food Chem 56:1942–1948

Svetoch EA, Levchuk VP, Pokhilenko VD, Eruslanov BV, Mitsevich EV, Mitsevich IP, Perelygin VV, Stepanshin YG, Stern NJ (2008b) Inactivating methicillin-resistant *Staphylococcus aureus* and other pathogens by use of bacteriocins OR-7 and E 50-52. J Clin Microbiol 46:3863–3865

Tabasco R, Garcia-Cayuela T, Pelaez C, Requena T (2009) *Lactobacillus acidophilus* La-5 increases lactacin B production when it senses live target bacteria. Int J Food Microbiol 132:109–116

Tagg JR (2004) Prevention of streptococcal pharyngitis by anti-Streptococcus pyogenes bacteriocin-like inhibitor substances (BLIS) produced by Streptococcus salivarius. Indian J Med Res 119 Suppl:13–16

Tahara T, Oshimura M, Umezawa C, Kanatani K (1996) Isolation, partial characterization, and mode of action of acidocin J1132, a two-component bacteriocin produced by *Lactobacillus acidophilus* JCM 1132. Appl Environ Microbiol 62:892–897

Takahashi H, Fujita T, Suzuki Y, Benno Y (2006) Monitoring and survival of *Lactobacillus gasseri* SBT2055 in the human intestinal tract. Microbiol Immunol 50:867–870

Ten Brink B, Minekus M, van der Vossen JM, Leer RJ, Huis in't Veld JH (1994) Antimicrobial activity of lactobacilli: preliminary characterization of production of acidocin B, a novel bacteriocin produced by Lactobacillus acidophilus M46. J Appl Bacteriol 77:140–148

Toba T, Samant SK, Yoshioka E, Itoh T (1991) Reutericin-6, a new bacteriocin produced by *Lactobacillus reuteri* LA-6. Lett Appl Microbiol 13:281–286

Trejo FM, Minnaard J, Perez PF, De Antoni GL (2006) Inhibition of *Clostridium difficile* growth and adhesion to enterocytes by *Bifidobacterium* supernatants. Anaerobe 12:186–193

Van Reenen CA, Dicks LMT, Chikindas ML (1998) Isolation, purification and partial characterization of plantaricin 423, a bacteriocin produced by *Lactobacillus plantarum*. J Appl Microbiol 84:1131–1137

von Buenau R, Jaekel L, Schubotz E, Schwarz S, Stroff T, Krueger M (2005) *Escherichia coli* strain Nissle 1917: significant reduction of neonatal calf diarrhoea. J Dairy Sci 88:317–323

Walsh MC, Gardiner GE, Hart OM, Lawlor PG, Daly M, Lynch B, Richert BT, Radcliffe S, Giblin L, Hill C, Fitzgerald GF, Stanton C, Ross P (2008) Predominance of a bacteriocin-producing *Lactobacillus salivarius* component of a five-strain probiotic in the porcine ileum and effects on host immune phenotype. FEMS Microbiol Ecol 64:317–327

Wenus C, Goll R, Loken EB, Biong AS, Halvorsen DS, Florholmen J (2008) Prevention of antibiotic-associated diarrhoea by a fermented probiotic milk drink. Eur J Clin Nutr 62:299–301

Wilding I (2000) Site-specific drug delivery in the gastrointestinal tract. Crit Rev Ther Drug Carr Sys 17:557–620

Willey JM, van der Donk WA (2007) Lantibiotics: peptides of diverse structure and function. Ann Rev Microbiol 61:477–501

Yan F, Cao HW, Cover TL, Whitehead R, Washington MK, Polk DB (2007) Soluble proteins produced by probiotic bacteria regulate intestinal epithelial cell survival and growth. Gastroenterology 132:562–575

Yildirim Z, Johnson MG (1998) Characterization and antimicrobial spectrum of bifidocin B, a bacteriocin produced by *Bifidobacterium bifidum* NCFB 1454. J Food Protect 61:47–51

Yildirim Z, Winters DK, Johnson MG (1999) Purification, amino acid sequence and mode of action of bifidocin B produced by *Bifidobacterium bifidum* NCFB 1454. J Appl Microbiol 86:45–54

Zhu WM, Liu W, Wu DQ (2000) Isolation and characterization of a new bacteriocin from *Lactobacillus gasseri* KT7. J Appl Microbiol 88:877–886

Zorič Peternel M (2007) Activity of bacteriocin complex of *Lactobacillus gasseri* K7 and *Lactobacillus gasseri* LF221 strains in different expression systems: doctoral dissertation. University of Ljubljana, Ljubljana, p 138

Zou J, Dong J, YU XF (2009) Meta-analysis: lactobacillus containing quadruple therapy versus standard triple first-line therapy for *Helicobacter pylori* eradication. Helicobacter 14:97–107

Chapter 15
Probiotics in Clinical Practice as Therapeutics Against Enteric Disorders

Ouafae Karimi and A. S. Peña

15.1 Introduction

Probiotics are useful in mild enteric disorders that are limited to the presence of individual symptoms such as diarrhoea, constipation, and bloating. More controversial is their use in colonic diverticular disease or in one of the most common disorders, namely the irritable bowel syndrome. With few exceptions and in a limited well defined patients' group, no evidence exists that probiotics induce and maintain remission in patients with inflammatory bowel disease. In this chapter we review the current evidence of the value of probiotics in clinical practice of the following enteric disorders as well as the formulations and compositions of specific probiotics. It will be clear that most of the compositions used are not classified as medicaments, and most of them are free to obtain as "over-the-counter" preparations administered without medical prescriptions. Therefore, their use as therapeutics is very limited at present. This situation is likely to change in coming years. As shown in other chapters of this book a multidisciplinary approach is bringing scientific protocols to the field of probiotics and new technology to study the complicated field of gut microbiology and ecology is being applied to the study of functional and inflammatory diseases of the gut. These advances will permit the design of specific, tailor-made probiotics to be used in specific clinical situations. Antibiotics are known to disrupt the normal gut microbiology of the individual. For example, patients who received antibiotic treatment during initial acute infectious diarrhoea had significantly more and longer lasting IBS compared to those with a natural course (Barbara et al. 2009; Stermer et al. 2006; Mearin et al. 2005) and patients with the so-called postinfectious irritable syndrome treated with antibiotics during initial acute infections have an increased severity and duration of diarrhoea. Thus demonstrating that during antibiotic treatment the intestinal microbiota is damaged additionally. In some cases patients suffer from overgrowth of intestinal flora and

A. S. Peña (✉)
Department of Pathology, VU University Medical Center, Laboratory of Immunogenetics,
P.O. Box 7057, 1007 MB Amsterdam, The Netherlands
e-mail: pena.as@gmail.com

J. J. Malago et al. (eds.), *Probiotic Bacteria and Enteric Infections*,
DOI 10.1007/978-94-007-0386-5_15,

in this case the antibiotics are effective in reducing the overgrowth (Posserud et al. 2007). In many of these clinical situations probiotics may turn out to be the appropriate agents to restore the normal microbiota.

The current indications of probiotics in the management of the following intestinal disorders are briefly reviewed with an emphasis on data obtained in clinical trials: The established indications such as probiotics in diarrhoea as in acute episodes following bacterial, fungal or viral infections; possibly indications under study, such as in travellers' diarrhoea or in antibiotic-associated, and diarrhoea and complications produced by *Clostridium difficile*. We also review studies of the usefulness of probiotics in constipation and bloating, in irritable bowel syndrome, and finally in inflammatory bowel disease, ulcerative colitis, pouchitis and in Crohn's disease.

15.2 Probiotics in Acute Diarrhoea

Multiple studies in children have shown that *Lactobacillus*, administered orally, may have antidiarrhoeal properties. To determine the effect of *Lactobacillus* GG on the course of acute diarrhoea in hospitalized children, a prospective, and placebo-controlled, triple-blind clinical trial was carried out in Pakistan. Forty children (mean age, 13 months) received either oral *Lactobacillus* GG (n=21) or placebo (n=19) twice daily for 2 days, after rehydration in addition to the usual diet. The clinical course of diarrhoea was followed during the treatment period. The features for admission into the study groups were similar and were characterized by severe diarrhoea, malnutrition and inappropriate management before presentation. Response was evident on day 2, when the frequency of both vomiting and diarrhoea was less in the *Lactobacillus* group. In those patients with acute non-bloody diarrhoea (n=32), the percentage of children with persistent watery diarrhoea at 48 hours was significantly lower in the *Lactobacillus* group (31% versus 75%). No significant difference was observed after 48 hours in those with bloody diarrhoea (Raza et al. 1995).

Van Niel et al. (2002) conducted a meta-analysis of randomized, controlled studies to assess whether treatment with *Lactobacillus* improved clinical outcome in children with acute infectious diarrhoea. They conducted a search in bibliographic databases of traditional biomedical as well as complementary and alternative medicine literature published between 1966 and 2000. The original search yielded 26 studies, nine of which met the criteria. A reduction of 0.7 days in diarrhoea duration and a reduction of 1.6 stools for diarrhoea frequency were attained on day 2 of treatment in the participants who received *Lactobacillus* compared to those who received placebo. A preplanned subanalysis suggested a dose-effect relationship. The results of this meta-analysis suggested that *Lactobacillus* is safe and effective as a treatment for children with acute infectious diarrhoea.

Campylobacter jejuni (*C. jejuni*) is an important cause of bacterial-induced enterocolitis in humans in the developed world, caused by consuming infected food (Young et al. 2007). Ternhag et al. studied 101,855 patients who had an episode of microbiologically confirmed gastrointestinal infections in the period of 1997–2004, in Sweden. In 56% of cases of acute diarrhoea *C. jejuni* was isolated (Ternhag et al. 2008). Acute intestinal infection with this pathogen may involve extraintestinal manifestations and lead to complications and in some cases chronic disease, like reactive arthritis and Guillain–Barré syndrome (Crushell et al. 2004), irritable bowel syndrome and chronic inflammatory bowel diseases (Garcia et al. 2006; Marshall et al. 2006). Some studies show that pathogen virulence and disease severity determine the ability of *C. jejuni* to invade the cytosol of human cells has been demonstrated (Byrne et al. 2007; Kalischuk et al. 2007). According to Wine et al. (2008) *C. jejuni* disruption of monolayers is mediated by invasion. The ability of this pathogen to invade epithelial cells is cell-type dependent. These findings provide new insight in the pathogen-host epithelial barrier interaction and offer potential mechanisms of intestinal injury and chronic immune stimulation. Wine et al. determined the ability of lactobacilli to inhibit *C. jejuni* invasion. *Lactobacillus helveticus* adheres to both T84 and intestine 407 cells. Protection of cells from *C. jejuni* invasion by lactobacilli seems to depend on strain specificity of both pathogen and probiotics (Wine et al. 2009).

The effect of the yeast *Saccharomyces boulardii* on acute diarrhoea was described by Szajewska et al. in a metaanalysis showing clinical benefits in those using probiotics above the control group by shortening the duration of diarrhoea. However, all trials included had methodological limitations (Szajewska et al. 2007a). The same group conducted a metaanalysis for the effect of LGG for treating diarrhoea in children. They found moderate clinical benefits for LGG in the treatment of acute diarrhoea in children. Also this metaanalysis discusses several methodological limitations and heterogeneity (Szajewska et al. 2007b).

Rotavirus was discovered in children with gastroenteritis by Bishop et al. in 1973 (Bishop 1999). This agent causes widespread morbidity and 870,000 deaths worldwide each year. Bishop said: “after doing a lot of background reading, it became clear that there probably was an infectious agent but we could not get anything to grow in culture”. Bishop (1999) participated in the development of vaccines against rotavirus, the first of which was licensed for use in the USA in 1998.

The effect of orally administered *lactobacilli* on acute rotavirus diarrhoea was tested by Isolauri et al. (1994) in 42 well-nourished children aged 5–28 months. After oral rehydration, the patients received human *L. casei* strain GG 10^{10} cfu twice daily for 5 days. The control group was not given *lactobacilli*. *Lactobacillus* GG was found in the faeces of 83% of the group with *L. casei* strain GG. The diarrhoeal phase was shortened in that group. The dietary supplementation with *lactobacilli* significantly influenced the bacterial enzyme profile. Urease activity during diarrhoea transiently increased in the control group but not in the group receiving *L. casei* strain GG. No intergroup differences were found in a-glucuronidase, a-glucosidase, and glycocholic acid hydrolase levels. Therefore, Isolauri et al. suggested that rotavirus infection gives rise to biphasic diarrhoea, the first phase being an osmotic diarrhoea and the second associated with overgrowth of specifically urease-producing bacteria. Oral bacteriotherapy appears to be a promising means to counteract the disturbed microbial balance.

To evaluate the ingested strain's adherent properties and ability to inhibit murine rotavirus infection, Duffy et al. (1994) administered human *Bifidobacterium* sp. strain *bifidum* to BALB/c lactating mice (n=58) and their litters (n=327 pups). ELISA and anaerobic bacteriologic techniques were used to measure murine rotavirus shedding and colonization of *Bifidobacterium* in the small intestine. At 1,316 days of gestation, pregnant dams (and their expected litters) were randomly assigned to one of four experimental groups as follows: normal controls; *B. bifidum*-treated only; murine rotavirus-infected only; and *B. bifidum*-treated plus murine rotavirus-infected dams and litters. During the acute phase of diarrhoea, 80% of small-intestine cultures in *B. bifidum*-treated litters were positive for the ingested *B. bifidum* strain compared to 24% of faecal cultures. The examination of tissue cross sections under electron microscopy revealed that the ingested *B. bifidum* strain survived passage through the upper gastrointestinal tract and adhered to the small-intestine epithelium. After the administration of the high dose of virus, diarrhoea developed in all pups, but onset was significantly delayed in *B. bifidum*-treated plus murine rotavirus-infected litters compared to litters infected with murine rotavirus only. *B. bifidum*-treated plus murine rotavirus-infected pups demonstrated a significant reduction in murine rotavirus shedding compared with litters challenged with murine rotavirus only at day 2–10 after inoculation. More direct studies are needed to assess the mechanisms by which this anaerobe may modify the course of murine rotavirus infection at the level of gut epithelium.

Qiao et al. (2002) evaluated the potential synergistic effects of *Bifidobacterium* spp. (*B. bifidum* and *B. infantis*), with or without prebiotic compounds (arabino-galactan, short-chain fructo-oligosaccharide, iso-malto-dextrins), on modulating the course of rhesus rotavirus infection, as well as their ability to mediate the associated mucosal and humoral immune responses. Therefore, they fed these species orally to Balb/c pups. Rotavirus-specific IgA and IgG in serum, rotavirus antigen, and specific IgA in faeces were measured by ELISA. Mucosal total IgA and IgG levels were determined in Peyer's patches by flow cytometry. Significantly delayed onset and early resolution of diarrhoea were observed in bifidobacteria-treated, rhesus rotavirus-infected mice compared with rhesus rotavirus-infected control mice. They saw that supplementation with prebiotic compounds did not shorten the clinical course of diarrhoea more than that observed with bifidobacteria treatment alone. Rotavirus-specific IgA in faeces was elevated 16-fold on day 5 postinfection in bifidobacteria-treated, rhesus rotavirus-infected mice compared with the rhesus rotavirus-infected only group. In addition, the level of rotavirus-specific IgA in serum was fourfold higher in bifidobacteria-treated, rhesus rotavirus-infected litters versus mice challenged with rhesus rotavirus alone on 28 and 42 days postinfection. They found no enhancement of the immune response in rhesus rotavirus-infected mice that were treated with both bifidobacteria and prebiotic compounds over those treated with bifidobacteria alone. These findings suggested that bifidobacteria may act as an adjuvant by modulating early mucosal and strong humoral rotavirus-specific immune responses, and mitigate the severity of rotavirus-induced diarrhoea (Qiao et al. 2002).

15.3 Antibiotic-Associated Diarrhoea

Diarrhoea is a common side-effect of both the short- and long-term use of antibiotics. Several reports exist on the benefits of probiotics in this common complication. For example, in a multicenter study the value of a probiotic *Enterococcus* SF68 or placebo was assessed in the prevention of antibiotic-associated diarrhoea. 45 patients treated with antibiotics were given, concurrently, one capsule of either *Enterococcus* SF68 or placebo for 7 days. This probiotic was effective in reducing the incidence of antibiotic-associated diarrhoea compared to placebo (8.7% compared to 27.2%, respectively) (Wunderlich et al. 1989). Unfortunately the *Enterococcus* SF68 has been withdrawn because of the risk of transfer of antibiotic resistance. Vanderhoof et al. showed a significant reduction in antibiotic associated diarrhoea, using *Lactobacillus* GG in children (Vanderhoof et al. 1999).

Clostridum difficile (*C. difficile*) is an anaerobic, spore-forming bacterium which can cause a primarily nosocomial disease ranging from mild diarrhoea to severe, life-threatening pseudomembranous colitis. Infections with *C. difficile* in hospitalised patients are an increasing worldwide problem. In double-blind, placebo-controlled study, performed in a high-risk group of 193 hospitalized patients receiving a new prescription for a β-lactam antibiotic and having no acute diarrhoea on enrollment it was shown that *Saccharomyces boulardii* (*S. boulardii*) 1 g/day caused a significant decrease in antibiotic associated diarrhoea. (7.2% in patients receiving *S. boulardii* compared to14.6% with placebo) $p=0.02$) (McFarland et al. 1995). In another study McFarland et al. measured the recurrence of active *C. difficile*-associated disease using a combination of *S. boulardii* and standard antibiotics compared to placebo. Therapy with *S. boulardii* showed a significant efficacy in recurrent *C. difficile*-associated disease compared to placebo (recurrence rate 34.6% in probiotic group compared to 64.7% in the placebo group). However, no benefit was found when *S. boulardii* was used to treat primary infection with *C. difficile* (recurrence rate 19.3% compared to 24.2% respectively; $P=0.86$) (McFarland et al. 1994).

Lactobacillus GG was successfully used to treat a group of patients with recurrent diarrhoea caused by *C. difficile* (Gorbach et al. 1987).

In summary, probiotics can be utilized to restore the normal gut function and to reduce the duration of acute gastroenteritis (Cuomo et al. 2007). Results so far are encouraging but the most effective dose and type of strain needs to be elucidated.

15.4 Probiotics in Constipation and Bloating

Constipation is a common heterogeneous gastrointestinal disease affecting up to 27% of the western population (Pare et al. 2001; Sonnenberg and Koch 1989; Stewart et al. 1999). Although there is evidence supporting the fact that probiotics favourably modify the intestinal function, placebo-controlled studies on the possible treatment using probiotics are very rare.

De Paula et al. (2008) investigated the effect of a probiotics in 266 females with functional constipation (according to Rome II criteria), randomized to receive either a mixture of *Bifidobacterium animalis* (DN-173 010) and prebiotic fructooligosaccharide (FOS) twice a day for 2 weeks or a lacteous dessert. The results show a 22% increase in the number of bowel movements per week and a slight increase in stool quality as assessed by the Bristol Stool Questionnaire when compared to placebo. Perception of pain and straining during defecation were significantly reduced in the probiotics group. Koebnick et al. (2003) showed a significant improvement in self reported severity of constipation 70 adults after the ingestion of *L. casei* shirota versus placebo during a period of 4 weeks.

According to Ouwehand et al. (2002) administering L. rhamnosus/*Propionibacterium freudenreichii* supplemented huice increases the defecation frequency by 24%. Nevertheless, they observed no reduction in laxative use.

15.5 Probiotics in Colonic Diverticular Disease

Colonic diverticulosis, characterized by sac-like protrusions, due to hernition of the colonic mucosa and sub-mucosa through defects in the muscular layer of the colon wall (Comparato et al. 2007; Jun and Stollman 2002) is highly prevalent in western countries, and rare in the developing world (Bogardus 2006; Hjern et al. 2006).

Recent data have shown that chronic inflammation and abnormal colonic microflora play an important role in the pathogenesis of diverticular disease, suggesting that diverticular disease is an inflammatory mucosal disease, similar to inflammatory bowel diseases (Peppercorn 2004; Ludeman et al. 2002). Therefore, normalizing the intestinal flora as well as administering an anti-inflammatory agent that has already proven effectiveness in inflammatory bowel disease (IBD), may help treat the symptoms of diverticular disease, prevent the onset of acute diverticulitis and reduce the risk of symptomatic recurrence.

Tursy et al. studied the efficacy of *Lactobacillus casei* DG VSL#3 (VSL Pharmaceuticals, Inc., Fort Lauderdale, FL, USA) in combination with 5-ASA (a pH-dependent formulation of mesalazine (Pentacol, SOFAR S.p.A, Trezzano Rosa (MI), Italy) or balsalazide, respectively] in patients with symptomatic, uncomplicated diverticular disease in remission. This probiotic/5-ASA combination performed better in preventing disease relapses and improving symptoms than the single-agent regimens (Tursi et al. 2006).

In another study, nonpathogenic *E. coli* (Nissle strain) combined with antibiotic therapy (dichlorchinolinol) and an intestinal absorbent (active coal) resulted in greater symptomatic improvement and longer periods of remission than with the combination of an antibiotic and absorbent regimen alone (Fric and Zavoral 2003).

15.6 Probiotics in Irritable Bowel Syndrome

Irritable bowel syndrome is a widespread and multifactorial functional disorder of the digestive tract (Niedzielin et al. 2001). It affects 8–22% of the population with a higher prevalence in women. It accounts for 20–50% of referrals to gastroenterology clinics and is characterized by abdominal pain, excessive flatus, variable bowel habit and abdominal bloating for which there is no evidence of detectable organic disease. Suggested aetiologies include gut motility and psychological disorders as well as psychophysiological phenomena and colonic fermentation (Madden and Hunter 2002).

A large proportion of patients have periods characterized by sudden and unforeseeable changes in the two main symptoms, constipation and diarrhoea, even within a few days (Madden and Hunter 2002). It is very likely that the syndrome represents different groups of patients with probably different pathogenesis. Irritable bowel syndrome may follow gastroenteritis and may be associated with an abnormal gut flora and with food intolerance (King et al. 1998). The faecal microflora in some of these patients has been shown to be abnormal with higher numbers of facultative organisms and low numbers of *lactobacilli* and *bifidobacteria* (Madden and Hunter 2002). Bacteria are the major component of formed stools and are influenced by substrates arriving with the ileal affluent. Stool production is related to quantitative and qualitative aspects of the colonic microflora and nearly 80% of the faecal dry weight consists of bacteria, 50% of which are viable.

Although there is no evidence of food allergy in irritable bowel syndrome, food intolerance has been identified and exclusion diets are beneficial to many of these patients. Food intolerance may be caused by an abnormal fermentation of food residues in the colon, as a result of disruption of the normal flora (Madden and Hunter 2002).

Some reports suggest that probiotics play a role in regulating the motility of the digestive tract (Niedzielin et al. 2001). This may result in improvements in pain and flatulence in response to probiotic administration (Madden and Hunter 2002).

To assess whether preceding gastroenteritis or food intolerance were associated with colonic malfermentation, King et al. (1998) conducted a crossover controlled trial with a standard diet and an exclusion diet matched for macronutrients in six female patients with irritable bowel syndrome and six female controls. In this study faecal excretion of fat, nitrogen, starch, and nonstarch polysaccharide was measured during the last 72 hours of each diet. The total excretion of hydrogen and methane were collected over 24 hours in purpose-built 1.4 m^3 whole body calorimeter. Breath hydrogen and methane excretion were measured for 3 hours after 20 g oral lactulose. The maximum rate of gas excretion was significantly greater in patients than in controls. The total gas production in patients was not greater than in controls, whereas hydrogen production was higher. After lactulose, breath hydrogen was greater on the standard than on the exclusion diet. This means that colonic-gas production, particularly of hydrogen, is greater in patients with irritable bowel disease than in controls, and both symptoms and gas production are reduced by an exclusion diet. This reduction may be associated with alterations in the activity of hydrogen-consuming bacteria. It was therefore concluded that fermentation may be an important factor in the pathogenesis of this syndrome (King et al. 1998).

Spiller et al. (2000) studied the intestinal permeability (lactulose/mannitol ratio) and histological and immunological features in rectal biopsy specimens in 21 patients who had acute *Campylobacter* enteritis, 10 patients with postdysenteric irritable bowel syndrome and 12 asymptomatic controls. They found that the increased enteroendocrine cell counts, T lymphocytes, and gut permeability, which may survive for more than a year after Campylobacter enteritis, contribute to post-dysenteric irritable bowel syndrome (Spiller et al. 2000), thus offering a rationale to use probiotics for several months after the infectious episode.

To determine the efficacy of *Lactobacillus rhamnosus* GG (LGG) in the management of functional abdominal pain disorders in children, Gawronska et al. performed a RCT in which 104 children with Rome-II criteria for functional dyspepsia, irritable bowel syndrome, or functional abdominal pain were enrolled. Fifty two patients received LGG for 4 weeks whereas the other 52 subjects received a placebo. The results show a benefit for those receiving LGG in all groups (Gawronska et al. 2007).

The effect of the probiotics was studied by Brigidi et al. (2001) in a clinical trial in which 10 patients suffering from this syndrome were administered the VSL#3 probiotic preparation. The results indicated that the administration of VSL#3 improved the clinical picture and changed the composition and biochemistry of faecal microbiota. The exact mechanisms of the positive effects are not known. The selection of patients may have had an important role in detecting the positive effects. Whether the induction of a significant increase in *lactobacilli, bifidobacteria,* and *S. thermophilus* contributed to the regulation of the motility disorders or the increase in faecal beta-galactosidase with a decrease in urease content indicate that a good response requires further study. The importance of this study is that it showed that the measurement of specific parameters and changes in the specific microflora was possible.

Kim et al. (2003) investigated the effects of VSL#3 on gastrointestinal transit and symptoms of patients with irritable bowel syndrome diagnosed with the criteria established by Rome II and with predominant diarrhoea. Twenty-five patients with diarrhoea-predominant irritable bowel syndrome were randomly assigned to receive VSL#3 powder (450 billion lyophilized bacteria/day) or matching placebo twice daily for 8 weeks after a 2-week run-in period. Pre- and post-treatment gastrointestinal transit measurements were performed in all patients. The patients recorded their bowel function and symptoms daily in a diary during the 10-week study, which was powered to detect a 50% change in the primary colonic transit endpoint. There were no significant differences in mean gastrointestinal transit measurements, bowel function scores or satisfactory global symptom relief between the two treatment groups, pre- or post-therapy. The differences in abdominal bloating scores between treatments were borderline significant. Abdominal bloating was reduced with VSL#3, but not with placebo. Furthermore, VSL#3 had no effects on individual symptoms such as abdominal pain, gas and urgency. VSL#3 was well tolerated by all patients, and thus it seems to relieve the abdominal bloating in patients with diarrhoea-predominant irritable bowel syndrome (Kim et al. 2003).

15.7 Probiotics in Inflammatory Bowel Disease

The term chronic inflammatory bowel disease includes three disease types: ulcerative colitis, Crohn's disease, and an intermediate form (about 10%). Crohn's disease is defined as a chronic granulomatous inflammation of the digestive tract that most commonly involves the distal ileum, colon and anus. Less often, the disease affects the mouth, esophagus, stomach and duodenum. Occasionally, extraintestinal sites are affected and it is referred to as: "metastatic Crohn's disease". In ulcerative colitis, the colon is affected and the disease usually starts in the rectum and progresses proximally, although sometimes the first manifestation may be the involvement of the whole colon and rectum (panproctocolitis).

Ulcerative colitis is slightly more common than Crohn's disease. In Western Europe and North of America, there are 3,000–5,000 new cases of Crohn's disease and 8,000–10,000 new cases of ulcerative colitis. The incidence and prevalence of Crohn's disease have been increasing five times faster than that of ulcerative colitis. Young people are more likely to be more affected by inflammatory bowel disease than older people, with a peak incidence at the age of 15–30 years.

The etiology of this disease is unknown. An infectious hypothesis has been considered for years, and *Mycobacterium paratuberculosis* has been mainly isolated from patients with Crohn's disease. However, some patients with ulcerative colitis and controls harbour this pathogen. Viruses have also been involved in the pathogenesis. Several factors other than infectious agents have been postulated as the cause of the disease. These different factors are immunologic, genetic and psychological. The chronic inflammatory nature of these diseases may indicate the presence of an infectious cause or the presence of a dysregulatory abnormality in the control of inflammation.

An increasing number of both clinical and laboratory observations support the importance of the ubiquitous luminal bacteria in the inflammatory responses of these disorders (Campieri and Gionchetti 1999). Bacteria are present throughout the gastrointestinal tract but are not evenly distributed and their diversity and numerical importance vary in the different sections of the gastrointestinal tract (Campieri and Gionchetti 1999; Kim et al. 2003; Simmering and Blaut 2001). In the stomach and duodenum there are facultative anaerobic bacteria (*Lactobacillus* spp. and *enterobacteriaceae*), with a small number of bacteria that are predominantly Gram-positive and aerobic (Campieri and Gionchetti 1999). In the lower distal part of the intestine there is a large variety of bacteria, mostly anaerobic bacteria belonging to *Bacteroides, Bifidobacterium, Clostridium, Fusobacterium, Peptostreptococcus* and *Ruminococcus* (Simmering and Blaut 2001). There is a transition to higher concentrations of bacteria and increasing number of Gram-negative bacteria in the distal ileum. Across the ileocecal valve there is a dramatic increase in bacterial concentration and more anaerobes than aerobes (Campieri and Gionchetti 1999).

Enteric bacteria have been detected in patients with Crohn's disease and in those with pouchitis. These patients may be effectively treated with antibiotics. Purified bacterial products may initiate and perpetuate experimental colitis. The inflammation is due to loss of normal tolerance to the commensal flora (Campieri and Gionchetti 1999).

The onset of inflammation is associated with an imbalance in the intestinal microflora with relative predominance of "aggressive" bacteria and an insufficient concentration of "protective" species. Reconditioning the flora through either direct supplementation with protective bacteria or by indirect stimulation plays a protective role in inflammatory bowel disease (Campieri and Gionchetti 1999).

Antioxidant properties, the ability to increase prostacyclin and crampy in endothelial cell cultures and the ability to modulate adhesion molecule expression on human lymphocytes are all effects which are relevant for the use of probiotics in the treatment of immunological disorders such as inflammatory bowel disease.

15.8 Probiotics in Ulcerative Colitis

Few data are available on the role of probiotics in human ulcerative colitis. Two studies have shown a significant decrease in lactobacilli concentration in colonic biopsies in patients with ulcerative colitis. Preventing or controlling the colitis is reported when the concentration of *Lactobacillus* was modulated through dietary supplementation with lactulose (prebiotic). This is a nondigestible food ingredient that affects the host by selectively stimulating the growth and activity of one or more "probiotic" bacteria, such as *Bifidobacterium* and *Lactobacillus* that have health-promoting properties (Gibson and Roberfroid 1995).

Ulcerative colitis is a chronic inflammation of the rectal and colonic mucosa, with a poorly defined etiology. Its characteristics are bloody diarrhoea and mucus associated with a negative stool culture for bacteria, ova, or parasites. There is also faecal stasis with bacterial overgrowth and mucosal ischemia. The therapeutic role of probiotics is shown through two studies; in one of these, oral administration of *Lactobacillus* GG caused an increase in intestinal IgA immune response in patients with Crohn's disease. In the other study, exogenous administration of *L. reuteri* (pure bacterial suspension or as fermented oatmeal soup) prevented acetic acid-induced colitis or methotrexate-induced colitis in rats.

These studies showed a significant decrease in lactobacilli concentration in patients with active ulcerative colitis. The results showed that *L. plantarum* was more effective in methotrexate-induced colitis, and *Lactobacillus* treatment prevented development of spontaneous colitis in IL-10 gene-deficient mice.

In an open label study with 20 patients, intolerant or allergic to 5-aminosalicylic acid (5-ASA), a treatment consisting of 6 g VSL#3 (1,800 billion bacteria)/day for 12 months was instituted. Clinical, endoscopic assessment and stool culture and faecal pH determination were recorded (Venturi et al. 1999). Nineteen patients completed the trial and 15 were in remission for the whole year. Faecal concentrations of bifidobacteria, lactobacilli, and *S. salivarius spp. Thermophilus* were significantly increased in all patients and remained stable throughout the study. No changes were noted in the concentrations of total aerobic this suggesting that the beneficial effects of VSL#3 were not related to suppression of endogenous luminal flora. The treatment was well-tolerated with no reported significant side effects like those seen in the treatment with 5-ASA oral compounds. This shows that the probiotic preparation was able to colonize the intestine and suggested its possible usefulness in maintaining remission in ulcerative colitis patients intolerant or allergic to 5-ASA (Venturi et al. 1999). The hypothesis from these studies is that the intestinal environment may contribute to the pathophysiology of ulcerative colitis.

Guslandi et al. (2003) studied the efficacy of *S. boulardii* in ulcerative colitis patients. Twenty-five patients with a mild to moderate clinical flare-up of ulcerative colitis received additional treatment with *S. boulardii* 250 mg three times a day for 4 weeks during maintenance treatment with mesalamine (mesalazine, 5-ASA). These patients were unsuitable for steroid therapy. Rachmilewitz's clinical activity index was calculated before and after the treatment. Of the 24 patients who completed the study, 17 attained clinical remission; this was endoscopically confirmed. The preliminary results suggested that *S. boulardii* may be effective in the treatment of ulcerative colitis.

Kruis et al. (2004) compared the efficacy of the probiotic preparation *Escherichia coli* Nissle 1917 and established therapy with mesalazine in maintaining remission in patients with ulcerative colitis. Three hundred and twenty seven patients received either probiotics 200 mg once daily ($n=162$) or mesalazine 500 mg three times daily ($n=165$) during a period of 12 months. Assessment was performed by clinical and endoscopic activity indices (Rachmilewitz) and histology. The results show relapses in 40 out of 110 (36.4%) and 38 out of 112 (33.9%) patients in the probiotic and mesalazine group respectively ($p=0.003$). Hereby, they show efficacy and safety of *E. coli Nissle* in maintaining remission equivalent to the gold standard mesalazine in patients with ulcerative colitis.

15.9 Probiotics in Pouchitis

Pouchitis is a nonspecific inflammation of the ileal reservoir that may appear after surgery for ulcerative colitis, and results in various clinical symptoms. It is a well-recognized long-term complication of restorative proctocolectomy. The risk of pouchitis increases in patients with a history of extraintestinal manifestations, primary sclerosing cholangitis, positive serology for perinuclear antineutrophil cytoplasmic antibodies, and backwash ileitis (Katz 2003).

Pouchitis is associated with bacterial overgrowth and dysbiosis, and antibiotics represent the first-choice treatment. The distal ileum and the large bowel, the sites with the highest bacterial concentration, are the most frequently affected by inflammation. Enteric bacteria or their products have been detected within the inflamed mucosa. A significant decrease of lactobacilli and bifidobacteria concentrations has been found in ulcerative colitis, Crohn's disease and pouchitis. Lactobacilli as maintenance showed less frequent relapses of pouchitis than those using placebo. Diversion of the faecal stream in the small and large intestine reduces the activity of the inflammation. The luminal contents and purified bacterial products added to isolated intestinal loops trigger systemic and local signs of inflammation.

In a study by Campieri and Gionchetti (1999), seven patients, after clinical, endoscopic, and histological diagnoses of inflammation of the ileal pouch anal anastomosis with a pouchitis disease activity index (PDAI) >7, were treated with 2 g/day of rifaximin (a non absorbable antibiotic) and 1 g/day of ciprofloxacin for 1 month. All patients went into remission during this month, as judged by clinical, endoscopic and histological examination. After remission, all seven patients were treated with the highly concentrated probiotic mixture VSL#3 (a mixture of four strains of lactobacilli (*Lactobacillus casei, L. plantarum, L. acidophilus, L. delbrueckii subsp. bulgaricus*), three strains of bifidobacteria (*Bifidobacterium longum, B. breve, B. infantis*) and one strain of *Streptococcus salivarius subsp. thermophilus*) for nine months. No patient had a relapse in this period. All patients who received a placebo had a relapse.

15.10 Probiotics in the Maintenance of Remission of Chronic Pouchitis

Gionchetti et al. (2000) evaluated the efficacy of VSL#3 in the maintenance of remission of chronic pouchitis. Forty patients in clinical and endoscopic remission were randomized to receive either VSL#3 6 g/day, or an identical placebo for 9 months. The patients were assessed clinically every month and endoscopically and histologically every 2 months or in the event of relapse. Three patients (15%) in the VSL#3 group had relapses within the 9-month follow-up period, compared with 20 (100%) in the placebo group ($p<0.001$). In the VSL#3-treated group, the faecal concentration of lactobacilli, bifidobacteria, and *S. thermophilus* increased significantly from baseline levels. These results suggested that oral administration of this new probiotic preparation is effective in preventing flare-ups of chronic pouchitis. In another RCT, 36 subjects were administered 6 g VSL#3 daily or a placebo for 12 months. A relapse was observed in 15% of the patients receiving VSL#3 versus 94% of those on placebo ($p<0.0001$) (Mimura et al. 2004).

15.11 Probiotics in Preventing the Onset of Pouchitis

A positive effect of VSL#3 in the prevention of pouchitis has been reported by Gionchetti et al. (2003), who compared probiotic therapy with VSL#3 versus placebo in the ability to prevent the onset of acute pouchitis during the first year after ileal pouch-anal anastomosis. Forty patients who underwent ileal pouch-anal anastomosis for ulcerative colitis were randomized to receive either VSL#3 or an identical placebo immediately after ileostomy closure for 1 year. Both groups consisted of 20 patients. The patients were assessed clinically, endoscopically, and histologically after 1, 3, 6, 9 and 12 months. Health-related quality of life was assessed using the Inflammatory Bowel Disease Questionnaire (IBDQ). Two of the 20 patients (10%) treated with VSL#3 had an episode of acute pouchitis compared with eight of the 20 patients (40%) treated with placebo ($p<0.01$). Treatment with VSL#3 determined a significant improvement in IBDQ score, which was not the case with placebo. During treatment with VSL#3, faecal concentration of lactobacilli, bifidobacteria, and *S. salivarius* increased significantly. The faecal concentration of *Bacteroides,* clostridia, coliforms, and enterococci were not modified. This suggested that the beneficial effect was not mediated by the suppression of the endogenous flora. The treatment with VSL#3 was effective in the prevention of the onset of acute pouchitis and improved the quality of life of patients with ileal pouch anal anastomosis.

As mentioned above, probiotic bacteria do not survive for long, and rapidly disappear as soon as the treatment is stopped. Therefore, prophylactic probiotic therapy of pouchitis might require long-term treatment and might not be indicated for all patients. For this reason, VSL#3 would be highly beneficial for patients at high risk of chronic pouchitis. In these cases prophylactic probiotics may be administered, pouch function improved and their quality of life after ileal pouch anal anastomosis could be maintained. Katz (2003) suggested that probiotics should be used for maintaining remission in chronic pouchitis and as prophylaxis against pouch inflammation in high-risk patients. Although preliminary results suggest that high doses VSL#3 may improve active pouchitis, probiotic therapy seems to be more effective to prevent mucosal inflammation than to treat it.

Kuisma et al. (2003) investigated the efficacy of *Lactobacillus* GG supplementation as primary therapy for pouchitis and its effect on the microbial flora. Twenty patients, with a previous history of pouchitis and endoscopic inflammation were recruited for a prospective, randomized, double-blind, placebo-controlled trial of *Lactobacillus* GG supplementation. Ten patients received *Lactobacillus* GG and 10 placebo, in two gelatin capsules b.i.d. for 3 months. Quantitative bacterial culture of fresh faecal samples and biopsies taken from the pouch and afferent limb was performed before and after supplementation. *Lactobacillus* GG supplementation was found to change the pouch intestinal flora by increasing the ratio of total faecal lactobacilli to total faecal anaerobes and enhancing the frequency of lactobacilli-positive cultures in the pouch and afferent limb mucosal biopsy samples. Only 40% of patients were colonized with *Lactobacillus* GG, and no differences were observed between the groups with regard to the mean PDAI or the total anaerobes or aerobes of faecal or tissue biopsy samples. Thus, a single-strain probiotic bacterium supplement of *Lactobacillus* GG changed the pouch intestinal bacterial flora, but was ineffective as primary therapy for a clinical or endoscopic response. More clinical trials are needed to evaluate the right placement and dosage of probiotics within a treatment regimen for pouchitis.

15.12 Probiotics in Crohn's Disease

The therapeutic role of probiotics in the prevention of postoperative recurrence of Crohn's disease has been reported in some studies. Campieri et al. (2000) studied the effects of VSL#3 in a randomized, investigator-blind trial. Forty patients with curative resection randomized within 1 week post surgery were divided into two groups of 20 patients. One group received mesalazine 4 g/day for 1 year and the other group received rifaximin 1.8 g/day for 3 months followed by VSL#3 6 g/day for 9 months. The endoscopic activity was assessed after 3 and 12 months. In the mesalazine group, eight patients had severe endoscopic recurrence after 3 months as well as after 12 months, whereas in the group with rifaximin and VSL#3, two patients had a severe recurrence after 3 months and two patients after 12 months. These results suggested the efficacy of the combination of a nonabsorbable antibiotic with a highly concentrated probiotic preparation in the prevention of severe endoscopic recurrence of Crohn's disease after surgical resection.

In a pilot study Guandalini (2002) investigated the possible effect of *Lactobacillus* GG in children with active Crohn's disease. Four male patients with a median age of 14.5 years (range 10–18) were enrolled. In terms of clinical outcome, the patients showed significant improvement. In three patients receiving *Lactobacillus* GG, it was possible to taper the dose of steroids.

In a third published study using *Lactobacillus* GG this effect could not be confirmed. Forty-five patients were randomized to receive *Lactobacillus* GG 12 billion cfu/day (23 patients) or placebo (22 patients). A clinical remission after 52 weeks was seen in 15 of the 23 patients with *Lactobacillus* GG (83.3%) and in 17 of the 22 patients with placebo (89.4%). Mild endoscopic activity was seen in nine of the 15 patients with remission in the *Lactobacillus* GG group (60%) and in six of the 17 patients with remission in the placebo group. This study failed to show effectiveness in the postoperative prevention on Crohn's disease (Prantera and Scribano 2002).

In a double blind RCT Marteau et al. administered 98 patients after surgery for CD *L. johnsonii* LA1 (4×10^9 cfu/day) or a placebo for 6 months. Endoscopic recurrence was observed in 49% of patients on Probiotics versus 64% of patients in the placebo group (Marteau et al. 2006). The same strain was studied by van Gossum et al. in a higher dose 10^{10} cfu/day in 70 postoperative patients in a multi center study (Van Gossum et al. 2007). They found *L. johnsonii* LA1 not effective in the prevention of recurrence. Twenty one and 15% in the probiotics versus placebo group developed severe endoscopic lesions ($p=0.33$). The percentages with clinical relapse was 15 and 13% in the probiotics and the placebo group respectively ($p=0.79$).

The limited experience indicates that different probiotics have different capacity to prevent intestinal inflammation. More studies are therefore necessary.

15.13 Conclusions and Perspectives

In this exploding era of clinical research, there is evidence for a strong link between the intestinal microbiota and intestinal disorders, both in the pathogenesis of inflammatory and regulatory pathways. Using probiotics as ecological treatment for these disorders have proven efficacy as is the case for specific forms of IBD as well as other intestinal diseases. Although the exact mechanisms by which probiotics exert their beneficial effects *in vivo* have not yet been fully clarified, the luminal bacterial flora appears to play a major role in the initiation and perpetuation of chronic inflammatory bowel diseases in animal and human models. Increasing evidence shows that probiotics modify the gastrointestinal microflora in such a way that the bacterial activities are advantageous to the health of the host without colonizing the gastrointestinal tract, suppress gastrointestinal inflammation, and modulate the inflammatory response (Simmering and Blaut 2001).

Probiotics appear to have multiple modes of action through the direct or indirect modulation of the endogenous flora or the immune system. This is based on cross-talk between luminal bacteria and the epithelial cells. Inhibition of pathogens includes competition for colonization sites and nutrients, production of toxic compounds, and stimulation of the immune system. These mechanisms are not mutually exclusive, and inhibition may include one, several, or all of these mechanisms (Patterson and Burkholder 2003). Modulation of the endogenous flora, enhancement of the intestinal barrier, and immunomodulator effects of down regulating inflammation are other mechanisms by which probiotics exert their effects (Karimi and Pena 2003).

Current evidence provides support for the consideration of probiotics therapy for intestinal diseases, keeping in mind that efficacy of probiotics is strain and disease specific. The variety of studies carried out with distinct strains of probiotics bacteria has suggested heterogeneous and strain-specific effects.

Limitations of most studies conducted with probiotics, either with regard to the power of the study, deficit of human studies, randomization, use of different strains and lack of standardized methodology it remains difficult to draw firm conclusions from the current trials.

Furthermore, discordance between recent meta-analyses (Szajewska et al. 2007a, b; Cremonini et al. 2002; McFarland 2006) and systematic reviews (Dendukuri et al. 2005; Hoveyda et al. 2009; Pillai and Nelson 2008) show important differences in the different trials.

Probiotics differ from one another, making results obtained with one strain, or a cocktail of strains, not easily extrapolated to another strain or cocktail of strains. The results from trials done on one probiotics strain in a specific patients group cannot be applied to other probiotics or other patient groups. Also the dose response studies are too few to establish the exact dose needed for optimal treatment. Investigation of probiotic application as therapeutic for different intestinal disorders increases the understanding of the role of gastrointestinal milieu in the pathogenesis of IBD and other intestinal diseases. Further validation of probiotic properties in humans and clarification of their mechanisms of action are needed to better understand the role of probiotics in promoting human health, and for better definition and application of the potential use of probiotics in different clinical settings. Therefore more and larger controlled randomized clinical trials are necessary to investigate the as yet unresolved issues with regard to efficacy based on immunological and microbiological analysis of colonic mucosa and stool, dose, and duration of use, and the exact strains for specific diseases and/or single or multistrain formulation are to prove the beneficial effects.

With regard to the described differences in the immunological and antipathogenic effects exerted by different probiotics, clinicians for the time being should advise those probiotic strains proven to be efficacious in relevant patient groups and encourage further clinical research studies in order to define the proper place in the management of infectious, functional and inflammatory disorders of the gastrointestinal tract.

References

Barbara G, Cremon C, Pallotti F, De Giorgio R, Stanghellini V, Corinaldesi R (2009) Postinfectious irritable bowel syndrome. J Pediatr Gastroenterol Nutr 48(Suppl 2):S95–97 (Apr)

Bishop R (1999) Ruth Bishop: rotaviruses and vaccines. Interview by Amanda Tattam. Lancet 353(9167):1860 (May)

Bogardus ST Jr (2006) What do we know about diverticular disease? A brief overview. J Clin Gastroenterol 40(Suppl 3):S108–S111 (Aug)

Brigidi P, Vitali B, Swennen E, Bazzocchi G, Matteuzzi D (2001) Effects of probiotic administration upon the composition and enzymatic activity of human faecal microbiota in patients with irritable bowel syndrome or functional diarrhoea. Res Microbiol 152(8):735–741 (Oct)

Byrne CM, Clyne M, Bourke B (2007) Campylobacter jejuni adhere to and invade chicken intestinal epithelial cells in vitro. Microbiology 153(Pt 2):561–569 (Feb)

Campieri M, Gionchetti P (1999) Probiotics in inflammatory bowel disease: new insight to pathogenesis or a possible therapeutic alternative? Gastroenterology 116(5):1246–1249 (May)

Campieri M, Rizzello F, Venturi A, Poggioli G, Ugolini F, Helwig U et al (2000) Combination of antibiotic and probiotic treatment is efficacious in prophylaxis of post-operative recurrence of crohn's disease: a randomized controlled study vs Mesalamine. Gastroenterology 118(4179):A781

Comparato G, Pilotto A, Franze A, Franceschi M, Di Mario F (2007) Diverticular disease in the elderly. Dig Dis 25(2):151–159

Cremonini F, Di Caro S, Nista EC, Bartolozzi F, Capelli G, Gasbarrini G et al (2002) Meta-analysis: the effect of probiotic administration on antibiotic-associated diarrhoea. Aliment Pharmacol Ther 16(8):1461–1467 (Aug)

Crushell E, Harty S, Sharif F, Bourke B (2004) Enteric campylobacter: purging its secrets? Pediatr Res 55(1):3–12 (Jan)

Cuomo R, Savarese MF, Gargano R (2007) Almost all irritable bowel syndromes are post-infectious and respond to probiotics: consensus issues. Dig Dis 25(3):241–244

Dendukuri N, Costa V, McGregor M, Brophy JM (2005) Probiotic therapy for the prevention and treatment of Clostridium difficile-associated diarrhoea: a systematic review. Can Med Assoc J 173(2):167–170 (19 July)

De Paula JA, Carmuega E, Weill R (2008) Effect of the ingestion of a symbiotic yogurt on the bowel habits of women with functional constipation. Acta Gastroenterol Latinoam 38(1):16–25 (Mar)

Duffy LC, Zielezny MA, Riepenhoff-Talty M, Dryja D, Sayahtaheri-Altaie S, Griffiths E et al (1994) Effectiveness of Bifidobacterium bifidum in mediating the clinical course of murine rotavirus diarrhoea. Pediatr Res 35(6):690–695 (June)

Fric P, Zavoral M (2003) The effect of non-pathogenic Escherichia coli in symptomatic uncomplicated diverticular disease of the colon. Eur J Gastroenterol Hepatol 15(3):313–315 (Mar)

Garcia Rodriguez LA, Ruigomez A, Panes J (2006) Acute gastroenteritis is followed by an increased risk of inflammatory bowel disease. Gastroenterology 130(6):1588–1594 (May)

Gawronska A, Dziechciarz P, Horvath A, Szajewska H (2007) A randomized double-blind placebo-controlled trial of Lactobacillus GG for abdominal pain disorders in children. Aliment Pharmacol Ther 25(2):177–184 (15 Jan)

Gibson GR, Roberfroid MB (1995) Dietary modulation of the human colonic microbiota: introducing the concept of prebiotics. J Nutr 125(6):1401–1412 (June)

Gionchetti P, Rizzello F, Venturi A, Brigidi P, Matteuzzi D, Bazzocchi G et al (2000) Oral bacteriotherapy as maintenance treatment in patients with chronic pouchitis: a double-blind, placebo-controlled trial. Gastroenterology 119(2):305–309 (Aug)

Gionchetti P, Rizzello F, Helwig U, Venturi A, Lammers KM, Brigidi P et al (2003) Prophylaxis of pouchitis onset with probiotic therapy: a double-blind, placebo-controlled trial. Gastroenterology 124(5):1202–1209 (May)

Gorbach SL, Chang TW, Goldin B (1987) Successful treatment of relapsing Clostridium difficile colitis with Lactobacillus GG. Lancet 2(8574):1519 (26 Dec)

Guandalini S (2002) Use of Lactobacillus-GG in paediatric Crohn's disease. Dig Liver Dis 34(Suppl 2):S63–S65 (Sept)

Guslandi M, Giollo P, Testoni PA (2003) A pilot trial of Saccharomyces boulardii in ulcerative colitis. Eur J Gastroenterol Hepatol 15(6):697–698 (June)

Hjern F, Johansson C, Mellgren A, Baxter NN, Hjern A (2006) Diverticular disease and migration—the influence of acculturation to a Western lifestyle on diverticular disease. Aliment Pharmacol Ther 23(6):797–805 (15 Mar)

Hoveyda N, Heneghan C, Mahtani KR, Perera R, Roberts N, Glasziou P (2009) A systematic review and meta-analysis: probiotics in the treatment of irritable bowel syndrome. BMC Gastroenterol 9:15

Isolauri E, Kaila M, Mykkanen H, Ling WH, Salminen S (1994) Oral bacteriotherapy for viral gastroenteritis. Dig Dis Sci 39(12):2595–2600 (Dec)

Jun S, Stollman N (2002) Epidemiology of diverticular disease. Best Pract Res Clin Gastroenterol 16(4):529–542 (Aug)

Kalischuk LD, Inglis GD, Buret AG (2007) Strain-dependent induction of epithelial cell oncosis by Campylobacter jejuni is correlated with invasion ability and is independent of cytolethal distending toxin. Microbiology 153(Pt 9):2952–2963 (Sept)

Katz JA (2003) Prevention is the best defense: probiotic prophylaxis of pouchitis. Gastroenterology 124(5):1535–1538 (May)

Kim HJ, Camilleri M, McKinzie S, Lempke MB, Burton DD, Thomforde GM et al (2003) A randomized controlled trial of a probiotic, VSL#3, on gut transit and symptoms in diarrhoea-predominant irritable bowel syndrome. Aliment Pharmacol Ther 17(7):895–904 (1 Apr)

King TS, Elia M, Hunter JO (1998) Abnormal colonic fermentation in irritable bowel syndrome. Lancet 352(9135):1187–1189 (10 Oct)

Karimi O, Pena AS (2003) Probiotics: isolated bacteria strain or mixtures of different strains? Two different approaches in the use of probiotics as therapeutics. Drugs Today 39(8):565–597 (Aug)

Koebnick C, Wagner I, Leitzmann P, Stern U, Zunft HJ (2003) Probiotic beverage containing Lactobacillus casei Shirota improves gastrointestinal symptoms in patients with chronic constipation. Can J Gastroenterol 17(11):655–659 (Nov)

Kruis W, Fric P, Pokrotnieks J, Lukas M, Fixa B, Kascak M et al (2004) Maintaining remission of ulcerative colitis with the probiotic Escherichia coli Nissle 1917 is as effective as with standard mesalazine. Gut 53(11):1617–1623 (Nov)

Kuisma J, Mentula S, Jarvinen H, Kahri A, Saxelin M, Farkkila M (2003) Effect of Lactobacillus rhamnosus GG on ileal pouch inflammation and microbial flora. Aliment Pharmacol Ther 17(4):509–515 (15 Feb)

Ludeman L, Warren BF, Shepherd NA (2002) The pathology of diverticular disease. Best Pract Res Clin Gastroenterol 16(4):543–562 (Aug)

Madden JA, Hunter JO (2002) A review of the role of the gut microflora in irritable bowel syndrome and the effects of probiotics. Br J Nutr 88 (Suppl 1):S67–S72 (Sept)

Marshall JK, Thabane M, Garg AX, Clark WF, Salvadori M, Collins SM (2006) Incidence and epidemiology of irritable bowel syndrome after a large waterborne outbreak of bacterial dysentery. Gastroenterology 131(2):445–450, quiz 660 (Aug)

Marteau P, Lemann M, Seksik P, Laharie D, Colombel JF, Bouhnik Y et al (2006) Ineffectiveness of Lactobacillus johnsonii LA1 for prophylaxis of postoperative recurrence in Crohn's disease: a randomised, double blind, placebo controlled GETAID trial. Gut 55(6):842–847 (June)

McFarland LV (2006) Meta-analysis of probiotics for the prevention of antibiotic associated diarrhoea and the treatment of Clostridium difficile disease. Am J Gastroenterol 101(4):812–822 (Apr)

McFarland LV, Surawicz CM, Greenberg RN, Fekety R, Elmer GW, Moyer KA et al (1994) A randomized placebo-controlled trial of Saccharomyces boulardii in combination with standard antibiotics for Clostridium difficile disease. J Am Med Assoc 271(24):1913–1918 (June)

McFarland LV, Surawicz CM, Greenberg RN, Elmer GW, Moyer KA, Melcher SA et al (1995) Prevention of beta-lactam-associated diarrhoea by Saccharomyces boulardii compared with placebo. Am J Gastroenterol 90(3):439–448 (Mar)

Mearin F, Perez-Oliveras M, Perello A, Vinyet J, Ibanez A, Coderch J et al (2005) Dyspepsia and irritable bowel syndrome after a Salmonella gastroenteritis outbreak: one-year follow-up cohort study. Gastroenterology 129(1):98–104 (June)

Mimura T, Rizzello F, Helwig U, Poggioli G, Schreiber S, Talbot IC et al (2004) Once daily high dose probiotic therapy (VSL#3) for maintaining remission in recurrent or refractory pouchitis. Gut 53(1):108–114 (Jan)

Niedzielin K, Kordecki H, Birkenfeld B (2001) A controlled, double-blind, randomized study on the efficacy of Lactobacillus plantarum 299 V in patients with irritable bowel syndrome. Eur J Gastroenterol Hepatol 13(10):1143–1147 (Oct)

Ouwehand AC, Lagstrom H, Suomalainen T, Salminen S (2002) Effect of probiotics on constipation, faecal azoreductase activity and faecal mucin content in the elderly. Ann Nutr Metab 46(3–4):159–162.

Pare P, Ferrazzi S, Thompson WG, Irvine EJ, Rance L (2001) An epidemiological survey of constipation in Canada: definitions, rates, demographics, and predictors of health care seeking. Am J Gastroenterol 96(11):3130–3137 (Nov)

Patterson JA, Burkholder KM (2003) Application of prebiotics and probiotics in poultry production. Poult Sci 82(4):627–631 (Apr)

Peppercorn MA (2004) The overlap of inflammatory bowel disease and diverticular disease. J Clin Gastroenterol 38(5, Suppl 1):S8–S10 (May–June)

Pillai A, Nelson R (2008) Probiotics for treatment of Clostridium difficile-associated colitis in adults. Cochrane Database Syst Rev 2008(1):CD004611

Posserud I, Stotzer PO, Bjornsson ES, Abrahamsson H, Simren M (2007) Small intestinal bacterial overgrowth in patients with irritable bowel syndrome. Gut 56(6):802–808 (June)

Prantera C, Scribano ML (2002) Probiotics and Crohn's disease. Dig Liver Dis 34(Suppl 2):S66–S67 (Sept)

Qiao H, Duffy LC, Griffiths E, Dryja D, Leavens A, Rossman J et al (2002) Immune responses in rhesus rotavirus-challenged BALB/c mice treated with bifidobacteria and prebiotic supplements. Pediatr Res 51(6):750–755 (June)

Raza S, Graham SM, Allen SJ, Sultana S, Cuevas L, Hart CA (1995) Lactobacillus GG promotes recovery from acute nonbloody diarrhoea in Pakistan. Pediatr Infect Dis J 14(2):107–111 (Feb)

Simmering R, Blaut M (2001) Pro- and prebiotics—the tasty guardian angels? Appl Microbiol Biotechnol 55(1):19–28 (Jan)

Sonnenberg A, Koch TR (1989) Physician visits in the United States for constipation: 1958 to 1986. Dig Dis Sci 34(4):606–611 (Apr)

Spiller RC, Jenkins D, Thornley JP, Hebden JM, Wright T, Skinner M et al (2000) Increased rectal mucosal enteroendocrine cells, T lymphocytes, and increased gut permeability following acute Campylobacter enteritis and in post-dysenteric irritable bowel syndrome. Gut 47(6):804–811 (Dec)

Stermer E, Lubezky A, Potasman I, Paster E, Lavy A (2006) Is traveler's diarrhoea a significant risk factor for the development of irritable bowel syndrome? A prospective study. Clin Infect Dis 43(7):898–901 (Oct)

Stewart WF, Liberman JN, Sandler RS, Woods MS, Stemhagen A, Chee E et al (1999) Epidemiology of constipation (EPOC) study in the United States: relation of clinical subtypes to sociodemographic features. Am J Gastroenterol 94(12):3530–3540 (Dec)

Szajewska H, Skorka A, Dylag M (2007a) Meta-analysis: saccharomyces boulardii for treating acute diarrhoea in children. Aliment Pharmacol Ther 25(3):257–264 (Feb)

Szajewska H, Skorka A, Ruszczyński M, Gieruszczak-Bialek D (2007b) Meta-analysis: lactobacillus GG for treating acute diarrhoea in children. Aliment Pharmacol Ther 25(8):871–881 (Apr)

Ternhag A, Torner A, Svensson A, Ekdahl K, Giesecke J (2008) Short- and long-term effects of bacterial gastrointestinal infections. Emerg Infect Dis 14(1):143–148 (Jan)

Tursi A, Brandimarte G, Giorgetti GM, Elisei W (2006) Mesalazine and/or Lactobacillus casei in preventing recurrence of symptomatic uncomplicated diverticular disease of the colon: a prospective, randomized, open-label study. J Clin Gastroenterol 40(4):312–316 (Apr)

Vanderhoof JA, Whitney DB, Antonson DL, Hanner TL, Lupo JV, Young RJ (1999) Lactobacillus GG in the prevention of antibiotic-associated diarrhoea in children. J Pediatr 135(5):564–568 (Nov)

Van Gossum A, Dewit O, Louis E, de Hertogh G, Baert F, Fontaine F et al (2007) Multicenter randomized-controlled clinical trial of probiotics (Lactobacillus johnsonii, LA1) on early endoscopic recurrence of Crohn's disease after Ileo-caecal resection. Inflamm Bowel Dis 13(2):135–142 (Feb)

Van Niel CW, Feudtner C, Garrison MM, Christakis DA (2002) Lactobacillus therapy for acute infectious diarrhoea in children: a meta-analysis. Pediatrics 109(4):678–684 (Apr)

Venturi A, Gionchetti P, Rizzello F, Johansson R, Zucconi E, Brigidi P et al (1999) Impact on the composition of the faecal flora by a new probiotic preparation: preliminary data on maintenance treatment of patients with ulcerative colitis. Aliment Pharmacol Ther 13(8):1103–1108 (Aug)

Wine E, Chan VL, Sherman PM (2008) Campylobacter jejuni mediated disruption of polarized epithelial monolayers is cell-type specific, time dependent, and correlates with bacterial invasion. Pediatr Res 64(6):599–604 (Dec)

Wine E, Gareau MG, Johnson-Henry K, Sherman PM (2009) Strain-specific probiotic (Lactobacillus helveticus) inhibition of Campylobacter jejuni invasion of human intestinal epithelial cells. FEMS Microbiol Lett 300(1):146–152 (Nov)

Wunderlich PF, Braun L, Fumagalli I, D'Apuzzo V, Heim F, Karly M et al (1989) Double-blind report on the efficacy of lactic acid-producing Enterococcus SF68 in the prevention of antibiotic-associated diarrhoea and in the treatment of acute diarrhoea. J Int Med Res 17(4):333–338 (July–Aug)

Young KT, Davis LM, Dirita VJ (2007) Campylobacter jejuni: molecular biology and pathogenesis. Nat Rev Microbiol 5(9):665–679 (Sept)

Chapter 16
Potential Mechanisms of Enteric Cytoprotection by Probiotics: Lessons from Cultured Human Intestinal Cells

Vanessa Liévin-Le Moal and Alain L. Servin

16.1 Introduction

Probiotic lactic acid strains are defined as live microorganisms which, when consumed in appropriate amounts in food, confer a health benefit on the host (FAO/WHO 2001, 2002). These strains include lactobacilli and bifidobacteria, which are commercialized as food or dietary supplements of living bacteria, and have attracted the interest of scientists as well as consumers (Gorbach 2000; Mercenier et al. 2003; Picard et al. 2005; Saxelin et al. 2005; Reid and Bruce 2006; Blandino et al. 2008; Koninkx and Malago 2008; Preidis and Versalovic 2009). One of the advantages of repeatedly consuming probiotic lactic acid bacteria in appropriate amounts could be that they maintain a higher, albeit generally transient, bacterial density in the upper part of the intestinal tract, a location that normally hosts very few resident microbiota bacteria, and in which enteric Gram-negative pathogens produce their harmful effects. The number and activity of live bacteria present in probiotic preparations sometimes declines if they are stored for inappropriate times and/or at inappropriate temperatures. Pharmaceutical preparations used medicinally to combat gastrointestinal diseases contain several inactivated or heat-killed lactic acid bacteria with or without their spent culture media. These pharmaceutical preparations have the advantage of providing a controlled and stable quantity of whole-cell bacteria and spent culture medium containing substances retaining some of the probiotic activities of the parental live strains. Although they cannot strictly be considered to be probiotics (Sanders et al. 2007), they have been defined as "biotherapeutic agents" (Servin 2004). It is important to note that a limited number of probiotic strains conserve their probiotic properties after heating.

A. L. Servin (✉)
Faculté de Pharmacie, Inserm Unité 756, 92296 Châtenay-Malabry, France
Tel.: +33-146-835661
e-mail: alain.servin@u-psud.fr

J. J. Malago et al. (eds.), *Probiotic Bacteria and Enteric Infections*,
DOI 10.1007/978-94-007-0386-5_16,

16.2 The Mucosal Barrier

The mucosal surface of the intestinal tract is the largest body surface in contact with the external environment (200–300 m^2). The gastrointestinal tract is a complex ecosystem that combines resident microbiota with cells of various phenotypes lining the epithelial wall and expressing complex metabolic activities. The surface of the intestine is lined by a simple columnar epithelium that is folded to form a number of invaginations, or crypts, which are embedded in the connective tissue. In the intestine, the polarized epithelial cells that form the epithelium separate two very different compartments (Fig. 16.1).

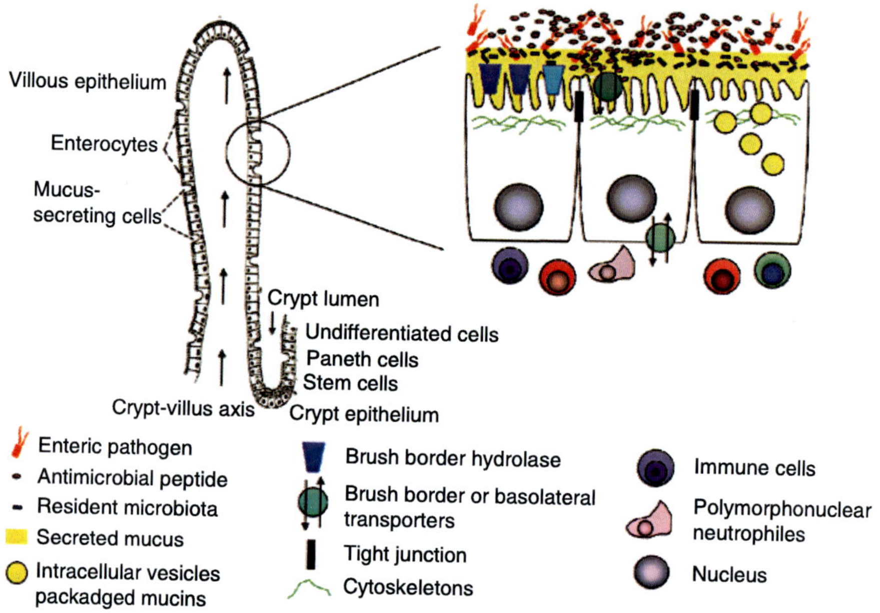

Fig. 16.1 The intestinal epithelial barrier.

Intestinal epithelium consists of an epithelial stem cell population with a high turnover rate and is located at the base of the crypt. It differentiates into multiple intestinal cell types including enterocytes, mucus-secreting and enteroendocrine cells during migration along the crypt-villus axis.

The intestinal mucosa normally consists of a single layer of polarized epithelial cells. The tight junction, a component of the apical junctional complex, seals the paracellular space between epithelial cells. The brush border expresses hydrolases, and both the apical and basolateral cell domains contain transporters. The intestinal epithelial barrier plays an essential role in maintaining immune homeostasis. The lamina propria, located beneath the basement membrane, contains immune cells and neutrophils. Intestinal microbiota resides in the lumen, outside the mucus layer. Intestinal functions are modulated by quorum sensing-dependent cross-talk between the microbiota and host cells. Secreted mucins, combined with membrane-bound mucins, act as a physicochemical barrier and protect the epithelial cell surface against undesirable and harmful pathogens. Secreted antimicrobial peptides form the first chemical defense system against unwanted enteric pathogens

16.2.1 Extracellular Components of the Barrier

The intestinal mucosa has a surface coating of mucus that is secreted by the specialized mucus-secreting cells (Laboisse et al. 1995). A discontinuous, thinner layer of mucus gel covers the epithelial cells that line the epithelium of the small intestine. The thickness of the mucus varies in the large intestine, gradually increasing from the colon to the rectum, and Peyer's patches apparently have no mucus covering. This mucus gel can be useful to enteric bacteria in at least two ways. First, mucins provide a source of energy by producing the saccharides used for the sustained growth of both the indigenous enteric microbiota and the pathogens adhering to the mucus. The mucus layer also creates a physical barrier that protects the host against enteric microbial pathogens (McCracken and Lorenz 2001; Liévin-Le Moal and Servin 2006), and so plays a pivotal cytoprotective role. Mucins can be divided into two distinct sub-families on the basis of their structure: forming secreted and membrane-bound mucins. Seventeen human mucin-type glycoproteins have so far been officially assigned to the *MUC* gene family with the approval of the Human Genome Organization Gene Nomenclature Committee (HUGO/GNC; http://www.gene.ucl.ac.uk/nomenclature). The secreted mucins consist of the MUC2, MUC5B, MUC5AC and MUC6 mucins. The membrane-associated mucins MUC1, MUC3A, MUC3B, MUC4, MUC11, MUC12, MUC13, MUC16, and MUC17 are associated with the cell membrane by an integral transmembrane domain, and are characterized by having relatively short cytoplasmic tails that are linked to cell cytoskeletal proteins. Moreover, the membrane-bound mucins are linked to the secreted mucins by both covalent and non-covalent bonds in order to create a high local concentration of specific molecular structures, and to develop functions including binding sites for lectins, selectin and adhesion molecules, stoichiometric power that enables them to exclude larger molecules and microorganisms, hygroscopic effects that influence the degree of hydration at the cell surface, ion-exchange effects; and an area in which growth factors, cytokines and chemokines are sequestered. Associated with the mucus lining, the intestinal epithelium secreted immunoglobulin A (Fagarasan and Honjo 2004) contributes to defending the intestine against luminal pathogens (Fagarasan and Honjo 2003).

Antimicrobial peptides (Ganz 2001; Selsted and Ouellette 2005) and angiogenins (Hooper et al. 2003) constitute the first line of chemical defenses of the host in intestine (McCracken and Lorenz 2001; Liévin-Le Moal and Servin 2006). One of their functions is to control the bacterial milieu in the intestine. Under physiological conditions, the continual release of preformed AMPs allows the chemical defense system to contribute directly to the innate immunity of the crypt microenvironment, and it probably does this by diffusing the secreted peptides into the lumen. AMPs are small peptides of 20–40 amino acids in length. Two major families of AMPs have been identified: the defensins and the cathelicidins. Defensins were first identified in mouse small-intestinal cells. Defensins have been classified on the basis of their secondary structure as AMPs containing an α-helical structure (α-defensins), or a β-sheet that contains three disulfide bonds (β-defensins), and circular AMP (θ-defensins).

The adult human intestine contains trillions of bacteria, representing hundreds of species and thousands of subspecies (Xu et al. 2007; Ley et al. 2008; Camp et al. 2009; Costello et al. 2009). Microbes resident in the intestinal microbiota share similar signature molecules known as microbe-associated molecular patterns (MAMPs), which include peptidoglycans, lipoproteins, and lipopolysaccharide (LPS) from Gram-negative bacteria, and teichoic acids from Gram-positive bacteria (Didierlaurent et al. 2002). Following infection, the host's innate mucosal immunity response is activated mainly as a result of the specific recognition by pattern recognition receptors of conserved "non-self" molecular structures found in large groups of pathogens, known as pathogen-associated molecular patterns (PAMPs) (Didierlaurent et al. 2002). Intestinal epithelial cells sense the environment within the gut by means of their pattern-recognition receptors (PRRs), which include Toll-like receptors (TLRs) and the NOD (nucleotide-binding oligomerization domain) proteins. TLRs are evolutionarily conserved proteins that are characterized by having an extracellular leucine-rich repeat domain involved in ligand recognition and an intracellular Toll/IL-1 receptor-like domain involved in signal transduction (Medzhitov 2001; Akira and Takeda 2004). It is well-established that human intestinal cells have low responsiveness to MAMPs such as LPS as a result of low TLR4 expression and no CD14 or MD-2 expression (Abreu et al. 2005; Cario and Podolsky 2005; Lenoir et al. 2008). Moreover, two mammalian nucleotide-binding leucine-rich repeat (NBS-LRR) proteins (NOD1 and NOD2) function as intracellular sensors of bacterial products in the induction of inflammatory responses (Franchi et al. 2009; Langefeld et al. 2009). In addition, the human intestinal microbiota (Pedron and Sansonetti 2008; Sansonetti 2008) has been associated with the immune modulation that plays a significant role in maintaining intestinal immune homeostasis and preventing inflammation. Importantly, the cells lining the gastrointestinal epithelium and the resident microbiota act independently and/or synergistically to promote an efficient host defense system (McCracken and Lorenz 2001; Liévin-Le Moal and Servin 2006).

16.2.2 Cellular Components of the Mucosal Barrier

The intestinal epithelium contains five highly specialized cell phenotypes: the fluid-transporting, neuroendocrine, mucus-secreting, Paneth and M cells. The intestinal epithelium is a model of tissue renewal since intestinal cells are constantly generated from a source of multipotent stem cells located in the crypts of Lieberkühn, and these provide new precursor cells that permit a high rate of cell turnover (Louvard et al. 1992). Intestinal cell differentiation results from a process with multiple stages, where the undifferentiated cells located at the base of the crypt, migrate along the crypto-villus axis, and during the final step of migration the fully differentiated cells localize at the tip of the villus. Absorptive, neuroendocrine, mucus-secreting cells and Paneth cells have a distinctive polarized organization. The polarized cellular architecture of these intestinal cells combines a complex cytoskeleton with actin microfilaments, intermediate filaments and microtubules, and two membrane domains. Sealing of these two membrane domains is triggered by the junctional complexes. In fluid-transporting cells, also known as enterocytes, the two membrane domains are structurally, biochemically and physiologically distinct. These systems allow enterocytes to carry out digestion and vectorial transport of nutrients. The apical membrane facing the external compartment contains membrane proteins, hydrolases and transporters; whereas the basal membrane domain facing the internal environment includes membrane proteins that interact with the basal lamina. The principal function of mucus-secreting cells, also known as goblet cells, is to produce secreted and membrane-bound mucins, and to secrete mucins that form the mucus layer at the surface of the intestinal epithelium (Laboisse et al. 1995). Mucins are stored intracellularly in granules, which are guided to the cell surface by microtubule network. The viscous mucus contained in granules is extruded after the granules have fused with plasma membranes and a fusion pore is formed, by a process that requires an expulsionary force. There are two secretory pathways in mucin-secreting cells. The steady vesicular constitutive pathway does not include any storage, since the vesicles are transported directly to the plasma membrane where they immediately undergo exocytosis. Additionally, this pathway does not require a signal to release the vesicle contents. The second pathway involves the packaging and storage of mucins in inert granules within the cell, and the release of the mucins which is regulated by specific stimuli involving transducing signal molecules. Paneth cells (Ayabe et al. 2004), which are located in the basal portion of intestinal crypts, underlying the zone of intestinal epithelial cell division, are specialized cells that produce antimicrobial substances. Paneth cells are pyramidal, columnar, exocrine cells identifiable within a few days after birth. The ultrastructure of Paneth cells shows that they have a basally-located nucleus with a nucleolus, a perinuclear region containing the rough endoplasmic reticulum and Golgi apparatus, and a supranuclear region containing numerous high-electron density, apically-located, eosinophilic secretory granules containing AMPs and other antimicrobial substances, including lysozyme, phospholipase A_2, α_1-antitrypsin, and AMPs. Finally, M cells are located in the epithelia above mucosa-associated lymphoid tissues, such as Peyer's patches, where they function as the antigen sampling cells of the mucosal immune system (Kraehenbuhl and Neutra 2000).

The intestinal cells that make up the epithelium provide a physical barrier that protects the host against the unwanted intrusion of microorganisms into the gastrointestinal microbiota, and against the penetration of harmful microorganisms which can hijack the cellular molecules and signaling pathways of the host to become pathogenic. The integrity of the intestinal epithelium is maintained by intercellular junctional complexes composed of tight junctions (TJs), adherent junctions, and desmosomes, whereas gap junctions allow intercellular communication to occur. TJs, the most apical components of the junctional complex, create a semipermeable diffusion barrier between individual cells, which can be regulated and acts as the permeability barrier (Matter and Balda 2003).

The mucosa of the intestinal tract is exposed to a high number of bacterial species present in the resident microbiota and to the unwanted intrusion of various enteric microbial pathogens. Three cell phenotypes, i.e., fluid-transporting, mucus-secreting and M cells, have been identified as the targets of these enterovirulent pathogens. Harmful enteric microorganisms can hijack the host cell systems by means of pathogenic factors. In the first step of infectious process, some enteric bacterial pathogens adhere to the brush border of enterocytes and goblet cells (Torres et al. 2005), enabling them to exploit the underlying signaling pathways (Kaper et al. 2004). Moreover, some invasive enteric microbial pathogens, including *Salmonella* (Patel and Galan 2005; Coburn et al. 2007), *Listeria* (Cossart and Toledo-Arana 2008), and *Shigella* (Sansonetti 2006), have developed specialized systems that, after this essential adhesion step, produce virulence factors that subvert host cell signaling pathways, enabling the pathogen to cross the epithelial barrier. The host cell cytoskeleton is commonly targeted by enteric microbial pathogens during the cell penetration step; it is exploited for purposes that include gaining entry into cells, moving within and between cells, and forming and remodeling vacuoles in order to create a specialized niche, which enhances the pathogen's chances of survival. Many pathogenic enteric bacteria target and exploit the TJ domain to accomplish their pathogenic strategies by modulating intestinal permeability (Turner 2009; Viswanathan et al. 2009). It has also been shown that some enteric pathogens use the M cells that lie over the organized mucosal-associated lymphoid system (MALT) as a route of invasion, and that after passing through these cells, the bacteria confront phagocytic cells, particularly the macrophages that are present in the follicle dome (Clark and Jepson 2003).

16.3 Enteric Cytoprotective Effects of Probiotics

The aim of this review is to assess the cytoprotective effects of lactic acid strains against structural and functional injuries produced by enteric pathogens and non-pathogenic agents including chemicals and drugs in the different phenotypes of intestinal cells that form the intestinal barrier. We have chosen to focus our analysis on reports concerning the cytoprotective effects and mechanisms of action of identified lactic acid strains using cultured human intestinal cells. Several cultured human intestinal cell lines expressing different specific characteristics of the cell phenotypes lining the intestinal epithelium, and by forming junctional complexes, constitute a monolayer that mimics the intestinal epithelial barrier situation. These cellular models have been extensively used to study specific intestinal functions, host-cell interactions of intestinal pathogens, and the role of probiotic strains in combating enteric infections. T84 cells (Madara et al. 1987) are known to exhibit intestinal crypt cell morphology, including sparse apical microvilli, that express intestinal hydrolases and transporters, and form monolayers with a well-organized junctional domain including highly regulated TJs (Zweibaum et al. 1991). T84 cells have been used to investigate intestinal functions including, in particular, the barrier function and transport systems of the intestine, and the impact of enteric pathogens on these functions and on the pro-inflammatory responses triggered by infection, such as the polarized production of pro-inflammatory cytokines and transepithelial migration of polymorphonuclear leukocytes. A cloned cell (HT29-Cl.19A) (Rouyer-Fessard et al. 1989), or HT-29 glc^{-} (Zweibaum et al. 1985), HT-29 glc$^{-/+}$ (Lesuffleur et al. 1991), HT29-D4 (Fantini et al. 1986), and HT29-18-C1 (Huet et al. 1987) cell subpopulations have been established, all of which were derived from the parental undifferentiated HT-29 cell line (Fogh et al. 1977) selected depending on the culture conditions. Cloned Caco-2$_{BB2}$ and Caco-2/TC7 (Chantret et al. 1994).) cells have been established from the parental Caco-2 cell line (Pinto et al. 1983), which spontaneously differentiates in culture. The parental Caco-2 cells, and the above cited clones and cell subpopulations all exhibit morphological and functional characteristics of the mature enterocytes of the small intestine, including a well-organized and dense apical brush border expressing intestinal hydrolases and transporters, and formed monolayers with a well-organized junctional domain TJs (Zweibaum et al. 1991). Caco-2 cells and clones have been used to investigate the cytoskeletal organization of polarized intestinal cells, brush border-associated intestinal functions, and barrier function, as well as the impact of enteric pathogens on the organization and function of these structures. Some other HT-29 cell subpopulations, HT29-MTX (Lesuffleur et al. 1991), and the clones HT29-Cl.16E (Augeron et al. 1992), HT-29/B6 (Kreusel et al. 1991), and HT29-18-N2 (Huet et al. 1987), displayed the structural and functional characteristics of intestinal goblet cells. These polarized cells, which form monolayers, have been extensively used to examine the production of secreted and membrane-bound intestinal mucins packaged into large vesicles centrally the cytoplasm, and the apical secretion of mucus which develops by exocytosis as the result of signaling that regulates the intracellular traffic of mucus-containing-vesicles. In addition, it was interesting to note that the HT29-FU cell subpopulation (Lesuffleur et al. 1991) is a mixed population of enterocyte-like cells (75%) and mucus-secreting cells (25%) which closely mimics the intestinal situation.

16.3.1 Preservation of the Barrier Function

In human intestinal epithelial HT29-Cl.19A and Caco-2 cell lines infected with the enteroinvasive *Escherichia coli* strain 029:NM, the decrease in transepithelial resistance (TER), elevated chloride secretory responses, and alterations in cytoskeletal and TJ proteins produced by *E. coli* were corrected in the presence of *Lactobacillus acidophilus* strain ATCC4356 (Resta-Lenert and Barrett 2003). Pretreating T84 cells with lactic acid-producing bacteria reduced the reduction in TER induced by enterohemorragic *E. coli* (EHEC) O157:H7- and *E. coli* O127:H6 (Sherman et al. 2005). *L. plantarum* MF1298 and *L. salivarius* DC5 induced an increase in TER of polarized monolayers of Caco-2 cells, and attenuated the decrease in TER induced by *Listeria monocytogenes* (Klingberg et al. 2005). In the human colon crypt-like T84 cell line, *L. casei* DN-114 001 is able to abrogate in a dose-dependent-manner, the increase in paracellular permeability and redistribution of zonula occludens-1

(ZO-1) induced by enteropathogenic *E. coli* (EPEC) strain E2348/69 (Parassol et al. 2005). EHEC strain O157:H7-impaired tight junction integrity, and the increase in production of cyclooxygenase 2 (Cox-2) are both antagonized by *L. acidophilus* NCFM and *L. salivarius* Ls-33 (Putaala et al. 2008). Johnson-Henry et al. (2008) observed that the *L. rhamnosus* strain GG antagonized the EHEC-induced changes in paracellular permeability in polarized MDCK-I and T84 epithelial cell monolayers, affecting parameters including the electrical resistance, dextran permeability, and the distribution and expression of claudin-1 and ZO-1. In contrast to the live strain, heat-inactivated *L. rhamnosus* GG has no effect on the EHEC-induced brush border lesions or disruption of the barrier function in T84 cells (Johnson-Henry et al. 2008). A similar effect was observed in Caco-2 cells with a *L. plantarum* strain that inhibited the increase in the paracellular passage of dextran, and the rearrangements of TJ proteins, claudin-1, occludin, JAM-1 and ZO-1 induced by enteroinvasive *E. coli* (EIEC) (Qin et al. 2009). The delocalization of the TJ-associated proteins, claudin-1 and occludin, promoted by *Shigella dysenteria* infection has been shown to be antagonized in the presence of *L. rhamnosus* and *L. acidophilus* strains (Moorthy et al. 2009).

Bifidobacterium infantis-conditioned medium applied to crypt-like T84 human epithelial intestinal cells prevents tumour necrosis factor-alpha (TNF-α)- and interferon-gamma (IFN-γ)-induced falls in TER and rearrangement of TJ proteins, ZO-1, claudins 1, 2, 3, and 4, and occludin and increase of p38 and extracellular regulated kinase (Erk)1/2 mitogen-activated kinases (MAPKs) activity (Ewaschuk et al. 2007). Putaala et al. (2008) have reported that the *B. lactis* 420 and *B. lactis* HN019 bacteria, and the cell-free culture supernatants (CFCS) of *Bifidobacterium lactis* 420, have the capacity to antagonize the damage caused to tight junction integrity induced by EHEC strain O157:H7, and block the overexpression of Cox-2 in Caco-2 cells.

Non-pathogen-induced lesions in barrier function can be also corrected by probiotic strains. A TNF-α-induced decrease in transepithelial electrical resistance was inhibited by a strain of *L. plantarum* that in parallel inhibits the activation of the MAP kinase, Erk1/2 and the degradation of the inhibitor of NF-κB (I-κB)-α in Caco-2 cell monolayers (Ko et al. 2007). *L. acidophilus* strain ATCC4356 has the ability to antagonize the TNF-α- and IFN-γ-induced decrease in TER, and increase in epithelial permeability in both enterocyte-like Caco-2 and HT29-Cl.19A cells (Resta-Lenert and Barrett 2006). Wheat gliadin, which induces severe intestinal symptoms and small-bowel mucosal damage, induces a loss in TER and rearrangements of ZO-1 expression in Caco-2 cells that can be protected by strains of *L. fermentum* or *B. lactis* (Lindfors et al. 2008). The effect of *L. rhamnosus* GG-produced soluble proteins (p40 and p75) on the hydrogen peroxide-induced disruption of tight junctions and barrier function in Caco-2 cell monolayers has been investigated (Seth et al. 2008). Pretreatment of cell monolayers with p40 or p75 attenuated the hydrogen peroxide-induced decrease in TER and increase in inulin permeability in a manner that was both time- and dose-dependent. Moreover, p40 and p75 also prevented hydrogen peroxide-induced redistribution of occludin, ZO-1, E-cadherin, and beta-catenin from the junctional domains, and their dissociation from the detergent-insoluble fractions. Interestingly, the inhibitory effect was accompanied by a rapid increase in the membrane translocation of protein-kinase C (PKC)-βI and PKC-epsilon. Abdominal pain is common in the general population and particularly so in patients with irritable bowel syndrome due to visceral hypersensitivity. Rousseaux et al. (2007) have recently published interesting report showing that after cell contact, *Lactobacillus* strains induce expression of opioid and cannabinoid receptors in cultured HT-29 cells, and mediate analgesic functions in the gut-similar to the effects of morphine.

Liévin-Le Moal et al. (2007) have shown that the prototype pharmaceutical preparation containing the heat-killed *L. acidophilus* strain LB and its spent culture medium (Lacteol Fort®, Axcan Pharma Ltd) antagonize the secreted autotransporter toxin (Sat)-induced increase in paracellular permeability and fluid-formed domes in intestinal Caco-2/TC7 cell monolayers. Moreover, this pharmaceutical preparation also inhibits the delocalization of the TJ-associated protein, ZO-1 by aspirin treatment in HT-29 cells (Montalto et al. 2004).

16.3.2 Preservation of Intestinal Cell Functions

Cross-talk between the species of the resident microbiota and intestinal cells occurs to modulate the functions of intestinal cells (Hooper et al. 2002; Zocco et al. 2007). Studies of the *Bacteroides thetaiotaomicron* strain VPI-5482, a component of the intestinal microflora of mice and humans (Xu et al. 2003), have documented this phenomenon (Hooper et al. 1999, 2000). For example, this commensal bacterium modulates the expression of genes involved in several important intestinal functions, including nutrient absorption, mucosal barrier fortification, xenobiotic metabolism, angiogenesis, and postnatal intestinal maturation (Hooper et al. 2001), mucosal defence (Lopez-Boado et al. 2000), and the glycosylation process via a mechanism that involves a soluble factor (Freitas et al. 2001). Since the *B. thetaiotaomicron* strain displays a considerable capacity to utilize polysaccharides, a property that has also been observed with a *L. casei* strain (Sonnenburg et al. 2006), it has been postulated that probiotic strains originating in the intestinal microbiota could also develop similar properties (Freitas et al. 2003).

It has been reported that several lactic acid strains have the capacity to up-regulate functional molecules present at the membrane of cultured human enterocyte-like cells. In HT-29 cells grown in the continuous presence of milks fermented by one of the following bacterial populations: *L. helveticus, L. acidophilus, Bifidobacterium,* increased activity is observed of brush border enzymes including sucrase-isomaltase (SI), dipeptidyl peptidase IV (DPP IV), aminopeptidase N and alkaline phosphatase (AP) (Baricault et al. 1995). The *L. plantarum* strain 299v inhibits the increase in short circuit current and secretory changes in Caco-2 cell monolayers after EPEC E2348/69 infection (Michail and Abernathy 2002). An *L. casei* strain increases the activity of the peptide transporter hPEPT1 in Caco-2 cells without modifying the hPEPT1 mRNA levels (Neudeck et al. 2004). Live but not heat-killed,

L. acidophilus and *L. casei* strains in Caco-2 cells display PI3-kinase-dependent increased surface expression of the apical anion exchanger, SLC26A3, which is correlated with an increase in Cl^-/OH^- exchange activity (Borthakur et al. 2008). Interestingly, this effect is mimicked by the cell-free culture supernatants (CFCSs) of both strains. *L. rhamnosus* GG induced a tyrosine kinase- and p38 MAP kinase-dependent Cox-2 expression in T84 cells (Korhonen et al. 2004). Tao et al. (2006) identified a low-molecular-weight, heat-stable peptide present in conditioned media of the *L. rhamnosus* strain GG, as the factor inducing the expression of cytoprotective heat shock protein (Hsp) 25 (Ropeleski et al. 2003) and 72 (Liu et al. 2003) in intestinal epithelial cells in a time- and concentration-dependent manner, and which is blocked by inhibitors of p38 and JNK MAPKs. Soluble effector molecules secreted by *L. acidophilus* strain ATCC 4356 reduced Niemann-Pick C1-like 1 (NPC1L1) gene expression via the liver X receptors (LXR), and inhibited the cellular uptake of micellar cholesterol in Caco-2 cells (Huang and Zheng 2009). Few reports have documented the antagonistic activity of lactic acid strains against the structural and functional damages triggered by enteric pathogens in cultured human enterocyte-like cells. For Liévin-Le Moal et al. (2002), the cellular structural and functional injuries in enterocyte-like Caco-2/TC7 cells including the disorganization of the brush border F-actin cytoskeleton and the down-expression of SI, DPP IV, AP and fructose transporter, promoted by the DAEC strain C1845, were antagonized by the *L. acidophilus* strain LB and its CFCS.

It has been reported that several lactobacilli strains are able to increase the production of extracellular components of the intestinal barrier. By incubating *L. plantarum* 299v with HT-29 cells, an increase in the expression levels of the mRNAs of mucins MUC2 and MUC3 has been observed (Mack et al. 1999, 2003). Moreover, *L. casei rhamnosus* GG mediates the up-regulation of epithelial MUC2 (Mattar et al. 2002). The *Lactobacillus* species present in the VSL#3 cocktail of probiotic strains, increased MUC2 gene expression in LS 174T colonic epithelial cells, an effect triggered by a heat-resistant soluble compound present in the CFCSs (Caballero-Franco et al. 2007). Schlee et al. (2008) have recently observed a MAPK Erk1/2-, p38- and c-Jun N-terminal kinase-dependent increase in the expression of human beta defensin-2 (hBD-2) gene in Caco-2 cells subjected to concentration-dependent treatment with heat-killed *L. acidophilus* PZ 1138 or *L. fermentum* PZ 1162 strains. Caco-2 cells exposed to *L. plantarum* bacteria significantly induce HBD-2 mRNA expression and the secretion of HBD-2 but not of HBD-3, in a dose- and time-dependent manner (Paolillo et al. 2009).

Clostridium difficile (CD) is the most common cause of antibiotic-associated diarrhoea. Its secreted toxins trigger deleterious effect on colon cells. Banerjee et al. (2009) have reported that the CD-conditioned-medium that induces cytotoxic effects on Caco-2 cells can be totally antagonized by the CFCS of *L. delbrueckii ssp. bulgaricus* strain B-30892. Interestingly, this effect is strain-specific, since other probiotic strains do not have this effect, and the antagonistic effect of CFCS-B-30892 is mediated by molecule(s), which probably inactivated the CD toxin(s).

The effect of probiotic strains against the deleterious effects produced by chemical molecules has also been evaluated. Resta-Lenert and Barrett (2006) have investigated the effect of the *L. acidophilus* strain ATCC4356 in Caco-2 and HT29-Cl.19A cells exposed basolaterally to IFN-γ or TNF-α. They found that the probiotic strain had the ability to restore chloride secretion through a mechanism involving MAPKs and PI3K, and to block the IFN-γ-induced down-regulation of the cystic fibrosis transmembrane conductance regulator (CFTR) and the NKCC1 cotransporter. In addition, probiotic-treated cells exposed to cytokines exhibited reduced activation of SOCS3 and STAT1,3. Interestingly, authors noted that the commensal, *B. thetaiotaomicron* has the same effect, but to a lesser extent.

Emerging evidence suggests that enteric glial cells, a major constituent of the enteric nervous system (ENS), are key regulators of intestinal epithelial barrier functions (Van Landeghem et al. 2009; Neunlist et al. 2003a, p. G1049). Moreover, it has been recently proposed that intestinal microbiota may play a role in the proposed two-way gut-brain axis (Collins and Bercik 2009). The regulatory function of ENS neurons has recently been investigated using a coculture model composed of human submucosa explants containing the submucosal neuronal network, and human enterocyte-like cell lines, primary cultures of rat ENS or human neuroblastoma SH-SY5Y cells (Neunlist et al. 2003b; Moriez et al. 2009). Immunohistochemistry and quantitative PCR examinations of enteric neurons have revealed that the enteric pathogen *S. flexneri* induces rapid neuronal morphological alterations, suggesting that cell death occurs in enteric submucosal neurons (Coron et al. 2009), whereas *Clostridium difficile* toxin B activates vasoactive intestinal peptide-immunoreactive neurons (VIP-neurons) (Neunlist et al. 2003a). This model looks attractive for finding out whether probiotic strains influence the ENS or correct the pathogen-induced changes in ENS functions and in turn those affecting intestinal functions.

16.3.3 Regulation of Pro-inflammatory Immune Responses

The colonic epithelium provides an interface between the host and micro-organisms colonizing the gastrointestinal tract. Molecular recognition of bacteria is facilitated through TLR, and the colonic epithelial cells express TLR2 TLR3; TLR4 proteins at different levels (Medzhitov 2001; Akira and Takeda 2004). The levels of expression of TLR1-4, CD14, and/or MD-2 differ in HT29 and Caco-2 cells (Abreu et al. 2005; Cario and Podolsky 2005; Furrie et al. 2005; Lenoir et al. 2008). The human intestinal microbiota (Pedron and Sansonetti 2008; Sansonetti 2008) has been associated with immune modulation; playing a significant role in the development and maturation of cells composing the epithelial barrier, maintaining intestinal immune homeostasis and preventing inflammation. Pro-inflammatory cytokines, many of the enteric pathogens, and deregulation of the host-symbiont interaction in the gut result in activation of the canonical NF-κB signaling pathway that plays a pivotal role in gut homeostasis (Pasparakis 2009). The role of probiotic strains to counteract or regulate this deleterious activation has attracted the attention of researchers. Studies in recent years have provided evidence that selected probiotic strains are able to antagonize or diminish experimental intestinal inflammation by their effects on mucosal immune functions and immune cells (Schiffrin and Blum 2002; Kelly et al. 2005; Boirivant and Strober 2007; Round and Mazmanian 2009).

Lipoteichoic acid is involved in the adhesion of *L. johnsonii* La1 to Caco-2 cells, but not in that of *L. acidophilus* La10 (Granato et al. 1999). When lipoteochoic acid is isolated from these two strains, it no longer stimulates HT-29 cells to produce cytokines, but inhibits the sCD14-mediated, LPS-induced response resulting from the production of pro-inflammatory cytokines, IL-8 and TNF-α (Vidal et al. 2002). When the probiotic *L. plantarum* strain BFE 1685, isolated from a child's faeces, and the probiotic *L. rhamnosus* strain GG were investigated for their ability to influence the innate immune response of undifferentiated HT-29 intestinal epithelial cells, they were shown to be able to stimulate transcription levels of TLR2 and TLR9, but not of TLR4 (Vizoso Pinto et al. 2009). Interestingly Kelly et al. (2004) observed that the symbiont *B. thetaiotaomicron* attenuates the inflammatory response in Caco-2 cells by enhancing the nuclear shuttling of peroxisome proliferator activated receptor-gamma (PPAR-γ), which is a major player in maintaining intestinal mucosa homeostasis, and of NF-κB subunit RelA. Voltan et al. (2008) identified the first such mechanism for a lactic acid strain, since the hydrogen peroxide-producing *L. crispatus* strain M247 induced PPAR-γ nuclear translocation, and enhanced transcriptional activity in epithelial CMT-93 cells via a hydrogen peroxide-dependent mechanism.

It has been reported that certain strains of *L. rhamnosus, L. delbrueckii,* and *L. acidophilus* suppress the production of the chemokine RANTES and interleukin-8 (IL-8) by stimulating HT-29 cells, that the extent of this suppression depends on the nature of the bacterial growth medium, and that the suppression of TNF-α and TGF-β production is a strain-dependent effect (Wallace et al. 2003). *L. reuteri* inhibits the constitutive synthesis by T84 and HT29 cells of IL-8 and inhibits IL-8 synthesis induced by *Salmonella enterica* serovar Typhimurium (Ma et al. 2004). The same effect has been reported for the *L. acidophilus* strain LB against the *S.* Typhimurium-induced IL-8 production in Caco-2/TC7 cells (Coconnier et al. 2000; Coconnier-Polter et al. 2005). The *L. plantarum* BFE 1685 and *L. johnsonii* 6128 strains significantly increased IL-8 chemokine production by HT-29 cells without modifying other cytokines, such as IL-1, IL-6, IL-10, monocyte chemoattractant protein-1 (MCP-1), TNF-α, and TGF-β (Vizoso Pinto et al. 2007). When HT-29 cells were first treated with lactobacilli and then infected with *S. enterica* serovar Typhimurium, the IL-8 levels were significantly increased, indicating that HT29 cells had been sensitized by the lactobacilli (Vizoso Pinto et al. 2009). Examining different lactic acid strains, Wallace et al. (2003) concluded that *L. rhamnosus* R0011 is the strain that has the most extensive effects on cytokine production in HT-29 cells. *L. plantarum* strain 299v inhibits IL-8-dependent, EPEC E2348/69-induced transepithelial neutrophil migration in Caco-2 cell monolayers (Michail and Abernathy 2003). O'Hara et al. (2006) investigated immune cell responses to commensal and pathogenic bacteria, and have reported that incubating human undifferentiated HT-29 cells with *S. typhimurium* strain UK1 and *L salivarius* strain UCC118, or pretreating the pathogen with the probiotic results in a decrease in the activation of inflammatory gene expression, the activation of NF-κB, and the secretion of IL-8. *L. rhamnosus* GG, which does not induce IL-8 production in human intestinal HT29-Cl.19A cells (Lammers et al. 2002) has the ability in Caco-2 cells to strongly inhibit the upregulation of IL-8 induced by enterotoxigenic *E. coli* strain K88, neutrophil transmigration, the expression of growth-related oncogene-alpha and of the epithelial neutrophil-activating peptide-78 gene (Roselli et al. 2006). *L. acidophilus* strain Bar13 reduced the production of IL-8 by HT-29 cells (Candela et al. 2008). In Caco-2, *L. plantarum* bacteria inhibited the LPS-induced increase in IL-23 secretion and IL-23 mRNA expression that occurs via a TLR2-dependent mechanism (Paolillo et al. 2009). Exposure of HT-29 cells to *Vibrio cholerae* results in NF-κB and p38 MAP kinase activation, and in upregulated gene expression for the neutrophil chemoattractant CXCL chemokines, IL-8, while downregulating the expression of macrophage-attracting C-C chemokines. *L. rhamnosus* strain GG specifically modulates the *V. cholerea*-induced IL-8, but not the other chemokine gene changes (Nandakumar et al. 2009). Interestingly, Tien et al. (2006) have reported that a strain of *L. casei* was able to manipulate the ubiquitin/proteasome pathway upstream of I-κBα in Caco-2 cells infected with *S. flexneri* promoting the down-regulation of the transcription of a number of genes encoding pro-inflammatory effectors, such as cytokines, chemokines and adherence molecules induced by invasive *S. flexneri*. *L. rhamonsus* strain GG prevents the K88-induced increased expression of IL-1β and TNF-α and decreased TGF-α in Caco-2 cells, and interestingly, the GG CFCS, either undigested or digested with proteases, displayed the same activity as the GG culture (Roselli et al. 2006). In HT-29 cells, the *L. rhamnosus* strain GG secretes proteins p75 and p40, which inhibit the activation of the pro-apoptotic p38 MAPK by TNF-α, IL-1α, or INF-γ (Yan and Polk 2002), and prevent cytokine-induced apoptosis by activating the anti-apoptotic Akt/protein kinase B regulating signaling pathway (Yan et al. 2007). In Caco-2 cell monolayers, the upregulation of the expressions of CCL20 and CXCL10 induced by *E. coli*, peptidoglycan or flagellin is antagonized by *L. rhamnosus* GG and *Lactobacillus casei* strains (Toki et al. 2009). It was noted that after heat-killing a strain of *L. reuteri* has been shown to lose its capacity to inhibit a pro-inflammatory response in T84 and HT-29 cells (Ma et al. 2004).

The *B. animalis* MB5 culture and its CFCS, either undigested or digested with proteases, blocks the upregulation of IL-8, neutrophil transmigration, the increased expression of IL-1β and TNF-α and the decrease in TGF-α, and in the growth-related oncogene-alpha and epithelial neutrophil-activating peptide-78 gene expression in Caco-2 cells induced by enterotoxigenic *E. coli* strain K88- (Roselli et al. 2006). IL-8 production by HT-29 cells is reduced by *B. longum* Bar33 (Candela et al. 2008). *B. breve* strain Yakult BbrY and *B. bifidum* strain Yakult BbiY inhibits IL-8 secretion in HT-29 cells (Imaoka et al. 2008). *B. longum* suppresses the TNFα-induced, NF-κB-dependent secretion of interleukin-8 in HT-29 cells (Bai et al. 2004). *B. breve* strain C50 conditioned medium containing soluble factors induces a time- and dose-dependent inhibition of the secretion of CXCL8 by HT29-Cl.19A cells driven by both AP-1 and NF-κB transcription pathways, and implying decreased phosphorylation of p38-MAPK and I-κBα (Heuvelin et al. 2009). Like those of *Lactobacillus* strains, the effects of *Bifidobacterium* strains

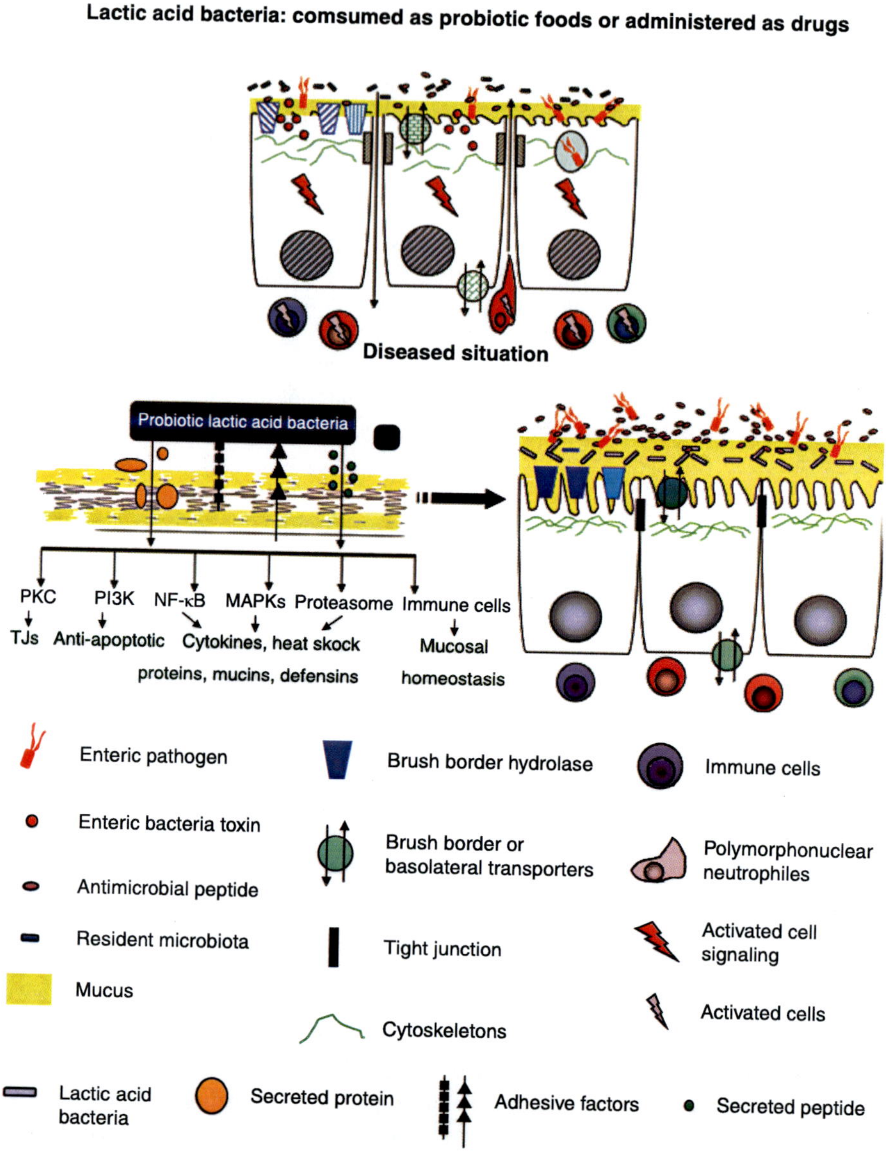

Fig. 16.2 How do lactic acid bacteria, including *Lactobacillus* and *Bifidobacterium*, consumed as probiotic foods or administered as drugs, produce their probiotic efffects on the intestinal epithelial barrier?

Enteric pathogens use their adhesive factors to interact with the brush border membrane and hijack membrane-bound molecules as receptors for triggering signaling events. Invasive pathogens cross the epithelial cell membrane, penetrate into the host cells and pursue sophisticated intracellular lifestyles either in vacuoles or directly within the cell cytoplasm. Enteric toxins are secreted by both adhering and non-adhering pathogens, and either bind to membrane-bound receptors, thus activating signaling pathways, or are internalized. Enteric pathogens, acting either directly or via their toxins, disrupt the epithelial barrier, increasing paracellular permeability and inducing proare strain-specific and concentration-dependent, since six out of eight bifidobacteria strains reversed the LPS-induced NF-κB activation, decreased IL-8 secretion and the mRNA levels for IL-8, TNF-α, Cox-2, and intercellular adhesion molecule 1 (ICAM-1) in HT-29 cells (Riedel et al. 2006). Menard et al. (2004) have reported that soluble factors present in conditioned media of a probiotic *B. breve,* and an intestinal commensal *B. bifidum* strains, were able to cross filter-grown, human intestinal epithelial HT29-Cl.19A cell monolayers using an active intestinal transport mechanism, and have conserved their inhibitory activity against the LPS-induced secretion of TNF-α in peripheral blood mononuclear cells or the THP-1 cell line. This is important, firstly because it demonstrates that after passage across the epithelium, secreted products of a lactic acid bacterium were able to exert an anti-inflammatory effect on immune cells located down-stream of the mucosal epithelial barrier.

16.4 Closing Remarks

Significant progress has been made in understanding how specific probiotic strains and biotherapeutic agents can be used to treat intestinal disorders caused by pathogens and non-pathogens. The activities have been shown to be strain-specific, and either triggered by whole-cell bacteria or related to particular cell surface molecules or secreted molecules. The mechanisms of action of a few number of selected lactic acid strains used as probiotics or biotherapeutic agents, have been elucidated during the last 20 years (Servin 2004; Lebeer et al. 2008) (Fig. 16.2), and have provided insights relevant to their therapeutic use for the management of diarrhoea. However, a large number of the studies conducted to evaluate their clinical efficacy have been conducted in open-label trials enrolling small numbers of patients. To mitigate this problem, meta-analyses of these clinical studies have been performed, even though the studies are not homogeneous and differ with regard to the age of the patients, the concentrations of lactic acid bacteria administered, the vehicles, the duration of administration, and the criteria used to evaluate the therapeutic efficacy, e.g., the volume of stools, the duration of diarrhoea, or the time to the production of a consistent stool. The conclusion of these meta-analyses is that several single lactic acid strains are apparently able to reduce the duration of diarrhoea. The use of lactic

---

inflammatory responses. Adhering and invading enteric pathogens trigger structural and functional lesions in epithelial cells, and even cell death.

Lactic acid-producing strains, including lactobacilli and bifidobacteria, whether commercialized as food or dietary supplements containing live bacteria, or used as "drugs" together with their spent culture media after being heat-killed, modulate intestinal function and/or correct the pathogen-induced structural and functional lesions in the epithelial barrier. Lactic acid triggers probiotic effects by activating cell signaling pathways after the engagement of their cell surface components with membrane-bound receptors. In addition, secreted proteins or peptides also activate cell signaling pathways acid bacteria as therapeutic agents against diarrhoea induced by enteric bacteria or rotaviruses or associated with antibiotics needs to be assessed by rigorous, double-blind, placebo-controlled, and randomized Phase III clinical trials. Such studies are currently in progress (Viswanathan et al. 2009).

The existence of complicated mechanisms underlying inflammatory bowel disease (IBD) is clearly indicated by the presence of many host susceptibility genes, and by the cooperative intervention of intestinal epithelial cells, immune cells and microbiota (Sands 2007; Xavier and Podolsky 2007). It is currently hoped that it may be possible to use probiotics to treat chronic IBD, comprising Crohn's disease and ulcerative colitis. However, Vanderpool et al. (2008) recently analyzed clinical studies conducted with single probiotic strains to treat IBD, and showed that most of them are less than promising. Only the VSL#3® cocktail of probiotic strains containing strains of *B. breve, B. longum, B. infantis, L. acidophilus, L. plantarum, L. paracasei, L. bulgaricus,* and *Streptococcus thermophilus* appeared to have clinical efficacy in an open-label trial in ulcerative colitis. This cocktail of strains may provide a simplified "normal microbiota" in a clinical context in which the presence of an activated coligenetic gut flora is suspected (Garrett et al. 2007). *In vivo* and *in vitro* methods have been used in attempts to elucidate the mechanisms of action at the cellular and molecular levels by which some lactic acid strains regulate pro-inflammatory responses at the mucosal barrier. Animal models, including genetically engineered mice and acute intestinal injury induced in mice by local administration of dextran sulfate sodium, each have both advantages and disadvantages (Mizoguchi and Mizoguchi 2008). If we are to use *in-vitro* methods to improve our knowledge about the mechanisms by which IBD develops, and to test the effects of probiotic strains *in vitro* cellular models integrating the different cell populations involved in the development of IBD will have to be devised. Moreover, as for diarrhoea, the use of probiotics to treat IBD needs to be assessed by rigorous, double-blind, placebo-controlled, and randomized Phase III clinical trials.

References

Abreu MT, Fukata M, Arditi M (2005) TLR signaling in the gut in health and disease. J Immunol 174:4453–4460

Akira S, Takeda K (2004) Toll-like receptor signalling. Nat Rev Immunol 4:499–511

Augeron C, Voisin T, Maoret JJ, Berthon B, Laburthe M, Laboisse CL (1992) Neurotensin and neuromedin N stimulate mucin output from human goblet cells (Cl.16E) via neurotensin receptors. Am J Physiol 262:G470–G476

Ayabe T, Ashida T, Kohgo Y, Kono T (2004) The role of Paneth cells and their antimicrobial peptides in innate host defense. Trends Microbiol 12:394–398

Bai AP, Ouyang Q, Zhang W, Wang CH, Li SF (2004) Probiotics inhibit TNF-alpha-induced interleukin-8 secretion of HT29 cells. World J Gastroenterol 10:455–457

Banerjee P, Merkel GJ, Bhunia AK (2009) *Lactobacillus delbrueckii* ssp. bulgaricus B-30892 can inhibit cytotoxic effects and adhesion of pathogenic *Clostridium difficile* to Caco-2 cells. Gut Pathog 1:8

Baricault L, Denariaz G, Houri JJ, Bouley C, Sapin C, Trugnan G (1995) Use of HT-29, a cultured human colon cancer cell line, to study the effect of fermented milks on colon cancer cell growth and differentiation. Carcinogenesis 16:245–252

Blandino G, Fazio D, Di Marco R (2008) Probiotics: overview of microbiological and immunological characteristics. Expert Rev Anti Infect Ther 6:497–508

Boirivant M, Strober W (2007) The mechanism of action of probiotics. Curr Opin Gastroenterol 23:679–692

Borthakur A et al (2008) The probiotic *Lactobacillus acidophilus* stimulates chloride/hydroxyl exchange activity in human intestinal epithelial cells. J Nutr 138:1355–1359

Caballero-Franco C, Keller K, De Simone C, Chadee K (2007) The VSL#3 probiotic formula induces mucin gene expression and secretion in colonic epithelial cells. Am J Physiol Gastrointest Liver Physiol 292:G315–G322

Camp JG, Kanther M, Semova I, Rawls JF (2009) Patterns and scales in gastrointestinal microbial ecology. Gastroenterology 136:1989–2002

Candela M et al (2008) Interaction of probiotic *Lactobacillus* and *Bifidobacterium* strains with human intestinal epithelial cells: adhesion properties, competition against enteropathogens and modulation of IL-8 production. Int J Food Microbiol 125:286–292

Cario E, Podolsky DK (2005) Intestinal epithelial TOLLerance versus inTOLLerance of commensals. Mol Immunol 42:887–893

Chantret I et al (1994) Differential expression of sucrase-isomaltase in clones isolated from early and late passages of the cell line Caco-2: evidence for glucose-dependent negative regulation. J Cell Sci 107(Pt 1):213–225

Clark MA, Jepson MA (2003) Intestinal M cells and their role in bacterial infection. Int J Med Microbiol 293:17–39

Coburn B, Grassl GA, Finlay BB (2007) *Salmonella*, the host and disease: a brief review. Immunol Cell Biol 85:112–118

Coconnier MH, Liévin V, Lorrot M, Servin AL (2000) Antagonistic activity of *Lactobacillus acidophilus* LB against intracellular *Salmonella enterica* serovar Typhimurium infecting human enterocyte-like Caco-2/TC-7 cells. Appl Environ Microbiol 66:1152–1157

Coconnier-Polter MH, Liévin-Le Moal V, Servin AL (2005) A *Lactobacillus acidophilus* strain of human gastrointestinal microbiota origin elicits killing of enterovirulent *Salmonella enterica* Serovar Typhimurium by triggering lethal bacterial membrane damage. Appl Environ Microbiol 71:6115–6120

Collins SM, Bercik P (2009) The relationship between intestinal microbiota and the central nervous system in normal gastrointestinal function and disease. Gastroenterology 136:2003–2014

Coron E et al (2009) Characterisation of early mucosal and neuronal lesions following *Shigella flexneri* infection in human colon. PLoS One 4:e4713

Cossart P, Toledo-Arana A (2008) *Listeria monocytogenes*, a unique model in infection biology: an overview. Microbes Infect 10:1041–1050

Costello EK, Lauber CL, Hamady M, Fierer N, Gordon JI, Knight R (2009) Bacterial community variation in human body habitats across space and time. Science 326(5960):1694–1697

Didierlaurent A, Sirard JC, Kraehenbuhl JP, Neutra MR (2002) How the gut senses its content. Cell Microbiol 4:61–72

Ewaschuk JB, Backer JL, Churchill TA, Obermeier F, Krause DO, Madsen KL (2007) Surface expression of Toll-like receptor 9 is upregulated on intestinal epithelial cells in response to pathogenic bacterial DNA. Infect Immun 75:2572–2579

Fagarasan S, Honjo T (2003) Intestinal IgA synthesis: regulation of front-line body defences. Nat Rev Immunol 3:63–72

Fagarasan S, Honjo T (2004) Regulation of IgA synthesis at mucosal surfaces. Curr Opin Immunol 16:277–283

Fantini J et al (1986) Spontaneous and induced dome formation by two clonal cell populations derived from a human adenocarcinoma cell line, HT29. J Cell Sci 83:235–249

FAO/WHO (2001) Joint FAO/WHO expert consultation on evaluation of health and nutritional properties of probiotics in food including powder milk with live lactic acid bacteria. Cordoba, Argentina, 1–4 Oct 2001. ftp://ftp.fao.org/docrep/fao/meeting/009/y6398e.pdf

FAO/WHO (2002) Guidelines for the evaluation of probiotics in food. Food and agriculture organization of the unites nations and world health organization working group report. ftp://ftp.fao.org/es/esn/food/wgreport2.pdf

Fogh J, Fogh JM, Orfeo T (1977) One hundred and twenty-seven cultured human tumor cell lines producing tumors in nude mice. J Natl Cancer Inst 59:221–226

Franchi L, Warner N, Viani K, Nunez G (2009) Function of Nod-like receptors in microbial recognition and host defense. Immunol Rev 227:106–128

Freitas M, Cayuela C, Antoine JM, Piller F, Sapin C, Trugnan G (2001) A heat labile soluble factor from *Bacteroides thetaiotaomicron* VPI-5482 specifically increases the galactosylation pattern of HT29-MTX cells. Cell Microbiol 3:289–300

Freitas M, Tavan E, Cayuela C, Diop L, Sapin C, Trugnan G (2003) Host-pathogens cross-talk. Indigenous bacteria and probiotics also play the game. Biol Cell 95:503–506

Furrie E, Macfarlane S, Thomson G, Macfarlane GT (2005) Toll-like receptors-2, -3 and -4 expression patterns on human colon and their regulation by mucosal-associated bacteria. Immunology 115:565–574

Ganz T (2001) Antimicrobial proteins and peptides in host defense. Semin Respir Infect 16:4–10

Garrett WS et al (2007) Communicable ulcerative colitis induced by T-bet deficiency in the innate immune system. Cell 131:33–45

Gorbach SL (2000) Probiotics and gastrointestinal health. Am J Gastroenterol 95:S2–S4

Granato D et al (1999) Cell surface-associated lipoteichoic acid acts as an adhesion factor for attachment of *Lactobacillus johnsonii* La1 to human enterocyte-like Caco-2 cells. Appl Environ Microbiol 65:1071–1077

Heuvelin E, Lebreton C, Grangette C, Pot B, Cerf-Bensussan N, Heyman M (2009) Mechanisms involved in alleviation of intestinal inflammation by *Bifidobacterium breve* soluble factors. PLoS One 4:e5184

Hooper LV, Xu J, Falk PG, Midtvedt T, Gordon JI (1999) A molecular sensor that allows a gut commensal to control its nutrient foundation in a competitive ecosystem. Proc Natl Acad Sci U S A 96:9833–9838

Hooper LV, Falk PG, Gordon JI (2000) Analyzing the molecular foundations of commensalism in the mouse intestine. Curr Opin Microbiol 3:79–85

Hooper LV, Wong MH, Thelin A, Hansson L, Falk PG, Gordon JI (2001) Molecular analysis of commensal host-microbial relationships in the intestine. Science 291:881–884

Hooper LV, Midtvedt T, Gordon JI (2002) How host-microbial interactions shape the nutrient environment of the mammalian intestine. Ann Rev Nutr 22:283–307

Hooper LV, Stappenbeck TS, Hong CV, Gordon JI (2003) Angiogenins: a new class of microbicidal proteins involved in innate immunity. Nat Immunol 4:269–273

Huang Y, Zheng Y (2009) The probiotic *Lactobacillus acidophilus* reduces cholesterol absorption through the down-regulation of Niemann-Pick C1-like 1 in Caco-2 cells. Br J Nutr 103(4):473–478

Huet C, Sahuquillo-Merino C, Coudrier E, Louvard D (1987) Absorptive and mucus-secreting subclones isolated from a multipotent intestinal cell line (HT-29) provide new models for cell polarity and terminal differentiation. J Cell Biol 105:345–357

Imaoka A et al (2008) Anti-inflammatory activity of probiotic *Bifidobacterium:* enhancement of IL-10 production in peripheral blood mononuclear cells from ulcerative colitis patients and inhibition of IL-8 secretion in HT-29 cells. World J Gastroenterol 14:2511–2516

Johnson-Henry KC, Donato KA, Shen-Tu G, Gordanpour M, Sherman PM (2008) *Lactobacillus rhamnosus* strain GG prevents enterohemorrhagic *Escherichia coli* O157:H7-induced changes in epithelial barrier function. Infect Immun 76:1340–1348

Kaper JB, Nataro JP, Mobley HL (2004) Pathogenic *Escherichia coli.* Nat Rev Microbiol 2:123–140

Kelly D et al (2004) Commensal anaerobic gut bacteria attenuate inflammation by regulating nuclear-cytoplasmic shuttling of PPAR-gamma and RelA. Nat Immunol 5:104–112

Kelly D, Conway S, Aminov R (2005) Commensal gut bacteria: mechanisms of immune modulation. Trends Immunol 26:326–333

Klingberg TD, Pedersen MH, Cencic A, Budde BB (2005) Application of measurements of transepithelial electrical resistance of intestinal epithelial cell monolayers to evaluate probiotic activity. Appl Environ Microbiol 71:7528–7530

Ko JS, Yang HR, Chang JY, Seo JK (2007) *Lactobacillus plantarum* inhibits epithelial barrier dysfunction and interleukin-8 secretion induced by tumor necrosis factor-alpha. World J Gastroenterol 13:1962–1965

Koninkx JF, Malago JJ (2008) The protective potency of probiotic bacteria and their microbial products against enteric infections—review. Folia Microbiol 53:189–194

Korhonen R, Kosonen O, Korpela R, Moilanen E (2004) The expression of COX2 protein induced by *Lactobacillus rhamnosus* GG, endotoxin and lipoteichoic acid in T84 epithelial cells. Lett Appl Microbiol 39:19–24

Kraehenbuhl JP, Neutra MR (2000) Epithelial M cells: differentiation and function. Ann Rev Cell Dev Biol 16:301–332

Kreusel KM, Fromm M, Schulzke JD, Hegel U (1991) Cl-secretion in epithelial monolayers of mucus-forming human colon cells (HT-29/B6). Am J Physiol 261:C574–C582

Laboisse C, Jarry A, Branka JE, Merlin D, Bou-Hanna C, Vallette G (1995) Regulation of mucin exocytosis from intestinal goblet cells. Biochem Soc Trans 23:810–813

Lammers KM et al (2002) Effect of probiotic strains on interleukin 8 production by HT29/19A cells. Am J Gastroenterol 97:1182–1186

Langefeld T, Mohamed W, Ghai R, Chakraborty T (2009) Toll-like receptors and NOD-like receptors: domain architecture and cellular signalling. Adv Exp Med Biol 653:48–57

Lebeer S, Vanderleyden J, De Keersmaecker SC (2008) Genes and molecules of lactobacilli supporting probiotic action. Microbiol Mol Biol Rev 72:728–764 (Table of Contents)

Lenoir C et al (2008) MD-2 controls bacterial lipopolysaccharide hyporesponsiveness in human intestinal epithelial cells. Life Sci 82:519–528

Lesuffleur T et al (1991) Increased growth adaptability to 5-fluorouracil and methotrexate of HT- 29 sub-populations selected for their commitment to differentiation. Int J Cancer 49:731–737

Ley RE et al (2008) Evolution of mammals and their gut microbes. Science 320:1647–1651

Liévin-Le Moal V, Amsellem R, Servin AL, Coconnier MH (2002) *Lactobacillus acidophilus* (strain LB) from the resident adult human gastrointestinal microflora exerts activity against brush border damage promoted by a diarrhoeagenic *Escherichia coli* in human enterocyte-like cells. Gut 50:803–811

Liévin-Le Moal V, Servin AL (2006) The front line of enteric host defence against unwelcome intrusion of harmful microorganisms: mucins, antimicrobioal peptides and microbiota. Clin Microbiol Rev 19(2):315–337

Liévin-Le Moal V, Sarrazin-Davila LE, Servin AL (2007) An experimental study and a randomized, double-blind, placebo-controlled clinical trial to evaluate the antisecretory activity of *Lactobacillus acidophilus* strain LB against nonrotavirus diarrhoea. Pediatrics 120:e795–e803

Lindfors K et al (2008) Live probiotic *Bifidobacterium lactis* bacteria inhibit the toxic effects induced by wheat gliadin in epithelial cell culture. Clin Exp Immunol 152:552–558

Liu TS et al (2003) Protective role of HSP72 against *Clostridium difficile* toxin A-induced intestinal epithelial cell dysfunction. Am J Physiol Cell Physiol 284:C1073–C1082

Lopez-Boado YS, Wilson CL, Hooper LV, Gordon JI, Hultgren SJ, Parks WC (2000) Bacterial exposure induces and activates matrilysin in mucosal epithelial cells. J Cell Biol 148:1305–1315

Louvard D, Kedinger M, Hauri HP (1992) The differentiating intestinal epithelial cell: establishment and maintenance of functions through interactions between cellular structures. Ann Rev Cell Biol 8:157–195

Ma D, Forsythe P, Bienenstock J (2004) Live *Lactobacillus reuteri* is essential for the inhibitory effect on tumor necrosis factor alpha-induced interleukin-8 expression. Infect Immun 72:5308–5314

Mack DR, Michail S, Wei S, McDougall L, Hollingsworth MA (1999) Probiotics inhibit enteropathogenic *E. coli* adherence in vitro by inducing intestinal mucin gene expression. Am J Physiol 276:G941–G950

Mack DR, Ahrne S, Hyde L, Wei S, Hollingsworth MA (2003) Extracellular MUC3 mucin secretion follows adherence of *Lactobacillus* strains to intestinal epithelial cells in vitro. Gut 52:827–833

Madara JL, Stafford J, Dharmsathaphorn K, Carlson S (1987) Structural analysis of a human intestinal epithelial cell line. Gastroenterology 92:1133–1145

Mattar AF, Teitelbaum DH, Drongowski RA, Yongyi F, Harmon CM, Coran AG (2002) Probiotics up-regulate MUC-2 mucin gene expression in a Caco-2 cell-culture model. Pediatr Surg Int 18:586–590

Matter K, Balda MS (2003) Functional analysis of tight junctions. Methods 30:228–234

McCracken VJ, Lorenz RG (2001) The gastrointestinal ecosystem: a precarious alliance among epithelium, immunity and microbiota. Cell Microbiol 3:1–11

Medzhitov R (2001) Toll-like receptors and innate immunity. Nat Rev Immunol 1:135–145

Menard S, Candalh C, Bambou JC, Terpend K, Cerf-Bensussan N, Heyman M (2004) Lactic acid bacteria secrete metabolites retaining anti-inflammatory properties after intestinal transport. Gut 53:821–828

Mercenier A, Pavan S, Pot B (2003) Probiotics as biotherapeutic agents: present knowledge and future prospects. Curr Pharm Des 9:175–191

Michail S, Abernathy F (2002) *Lactobacillus plantarum* reduces the in vitro secretory response of intestinal epithelial cells to enteropathogenic *Escherichia coli* infection. J Pediatr Gastroenterol Nutr 35:350–355

Michail S, Abernathy F (2003) *Lactobacillus plantarum* inhibits the intestinal epithelial migration of neutrophils induced by enteropathogenic *Escherichia coli*. J Pediatr Gastroenterol Nutr 36:385–391

Mizoguchi A, Mizoguchi E (2008) Inflammatory bowel disease, past, present and future: lessons from animal models. J Gastroenterol 43:1–17

Montalto M et al (2004) *Lactobacillus acidophilus* protects tight junctions from aspirin damage in HT-29 cells. Digestion 69:225–228

Moorthy G, Murali MR, Devaraj SN (2009) Lactobacilli facilitate maintenance of intestinal membrane integrity during *Shigella dysenteriae* 1 infection in rats. Nutrition 25:350–358

Moriez R, Abdo H, Chaumette T, Faure M, Lardeux B, Neunlist M (2009) Neuroplasticity and neuroprotection in enteric neurons: role of epithelial cells. Biochem Biophys Res Commun 382:577–582

Nandakumar NS, Pugazhendhi S, Madhu Mohan K, Jayakanthan K, Ramakrishna BS (2009) Effect of *Vibrio cholerae* on chemokine gene expression in HT29 cells and its modulation by *Lactobacillus* GG. Scand J Immunol 69:181–187

Neudeck BL, Loeb JM, Faith NG (2004) *Lactobacillus casei* alters hPEPT1-mediated glycylsarcosine uptake in Caco-2 cells. J Nutr 134:1120–1123

Neunlist M et al (2003a) Toxin B of Clostridium difficile activates human VIP submucosal neurons, in part via an IL-1beta-dependent pathway. Am J Physiol Gastrointest Liver Physiol 285:G1049–G1055

Neunlist M et al (2003b) Human ENS regulates the intestinal epithelial barrier permeability and a tight junction-associated protein ZO-1 via VIPergic pathways. Am J Physiol Gastrointest Liver Physiol 285:G1028–G1036

O'Hara AM et al (2006) Functional modulation of human intestinal epithelial cell responses by *Bifidobacterium infantis* and *Lactobacillus salivarius*. Immunology 118:202–215

Paolillo R, Romano Carratelli C, Sorrentino S, Mazzola N, Rizzo A (2009) Immunomodulatory effects of *Lactobacillus plantarum* on human colon cancer cells. Int Immunopharmacol 9:1265–1271

Parassol N, Freitas M, Thoreux K, Dalmasso G, Bourdet-Sicard R, Rampal P (2005) *Lactobacillus casei* DN-114 001 inhibits the increase in paracellular permeability of enteropathogenic *Escherichia coli*-infected T84 cells. Res Microbiol 156:256–262

Pasparakis M (2009) Regulation of tissue homeostasis by NF-kappaB signalling: implications for inflammatory diseases. Nat Rev Immunol 9:778–788

Patel JC, Galan JE (2005) Manipulation of the host actin cytoskeleton by *Salmonella*—all in the name of entry. Curr Opin Microbiol 8:10–15

Pedron T, Sansonetti P (2008) Commensals, bacterial pathogens and intestinal inflammation: an intriguing menage a trois. Cell Host Microbe 3:344–347

Picard C, Fioramonti J, Francois A, Robinson T, Neant F, Matuchansky C (2005) Review article: bifidobacteria as probiotic agents—physiological effects and clinical benefits. Aliment Pharmacol Ther 22:495–512

Pinto M et al (1983) Enterocyte-like differentiation and polarization of the human colon carcinoma cell line Caco-2 in culture. Biol Cell 47:323–330

Preidis GA, Versalovic J (2009) Targeting the human microbiome with antibiotics, probiotics, and prebiotics: gastroenterology enters the metagenomics era. Gastroenterology 136:2015–2031

Putaala H et al (2008) Effect of four probiotic strains and *Escherichia coli* O157:H7 on tight junction integrity and cyclo-oxygenase expression. Res Microbiol 159:692–698

Qin H, Zhang Z, Hang X, Jiang Y (2009) *L. plantarum* prevents enteroinvasive *Escherichia coli*-induced tight junction proteins changes in intestinal epithelial cells. BMC Microbiol 9:63

Reid G, Bruce AW (2006) Probiotics to prevent urinary tract infections: the rationale and evidence. World J Urol 24:28–32

Resta-Lenert S, Barrett KE (2003) Live probiotics protect intestinal epithelial cells from the effects of infection with enteroinvasive *Escherichia coli* (EIEC). Gut 52:988–997

Resta-Lenert S, Barrett KE (2006) Probiotics and commensals reverse TNF-alpha- and IFN-gamma-induced dysfunction in human intestinal epithelial cells. Gastroenterology 130:731–746

Riedel CU, Foata F, Philippe D, Adolfsson O, Eikmanns BJ, Blum S (2006) Anti-inflammatory effects of bifidobacteria by inhibition of LPS-induced NF-kappaB activation. World J Gastroenterol 12:3729–3735

Ropeleski MJ, Tang J, Walsh-Reitz MM, Musch MW, Chang EB (2003) Interleukin-11-induced heat shock protein 25 confers intestinal epithelial-specific cytoprotection from oxidant stress. Gastroenterology 124:1358–1368

Roselli M, Finamore A, Britti MS, Mengheri E (2006) Probiotic bacteria *Bifidobacterium animalis* MB5 and *Lactobacillus rhamnosus* GG protect intestinal Caco-2 cells from the inflammation-associated response induced by enterotoxigenic *Escherichia coli* K88. Br J Nutr 95:1177–1184

Round JL, Mazmanian SK (2009) The gut microbiota shapes intestinal immune responses during health and disease. Nat Rev Immunol 9:313–323

Rousseaux C et al (2007) *Lactobacillus acidophilus* modulates intestinal pain and induces opioid and cannabinoid receptors. Nat Med 13:35–37

Rouyer-Fessard C, Augeron C, Grasset E, Maoret JJ, Laboisse CL, Laburthe M (1989) VIP receptors and control of short circuit current in the human intestinal clonal cell line Cl.19A. Experientia 45:1102–1105

Sanders ME, Hamilton J, Reid G, Gibson G (2007) A nonviable preparation of *Lactobacillus acidophilus* is not a probiotic. Clin Infect Dis 44:886

Sands BE (2007) Inflammatory bowel disease: past, present, and future. J Gastroenterol 42:16–25

Sansonetti PJ (2006) The bacterial weaponry: lessons from *Shigella*. Ann N Y Acad Sci 1072:307–312

Sansonetti PJ (2008) Host-bacteria homeostasis in the healthy and inflamed gut. Curr Opin Gastroenterol 24:435–439

Saxelin M, Tynkkynen S, Mattila-Sandholm T, de Vos WM (2005) Probiotic and other functional microbes: from markets to mechanisms. Curr Opin Biotechnol 16:204–211

Schiffrin EJ, Blum S (2002) Interactions between the microbiota and the intestinal mucosa. Eur J Clin Nutr 56(Suppl 3):S60–S64

Schlee M, Harder J, Koten B, Stange EF, Wehkamp J, Fellermann K (2008) Probiotic lactobacilli and VSL#3 induce enterocyte beta-defensin 2. Clin Exp Immunol 151:528–535

Selsted ME, Ouellette AJ (2005) Mammalian defensins in the antimicrobial immune response. Nat Immunol 6:551–557

Servin AL (2004) Antagonistic activities of lactobacilli and bifidobacteria against microbial pathogens. FEMS Microbiol Rev 28:405–440

Seth A, Yan F, Polk DB, Rao RK (2008) Probiotics ameliorate the hydrogen peroxide-induced epithelial barrier disruption by a PKC- and MAP kinase-dependent mechanism. Am J Physiol Gastrointest Liver Physiol 294:G1060–G1069

Sherman PM, Johnson-Henry KC, Yeung HP, Ngo PS, Goulet J, Tompkins TA (2005) Probiotics reduce enterohemorrhagic *Escherichia coli* O157:H7- and enteropathogenic *E. coli* O127:H6-induced changes in polarized T84 epithelial cell monolayers by reducing bacterial adhesion and cytoskeletal rearrangements. Infect Immun 73:5183–5188

Sonnenburg JL, Chen CT, Gordon JI (2006) Genomic and metabolic studies of the impact of probiotics on a model gut symbiont and host. PLoS Biol 4:e413

Tao Y et al (2006) Soluble factors from *Lactobacillus* GG activate MAPKs and induce cytoprotective heat shock proteins in intestinal epithelial cells. Am J Physiol Cell Physiol 290:C1018–C1030

Tien MT et al (2006) Anti-inflammatory effect of *Lactobacillus casei* on *Shigella*-infected human intestinal epithelial cells. J Immunol 176:1228–1237

Toki S et al (2009) *Lactobacillus rhamnosus* GG and *Lactobacillus casei* suppress *Escherichia coli*-induced chemokine expression in intestinal epithelial cells. Int Arch Allergy Immunol 148:45–58

Torres AG, Zhou X, Kaper JB (2005) Adherence of diarrhoeagenic *Escherichia coli* strains to epithelial cells. Infect Immun 73:18–29

Turner JR (2009) Intestinal mucosal barrier function in health and disease. Nat Rev Immunol 9:799–809

Vanderpool C, Yan F, Polk DB (2008) Mechanisms of probiotic action: Implications for therapeutic applications in inflammatory bowel diseases. Inflamm Bowel Dis 14:1585–1596

Van Landeghem L et al (2009) Regulation of intestinal epithelial cells transcriptome by enteric glial cells: impact on intestinal epithelial barrier functions. BMC Genomics 10:507

Vidal K, Donnet-Hughes A, Granato D (2002) Lipoteichoic acids from *Lactobacillus johnsonii* strain La1 and *Lactobacillus acidophilus* strain La10 antagonize the responsiveness of human intestinal epithelial HT29 cells to lipopolysaccharide and gram-negative bacteria. Infect Immun 70:2057–2064

Viswanathan VK, Hodges K, Hecht G (2009) Enteric infection meets intestinal function: how bacterial pathogens cause diarrhoea. Nat Rev Microbiol 7:110–119

Vizoso Pinto MG, Schuster T, Briviba K, Watzl B, Holzapfel WH, Franz CM (2007) Adhesive and chemokine stimulatory properties of potentially probiotic *Lactobacillus strains*. J Food Prot 70:125–134

Vizoso Pinto MG, Rodriguez Gomez M, Seifert S, Watzl B, Holzapfel WH, Franz CM (2009) Lactobacilli stimulate the innate immune response and modulate the TLR expression of HT29 intestinal epithelial cells in vitro. Int J Food Microbiol 133:86–93

Voltan S et al (2008) *Lactobacillus crispatus* M247-derived H_2O_2 acts as a signal transducing molecule activating peroxisome proliferator activated receptor-gamma in the intestinal mucosa. Gastroenterology 135:1216–1227

Wallace TD, Bradley S, Buckley ND, Green-Johnson JM (2003) Interactions of lactic acid bacteria with human intestinal epithelial cells: effects on cytokine production. J Food Prot 66:466–472

Xavier RJ, Podolsky DK (2007) Unravelling the pathogenesis of inflammatory bowel disease. Nature 448:427–434

Xu J et al (2003) A genomic view of the human-*Bacteroides thetaiotaomicron* symbiosis. Science 299:2074–2076

Xu J et al (2007) Evolution of symbiotic bacteria in the distal human intestine. PLoS Biol 5:e156

Yan F, Polk DB (2002) Probiotic bacterium prevents cytokine-induced apoptosis in intestinal epithelial cells. J Biol Chem 277:50959–50965

Yan F, Cao H, Cover TL, Whitehead R, Washington MK, Polk DB (2007) Soluble proteins produced by probiotic bacteria regulate intestinal epithelial cell survival and growth. Gastroenterology 132:562–575

Zocco MA, Ainora ME, Gasbarrini G, Gasbarrini A (2007) *Bacteroides thetaiotaomicron* in the gut: molecular aspects of their interaction. Dig Liver Dis 39:707–712

Zweibaum A et al (1985) Enterocytic differentiation of a subpopulation of the human colon tumor cell line HT-29 selected for growth in sugar-free medium and its inhibition by glucose. J Cell Physiol 122:21–29

Zweibaum A, Laburthe M, Grasset E, Louvard D (1991) Use of cultured cell lines in studies of intestinal cell differentiation and function. In: Schultz SJ, Field FM, Frizell RA (eds) Handbook of physiology, the gastrointestinal system, intestinal absorption and secretion. American Physiological Society, Bethesda, pp 223–255

Chapter 17
Probiotics and Enteric Cancers

Min-Tze Liong, Huey-Shi Lye, Siok-Koon Yeo, Joo-Ann Ewe, Lay-Gaik Ooi and Ting-Jin Lim

Abbreviations

ACF	Aberrant crypt foci
A/E	Attaching and effacing
AFTB	Aflatoxin B1
AMIQ	2-amino-3-methyl-3H-imidazoquinoline
AMPI	2-amino-3-methyl-9H-pyrido (3,3-6) indole
AOM	Azoxymethane
CD8	Cluster of differentiation 8
DMH	1,2-dimethylhydrazine
DNA	Deoxyribonucleic acid
EHEC	Enterohaemorrhagic *E. coli*
FNFA	2-(2-furyl)-3-(5-nitro-2-furyl) acrylamide
FOS	Fructooligosaccharides
GST	Glutathione-S-transferase
HCAs	Heterocyclic amines
HMA	Human microbiota-associated
IBD	Inflammatory bowel disease
IL-12	Interleukin-12
IQ	2-amino-3-methylimidazo[4,5-*f*]quinoline
MAM	Methylazoxymethanol
MeIQ	2-amino-3,4-dimethylimidazo[4,5-*f*]quinoline
MeIQX	2-amino-3,8-dimethylimidazo(4,5-*f*)quinoxaline
MHIQ	2-amino-3-methyl-3H-imidazo[4,5-*f*]quinoline
MNNG	*N*-methyl-*N*′-nitro-*N*-nitrosoguanidine
MUC	Mucins
NPD	4-nitro-*O*-phenylenediamine

M.-T. Liong (✉)
School of Industrial Technology, Universiti Sains Malaysia, 11800 Penang, Malaysia
Tel.: +604-653-2114
Fax: +604-657-3678
e-mail: mintze.liong@usm my

J. J. Malago et al. (eds.), *Probiotic Bacteria and Enteric Infections*,
DOI 10.1007/978-94-007-0386-5_17, © Springer Science+Business Media B.V. 2011

NQO	4-nitroquinoline-N-oxide
O^6-meG	*O*^{6}-methylguanine
ODC	Ornithine decarboxylase
OF	Oligofructose
PhIP	2-amino-1-methyl-6-phenylimidazo[4,5-b]pyridine
PhMIP	5-phenyl-2-amino-l-methylimidazo[4,5-*f*]pyridine
PPARs	Peroxisome proliferator-activated receptors
PPARγ	Peroxisome proliferator-activated receptors-gamma
SCFAs	Short chain fatty acids
TNF-α	Tumour necrosis factor alpha
Trp-P-1	3-amino-I,4-dimethy-^{5}H-pyrido(4,3-b)indole

17.1 Introduction

The amount of bacteria in the colon is much higher than the amount of human cells in our body. Most of the bacteria in our gastrointestinal tract are harmless and beneficial to our health. However, recent evidences have suggested that some harmful microflora in the gastrointestinal tract may play a role in the occurrence of gut cancer. Gut cancers occur mostly in the colon whereby billions of bacteria are in contact with the gastrointestinal surface. Although the exact origins of gut cancers remain to be elucidated, several factors have been identified to correlate directly to the incidences of cancer growth, namely carcinogenesis, mutagenesis, gut lesions, neoplasm and deoxyribonucliec acid (DNA) damage.

Carcinogenesis and mutagenesis are characterized by changes from normal mucosa through early and advanced adenomas to invasive carcinoma (Treyzon et al. 2006). The disease develops through a series of mutations which involve activations and inactivations of oncogenes and tumour suppressor genes (Fearon and Vogelstein 1990). Carcinogenesis and mutagenesis have been reported to be induced by procarcinogens and mutagens such as heterocyclic amines (HCAs) that are generated from diet and food processing namely, meat during cooking. These molecules have been shown to exert carcinogenicity and mutagenicity in mice, rats and monkeys via the induction of hepatic, intestinal and mammary tumours. In human diets, the major group of HCAs is aminoimidazoazaarenes. All these compounds require the presence of metabolic activation to become genotoxic (Terada et al. 1986).

Additionally, the initiation of colorectal cancer has also been found to be attributed to the transformation of normal epithelium cells to neoplastic cells. This modification can alter the receptiveness of the mutated cells to their microenvironment while providing them with a growth advantage relative to normal cells. Neoplastic cells are self-sufficiency in growth signals, insensitivity to growth inhibition, and evasion of apoptosis, limitless replicative capacity, angiogenesis and capable of activating tissue invasion (Pacifico and Leonardi 2006). Neoplasm has been associated with the incidences of colorectal cancers.

Gut lesions are any abnormal tissue found in the organs, usually damaged by diseases or trauma. Pretlow et al. (1991) previously reported that putative preneoplastic lesions are present in the colonic mucosa of subjects with colon cancer. Aberrant crypt foci (ACF) have been suggested not only morphologically but also genetically, as distinct lesions and are precursors of adenoma and cancer (Takayama et al. 1998). Interrupting the progression of these early lesions has been found to improve the proliferation of gut cancer cells.

Recent evidences have suggested that probiotics could act as cancer preventive agents and have been proven beneficial in the prevention and alleviation of symptoms of colon cancer, in the representation of new novel therapeutic options. Probiotics are viable microorganisms that confer health benefits to the host once consumed in adequate amounts, primarily by promoting the proliferation of beneficial gastrointestinal indigenous microflora. Various microorganisms have been found to possess such properties, although *Lactobacillus* and *Bifidobacterium* are the most common probiotic bacteria used as food adjuvants (Lye et al. 2009). Probiotics have the potential to impact significantly on the development, progression and treatment of colorectal cancer (Geier et al. 2006), either as whole cells or in probiotic fermented products. However, different strains of probiotics possess varying responses towards the prevention of colorectal cancer. Thus, this chapter will discuss the roles of probiotics based on references of specific strains. This chapter will highlight the mechanisms and evidences of probiotics on enteric cancers as whole and colonic cancers specifically, with emphasis on carcinogenesis, mutagenesis, gut lesions, neoplasm and DNA damage.

17.2 Effects of Probiotics on Carcinogens

17.2.1 Clinical Evidences

Cancer in general is induced via the activation of abnormal genes that control cell growth and division and such risks could be increased by various processes or exposures. Carcinogens (chemical exposures) are the most risky exposure among other potentially risky exposures. Carcinogens are the cancer-causing chemicals that can be ingested or synthesized by metabolic activity of the human gastrointestinal resident flora (Parvez et al. 2006). Recent evidences have suggested that probiotics could be used to prevent the growth of tumour cells induced by carcinogens.

Sivieri et al. (2008) conducted a placebo-controlled design trial involving 30 male Wistar SPF rats to evaluate the effects of a probiotic strain, *Enterococcus faecium* CRL 183 on the incidence of colorectal tumours induced by 1,2-dimethlhydrazine (DMH). The authors reported that rats administered with *E. faecium* CRL 183 (10^8 CFU/ml) for 42 weeks showed a 40% reduction of adenocarcinoma incidence and diminished mean tumour volumes compared to rats without the administration of probiotics. Additionally, Singh et al. (1997) conducted a placebo-controlled design trial to evaluate the effects of *B. longum* on 60 male F344 azoxymethane (AOM)-induced colon carcinogenesis rats. The rats were fed a modified AIN-76A diet containing 0 or 2% lyophilized cultures of *B. longum* (4×10^{10} live cells/g diet) and administered AOM dissolved in normal saline, once weekly for 2 weeks and killed on 40 weeks after second AOM injection to evaluate the incidences of colon tumour. The authors revealed that the administration of *B. longum* significantly reduced the incidence of colon adenocarcinomas, colon tumour multiplicity in terms of tumours/animal and tumours/tumour-bearing animal compared to those on the control diet.

The consumption of probiotics or probiotic fermented products not only reduced the incidence of tumours caused by carcinogens but also change the faecal enzyme activities in the subjects. Lidbeck et al. (1991) studied the effect of *L. acidophilus* fermented milk on faecal microflora and β-glucuronidase activity in 14 colon cancer patients. The authors reported that the feeding of *L. acidophilus* (10^{11} CFU/day) for six weeks caused a reduction of *E. coli* and increased the number of lactobacilli in the faeces that subsequently led to a 14% reduction of β-glucuronidase activity, an enzyme which generates carcinogens in the digestive system of humans.

This was supported by Ling et al. (1994) that studied the effect of *Lactobacillus* strain GG on the faecal enzyme activity in 64 subjects. The authors found that the consumption of yoghurt containing viable *Lactobacillus* strain GG (10^{11} CFU/l) not only decreased faecal β-glucuronidase but also other faecal enzyme activities such as nitroreductase and glycocholic acid hydrolase activities ($P<0.05$) after consumption of yoghurt containing probiotic for 4 weeks. Same observation was reported by Marteau et al. (1990) that the nitroreductase activity was significantly reduced ($P<0.05$) in 9 healthy volunteers after administration of 100 g/day of fermented milk product containing *L. acidophilus* (10^7 CFU/g), *B. bifidum* (10^8 CFU/g), *Streptoccoccus* (*Lactococcus*) *lactis* (10^8 CFU/g) and *S. cremoris* (*Lactococcus lactis* subsp. *cremoris*) *lactis* (10^8 CFU/g) for 3 weeks. The authors also found that the β-glucosidase activity was significantly increased ($P<0.05$) after the consumption of probiotic fermented milk.

In another study, Goldin and Gorbach (1984) evaluated the effects of milk containing *L. acidophilus* on faecal enzyme activity in 16 women and 5 men. The authors found that the oral administration of *L. acidophilus* (2×10^6 CFU/ml) for 4 weeks significantly reduced ($P<0.05$) most of the faecal enzyme activities such as β-glucuronidase, nitroreductase and azoreductase with 2- to 4-fold reductions during the period of lactobacilli feeding.

17.2.2 Mechanisms

Although the mechanisms of anticarcinogenic effects of probiotics are remaining unclear, however, several mechanisms of cancer prevention by probiotics have been proposed and investigated. Past studies have reported that the administration of lactic acid bacteria and fermented dairy products could affect the gut flora enzyme activities which associated with colon carcinogenesis (Wollowski et al. 2001). The intestinal microflora composes of different microbial species that form a metabolically complex ecosystem. Some of colonic microbial may also convert harmless compounds into metabolites that cause inflammation or tumourigenesis (Huycke and Gaskins 2004). This intestinal microflora could influence the carcinogenesis by producing enzymes such as β-glucosidase, β-glucuronidase, azoreductase and nitroreductase (Rolfe 2000). *E. coli* and *Clostridium perfringens* are the common enteropathogens that produce β-glucuronidase (Anuradha and Rajeshwari 2005). This faecal enzyme could hydrolyze glucuronide, a compound that is required to detoxify foreign compounds and produce carcinogenic aglycones in the intestinal lumen (Rafter 2002).

On the other hand, bifidobacteria and lactobacilli have been reported to possess lower activities of these xenobiotic-metabolizing enzymes involved in carcinogen formation and metabolism compared to bacteroides, clostridia, and enterobacteriaceae. In addition, beta-glucosidase produced by probiotics has been found to prevent colon cancer by releasing flavonoids with anticancerogenic, antioxidative, antimutagenic and immune stimulatory effects (Wollowski et al. 2001).

Moreover, probiotic could also produce lactic and acetic acids that subsequently lower the intestinal pH and create a bactericidal environment for these putative enteropathogens, and thus develop a favourable microenvironment which modulates the bacterial enzymes (Anuradha and Rajeshwari 2005). Biasco et al. (1991) reported that the administration of both *L. acidophilus* and *B. bifidus* caused significant reduction in faecal pH of patients with colonic adenomas, thus reduces the infection of pathogens and heightened the anticarcinogenic effects of probiotic. Sreekumar and Hosono (2000) also found that *L. acidophilus* SBT2074 could efficiently inhibit the number of coliforms in the duodenum, jejunum and the ileum. Therefore, the consumption of probiotic or probiotic fermented products may affect the initiation phase of carcinogenesis as they may alter the balance of intestinal microflora, leading to reduced secretion of enzymes responsible for the conversion of procarcinogens into carcinogens and thus reduces the absorption of harmful carcinogens involved colon carcinogenesis.

Probiotics have also been found to bind and degrade carcinogenic compounds in the gastrointestinal tract. The cell wall of probiotics may be an important factor in determining the binding of free toxins in the intestine (Rafter 2002). In addition, different strains of probiotics have been found to exert different affinity towards different carcinogenic compounds. Nevertheless, the binding of carcinogen by probiotics has been shown to correlate well with the reduction of carcinogenesis (Rafter 2002). Haskard et al. (2001) reported that *L. rhamnosus* strain GG (ATCC 53103) and *L. rhamnosus* strain LC-705 (DSM 7061) could remove aflatoxin B1 (AFTB), which is a carcinogenic compound from liquid media via binding. Approximately 90% of bound AFTB was recovered from the bacteria upon solvent extraction. Similarly, Oatley et al. (2000) revealed that strains of bifidobacteria could also bind significant amounts of AFTB, ranging from 25 to 60%. Strains of *Lactobacillus* have been reported to bind to carcinogenic indoles such as indole-acetate and indole-propionate, that act as cancer promoters. Bolognani et al. (1997) demonstrated that probiotic could not only bind indole and AFTB but also other carcinogenic compounds such as benzo[a]pyrene, and other food- heat generated carcinogens such as 2-amino-3-methyl-3H-imidazo[4,5-*f*]quinoline (MHIQ), 2-amino-3,4-dimethylimidazo[4,5-*f*] quinoline (MeIQ), 2-amino-3,8-dimethylimidazo[4,5-*f*]quinoxaline (MeIQX) and 5-phenyl-2-amino-l-methylimidazo[4,5-*f*]pyridine (PhMIP). Probiotics have also been found to degrade nitosamines, a group of carcinogens well studied in animal models and have been detected in human faeces (Fotiadis et al. 2008). Therefore, probiotic could prevent or delay the development of tumour or carcinogenesis in human as the bound and/or degraded carcinogens are less available to absorb into the human leading to decreased levels in blood.

Probiotics have also been reported to deactivate carcinogens, via increasing the enzyme activities or processes that protect cells against carcinogen-induced damage. Wollowski et al. (1999) reported that *Lactobacillus delbrueckii* ssp. *bulgaricus* 191R and *Streptococcus salivarius* ssp. *thermophilus* CH3 could prevent initiation of carcinogenesis by producing antigenotoxic metabolites, which can deactivate and detoxify carcinogenic compounds such as *N*-methyl-*N'*-nitro-*N*-nitrosoguanidine (MNNG). The authors hypothesized that the thiol-containing breakdown products of proteins produced by bacterial proteases would be the compounds responsible to deactivate the MNNG or other carcinogens in the gut lumen that further reduces the damage of colonic mucosal cells.

17.3 Effects of Probiotics on Mutagens

17.3.1 In-vitro and In-vivo Evidences

The development of gut cancer is a multistage process due to mutation of protooncogenes and tumour suppressor genes which could initiate cancer. The mutation could be induced by several mutagens derived from food or environment. High consumption of high temperature cooking meat has been associated with the development of colon cancer. HCAs, a potent mutagen formed during high temperature cooking could be a credible risk factor that initiates colon carcinogenesis. Therefore, taking into consideration that the natural capabilities of human to defend against mutation is limited, the ability to prevent the absorption and the mutagenic activity of these compounds provides a possible alternative to suppress the development of gut cancer. Interestingly, probiotics have been studied extensively for their antimutagenic activity and thus could reduce the risk of gut cancer (Wollowski et al. 2001).

The antimutagenic effects of probiotics and probiotic-fermented products have received much attention and widely researched in the past decades. In 1980s, Hosono et al. (1986) reported that milk cultured with *Lactobacillus bulgaricus* and *Streptococcus thermophillus* exhibited antimutagenic effects against mutagens such as 2-(2-furyl)-3-(5-nitro-2-furyl) acrylamide (FNFA), 4-nitroquinoline-N-oxide

(NQO) and faecal mutagenic extract in an *in vitro* assay system using streptomycin-dependent strains of *Salmonella*. The authors postulated that the antimutagenic effect of probiotics against these mutagens was probably due to the production of certain heat labile antimutagenic factors during fermentation. In early 1990s, an *in vitro* assay using *Salmonella typhimurium* TA 100 was conducted to investigate the effects of milk cultured with 71 strains of lactic acid bacteria on the mutagenicity of MNNG (Hosoda et al. 1992). The study demonstrated that milk containing lactic acid bacteria belonging to the genus of *Lactobacillus, Streptococcus, Lactococcus*, and *Bifidobacterium* could inhibit mutagenicity of MNNG by 14.9–77%. MNNG is a potential mutagen which is strongly associated with the increased risk of colon cancers. The mutagen decomposes into highly reactive electrophiles which then attack the nucleophilic sites of DNA bases, leading to base substitution mutation (Reddy et al. 1974). Therefore, inhibition of this mutagen could reduce the risks of colon mutagenicity.

In another study, viable cells of *L. acidophilus* and *L. bulgaricus* isolated from yogurt were found to exert strong antimutagenic activity (64–98%) against a potent mutagen, 2-nitrofluorene (Mobarez et al. 2007). However, the antimutagenic activity of the probiotic cells was reduced by 36–47% upon heat treatment (100°C for 15 min) on the cells, indicating that the antimutagenic activity was strongly dependant on the survival of probiotics. 2-Nitrofluorene is a nitro-polycyclic aromatic hydrocarbon which lead to formation of direct acting mutagens in the intestine and is strongly associated with fore-stomach and intestinal carcinogenesis (Moller 1994). Thus, the overall antimutagenic effect of probiotics especially live cells against 2-nitrofluorene provide an alternative approach to reduce or prevent the incidence of gut cancers.

In vivo studies have also shown promising evidence that probiotics could exhibit antimutagenic activities against various mutagens including HCAs. Previous animal studies have shown strong correlations between probiotics and the reduced risks of mutagenicity and carcinogenesis induced by mutagens such as HCAs. Surono et al. (2009) conducted a placebo-controlled study involving 20 male rats administered with probiotic-cultured skim milk, unfermented skim milk or water (control) for 5 days. The authors found that the consumption of skim milk cultured with *Enterococcus faecium* IS-27526 (8.4×10^8 viable cells/ml) by rats significantly reduced faecal 3-amino-I,4-dimethy-^{5}H-pyrido(4,3-b)indole (Trp-P-1), which induces mutagenicity, compared to the control and unfermented skim milk. Considering that mutagenesis induced by this mutagen could lead to colon carcinogenesis, the *in vivo* antimutagenic activity of probiotics against this mutagen provided a strong indication that probiotics could reduce the risks of colon cancers.

Reddy and Rivenson (1993) have also demonstrated that a diet containing *B. longum* (2×10^{10} live bacterial cells/g) could inhibit colon carcinogenesis induced by 2-amino-3-methylimidazo[4,5-*f*]quinoline (IQ), a food mutagen. A total of 156 rats (78 female and 78 male) were fed with a control diet (high fat diet without containing *B. longum*) or experimental diet containing 0.5% lyophilized *B. longum* (2×10^{10} live bacterial cell/g) with or without IQ (125 ppm) for 58 weeks. This randomised study found that IQ induced gut carcinogenesis while dietary supplementation of *B. longum* significantly inhibited the colon and small intestinal tumour incidence in the rats.

Probiotics were also found useful to reduce colon DNA lesions induced by HCAs in rats (Tavan et al. 2002). In a randomised study, a total of 60 weanling males rats were induced with 250 mg of IQ, MeIQ and 2-amino-1-methyl-6-phenylimidazo[4,5-b]pyridine (PhIP) (ratio 1:1:1). These mutagens-induced rats were randomly assigned to four groups each fed with water, non-fermented milk, *B.animalis* DN-173 010 ($5.4 \pm 1 \times 10^8$ CFU/day) fermented milk or *St. thermophilus* DN-001 158 ($5.4 \pm 1 \times 10^8$ CFU/day) fermented milk for 7 weeks. The authors found that rats fed with both probiotics-fermented milk significantly decreased the incidence of ACF compared to rats fed with water and unfermented milk. Ingestion of *B.animalis* fermented milk and *St. thermophilus* fermented milk inhibited the incidence of colon ACF by 96 and 93%, respectively. In addition, the ingestion of this fermented milk also reduced the faeces mutagenicity where the faeces mutagenicity was insignificant with that of prior to the induction of HCAs. This study indicated that probiotic-fermented products could reduce gut mutagenicity.

The reduction of gut mutagenicity upon consumption of probiotic containing products was also observed in human subjects. Hosoda et al. (1996) demonstrated that ingestion of milk fermented with *L. acidophilus* LA-2 in six male subjects (26–38 years old) caused a remarkable decrease in the faeces mutagenicity as compared to initial faeces mutagenicity (decreased by 71.9%). The subjects were selected on the basis of high and stabilised faeces mutagenicity. All the subjects were given 100 g of probiotic fermented milk three times daily as supplement for seven consecutive days. The faeces mutagenicity of the subjects consuming their regular diet was evaluated before and during administration of tested diet. Similarly, Matsumoto and Benno (2004) conducted a cross over, placebo-controlled study and demonstrated that consumption of *B. lactis* containing yogurt reduced gut mutagenicity. The authors postulated that *B. lactis* LKM512 yogurt may be a promising food to help prevent colon tumourigenesis. It was reported that the consumption of 100 g/day of *B. lactis*—containing yogurt for 2 weeks significantly reduced the faeces mutagenicity of seven healthy subjects (6 females and 1 male) compared to the consumption of placebo yogurt without *B. lactis*). The faeces mutagenicity (evaluated based on *umu*-test) of the subjects upon consumption of *B. lactis*-yogurt was 60.1% lower than that of placebo.

17.3.2 Mechanisms

Several mechanisms have been proposed to explain the efficiency of probiotics in suppressing and preventing gut mutagenicity. One of the potential mechanisms is the removal of mutagens via binding ability of lactic acid bacteria to mutagen compounds. In fact, past studies have demonstrated that live probiotic cells could permanently inhibit the activity of mutagens via binding to the live cell (Lankaputhra and Shah 1998). The authors reported that various types of mutagens and promutagens such as MNNG, 2-nitroflourene, 4-nitro-*O*-phenylenediamine (NPD), NQO, AFTB, 2-amino-3-methyl-3H-imidazoquinoline (AMIQ), PhIP and 2-amino-3-methyl-9H-pyrido (3,3-6) indole (AMPI) could bind (10–85% bound) to live cells of *Lactobacillus* and *Bifidobacterium* spp. and the recoverability of the mutagens was very low upon extraction with dimethyl sulphoxide. This indicates that the mutagens were permanently bound to the live cells, leading to permanent inhibition of the activity of these mutagens.

In another study, Zhang and Ohta (1991) demonstrated that cell wall skeleton of probiotics was mainly responsible for the binding properties of the bacteria. Cell wall skeleton of *St. cremoris* 225, *L. acidophilus* IFO 13951 and *B. bifidum* IFO 14252 was reported to exhibit the ability to bind to various mutagen pyrolysates such as Trp-P-1, MeIQ and MeIQX. According to the authors, peptidoglycan and polysaccharides of the bacterial cell walls are the main components responsible for the binding of lactic acid bacteria to mutagenic pyrolysates. This was in agreement with Sreekumar and Hosono (1998) who investigated the HCAs (Trp-P-1)-binding receptor of *L. gasseri* and found that polysaccharides component of the cell wall contains binding receptors for the HCAs. Considering that mutagens (Trp-P-1, MeIQ, MeIQX) may be formed during food preparation and are exposed to the human colon leading to colorectal cancers (De Kok and van Maanen 2000), the inhibition of HCAs via binding by probiotics is a plausible alternative to prevent the risks of gut cancers.

In addition to the cell wall skeleton, mutagens could also bind to other parts of the probiotic cell. Tsuda et al. (2008) demonstrated that the capsular cell of *L. plantarum* could bind to food borne mutagens such as HCAs and the binding ability was due to exopolysaccharides of the bacteria cell surface. The authors postulated that the mutagen-binding mode of the exopolysaccharides may consist of cation-exchange and hydrophobic bond. In fact, the exopolysaccharides were stable in the presence of bile thus suggesting that exopolysaccharides of probiotics could possibly bind mutagens in the gastrointestinal condition of human, leading to the reduced risks of gut cancer in humans.

Another possible antimutagenicity mechanism by probiotics bacteria involved the production of metabolites which could inhibit the activity of mutagens. It is commonly known that probiotic bacteria could ferment carbohydrate to produce various organic acid such as lactic, acetic and butyric acid. The organic acid produced by probiotics could act as desmutagens and has been reported to reduce the mutagenic activity of various mutagens. Lankaputhra and Shah (1998) demonstrated that organic acid especially butyric acids produced by *L. acidophilus* and *Bifidobacterium* spp. was highly effective on inhibiting the activity of several mutagens and promutagens including MNNG, 2-nitroflourene, NPD, NQO, AFTB, AMIQ, PhIP and AMPI.

Additionally, fermentation of probiotics could leads to the biosynthesis of several bioactive compounds. Some of these bioactive compounds could potentially exert antimutagenic activities against mutagens. Chalova et al. (2008) suggested that the antimutagenic compound synthesis by *Lactobacillus* and *Bifidobacterium* exist extracellularly. It has been found that the compounds biosynthesised extracellularly into the supernatant could inhibit the mutagenic activity of benzo[a]pyrene and sodium azide. However, the exact compounds that are responsible for such activities have yet to be identified. Rhee and Park (2001) suggested that glycoprotein produced by probiotics are responsible for the antimutagenic activities and three glycoproteins which exert antimutagenic activities were successfully extracted and purified from the supernatant of *L. plantarum* cells. This purified fractions exhibited antimutagenic effect against MNNG on *Salmonella enterica* serovar Typhimurium cells.

Preventing the activation of promutagens by cytochrome P-450 is another possible mechanism of probiotics in preventing the incidence of gut cancers (Nadathur et al. 1996). The mutagenic effects of ingested mutagens such as HCAs are highly dependant on the metabolism of the compounds. These compounds require metabolic activation in order to cause a mutation (Cross and Sinha 2004). The promutagens must be first converted to potential genotoxic metabolites or direct acting mutagens by the hepatic cytochrome P-450 1A2 enzyme in order to cause genetic mutation, leading to the development of colorectal cancer. Therefore, it is important to terminate such pathways in order to prevent mutation and the risk of cancer. Interestingly, probiotics have been found to reduce hepatic cytochrome P-450 enzyme and was postulated to prevent the activation of promutagens. Tavan et al. (2002) demonstrated that the consumption of milk fermented by *B. animalis* ($5.4 \pm 1 \times 10^8$/day) and *St. thermophilus* ($5.4 \pm 1 \times 10^8$/day) for 7 weeks decreased the hepatic cytochrome P450 metabolic activity in 60 rats induced by HCAs (IQ:MeIQ:PhIP in the ratio of 1:1:1). Promutagens could undergo a metabolic activation pathway to directly form acting mutagens, and probiotics could prevent such activation (Fig. 17.1).

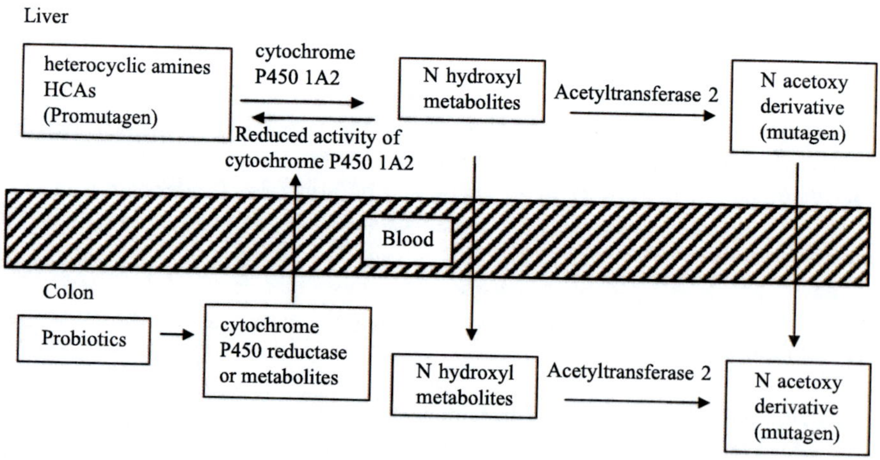

Fig. 17.1 Route of metabolic activation of promutagen to direct acting mutagen and inhibitory effect of probiotics on cytochrome P450 1A2 activity

17.4 Probiotics and Neoplasm

17.4.1 Anti-neoplastic Activities of Probiotics

As ordinary cancerous development, colon carcinogenesis can be divided into three stages, the initiation, promotion and progression. It is a multi step process involving cellular, molecular and morphological levels (Fodde 2002). The expansion of genetically altered initiated cells in the promotion and progression stage causes the growth of neoplasm. A neoplasm is an abnormal tissue that grows by cellular proliferation more rapidly than normal that resulted from neoplasia. The abnormal tissue would persist in the same excessive manner even after the stimuli that initiated the growth cease. Neoplasms show partial or complete lack of structural organization and functional coordination, with that of the normal tissues around it. The distinct mass of tissue formed may be benign (tumour), pre-malignant or malignant (cancer). Considering that the development of cancer is a multistage process, prevention should focus on stopping the carcinogenesis process at the earliest possible point in the pathway through mechanism-based approaches (Hursting and Kari 1999). Lately, probiotics have been advocated for the prevention and treatment of a diverse range of disorders, from acute gastroenteritis to intestinal neoplasia (Boyle et al. 2006). By possessing systemic anti-neoplastic activity, probiotics play a crucial role in the prevention of colorectal cancer (Fig. 17.2).

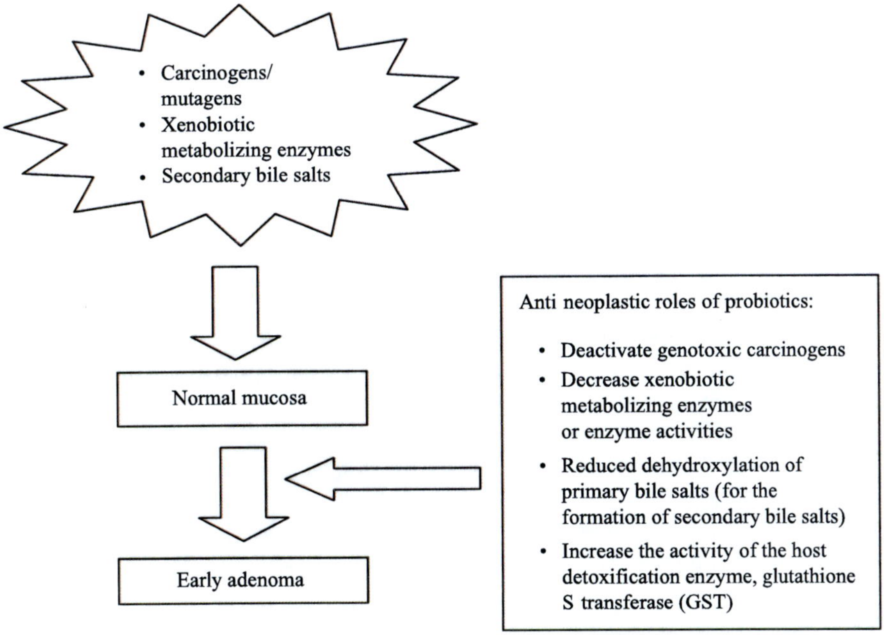

Fig. 17.2 The anti-neoplastic activities of probiotics in inhibiting the occurrence of colon carcinogenesis

The incidence of colorectal cancer in high proportion of the population is due to somatic mutations induced by carcinogens during the lifetime. Hence, by deactivating genotoxic carcinogens, probiotics could prevent the development of neoplastic cell which could finally lead to colorectal cancer. The initiating step involved in the development of colon cancer involves a carcinogen that causes an alteration in the DNA and most of the chemical carcinogens are genotoxic, causing DNA damage by reacting with DNA bases in the nucleus and mitochondria. DNA damage in proto-oncogenes and tumour suppressor genes could lead to mutation of normal cell. The mutated proto-oncogenes and tumour suppressor genes can cause uncontrolled cell growth that increase the probability of neoplastic transformation (Hursting and Kari 1999). The accumulation of mutations in these genes then results in cancer initiation. Therefore, probiotics particularly lactic acid bacteria have been investigated extensively in model systems for their ability to prevent mutations. Different mutagenic activities in the *Salmonella typhimurium* mutagenic assay have been performed by various genera of probiotics such as *Lactobacillus, Streptococcus, Lactococcus* and in fermented milk products containing *Bifidobacterium* (Renner and Münzner 1990; Hosoda et al. 1992; Abdelali et al. 1995a).

Probiotics also play a part in the inactivation of carcinogens that would cause DNA alteration in colonic epithelium by modification of toxifying and detoxifying enzymes. This may be the preceding step of anti-genotoxic activity possessed by probiotics. The variation in metabolic capacity reveals the vast numbers and diversity of the human intestinal microflora. The metabolic activities of the intestinal microflora have resulted in both detrimental and beneficial effects on the health of the host (Rowland and Gangolli 1999). The aetiology of colorectal cancer has been associated with endogenous toxic and genotoxic compounds related to enzyme activities of intestinal microflora. β-glucuronidase, nitroreductase and azoreductase are the bacterial enzymes released in the intestinal tract which are capable of converting indirect acting carcinogens to proximal carcinogens in the large bowel (Rowland 1991). Toxic compounds that are already detoxified in the liver would be regenerated and liberated in the intestinal lumen by the action of these enzymes. Interestingly, studies reported that both pathogenic and beneficial bacteria found in intestine usually differ in their enzyme activities (Mital and Garg 1995). Probiotics such as *L. acidophilus* have lower activities of these xenobiotic-metabolizing enzymes than do bacteroides, clostridia and enterobacteriaceae. Thus, a decrease in xenobiotic-metabolizing enzymes or enzyme activities is regarded as an advantage of health since less carcinogens would induce the occurrence of cell neoplascity.

In addition, another bacterial enzyme 7-α-dehydroxylase, has been reported to be involved in the dehydroxylation of primary bile salts leading to the formation of detrimental secondary bile salts, deoxycholic and lithocholic acids. These acids could exert a cytotoxic effect upon epithelial cells, increasing cell proliferation and leading to a higher probability of colon cancer development (Ling 1995). The activity of 7-α-dehydroxylase was reported to be affected through probiotic consumption. It was suggested that probiotics may involve modulation of intestinal microflora or bind physically with the bile salts, leading to reduced dehydroxylation.

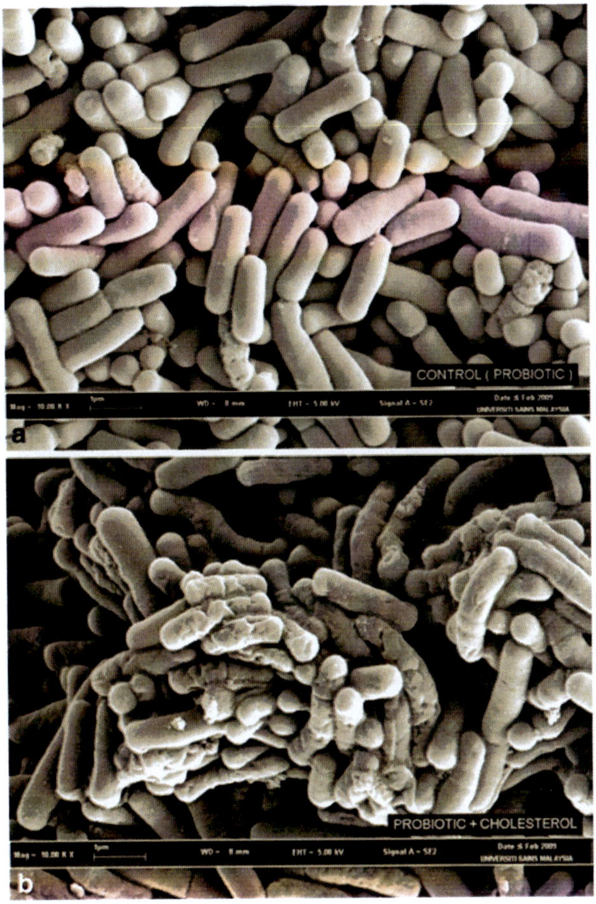

Fig. 17.3 SEM micrograph of *Lactobacillus bulgaricus* that grew in **a** media without cholesterol and **b** broth supplemented with cholesterol

Another hypothesis involves the unavailability of cholesterol to form new bile. *In vitro* evidence has shown that probiotic cells could bind cholesterol to cellular surface (Fig. 17.3). Cholesterol is the precursor for the de novo synthesis of new bile acids (Lye et al. 2009). Bound cholesterol is less available for the formation of new bile and may lead to reduced dehydroxylation of primary bile salts.

In addition to bacterial enzymes, probiotics were suggested to be able to suppress the host's enzyme-related cellular growth. Polyamines play an important role in cell proliferation, differentiation and participate in macromolecular synthesis. Ornithine decarboxylase (ODC; EC 4.1.1.17) is a crucial enzyme involved in the biosynthetic pathway of polyamine. In neoplastic human colons, ODC activity is often elevated compared to normal-appearing colonic mucosa (Singh et al. 1997). Lyophilized cultures of *B. longum* have been found to inhibit colon tumour via the inhibition of colonic mucosal cell proliferation, and the suppression of ODC activity in the colonic mucosa (Reddy 1999). Meanwhile, ingested probiotics was found able to increase the activity of the host detoxification enzyme glutathione-

S-transferase (GST) in the gut. The enzyme catalyses the conjugation of arene oxides as well as other substrates possessing electrophilic carbon atoms with glutathione, renders products with exceptionally stable thioether linkages. This inactivated product would then be secreted in the bile as GST conjugates or are further metabolized to mercapturic acid (Challa et al. 1997). Thus, by modifying toxic and detoxifying enzyme activities, probiotics may prevent carcinogens from inducing the development of neoplastic cell.

17.4.2 In-vitro and In-vivo Evidences

The ability of probiotic cultures to prevent damages and mutations have been demonstrated through *in vitro* and *in vivo* studies. An *in vitro* study evaluating the effects of lactic acid bacteria on the induction of mutations by carcinogens has been performed by Pool-Zobel et al. (1993) using the Ames *Salmonella* assay. The authors reported that both *L. casei* and *L. lactis* inhibited the mutagenic activity of nitrosated beef by more than 85%. Subsequently, an *in vivo* study involving *L. casei* has been shown to inhibit cellular damages in the colon of rats exposed to mutagen MNNG. The authors exposed the gastrointestinal tract of rats to MNNG and found that cellular damage developed in cells within 24 h. However, antigenotoxic property was detected when 10^{10} viable cells *L. casei* (in 10 ml of 0.9% NaCl/kg body weight) was administered upon oral ingestion of MNNG by the rats. The protective effect of the product was greatest when bacteria were applied 8 h before exposure to carcinogens (Pool-Zobel et al. 1993). A subsequent study by Pool-Zobel et al. (1996) involving other probiotics such as *L. acidophilus, L. confuses, L. gasseri, S. thermophilus, B. longum* and *B. breve* had also reached a similar positive finding.

Probiotic supplements have been shown to modify bacterial enzyme activities in various *in vivo* studies. 18 male Sprague-Dawley rats (4 weeks aged) were orally administered *B. longum* SPM 1205 or Duolac™ (consist of *L. casei, L. rhamnosus, L. lactis, L. plantarum* and *B. longum*) in 10^9 CFU/kg body weight/day for 4 weeks. The results obtained in this experiment showed that the activity of β-glucuronidase was potently inhibited by the administration of *B. longum* SPM 1205 or the commercial strains for 4 weeks (Choi et al. 2005). In another study, Villarini et al. (2008) administered *L. casei* suspension (10^9 bacteria/ml, 10 ml/kg body weight) to 20 male Sprague-Dawley rats (1-2 months old) for a period of 6 days; the control group was given saline. The authors reported that there was a significant reduction of β-glucuronidase activity in DMH induced probiotic-group compared to the control. Ouwehand et al. (2002) evaluated the effects of juice supplemented with probiotics on 28 elderly subjects for a period of 10 weeks. The subjects were divided into 3 groups: 1 group receiving only juice (control), 1 group receiving juice supplemented with *L. reutari,* and 1 group receiving *L. reutari* and *Propionibacterium freudenreichii.* Results showed that subjects receiving the *L. reutari* and *Propionibacterium freudenreichii* juice exhibited significantly reduced faecal azoreductase activity.

In another cross-over study, Ayebo et al. (1980) evaluated the consumption of non-fermented milk containing *L. acidophilus* in 2×10^6 CFU/g on faecal β-glucuronidase and β-glucosidase activity in three male and nine female healthy subjects for four weeks. Low fat milk was given as a control throughout the period. An increase of approximately one order of magnitude in feacal lactobacilli counts during the period of probiotic consumption was detected. The β-glucuronidase activity also decreased after four weeks of lactobacilli consumption. This was supported by Spanhaak et al. (1998) in a placebo-controlled trial involving 20 healthy male subjects (age 40–65 years old). The subjects were assigned to both control and treatment group. The treatment group received three times daily 100 ml of fermented milk containing 10^9 *L. casei* Shirota/ml, while the control group received unfermented milk for 4 weeks. The authors reported a significant increase in faecal levels of *Lactobacillus* and *Bifidobacterium* in the group receiving *L. casei* Shirota-fermented milk. In addition, this group also showed significant decreased activity of faecal β-glucuronidase.

17.5 Probiotics and Deoxyribonucleic Acid (DNA)

17.5.1 DNA-Induced Colonic Damage

DNA damages of the enteric cells have been highly associated with the occurrences of enteric cancers. Although various factors could cause such damages, microbial-induced DNA damages have attracted much attention lately. Pathogenic bacteria residing in the colons have been reported to produce reactive substances leading to the formation of cancer cells. *Enterococcus faecalis* is a pathogenic gut bacterium that can survive in the intestinal tract, via both respiration and fermentation. During fermentation conditions, *E. faecalis* releases superoxide and reactive oxygen species such as hydrogen peroxide and hydroxyl radicals through autoxidation of membrane-associated demethylmenaquinone. These by-products damage the DNA of colonic epithelial cells, leading to the formation of colon tumours (Huycke et al. 2002). DNA has a key role in predisposition to gut cancer and in its initiation and progression. In an *in vitro* study, Glei et al. (2007) studied the role of hydrogen peroxide, trans-2-hexenal, and 4-hydroxy-2-nonenal in causing the DNA damage in genes which contribute to the development of colon cancer. The authors reported that these compounds have equal genotoxic potency on DNA damage at varying concentrations (concentrations of hydrogen peroxide: from 0–150 μM; trans-2-hexenal: 0–1,600 μM; 4-hydroxy-2-nonenal: 0–250 μM).

Mucins (MUC) are high-molecular-weight glycoproteins synthesized and secreted by epithelial cells of gastrointestinal tracts (Mack et al. 1999), comprising of the major organic components of mucus, 20% of the protein and 70–80% of carbohydrate by weight (Juntunen et al. 2001). Several studies have evaluated the roles of MUC in the mucosal defense and down-regulation of gut cancer. In an *in vitro* study, Ogata et al. (1992) used Northern blot analysis to study the expression of MUC1, MUC2, MUC3, and MUC4 mRNA in paired normal and cancerous colonic tissues, and nine colon cancer cell lines. Northern blot analysis is frequently used in molecular biology research to study gene expression by detection of mRNA (Higo et al. 2002). The authors found that all four MUC genes (MUC1, MUC2, MUC3 and MUC4) were expressed rather weakly or not at all in colon cancer cell lines. These results indicated that the human colon expresses a broad repertoire of MUC genes which are differentially regulated in malignancy. In another *in vitro* study, Williams et al. (1999) have identified two novel MUC cDNAs (*dd*34 and *dd*29) which were designated MUC11 and MUC12, respectively, that are down-regulated in colorectal cancers. The expression patterns of MUC11 and MUC12 mRNAs were observed in a range of normal human tissues. The authors found that these two partial cDNAs (*dd*34 and *dd*29) were highly expressed in colonic epithelium. MUC12 mRNA was not expressed in the colorectal cancer cell lines analyzed and it was down-regulated in 40% tumours compared with matched normal colonic tissue. In the study, the authors also demonstrated that MUC11 mRNA was down-regulated in 80% tumours, comparing to the matched normal colonic tissue. Sequence homology occurred between the MUC12 epidermal growth factor-like domain and epidermal growth factor receptor-binding growth factors. These discoveries implied that these MUC acted as growth regulators in colonic epithelium. The production of MUC is important to minimize the interaction of pathogen microorganisms with intestinal mucosal cells which play a vital role in gut health. In an *in vitro* study, Juntunen et al. (2001) determined the adherence of probiotics to the isolated mucus MUC extracted from infant faecal sample. In the study, the authors were using five probiotics such as *Lactobacillus rhamnosus* GG, *Lactobacillus casei Shirota, Lactobacillus paracasei* F19, *Lactobacillus acidophilus* LA5 and *Bifidobacterium lactis* Bb12 for the adherence. The authors found that each probiotic strain examined had significantly different adherence to the mucus MUC extracted from faeces. The result of the study showed that the concentration of MUC did not diminish during the inflammation of pathogen microorganisms and the binding of the probiotics to the mucus MUC has the potential as a protector against the pathogen microorganisms along the gastrointestinal tract.

17.5.2 In-vitro and In-vivo Evidences

Past *in vitro* and *in vivo* studies have evaluated a number of proposed mechanisms by which probiotics exert their effects on DNA leading to the reduced risks of gut cancer. Some postulated mechanisms include prevention of oxidative DNA damage in colonic cells, MUC secretion, bioproduction of conjugated linoleic acid, and the production of short chain fatty acids (SCFAs) that modulate the induction of DNA damage. Different strains of probiotics have different effects on the intestinal luminal environment, epithelial and mucosal barrier function and the mucosal immune system (Hirayama et al. 2000).

Formation of reactive oxygen species contributes to the development of cancers, including gut cancers, mainly via the induction of DNA damage. In an *in vitro* study, Koller et al. (2008) evaluated the detoxification of reactive oxygen species by using several strains of probiotics strains, isolated from the gastrointestinal microflora and commercialized probiotics products. The authors incubated the human derived colon cell line (HT-29 cells) with 3×10^7 bacteria/ml of each probiotic strain and subsequently exposed it to 5-hydroxy-2-methyl-1,4-naphthoquinone and hydrogen peroxide which would contribute to DNA migration. The results obtained were positive, whereby probiotic strains such as *L. acidophilus* VM19, *L. casei casei* VM25, *L. d. lactis* VM33, *L. paraplantarum* VM35, *L. pentosus* VM36, *L. plantarum* VM38, *L. rhamnosus* VM40 and *S. thermophilus* VM44 contributed to reduced DNA damages caused by hydrogen peroxide and 5-hydroxy-2-methyl-1,4-naphthoquinone.

MUC1, MUC2, MUC3, MUC4 and MUC5AC are the MUC genes that have been identified to be expressed in the human colon. MUC2 is the predominant ileocolonic MUC and the gel-forming MUC of the gastrointestinal tract which is the main structure component of the mucus gel. The MUC play a role in the prevention of the enteropathogens attachment on epithelial cell in the gastrointestinal tract. This is achieved through the effects of competitive inhibition for attachment sites on MUC mimicking epithelial cell bacterial attachment sites. In an *in vitro* study, Mack et al. (1999) found that the incubation of a probiotic strain, *Lactoacbillus plantarum* 299v with human-derived colon cell line (HT-29 cells) increased MUC2 and MUC3 mRNA expression levels. The result indicated that the increased level of MUC2 and MUC3 mRNA expression in the presence of probiotics could contribute to increased intestinal MUC production, which would prevent the attachment of enteropathogens in the gastrointestinal tract. In an *in vivo* study evaluating the effect of probiotic on MUC secretion in gut, Caballero-Franco et al. (2007) administered 0.15 mg/kg (3×10^9) of probiotic formula in five rats for seven days. In this study, the authors found that there was a significant ($P < 0.003$) increase in MUC2 gene expression compared to control group. The results indicated that the probiotic induced colonic MUC secretion, enhancing the protective mucus barrier in animals that indirectly lead to improvement of gut function and further reduced the occurrence risk of gut cancer.

The production of conjugated linoleic acid by probiotics was also found to inhibit the growth of HT-29 (human colon carcinoma cells) and Caco-2 colon cancer cells (human epithelial colorectal adenocarcinoma cells). Conjugated linoleic acid is a ligand for the peroxisome proliferator-activated receptors-gamma (PPARγ) and an exposure of conjugated linoleic acid was found to repress the growth of colon cancer cells. The peroxisome proliferator-activated receptors (PPARs) are nuclear receptor proteins responsible as transcription factors which regulating the expression of genes such as in the regulation of cellular differentiation, development, and macronutrient metabolism and tumourigenesis of organisms (Liu et al. 2004). In addition, conjugated linoleic acid was found to reduce the mRNA ratio of *Bax/Bcl-2* in the colonic mucosa of animals, decrease cellular proliferation and induce apoptosis of the colonic mucosa. *Bax* is a protein group of the *Bcl-2* gene family which is a pro-apoptotic *Bcl-2* protein (a prototype for a gene and proteins) whereby it promotes apoptosis by competing with *Bcl-2* proper (Miquel et al. 2005). Erb33 is a protein that is implicated with the development of colon cancer. The expression of Erb33 was reduced in HT-29 cells during the exposure to conjugated linoleic acid. In an *in vitro* study, Ewaschuk et al. (2006) evaluated the ability of probiotics to exert anticarcinogenic effects through the production of conjugated linoleic acid. The authors found that the conditioned medium containing probiotic-produced conjugated linoleic acid reduced the viability and induced the apoptosis of HT-29 and Caco-2 cell. In addition, Ewaschuk et al. (2006) also evaluated the ability of probiotics to produce conjugated linoleic acid in an *in vivo* study. The study utilized four mice for a period of twelve weeks whereby the mice were fed 30 μl of several strains of probiotics (*L. casei, L. plantarum, L. acidophilus, L. delbrueckii subsp bulgaricus, B. infantis, B. breve, B. longum* and *S. salivarius subsp. thermophilus*). The authors found that the probiotics were capable of producing conjugated linoleic acid from linoleic acid. The author also observed the conversion of linoleic acid to conjugated linoleic acid in varying degrees depending upon the strains of probiotic. In this study, the authors also demonstrated that the probiotics strains have the capacity to induce the up-regulation of PPARγ which contributed to the reduction of cancer cell viability and the induction of apoptosis.

Butyrate, acetate and propionate are known as SCFAs. SCFAs are the main fermentation metabolites of probiotics in the gut, and are often utilized as energy sources by colonocytes. Additionally, they have been found to deliberate the proliferation of colonic tumour cells. Butyrate induces GST which is a detoxifying enzyme system which provides defenses against carcinogens and oxidative DNA damage. In an *in vitro* study, Abrahamse et al. (1999) investigated the potential of SCFAs in the prevention of DNA damage induced by hydrogen peroxide. In this study, the authors incubated freshly isolated colon cells with 6.25 mM of individual SCFAs (such as butyrate, acetate and propionate) and SCFAs mixture (containing acetate, propionate and butyrate), respectively and those cells were further exposed to 0, 100, 200 or 500 μM of hydrogen peroxide for 15 min. The results of the study showed that exposure to 100–500 μM of hydrogen peroxide caused a significant decrease in the cell damage in cells pre-treated with individual SCFAs such as butyrate and acetate. It was postulated that the treatment with SCFAs resulted in the reduction of vulnerability against hydrogen peroxide and further stimulation of DNA restoration and antioxidant defense systems of the colonic cells.

17.6 Probiotics and Gut Lesions

Lesions in the colons of mammals are often identified and appeared to be of putative pre-neoplastic origins (Bird 1987; Sandforth et al. 1988). ACF, one of the preneoplastic lesions, generally consist of large, thick crypts in methylene blue stained specimen of colon from mice treated with carcinogens (Bird 1987). Aberrant crypts may appear solely or as groups of aberrant crypts within single focus and these crypts will progress to polyps and eventually to tumours. Oral administrations of lactic acid bacteria (bifidobacteria and lactobacilli) have been found able to prevent and decrease the generation of putative pre-neoplastic lesions.

17.6.1 Carcinogen-Induced Lesions

Using AOM-induced colon ACF in rats, Arimochi et al. (1997) found that the number of ACF in overnight cultures of *L. acidophilus* treated group was significantly decreased to 73.6% compared to the control. Additionally, the half-life of O^6-methylguanine (O^6-meG) in the *L. acidophilus* treated group was reduced to 52.2% compared to 74.4% of that in the control. The authors suggested that the inhibition of ACF was attributed to the elimination of O^6-meG from the colon mucosal DNA. Formation of O^6-meG is a crucial pathway in the initiation stage of carcinogenesis (Zalidi et al. 1995). Thus, the inhibitory effect of probiotics in the prevention of the formation of ACF was due to the enhanced removal of O^6-meG from the colon mucosal DNA. In another study, Kulkarni and Reddy (1994) induced colonic ACF in male F344 rats by AOM and determined the effect of lyophilized cultures of *B. longum* on the formation of pre-neoplastic lesion in the colon. The administration of lyophilized cultures of *B. longum* significantly inhibited the development of ACF by 53% and the crypt multiplicity in the colon.

In another study, Yamazaki et al. (2000) evaluated the effect of *L. casei* strain Shirota on AOM-induced colonic ACF and they found that the large ACF had significantly decreased in the rats consuming diets containing *L. casei* strain Shirota compared to control group. The authors suggested that the preventive effect against colon carcinogenesis may attribute to the cluster of differentiation 8 (CD8) positive T-lymphocytes. In addition, Bolognani et al. (2001) examined the effects of *L. acidophilus* in a high fat diet on the formation of colon ACF in rats induced by AOM. In this study, 48 male Sprague Dawley rats were fed an experimental diets containing freeze-dried *L. acidophilus* (0.5% w/w, 5×10^9 viable cells/kg diet) for 4 weeks. The authors found that the number of ACF was significantly reduced by 38% compared to the high fat control diet.

Most of the suppression effects of probiotics on ACF have also been shown in DMH-induced pre-neoplastic lesions in rats. In a recent study, Park et al. (2007) assessed the effects of *B. polyfermenticus* on the antioxidant system and the development of ACF in F344 male rats. In this study, 30 rats were fed a high fat-low fiber experimental diet supplemented with freeze-dried *B. polyfermenticus* (3×108 CFU/1.3 g) for 10 weeks. The total numbers of ACF in the colon of DMH-treated rats supplemented with *B. polyfermenticus* were significantly decreased by 40% compared to the DMH group. The authors indicated that oxidant-antioxidant homeostasis was disrupted in DMH-treated rats. Plasma lipid peroxidation was significantly increased, while the plasma total antioxidant potential was significantly reduced in the DMH-treated rats compared to the control groups. The supplementation of *B. polyfermenticus* increased the plasma total antioxidant potential levels,

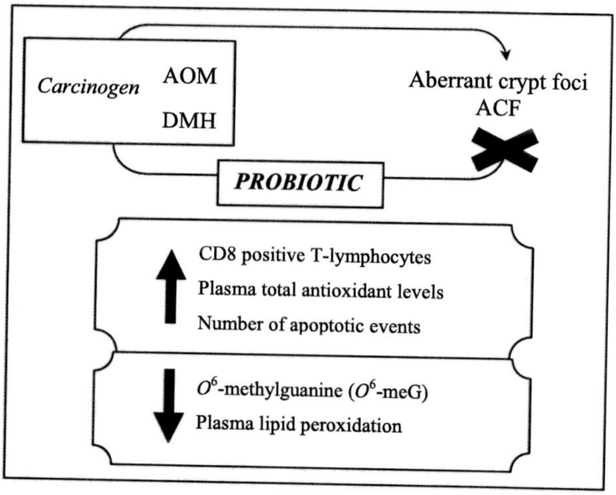

Fig. 17.4 Probiotic and inhibition of pre-neoplastic lesions, aberrant crypt foci (ACF)

as well as decreased in plasma lipid peroxidation compared to the DMH-treated rats. Thus, the postulated mechanism underlying the reduction of ACF by *B. polyfermenticus* may involve the putative antioxidant activity of *B. polyfermenticus*. In another study, Lan et al. (2008) evaluated the effects of *Propionibacterium freudenreichii* TL133 on colonic mucosal crypts in human microbiota-associated (HMA) rats treated with DMH. The administration of propionic acid bacteria increased the number of apoptotic events in all crypt zones in the mid and distal regions of the colon. The authors suggested that *P. freudenreichii* may act as a suppressing agent by inducing previously initiated cells to follow the apoptotic pathway and causing their deletion, thus preventing the expansion of pre-neoplastic lesions. Figure 17.4 illustrates the roles of probiotics in preventing and reducing the occurrence of ACF.

Numerous studies have positively indicated that the probiotics possess the ability to prevent the putative pre-neoplastic lesions induced by carcinogens. In order to increase the potency of any protective effects, the combined administration of a probiotic and an indigestible food ingredient (prebiotic) known to stimulate the probiotic numbers in the gut has been established via various *in vivo* trials (Table 17.1). Prebiotics are non-digestible food ingredients that can escape digestion under the harsh conditions of the upper gastrointestinal tract and reach the lower gut as substrates for the fermentation by selective indigenous gut microflora (Yeo et al. 2009).

17.6.2 Enzyme-Induced Lesions

Several studies have found that β-glucuronidase activity increases in incidences of cancers and inflammations (Kohno et al. 2002; Shimoi et al. 2001). The bacterial enzyme β-glucuronidase is one of the deconjugation enzymes that possess the ability to hydrolyse glucuronides, thus liberate the carcinogenic aglycones in the intesti-

Table 17.1 Effects of probiotics and prebiotics on the prevention of putative pre-neoplastic lesions aberrant crypt foci (ACF)

Intervention	ACF inducer	Animals	Treatment	Effects	References
Bifidobacteria + Oligofructose (OF)	1,2-dimethylhydrazine (DMH)	Male Wistar rats	Basal diet (AIN-76A) containing 10^8 *bifidobacteria* and 2% of OF	Significantly reduced the number of ACF relative to the control group	Gallaher and Khil (1999)
Bifidobacteria + Fructooligosaccharides (FOS)	1,2-dimethylhydrazine (DMH)	Male Wistar rats	Basal diet (AIN-76) containing 10^8 *bifidobacteria* and 2% FOS	Significantly fewer numbers of ACF in the Bifido-FOS group compared with the Skim-Basal and Bifido-Basal groups	Gallaher et al. (1996)
Bifidobacteria + Neosugar [Fructooligosaccharides (FOS)]	1,2-dimethylhydrazine (DMH)	Female CF1 mice	Bifidobacterium pseudolongum and 5% neosugar	Decreased numbers of ACF at 38 weeks after the last injection of the carcinogen	Koo and Rao (1991)
Bifidobacteria + Lactulose	azoxymethane (AOM)	61 male F344 weanling rats	Control diet (AIN-76A) containing 0.5% *B. longum* and 2.5% lactulose	Decreased in total number of ACF in the colons of rats treated with AOM compared to the control group	Challa et al. (1997)
Bifidobacteria + Inulin	azoxymethane (AOM)	60 male Sprague-Dawley rats	Control diet (CO25) containing 1.7% *B. longum* and 5% inulin	Decreased in total number of ACF by 74% and ACF with 1-3 crypts per focus by 80%	Rowland et al. (1998)

nal lumen (Hirayama and Rafter 2000). Takada et al. (1982) previously showed that the glucuronides will be converted into aglycone methylazoxymethanol (MAM) by the action of bacterial β-glucuronidase, where MAM is an active carcinogen readily and spontaneously reacts with cellular cells. Therefore, the findings suggested that the supplementation of probiotics reduced the bacterial β-glucuronidase, leading to the suppression of ACF.

In another study, Abdelali et al. (1995b) studied the effect of dairy products on the formation of putative pre-neoplastic lesions in rat colon. The incidence of ACF was significantly reduced by 49% upon consumption of skim milk fermented with *Bifidobacterium* sp. Bio (Danone strain 173010) compared to control diet. Although there was no difference in cecal pH between the rats fed with probiotics and the control diet, the diet incorporated with *Bifidobacterium* strain suspension significantly decreased cecal β-glucuronidase activity. The observed reduction of β-glucuronidase activity was associated with the ability of probiotics in preventing gut lesions through the inhibition of ACF. Goldin and Gorbach (1984) also previously showed that the activity of bacterial enzyme β-glucuronidase was reduced upon the consumption of *L. acidophilus*. In this study, the β-glucuronidase activity of 21 subjects with negative history of diarrhoea, constipation or bowel disorders were studied for 5 experimental periods. All subjects consumed 2 glasses of milk per day supplemented with and without *L. acidophilus* (2×10^6/ml) for 4 weeks. The β-glucuronidase activity upon ingestion of milk supplemented with *L. acidophilus* was significantly decreased compared to the control diet (milk without *L. acidophilus*).

17.6.3 Pathogen-Induced Lesions

Enteric pathogen may cause lesions at the mucosal surface due to direct adherence onto intestinal epithelial cells. Enterohemorrhagic *E. coli* (EHEC) produces attaching and effacing (A/E) lesions on host epithelial cells, thus reduces intestinal epithelial barrier function (Britton and Versalovic 2008). EHEC infections have caused sporadic outbreaks of hemorrhagic colitis throughout the world and hemorrhagic colitis due to *E. coli* O157 may deteriorate into advanced colorectal cancer (Ishikubo et al. 2003).

The applications of probiotics have successfully inhibited A/E lesion formation and enhancing *in vitro* barrier function in response to EHEC infection. Johnson-Henry et al. (2008) previously showed that EHEC O157:H7 attached to the intestinal epithelial monolayers and produced A/E lesions. Using the *L. rhamnosus* strain GG, the authors demonstrated that probiotics reduced the number of A/E lesions induced in response to EHEC O157:H7 infection in polarized MDCK-I and T84 cells. Pathogen binding and bacterium-induced A/E lesions were indicated by the formation of α-actinin foci. The authors revealed that adhesion of *L. rhamnosus* strain GG to MDCK-I cells showed an absence of α-actinin foci. In addition, pre-incubation of the epithelial cells with *L. rhamnosus* strain GG prior to *E. coli* O157:H7 infection inhibited the pathogen attachment and attenuated bacterium induced A/E lesion formation. It has been postulated that probiotics may suppress the formation of A/E lesions or interfere with the abilities of specific pathogens to adhere directly to the intestinal surface through the production of bacteriocins, hydrogen peroxide and biosurfactants (Boyle et al. 2006; Britton and Versalovic 2008).

Helicobacter pylori is a Gram-negative spiral-shaped, micro-aerophilic rod colonizing the human gastric mucosa and has recently been shown to be an important chronic gastritis and as a risk factor for gastric adenocarcinoma in human. It has also been postulated that *H. pylori* infection leads to stomach carcinoma (Gotteland et al. 2006; Rolfe 2000) and probiotics have been shown to be antagonistic to *H. pylori*. Brzozowski et al. (2006) previously investigated the effect of probiotics on the gastric mucosal impairment in *H. pylori*-infected Mongolian Gerbils and they found that treatment with Lacidofil containing probiotic bacteria *L. rhamnosus* and *L. acidophilus* significantly attenuated the lesion index. The authors suggested that Lacidofil exerts a beneficial effect against *H. pylori* via its attenuating effect on the *H. pylori* colonization, the mucosal inflammation, the impairment of gastric-somatostatin link and its ability to reverse the over-expression of COX-2 in the *H. pylori*-infected gastric mucosa. In another study, Pena et al. (2005) studied the potential role of probiotic *Lactobacillus* spp. in diminishing *Helicobacter hepaticus*-induced inflammatory bowel disease (IBD) *in vivo*. IBD such as Crohn's disease and ulcerative colitis are aggressive chronic inflammation of the gastrointestinal tract with characteristic patchy transmural lesions that predispose to the development of colorectal cancer (Tytgat et al. 1995; Kukitsu et al. 2008). In the study of Pena et al. (2005), 15 female mice were pre-colonized with an equal mixture (10^9 cells per dose) of *L. paracasei* 1602 and *L. reuteri* 6798. Using *L. paracasei* 1602 and *L. reuteri* 6798 for a period of 81 days, the authors observed a significant reduction in cecocolic lesion scores in female mice compared to animals that were mono-infected with *H. hepaticus*. Probiotics may diminish inflammation by inhibiting the virulence gene expression of *H. hepaticus* (Collado et al. 2009). The authors suggested that *L. paracasei* 1602 and *L. reuteri* 6798 reduced mucosal interleukin-12 (IL-12) expression and lipopolysaccharide-stimulated colonic tumour necrosis factor alpha (TNF-α) production. These proinflammatory colonic cytokine (TNF-α and IL-12) possibly facilitated the colonization of *H. hepaticus*. *L. paracasei* 1602 and *L. reuteri* 6798 provided protection against *H. hepaticus*-induced IBD-like lesions, thus prevent the occurrence of gastric inflammation that may lead to stomach carcinoma.

17.7 Conclusions

To this end, past *in vitro* and *in vivo* studies have provided strong experimental evidences to support the positive influences of probiotics on colon carcinogenesis. However, more clinical trials are needed to better understand the exact underlying mechanisms, and to strengthen the suggestions that probiotics could play a major role in preventing colonic cancers.

References

Abdelali H, Cassand P, Soussotte V, Koch-Bocabeilla B, Narbonne JF (1995a) Antimutagenicity of components of dairy products. Mutation Research 331:133–141

Abdelali H, Cassand P, Soussotte V, Daubeze M, Bouley C, Narbonne JF (1995b) Effect of dairy products on initiation of precursor lesions of colon cancer in rats. Nutr Cancer 24(2):121–132

Abrahamse SL, Pool-Zobel BL, Rechkemmer G (1999) Potential of short chain fatty acids to modulate the induction of DNA damage and changes in the intracellular calcium concentration by oxidative stress in isolated rat distal colon cells. Carcinogenesis 20:629–634

Anuradha S, Rajeshwari K (2005) Probiotics in health and disease. J Indian Acad Clin Med 6:67–72

Arimochi H, Kinouchi T, Kataoka K, Kuwahara T, Ohnishi Y (1997) Effect of intestinal bacteria on formation of azoxymethane-induced aberrant crypt foci in the rat colon. Biochem Biophys Res Commun 238:753–757

Ayebo AD, Angelo IA, Shahani KM (1980) Effect of ingesting Lactobacillus acidophilus milk upon faecal flora and enzyme activity in humans. Milchwissenschaft 35:730–733

Biasco G, Paganelli GM, Brandi G, Brillanti S, Lami F, Callegari C, Gizzi G (1991) Effect of Lactobacillus acidophilus and Bifidobacterium bifidum on rectal cell kinetics and faecal pH. Italian J Gastroenterol 23:142

Bird RP (1987) Observation and quantification of aberrant crypts in the murine colon treated with a colon carcinogen: preliminary findings. Cancer Lett 37:147–151

Bolognani F, Rumney CJ, Rowland IR (1997) Influence of carcinogen binding by lactic acid producing bacteria on tissue distribution and *in vivo* mutagenicity of dietary carcinogens. Food Chem Toxicol 35:535–545

Bolognani F, Rumney CJ, Pool-Zobel BL, Rowland IR (2001) Effect of lactobacilli, bifidobacteria and inulin on the formation of aberrant crypt foci in rats. Eur J Nutr 40:293–300

Boyle RJ, Robins-Browne RM, Tang MLK (2006) Probiotic use in clinical practice: what are the risks? Am J Clin Nutr 83:1256–1264

Britton RA, Versalovic J (2008) Probiotics and gastrointestinal infections. Interdiscipl Perspect Infect Dis 290769:1–10

Brzozowski T, Konturek PC, Mierzwa M, Drozdowicz D, Bielanski W, Kwiecien S, Konturek SJ, Stachura J, Pawlik WW, Hahn EG (2006) Effect of probiotics and triple eradication therapy on the cyclooxygenase (COX)-2 expression, apoptosis, and functional gastric mucosal impairment in *Helicobacter pyroli*-infected Mongolian Gerbils. Helicobacter 11:10–20

Caballero-Franco C, Keller K, De Simone C, Chadee K (2007) The VSL#3 probiotic formula induces mucin gene expression and secretion in colonic epithelial cells. Am J Physiol Gastrointest Liver Physiol 292:315–322

Challa A, Rao Ramkishan D, Chawan CB, Shackelford L (1997) *Bifidobacterium longum* and lactulose suppress azoxymethane-induced colonic aberrant crypt foci in rats. Carcinogenesis 18(3):517–521

Chalova VI, Lingbeck JM, Kwon YM, Ricke SC (2008) Extracellular antimutagenic activities of selected probiotic *Bifidobacterium* and *Lactobacillus* spp. as a function of growth phase. J Environ Sci Health Part B 43:193–198

Choi SS, Kang BY, Chung MJ, Kim SD, Park SH, Kim JS, Kang CY, Ha NJ (2005) Safety assessment of potential lactic acid bacteria *Bifidobacterium longum* SPM1205 isolated from healthy Koreans. J Microbiol 43:493–498

Collado MC, Isolauri E, Salminen S, Sanz Y (2009) The impact of probiotic on gut health. Curr Drug Metab 10:68–78

Cross AJ, Sinha R (2004) Meat-related mutagens/carcinogens in the etiology of colorectal cancer. Environ Mol Mutagene 44:44–55

De Kok TMCM, van Maanen JMS (2000) Evaluation of faecal mutagenicity and colorectal cancer risk. Mutat Res/Rev Mutat Res 463:53–101

Ewaschuk JB, Walker JW, Diaz H, Madsen KL (2006) Bioproduction of conjugated linoleic acid by probiotic bacteria occurs in vitro and in vivo in mice. J Nutr 136:1483–1487

Fearon ER, Vogelstein B (1990) A genetic model for colorectal tumourigenesis mice. Cell 61:759–767

Fodde R (2002) The APC gene in colorectal cancer. Eur J Cancer 38:867–871

Fotiadis CI, Stoidis CN, Spyropoulos BG, Zongrafos ED (2008) Role of probiotics, prebiotics and symbiotics in chemoprevention for colorectal cancer. World J Gastroenterol 14:6453–6457

Gallaher DD, Khil J (1999) The effect of synbiotics on colon carcinogenesis in rats. J Nutr 129:1483S–1487S

Gallaher DD, Stallings WH, Blessing LL, Busta FF, Brady LJ (1996) Probiotics, cecal microflora, and aberrant crypts in the rat colon. J Nutr 126:1362–1371

Geier MS, Butler RN, Howarth GS (2006) Probiotics, prebiotics and symbiotic: A role in chemoprevention for colorectal cancer? Cancer Biol Ther 5:1265–1269

Glei M, Schaeferhenrich A, Claussen U, Kuechler A, Liehr T, Weise A, Marian B, Sendt W, Pool-Zobel BL (2007) Comet fluorescence in situ hybridization analysis for oxidative stress–induced DNA damage in colon cancer relevant gene. Toxicologic Sci 96:279–284

Goldin BR, Gorbach SI (1984) The effect of milk and Lactobacillus feeding on human intestinal bacterial enzyme activity. Am J Clin Nutr 39:756–761

Gotteland M, Brunser O, Cruchet S (2006) Systematic review: are probiotics useful in controlling gastric colonization by *Helicobacter pylori?* Aliment Pharmacol Ther 23:1077–1086

Haskard CA, El-Nezami H, Kankaanoa PE, Salminen S, Ahokasa JT (2001) Surface binding of aflatoxin B1 by lactic acid bacteria. Appl Environ Microbiol 67:3086–3091

Higo N, Oishi T, Yamashita A, Matsuda K, Hayashi M (2002) Northern blot and in situ hybridization analyses of MARCKS mRNA expression in the cerebral cortex of the macaque monkey. Cereb Cortex 12:552–564

Hirayama K, Rafter J (2000) The role of probiotic bacteria in cancer prevention. Microbe Infect 2:681–686

Hirayama K, Baranczewski P, Åkerlund JE, Midtvedt T, Möller L, Rafter J (2000) Effects of human intestinal flora on mutagenicity of and DNA adduct formation from food and environmental mutagens. Carcinogenesis 21:2105–2111

Hosoda M, Hashimoto H, Morita H, Hosono A, Chiba M (1992) Antimutagenicity of milk cultured with lactic acid bacteria against N-methyl-N′-nitro-N-nitrosoguanidine. J Dairy Sci 75:976–981

Hosoda M, Hashimoto H, He F, Morita H, Hosono A (1996) Effect of administration of milk fermented with *Lactobacillus acidophilus* LA-2 on faecal mutagenicity and microflora in the human intestine. J Dairy Sci 79:745–749

Hosono A, Kashina T, Kada T (1986) Antimutagenic properties of lactic acid-cultured milk on chemical and faecal mutagens. J Dairy Sci 69:2237–2242

Hursting SD, Kari FW (1999) The anti-carcinogenic effects of dietary restriction mechanisms and future directions. Mutat Res 443:235–249

Huycke MM, Gaskins HR (2004) Commensal bacteria, redox stress, and colorectal cancer: Mechanisms and models. Exp Biol Med 229:586–597

Huycke MM, Abrams V, Moore DR (2002) *Enterococcus faecalis* produces extracellular superoxide and hydrogen peroxide that damages colonic epithelial cell DNA. Carcinogenesis 23:529–536

Ishikubo T, Katou T, Fujiwara N, Setoyama T, Teshigawara M, Tada M (2003) Hemorrhagic colitis due to *E. coli* O157 with advanced rectal cancer detected by total colonoscopy-a case study. Jichi Med Sch J 26:123–128

Johnson-Henry KC, Donato KA, Tu GS, Gordanpour M, Sherman PM (2008) *Lactobacillus rhamnosus* strain GG prevents enterohemorrhagic *Escherichia coli* O157:H7-induced changes in epithelial barrier function. Infect Immun 76(4):1340–1348

Juntunen M, Kirjavainen PV, Ouwehand AC, Salminen SJ, Isolauri E (2001) Adherence of probiotic bacteria to human intestinal mucus in healthy infants and during rotavirus infection. Clin Diagn Lab Immun 8(2):293–296

Kohno H, Tanaka T, Kawabata K, Hirose Y, Sugie S, Tsuda H, Silymarin MH (2002) A naturally occurring polyphenolic antioxidant flavonoid, inhibits azoxymethane-induced colon carcinogenesis in male F344 rats. Int J Cancer 101:461–468

Koller VJ, Marian B, Stidl R, Nersesyan A, Winter H, Simić T, Sontag G, Knasmüller S (2008) Impact of lactic acid bacteria on oxidative DNA damage in human derived colon cells. Food and Chem Toxicol 46:1221–1229

Koo M, Rao AV (1991) Long term effect of bifidobacteria and neosugar on precursor lesions of colonic cancer in mice. Nutr Cancer 16:249–257

Kukitsu T, Takayama T, Miyanishi K, Nobuoka A, Katsuki S, Sato Y, Takimoto R, Matsunaga T, Kato J, Sonoda T, Sakamaki S, Niitsu Y (2008) Aberrant crypt foci as precursors of the dysplasia-carcinoma sequence in patients with ulcerative colitis. Clin Cancer Res 14(1):48–54

Kulkarni N, Reddy BS (1994) Inhibitory effect of Bifidobacterium longum cultures on the azoxymethane-induced aberrant crypt foci formation and faecal bacterial beta-glucuronidase. Proc Soc Exp Biol Med 207(3):278–283

Lan A, Bruneau A, Bensaada M, Philippe C, Bellaud P, Rabot S, Jan G (2008) Increased induction of apoptosis by Propionibacterium freudenreichii TL133 in colonic mucosal crypts of human microbiota-associated rats treated with 1,2-dimethylhyrazine. Br J Nutr 100:1251–1259

Lankaputhra WEV, Shah NP (1998) Antimutagenic properties of probiotic bacteria and of organic acids. Mutat Res 397:169–182

Lidbeck A, Geltner Allinger U, Orrhage KM, Ottova L, Brismar B, Gustafsson JA, Rafter JJ, Nord CE (1991) Impact of *Lactobacillus acidophilus* supplements on the faecal microflora and soluble faecal bile acids in colon cancer patients. Microb Ecol Health Dis 4:81–88

Ling WH (1995) Diet and colonic microflora interaction in colorectal cancer. Nutr Res 15:439–454

Ling WH, Korpela R, Mykkanen H, Salminen S, Hanninen O (1994) *Lactobacillus* strain GG supplementation decreases colonic hydrolytic and reductive enzyme activities in healthy female adults. J Nutr 124:18–23

Liu Y, Zhu Y, Rannou F, Lee TS, Formentin K, Zeng L, Yuan X, Wang N, Chien S, Forman BM, Shyy JYJ (2004) Laminar flow activates peroxisome proliferator-activated receptor- γ in vascular endothelial cells. Circulation 110:1128–1133

Lye HS, Kuan CY, Ewe JA, Fung WY, Liong MT (2009) The Improvement of hypertension by probiotics: effects on cholesterol, diabetes, renin, and phytoestrogens. Int J Mol Sci 10:3755–3775

Mack DR, Michail S, Wei S, McDougall L, Hollingsworth MA (1999) Probiotics inhibit enteropathogenic *E. coli* adherence in vitro by intestinal mucin gene expression. Am J Physiol Gastrointest Liver Physiol 276:941–950

Marteau P, Pochart P, Flourie B, Pellier P, Santos L, Desjeux J-H, Rambaud J-C (1990) Effect of chronic ingestion of a fermented dairy product containing *Lactobacillus acidophilus* and *Bifidobacterium bifidum* on metabolic activities of the colonic flora in humans. Am J Clin Nutr 52:685–688

Matsumoto M, Benno Y (2004) Consumption of *Bifidobacterium lactis* LKM512 yogurt reduces gut mutagenicity by increasing gut polyamine contents in healthy adult subjects. Mutat Res/Fundam Mol Mech Mutagene 568:147–153

Miquel C, Borrini F, Grandjouan S, Aupérin A, Viguier J, Velasco V, Duvillard P, Praz F, Sabourin JC (2005) Role of *bax* mutations in apoptosis in colorectal cancers with microsatellite instability. Am J Pathol 123:562–570

Mital BK, Garg SK (1995) Anticarcinogenic, hypocholesterolemic, and antagonistic activities of *Lactobacillus acidophilus*. Crit Rev Microbiol 21:175–214

Mobarez AM, Doust RH, Hassan ZM, Kamali S (2007) Antimutagenic effect of *Lactobacillus acidophilus* and *Lactobacillus bulgaricus* isolated from Iranian yoghurt on 2-nitrofluorene. Res J Microbiol 2:524–529

Moller L (1994) *In vivo* metabolism and genotoxic effects of nitrated polycyclic aromatic hydrocarbons. Environ Health Perspect 102:139–146

Nadathur SR, Carney JR, Gould SJ, Bakalinsky AT (1996) Palmitic acid is the major fatty acid responsible for significant anti-N-methyl-N′-nitro-N-nitrosoguanidine (MNNG) activity in yogurt. Mutat Res 359:179–189

Oatley JT, Rarick MD, Ji GE, Linz JE (2000) Binding of aflatoxin B1 to bifidobacteria *in vitro*. J Food Prot 63:1133–1136

Ogata S, Uehara H, Chen A, Itzkowitz SH (1992) Mucin gene expression in colonic tissues and cell lines. Cancer Res 52:5971–59

Ouwehand AC, Lagstrom H, Suomalainen T, Salminen S (2002) Effect of probiotics on constipation, faecal azoreductase activity and faecal mucin content in the elderly. Annal Nutr Metab 46:159–162

Pacifico F, Leonardi A (2006) NF-KB in solid tumours. Biochem Pharmacol 72:1142–1152

Park E, Jeon GI, Park JS, Paik HD (2007) A probiotic strain of *Bacillus polyfermenticus* reduces DMH induced precancerous lesions in F344 male rat. Biol Pharm Bull 30(3):569–574

Parvez S, Malik KA, Ah Kang S, Kim HY (2006) Probiotics and their fermented food products are beneficial for health. J Appl Microbiol 100:1171–1185

Pena JA, Rogers AB, Ge ZM, Ng V, Li SY, Fox JG, Versalovic J (2005) Probiotic *Lactobacillus* spp. diminish *Helicobacter hepaticus*-induced inflammatory bowel disease in interleukin-10-deficient mice. Infect Immun 73(2):912–920

Pool-Zobel BL, Munzner R, Holzaapfel H (1993) Antigenotoxic properties of lactic acid bacteria in the *S. typhimurium* Mutagenicity assay. Nutr Cancer 20:261–270

Pool-Zobel BL, Neudecker C, Domizlaff I, Ji S, Schillinger U, Rumney C, Moretti M, Vilarini I, Scasellati-Sforzolini R, Rowland IR (1996) *Lactobacillus* and *Bifidobacterium* mediated antigenotoxicity in the colon of rats. Nutr Cancer 26:365–380

Pretlow TP, Barrow BJ, Ashton WS, O'Riordan MA, Pretlow TG, Jurcisek JA, Stellato TA (1991) Aberrant crypts: putative preneoplastic foci in human colonic mucosa. Cancer Res 51:1564–1567

Rafter J (2002) Lactic acid bacteria and cancer: Mechanistic perspective. Br J Nutr 88:S89–S94

Reddy BS (1999) Possible mechanisms by which pro- and prebiotics influence colon carcinogenesis and tumour growth. Am Soc Nutr Sci 129:1478S–1482S

Reddy BS, Rivenson A (1993) Inhibitory effect of Bifidobacterium longum on colon, mammary, and liver carcinogenesis induced by 2-amino-3-methylimidazo[4,5-f]quinoline, a food mutagen. Cancer Res 53:3914–3418

Reddy BS, Weisburger JH, Narisawa T, Wynder EL (1974) Colon carcinogenesis in germ-free rats with 1,2-dimethylhydrazine and N-methyl-N′-nitro-N-nitrosoguanidine. Cancer Res 34:2368–2372

Renner HW, Münzer R (1990) The possible role of probiotics as dietary antimutagens. Mutat Res 262:239–245

Rhee CH, Park HD (2001) Three glycoproteins with antimutagenic activity identified in Lactobacillus plantarum KLAB21. Appl Environ Microbiol 67:3445–3449

Rolfe RD (2000) The role of probiotic cultures in the control of gastrointestinal health. J Nutr 130:396S–402S

Rowland IR (1991) Nutrition and gut microflora metabolism. In: Rowland IR (ed) Nutrition, toxicity and cancer. CRC Press, Boston, pp 113–136

Rowland IR, Gangolli SD (1999) Role of gastrointestinal flora in the metabolic and toxicological activities of xenobiotics. In: Ballantyne B, Marrs TC, Syverson T (eds) General and applied toxicology, 2nd edn. Macmillan, London, pp 561–576

Rowland IR, Rumney CJ, Coutts JT, Lievense LC (1998) Effect of *Bifidobacterium longum* and inulin on gut bacterial metabolism and carcinogen-induced aberrant crypt foci in rats. Carcinogenesis 19(2):281–285

Sandforth F, Heimpel S, Balzer T, Gutschmidt S, Riecken EO (1988) Characterization of stereomicroscopically identified preneoplastic lesions during dimethylhydrazine-induced colonic carcinogenesis. Eur J Clin Investig 18:655–662

Shimoi K, Saka N, Nozawa R, Sato M, Amano I, Nakayama T, Kinae N (2001) Deglucuronidation of a flavonoid, luteolin monoglucuronide, during inflammation. Drug Metab Dispos 29:1521–1524

Singh J, Rivenson A, Tomita M, Shimamura S, Ishibashi N, Reddy BS (1997) *Bifidobacterium longum*, a lactic acid-producing intestinal bacterium inhibits colon cancer and modulates the intermediate biomarkers of colon carcinogenesis. Carcinogenesis 18:833–841

Sivieri K, Spinardi-Barbusan ALT, Barbisan LF, Bedani R, Pauly ND, Carlos IZ, Benzatti F, Vendramini RC, Rossi EA (2008) Probiotic *Enterococcus faecium* CRL 183 inhibit chemically induced colon cancer in male Wistar rats. Eur Food Res Technol 228:231–237

Spanhaak S, Havenaar R, Schaafsma G (1998) The effect of consumption of milk fermented by *Lactobacillus casei* strain Shirota on the intestinal microflora and immune parameters in humans. Eur J Clin Nutr 52:899–907

Sreekumar O, Hosono A (1998) The heterocyclic amine binding receptors of *Lactobacillus gasseri* cells. Mutat Res 421:65–72

Sreekumar O, Hosono A (2000) Immediate effect of *Lactobacillus acidophilus* on the intestinal flora and faecal enzymes of rats and the *in vitro* inhibition of *Escherichia coli* in coculture. J Dairy Sci 83:931–939

Surono IS, Pato U, Koesnandar, Hosono A (2009) *In vivo* antimutagenicity of dadih probiotic bacteria towards Trp-P1. Australasian J Anim Sci 22:119–123

Takada H, Hirooka T, Hiramatsu Y, Yamamoto M (1982) Effect of β-glucuronidase inhibitor on azoxymethane-induced colonic carcinogenesis in rats. Cancer Res 42:331–334

Takayama T, Katsuki S, Takahashi Y, Ohi M, Nojiri S, Sakamaki S, Kato J, Kogawa K, Miyake H, Niitsu Y (1998) Aberrant crypt foci of the colon as precursors of adenoma and cancer. New Engl J Med 339(18):1277–1284

Tavan E, Cayuela C, Antoine JM, Trugnan G, Chaugier C, Cassand P (2002) Effects of dairy products on heterocyclic aromatic amine-induced rat colon carcinogenesis. Carcinogenesis 23:477–483

Terada M, Nagao M, Nakayasu M, Sakamoto H, Nakasato F, Sugimura T (1986) Mutagenic activities of heterocyclic amines in chinese hamster lung cells in culture. Environ Health Perspect 67:117–119

Treyzon L, Ohning G, Heber D (2006) Colon cancer. In: Heber D, Blackburn GL, Go VLW, Milner J (eds) Nutritional oncology, 2nd edn. Academic Press, New York, pp 423–435

Tsuda H, Hara K, Miyamoto T (2008) Binding of mutagens to exopolysaccharide produced by *Lactobacillus plantarum* mutants strain 301102S. J Dairy Sci 91:2960–2966

Tytgat GNJ, Dhir V, Gopinath N (1995) Endoscopic appearance of dysplasia and cancer in inflammatory bowel disease. Eur J Cancer 31:1174–1177

Villarini M, Caldini G, Moretti M, Trotta F, Pasquini R, Cenci G (2008) Modulatory activity of a *Lactobacillus casei* strain on 1,2-dimethylhydrazine-induced genotoxicity in rats. Environ Mol Mutagenes 49:192–199

Williams SJ, McGuckin MA, Gotley DC, Eyre HJ, Sutherland GR, Antalis TM (1999) Two novel mucin genes down-regulated in colorectal cancer identified by differential display. Cancer Res 59:4083–4089

Wollowski I, Ji S-T, Bakalimsky AT, Neudecker C, Pool-Zobel BL (1999) Bacteria used for the production of yogurt inactivate carcinogens and prevent DNA damage in the colon of rats. J Nutr 129:77–82

Wollowski I, Rechkemmer G, Pool-Zobel BL (2001) Protective role of probiotics and prebiotics in colon cancer. Am J Clin Nutr 73:451S–455S

Yamazaki K, Tsunoda A, Sibusawa M, Tsunoda Y, Kusano M, Fukuchi K, Yamanaka M, Kushima M, Nomoto K, Morotomi M (2000) The effect of an oral administration of *Lactobacillus casei* strain shirota on azoxymethane-induced colonic aberrant crypt foci and colon cancer in the rat. Oncol Rep 7(5):977–982

Yeo SK, Ooi LG, Lim TJ, Liong MT (2009) Antihypertensive properties of plant-based prebiotics. Int J Mol Sci 10:3517–3530

Zalidi NH, Pretlow TP, O'Riordan MA, Dumenco LL, Allay E, Gerson SL (1995) Transgenic expression of human MGMT protects against azoxymethane-induced aberrant crypt foci and G to A mutations in the K-ras oncogene of mouse colon. Carcinogenesis 16:451–456

Zhang XB, Ohta Y (1991) Binding of mutagens by fractions of the cell wall skeleton of lactic acid bacteria on mutagens. J Dairy Sci 74:1477–1481

Part VII
The Future of Probiotics

Chapter 18
Designer Probiotics and Enteric Cytoprotection

Adrienne W. Paton, Renato Morona and James C. Paton

18.1 Introduction

Bacterial enteric pathogens cause disease in multi-step processes, typically comprising oral ingestion, resistance to innate barriers, competition with endogenous gut microflora, colonization or invasion of the intestinal epithelium, and generation of symptoms via release of toxins or triggering of host inflammatory responses. This is achieved through a dynamic and complex interplay between specific bacterial virulence factors and the innate and acquired defences of the host. Several stages of this process are amenable to modulation with "conventional" probiotics, such as *Lactobacillus, Lactococcus, Bifidobacterium* and *Saccharomyces* spp.

Conventional probiotic approaches have been extensively studied in the context of prevention of *Clostridium difficile* associated diarrhoea (CDAD). This is the leading cause of hospital-acquired diarrhoea and life-threatening pseudomembranous colitis, and is a major cost burden for health systems in developed countries. Furthermore, morbidity and mortality have increased in recent years with the spread of hypervirulent *C. difficile* strains (Voth and Ballard 2005; Parkes et al. 2009). The organism is commonly carried in the gut, but competition with resident microflora keeps its numbers low with no obvious associated symptoms. However, when the gut flora is disturbed by treatment with certain broad-spectrum antibiotics (particularly clindamycin, third generation cephalosporins and fluoroquinolones), overgrowth with *C. difficile* may occur, and disease results from production of two highly potent cytotoxins referred to as toxin A (TcdA) and toxin B (TcdB) (Voth and Ballard 2005). Treatment typically involves administration of antibiotics more specifically targeted at *C. difficile* (e.g. metronidazole or vancomycin), but relapse is common. The fact that CDAD occurs in a readily identifiable at-risk patient group has facilitated clinical trials of a variety of probiotic regimens for either primary

A. W. Paton (✉)
Research Centre for Infectious Diseases, School of Molecular and Biomedical Science, University of Adelaide, Adelaide, SA 5005, Australia
e-mail: adrienne.paton@adelaide.edu.au

J. J. Malago et al. (eds.), *Probiotic Bacteria and Enteric Infections*,
DOI 10.1007/978-94-007-0386-5_18,

or secondary prophylaxis, with or without adjunct antibiotic therapy. Although results have been mixed, at least some of these studies have demonstrated statistically significant benefit, either in terms of reduced incidence or relapse rate of CDAD, reduced carriage of *C. difficile*, or reduced levels of TcdA and TcdB in stool samples (Parkes et al. 2009). The mechanistic basis of these clinical benefits is uncertain, but may include non-specific effects such as inhibition of pathogen growth through alterations in short-chain fatty acid composition of the gut lumen, stimulation of colonic mucin production, and enhancement of secretory antibody responses (e.g. to TcdA and TcdB). At least in the case of *Saccharomyces boulardii,* degradation of the toxins by a secreted yeast protease is a potential cytoprotective mechanism (Parkes et al. 2009), while direct inhibitory effects of cell-free culture supernatants from a specific *Lactococcus delbrueckii* strain on the cytotoxins has also been described (Banerjee et al. 2009).

Although clinical trial data are not available, laboratory studies suggest that conventional probiotic approaches may also have benefit against Shiga toxigenic *Escherichia coli* (STEC) infections. STEC, including the notorious O157:H7 serotype, are an important cause of food-borne gastrointestinal disease in humans, which can progress to haemorrhagic colitis (HC) and the life-threatening haemolytic uraemic syndrome (HUS) (Paton and Paton 1998). Pathogenesis of human disease initially involves colonization of the gut by STEC, followed by production of one or more members of the Shiga toxin (Stx) family. These are potent AB_5 cytotoxins that are absorbed systemically and inhibit eukaryotic protein synthesis in remote tissues (particularly renal epithelium and the microvasculature of the kidney, gut and brain) that express the Stx receptor. *In vitro* studies have shown that co-incubation of STEC with a variety of *Lactobacillus, Pediococcus* and *Bifidobacterium* species resulted in specific down-regulation of Stx gene expression, as well as inhibition of growth of the pathogen. These effects were attributable to production of short-chain fatty acids by the probiotic species which lowered the pH of the medium (Carey et al. 2008). Another study reported that extracellular products of *Lactobacillus acidophilus* interfered with a quorum-sensing virulence gene regulatory network in STEC and reduced expression of genes required for generation of attaching-effacing lesions on enterocytes, which is a crucial gut colonization mechanism for certain STEC serotypes, including O157:H7 (Medellin-Peña et al. 2007). Certain non-pathogenic *E. coli* strains (including the well-studied Nissle 1917 strain) have been shown to inhibit growth and Stx production by STEC (Reissbrodt et al. 2009); similar inhibition of growth of enterotoxigenic *E. coli* by commensal *E. coli* strains has also been reported (Setia et al. 2009). Probiotics can also directly modulate inflammatory responses of the host to attack by pathogens. For example, *Lactococcus casei* has been shown to attenuate the pro-inflammatory signalling induced by invasion of the intestinal mucosa by *Shigella flexneri*, by inhibition of the NF-κB signalling pathway (Tien et al. 2006). More recently, Ryan et al. (2009) have reported that exposure of human gastric epithelial cells to certain *Lactococcus salivaris* strains modulates chemokine responses to infection with *Helicobacter pylori*, as well as down-regulating expression of virulence factors encoded by the Cag pathogenicity island.

18.2 "Designer" Probiotics that Target Bacterial Toxin-Receptor Interactions

Production of cytotoxins is central to the pathogenesis of many enteric infectious diseases, and interaction of these toxins with the host involves highly specific binding to cell surface receptors. In many cases, these receptors are oligosaccharides displayed on glycolipids or glycoproteins. Similar recognition of host oligosaccharides by bacterial adhesins may also be critical for colonization of the intestinal epithelium by the pathogen. Thus, the presence and distribution of key receptors is a principal determinant of a pathogen's host specificity, tissue tropism of toxins, and the resultant clinical spectrum of disease (Karlsson 1998). A number of studies have investigated the anti-infective potential of compounds that interfere with these interactions (Zopf and Roth 1996; Fan et al. 2000; Mulvey et al. 2001; Schengrund 2003; Bovin et al. 2004), and this has been facilitated by knowledge of the cognate receptor structures. Capacity to block adherence would prevent establishment of an infection altogether, while toxin neutralization would protect host tissues until the pathogen is eliminated by natural mucosal immune responses.

The simplest approach is to use synthetic oligosaccharides corresponding to a given receptor epitope to competitively inhibit ligand binding (Zopf and Roth 1996). However, free oligosaccharides may be vulnerable to glycosidases present in the small intestine, limiting their utility for treatment of most enteric infections. Oligosaccharide conformation may also be important and furthermore, optimal ligand binding often necessitates multivalent interactions. This may be critically important in the case of pathogens such as STEC, enterotoxigenic *E. coli* (ETEC) and *Vibrio cholerae*, which cause disease by release of AB_5 family cytotoxins, namely Stx, labile enterotoxin (LT) and cholera toxin (Ctx), respectively. Binding of these toxins to specific cell surface glycolipids is mediated by a pentamer of identical B subunits, and each of these is capable of engaging at least one cognate host glycan (Mulvey et al. 2001). The resultant cooperative, multivalent, high-avidity interaction between the toxin and the host cell surface triggers holotoxin internalization and intoxication of the cell. Experience has shown that multivalent oligosaccharide conjugates are much more effective than the respective free oligosaccharides at blocking AB_5 toxin-receptor interactions (Fan et al. 2000; Mulvey et al. 2001; Schengrund 2003; Bovin et al. 2004). Indeed, the most effective synthetic toxin-binding agents have been developed with the aid of crystal structures of the respective B subunit pentamer-receptor glycan complexes. This has enabled design of elaborate three dimensional scaffolds displaying multiple oligosaccharide epitopes in the optimal spatial orientation (Kitov et al. 2000; Merritt et al. 2002). However, notwithstanding the elegance of the underlying chemistry, these complex toxin-binding molecules are expensive to synthesise and this may limit their utility in regions where disease prevalence is highest.

We have conceived and developed an alternative recombinant probiotic strategy involving expression of oligosaccharides that mimic the receptors for AB_5 toxins

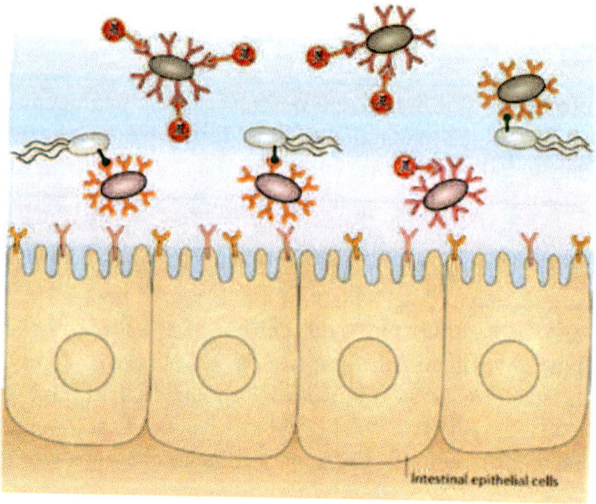

Fig. 18.1 The receptor mimic probiotic strategy. Recombinant bacteria expressing oligosaccharide receptor mimics on their surface (*black*) bind toxins in the gut lumen or adhesins on the pathogen surface (*white*), preventing interactions with receptors on intestinal epithelial cells. (Reproduced from Nat Reviews Microbiology (Paton et al. 2006a))

on the surface of a non-pathogenic *E. coli* strain. These organisms have the capacity to directly bind toxins in the gut lumen, or adhesins expressed on the surface of the pathogen, preventing them from interacting with their receptors on epithelial surface, as shown in Fig. 18.1. The *E. coli* host strains have mutations in the genes encoding the outer core region of the lipopolysaccharide (LPS) such that the resultant LPS is truncated. Introduction of additional genes encoding specific glycosyltransferases from heterologous bacteria resulted in expression of a chimeric LPS, the terminal structure of which was determined by the specificity of the introduced transferases (reviewed by Paton et al. 2006a). We capitalized on the knowledge that the LPS core structures of certain pathogenic Gram-negative bacteria (including *Haemophilus influenzae, Neisseria meningitidis, N. gonorrhoeae, Campylobacter jejuni* and *Helicobacter pylori*) mimic a range of host oligosaccharide structures, as part of a strategy to evade detection by the host immune system. Genome sequences are now available for one or more representatives of each of the above species, and in many cases, the genes encoding synthesis of the host-mimicking epitopes have been functionally characterised. This enabled us to select genes encoding appropriate specific glycosyltransferases, and additional sugar precursor synthases where required, to direct expression of the desired oligosaccharide structure on the surface of the recombinant *E. coli* strain. The resultant chimeric LPS molecules were assembled and exported to the outer face of the bacterial outer membrane via the usual route, enabling display of the receptor mimic structure as a high-density 2D array. Moreover, the capacity for lateral diffusion of LPS molecules in the outer membrane was expected to optimise multi-valent interactions of the oligosaccharide epitope with target toxin B subunit pentamers (Paton et al. 2006a). Applications of this receptor-mimic probiotic approach for treatment and prevention of specific toxin-mediated enteric infections is described below.

18.2.1 STEC Disease

The prototype receptor mimic probiotic was designed to treat or prevent STEC infections. The serious systemic complications of STEC disease and much of the gastrointestinal pathology are directly attributable to Stx-mediated endothelial injury, and there is a lag of up to 7–10 days between onset of the prodromal diarrhoeal symptoms and development of severe haemorrhagic colitis and the life-threatening systemic sequelae (Paton and Paton 1998). We had previously demonstrated in an outbreak setting that patients can be diagnosed with STEC disease by direct PCR analysis of stool cultures early in the prodromal phase (Paton et al. 1996). This provides a potential window of opportunity for therapeutic intervention with an agent capable of neutralizing free Stx in the gut lumen and preventing its systemic absorption. The receptor recognised by the B pentamer of all variants of Stx associated with disease in humans is the glycolipid globotriaosyl- (Galα1-4Galβ1-4Glc-) ceramide (Gb_3). Thus, we constructed an *E. coli* strain expressing a chimeric LPS that terminated in Galα1-4Galβ1-4Glc- and tested its capacity to bind and neutralize Stx family toxins. The expression host was a derivative of an *E. coli* R1 strain (CWG308) which has a *waaO* mutation that truncates its LPS, such that it comprises just the lipid A and inner oligosaccharide core components, terminating in glucose (Glc) (Heinrichs et al. 1998). The phase variable lipooligosaccharides (LOS) of both *N. gonorrhoeae* and *N. meningitidis* were previously known to mimic globotriose as part of their surface antigen repertoire, and this required expression of two active galactosyl transferases encoded by the *lgtC* and *lgtE* genes. Phase variable expression of the epitope was due to replicative slippage at an unstable poly-G tract within *lgtC* (Yang and Gotschlich 1996). We first "locked" the *lgtC* gene in the active configuration by site-directed mutagenesis of the poly-G tract and then introduced this stabilized gene, along with *lgtE* into the *E. coli* CWG308 host using a plasmid. The exogenous transferases then directed addition of Galα1-4Galβ1-4- to the terminal Glc acceptor displayed on the truncated LPS of CWG308, generating an exact mimic of the Stx receptor (Paton et al. 2000). This construct bound purified Stx1 or Stx2 with extraordinary avidity; one mg dry weight of recombinant bacteria neutralized over 150 μg of toxin. We then tested the capacity of the Gb3 mimic construct to protect mice from oral challenge with highly virulent STEC strains. Twice daily administration of the Gb3 mimic construct was 100% effective at preventing otherwise fatal disease (Paton et al. 2000). The model employed involves pre-treatment of mice with streptomycin to enable colonization with STEC, and maintains high numbers of the pathogen in the gut ($>10^9$ CFU per gram) throughout the course of the experiment. Unprotected mice do not get diarrhoea, but they die from Stx-induced acute renal tubular necrosis 4–7 days after challenge. Thus, the model mimics the systemic complications of human STEC disease, and strongly suggests that administration of the receptor mimic probiotic in the early stages of human disease (or prophylactically) could prevent HUS in humans.

Notwithstanding its spectacular efficacy in the mouse model, there are some issues regarding deployment of the above construct in humans. Firstly, it uses antibiotic selection (in this case kanamycin) to maintain the *lgtCE*-encoding plasmid during large scale growth. Secondly, the *E. coli* host strain CWG308 has not been subjected to human safety trials. We have addressed both of these issues in a second generation Gb3 construct. Extensive human safety data are available for *E. coli* K-12 and deletion of both the *waaO* and *waaB* genes in this background yields the same truncated, Glc-terminating LPS core structure as CWG308. We also deleted the *thyA* gene from this strain to make it thymine dependent, and replaced the kanamycin resistance gene in the plasmid used to express *lgtCE* with a *thyA* gene derived from *Salmonella*. This provides a nutritional selection system, which eliminates use of antibiotics during production, and ensures plasmid stability and receptor mimic expression *in vivo*, without any impact on its protective capacity (Pinyon et al. 2004). *E. coli* K-12 derivatives may not be ideal hosts for expression of receptor mimics *in vivo*, as their capacity to compete with the microbiota of the human gut may be poor (hence their early approval as recombinant DNA cloning hosts). Other more robust *E. coli* strains with good safety records, for example Nissle 1917, may provide advantages in terms of extended survival in the human gut, enabling reductions in dose and/or frequency of administration.

A subset of STEC strains that are of high human virulence have recently been shown to produce the prototype of a new family of AB_5 toxins called subtilise cytotoxin (SubAB) (Paton et al. 2004). Unlike all other AB_5 toxin families, the A subunit, SubA, is a subtilase-like serine protease with a unique substrate, the endoplasmic reticulum (ER) chaperone BiP/GRP78, which is the master regulator of the ER stress response (Paton et al. 2006b). Cleavage of BiP by the toxin induces massive ER stress and triggers apoptosis (Wolfson et al. 2008). The receptor specificity of the SubB pentamer was determined by glycan array analysis; SubB was highly specific for glycans terminating in α2-3-linked *N*-glycolylneuraminic acid. Binding was about 20-fold weaker when the terminal sugar was α2-3-linked *N*-acetylneuraminic acid (NeuAc), which differs by a single hydroxyl group (Byres et al. 2008). Nevertheless, significant *in vitro* neutralization of SubAB cytotoxicity could be achieved using a CWG308 derivative that displays a mimic of the ganglioside GM2 (GM2 contains NeuAc α2-3-linked to a subterminal Gal) (Paton et al. 2004). We engineered expression of the GM2 mimic by introducing the *Neisseria* β1-4galactosyltransferase gene *lgtE,* as well as an α2-3sialyltransferase (*cstII*), and a β1-4*N*-acetylgalactosaminyltransferase gene (*cgtA*) from *Campylobacter jejuni* NCTC11168, along with a UDP-*N*-acetylglucosamine-4-epimerase gene (*gne*) from *E. coli* O113 on two separate plasmids (Paton et al. 2005). The latter gene was required because CWG308 does not have the capacity to produce UDP-Gal*N*Ac, the activated sugar substrate for the Gal*N*Ac transferase CgtA. Although SubAB is highly lethal when injected into mice, causing massive thrombotic microangiopathy and multiple organ damage (Paton et al. 2004; Wang et al. 2007), its contribution to human disease is unproven. However, a combination of the Gb_3 and GM2 mimic probiotics is likely to be effective against those STEC strains that produce both Stx and SubAB.

Another distinct subgroup of STEC strains produce a variant Shiga toxin (Stx2e). These strains do not infect humans, but cause oedema disease in weanling piglets. Stx2e has a different receptor specificity to the other members of the Stx family, binding to globotetraosyl ceramide (Gb_4; GalNAcβ1-3Galα1-4Galβ1-4Glc-ceramide) preferentially over Gb_3 (DeGrandis et al. 1989). Piglets are most at risk of oedema disease immediately after weaning, and so routine supplementation of the weaning diet with a Gb_4-mimic probiotic would be a potentially cost-effective prophylactic. We therefore modified the Gb3-mimic bacterium by introducing an additional *N*-acetylgalactosamine transferase gene (*lgtD* from *N. gonorrhoeae*) as well as the *gne* gene from *E. coli* O113 referred to above. The new Gb_4-mimic bacterium had a reduced capacity to neutralize Stx1 and Stx2c *in vitro* (as predicted), but neutralized 98.4% of the cytotoxicity in crude Stx2e extracts (Paton et al. 2001a).

18.2.2 ETEC Disease

ETEC strains are a major global cause of diarrhoea, particularly in developing countries, where they are a significant killer of young children. They are also a common cause of diarrhoea in immunologically naïve visitors to regions where ETEC disease is endemic, and the condition is often referred to as "travellers' diarrhoea" (Nataro and Kaper 1998). ETEC strains produce two classes of enterotoxin, the AB_5 toxin LT (which is structurally related to Ctx), and a heat stable enterotoxin (STa). Strains that infect humans can produce either or both of these toxins. Like Ctx, LT binds via its B subunit to the ganglioside GM1 on the surface of gut epithelial cells, which then triggers internalisation, leading to A subunit-mediated ADP-ribosylation of the host cell G_S protein. This results in uncontrolled stimulation of adenylate cyclase, which interferes with ion transport, thereby causing watery diarrhoea (Nataro and Kaper 1998). STa, on the other hand, binds to the extracellular domain of guanylyl cyclase C (GC-C), thereby activating its cytoplasmic enzymic domain (Vaandrager 2002). This increases the intracellular concentration of cGMP, which also dysregulates cellular ion transport, resulting in secretory diarrhoea. The extracellular domain of GC-C is heavily glycosylated, and fucosylated glycolipids found in human milk have been reported to inhibit STa activity (Newburg 2009). However, other research suggests that STa binding may be independent of glycans, as non-glycosylated GC-C expressed in *E. coli* can bind STa *in vitro* (Nandi et al. 1996).

Travellers' diarrhoea has a high strike rate and is particularly amenable to prophylactic toxin-binding probiotics, because travel to regions where ETEC disease is endemic is usually planned well in advance. Thus, administration can be commenced prior to exposure. A recombinant bacterium expressing a mimic of GM1 should be highly efficacious against the effects of LT. However, strains of *C. jejuni* that mimic human gangliosides have been associated with autoimmune diseases such as Guillain Barré Syndrome (discussed later), and so there is at least a theoretical risk that a GM1 mimic probiotic might elicit auto-reactive antibodies. However, the receptor specificity of the LT B subunit is more relaxed than that of Ctx, and it is also capable of binding to other glycans including lacto-N-neotetraose (LNT;

Galβ1-4GlcNAcβ1-3Galβ1-4Glc) (Angstrom et al. 1994, 2000). LNT is one of the phase-variable LOS epitopes of *N. gonorrhoeae* and *N. meningitidis,* and can be expressed on the surface of *E. coli* CWG308 by introduction of the *lgtA, lgtB* and *lgtE* glycosyltransferase genes (Paton et al. 2005). This construct neutralized ≥93.8% of the LT activity in culture lysates of diverse ETEC strains and was capable of adsorbing approximately 5% of its own weight of purified toxin. Significant *in vivo* protection from LT-induced fluid secretion in rabbit ligated ileal loops was also achieved by pre-absorption with, or co-administration of the LNT mimic (Paton et al. 2005). However, an effective anti-ETEC probiotic would also need to cover STa-producing strains. This will necessitate defining the precise epitope on GC-C that is recognised by STa in order to design constructs that mimic either the oligosaccharide or peptide receptor.

18.2.3 Cholera

Cholera is another example of an essentially toxin-mediated enteric disease, which is endemic in Asia and causes epidemics in nearly all regions (Sack et al. 2004). Infection with the causative bacterium *Vibrio cholerae* usually results from ingestion of contaminated water or seafood. The organism colonises the small intestine and releases Ctx, the B subunit of which (like LT) targets the ganglioside GM1. A subunit-mediated dysregulation of ion transport causes the massive diarrhoea and electrolyte imbalance that is the hallmark of cholera. Without treatment, the case fatality rate is high, although this can be reduced substantially by oral rehydration therapy. However, treatment needs to be maintained until the organism is cleared from the gut and the Ctx-intoxicated epithelial cells are replaced, enabling restoration of electrolyte balance. *In situ* neutralization of Ctx in the gut should therefore speed recovery from an established infection with *V. cholerae,* or prevent the disease from developing if administered early in the infectious process. In spite of the structural and functional homology between the toxins, the LNT receptor mimic bacterium referred to above neutralized Ctx less efficiently than it did LT (Paton et al. 2005). A GM1 mimic was therefore constructed by expressing the genes required for expression of the GM2 oligosaccharide structure in CWG308 (*N. gonorrhoeae lgtE, E. coli* O113 *gne,* and *C. jejuni* NCTC11168 *cstII* and *cgtA*), as described above, as well as *cgtB* from *C. jejuni* NCTC11168, which encodes a β-1,3 galactosyltransferase (Focareta et al. 2006). The recombinant bacterium bound purified Ctx with high avidity, adsorbing >5% of its own weight of toxin *in vitro*. Administration of the GM1-expressing probiotic was highly protective against challenge with virulent *V. cholerae* in the infant mouse cholera model, even when treatment was delayed until after establishment of infection (Focareta et al. 2006). Importantly, the mimic was clearly capable of binding Ctx in the gut lumen in direct competition with the natural receptor on the epithelial surface. This augers well for potential efficacy in humans, in either the prophylactic or therapeutic setting. The GM1-mimic is stable for long periods in either liquid or dried form without refrigeration, and so might be conveniently formulated in sachets with oral rehydration salts, which are the current mainstay of cholera therapy (Sack et al. 2004).

18.3 Safety and Regulatory Issues

Gut microorganisms display a vast array of oligosaccharide structures on their surface, many of which may closely resemble host cell epitopes, such as the various human blood group antigens. Immune response to these microbial molecules explains why humans have antibodies to those blood group antigens that they do not express. Moreover, on-going exposure to antigens that they do express is believed to play an important role in self/non-self immune recognition and maintenance of immune tolerance (Moran et al. 1996). Thus, mucosal presentation of host oligosaccharide receptor mimics on the surface of recombinant probiotic bacteria, such as those described above, would not be expected to elicit inappropriate host immune responses. In the case of glycosphingolipids such as Gb3, Gb4 and LNT, the oligosaccharide components are also mimicked by mucosal pathogens such as *H. influenzae, N. meningitidis* and *N. gonorrhoeae.* In addition to causing serious invasive disease, these organisms can be carried asymptomatically on respiratory or genital mucosae for extended periods, but there is no known human autoimmune condition associated with either carriage or disease due to these organisms. However, the situation may not be so clear-cut for mimicry of host gangliosides such as GM1. A particular concern is that human neural tissue is rich in GM1 and related gangliosides, and antibodies capable of cross-reacting with these structures have been associated with autoimmune neuropathies, most notably Guillain-Barré syndrome (GBS) (Prendergast and Moran 2000; Willison and Yuki 2002; Hughes 2004). Many patients with GBS have had antecedent enteric infections with certain serotypes of *C. jejuni* which express GM1-related LPS structures on their surface. However, establishing a direct causal relationship is not straight forward, particularly since not all GBS-associated strains express GM1 mimics. Conversely, not all *C. jejuni* serotypes known to express GM1-like LPS have been associated with GBS. This suggests that other LPS structures, including those with multiple sialic acid groups, may be involved, as well as other yet to be characterized bacterial factors (Prendergast and Moran 2000; Nachamkin et al. 2002; Willison and Yuki 2002). Indeed, a small proportion of GBS-associated strains do not express sialated ganglioside mimics at all (Godschalk et al. 2007). Examination of the relationship between infection with a *C. jejuni* strain expressing a given LPS structure (or carrying genes encoding synthesis thereof) and subsequent development of GBS is extremely complicated, because expression of key glycosyltransferase genes such as *cgtB* is subject to bi-directional phase variation. This will generate populations of cells expressing multiple LPS types at any given time (Linton et al. 2000; Gilbert et al. 2002; Guerry and Szymanski 2008). Moreover, the *in vitro* LPS expression profile of a given strain may not reflect that which was previously expressed in the patient. An additional complication is that polymorphisms in the *cstII* gene resulting in single amino acid differences in the encoded sialyltransferase have been shown to determine the capacity to add sialic acid at either single or multiple positions in the LPS core structure (Godschalk et al. 2007; Gilbert et al. 2002; Guerry and Szymanski 2008). Nevertheless, there has been a report of immunization of rabbits with GM1-like *C. jejuni* LPS eliciting anti-GM1 antibodies and flaccid limb weakness in a proportion of animals. However, this required sub-cutaneous injection of LPS in the presence of complete Freund's adjuvant every three weeks for up to a year (Yuki et al. 2004). This finding suggests that if the right cytokine environment or immune stimulus is provided, it is possible to break immune tolerance and trigger anti-ganglioside responses in some individuals. Such circumstances may well arise in the context of the complex pattern of inflammatory responses that occurs in association with acute *C. jejuni* gastroenteritis. However, it seems much less likely that this would result as a consequence of passive presentation of a GM1 epitope on the surface of a non-invasive probiotic bacterium. Nevertheless, careful studies in animal models as well as phase I clinical trials will be needed to ensure that oral administration of recombinant GM1-mimic bacteria do not elicit auto-reactive antibodies.

A separate issue is that recombinant probiotics are, by definition, genetically modified organisms (GMOs), and oral administration to patients constitutes "planned release". Release of live GMOs is subject to very stringent regulatory approval processes in most countries. However, there are now examples of anti-viral and anti-malarial vaccines comprising live GMOs that have received approval for human trials in the United States, Britain and elsewhere, indicating that the regulatory hurdles are not insurmountable. Notwithstanding the potential benefits that might emanate from the deployment of live recombinant probiotics such as those described above, the (not unreasonable) stringency of the approval process, combined with market-place resistance to GMOs in general, has acted as an impediment to commercial involvement in their clinical development. One potential interim course of action would be to conduct initial human trials using appropriately killed receptor mimic bacteria, effectively side-stepping the GMO issue. There are a number of alternative processes for achieving this, including chemical modification, irradiation, pasteurisation, and use of phage-encoded lysins to generate bacterial ghosts, all of which are adaptable to commercial scale operations. We have previously shown that either formalin- or heat-killed Gb3-mimic probiotic bacteria were still capable of neutralizing Stx *in vitro*. The formalin-killed form was also evaluated in the mouse model and protected against lethal STEC challenge. However, it was necessary to increase the frequency of administration from two to three times daily in order to maintain the 100% protection achieved by the live construct, presumably due to more rapid clearance of the killed mimic from the mouse GI tract (Paton et al. 2001b). Apart from the advantage of a simpler large-scale production process, the capacity of the live GMO to undergo a limited degree of replication *in vivo* and persist in the gut for more than a day or so is likely to confer significant advantages in terms of dose and frequency of administration.

18.4 Future Prospects

There are a range of enteric pathogens that express toxins or adhesins that bind to specific host oligosaccharides, and so might be targeted by a similar receptor mimic probiotic approach to that used above for STEC, ETEC and *Vibrio cholerae.* Foremost amongst these is *C. difficile.* The two major toxins TcdA and TcdB are not AB_5 cytotoxins, but they do have multiple receptor-binding domains, suggesting the potential for high avidity multi-valent interactions with both host receptors and receptor mimic constructs. Attempts to develop toxin-binding agents have been hampered by uncertainty regarding the actual structure of the host receptor for either toxin. Recent glycan array data for TcdA available from the Consortium for Functional Glycomics (http://www functionalglycomics.org/glycomics/publicdata/selectedScreens.jsp) showed strong binding to fucosylated oligosaccharides resembling the Lewis[a] blood group antigen. However, similar analysis of TcdB detected no binding to over 300 distinct glycan structures represented on the array. This is a disappointment in the light of recent findings which indicate that TcdB may be much more important than TcdA in pathogenesis of disease (Lyras et al. 2009).

Knowledge of the cognate receptor oligosaccharide for a given toxin (or adhesin) enables selection of appropriate glycosyltransferase genes to direct expression of modified LPS outer core structures in *E. coli* expression hosts such as CWG308. However, to date only transferases derived from the LPS core regions of other Gram-negative bacteria have been able to recognise the truncated LPS core of CWG308 as an acceptor. Whilst there is substantial variation in LPS core structure within and between Gram-negative species, the diversity is not nearly as great as that which occurs in the O-antigen polysaccharide, and this may place limits on the repertoire of oligosaccharide structures that can be expressed in this system. However, it may be possible to exploit the diversity of Gram−ve O-antigen repeat units as ready-made glycan epitopes. These are assembled on the inner face of the cytoplasmic membrane on a lipid carrier (undecaprenyl pyrophosphate) and after translocation across the membrane and polymerization, the O-polysaccharide is attached to the LPS core by the O-antigen ligase (Raetz and Whitfield 2002). In *E. coli*, the ligase WaaL can transfer virtually any oligosaccharide onto the full LPS core, enabling considerable expansion of available structures for use as receptor mimics. Interestingly, much recent attention has been focused on the possibility of exploiting certain bacterial surface glycoproteins as scaffolds for display of heterologous oligosaccharides (Feldman et al. 2005; Wacker et al. 2006). Several bacterial species, including *Campylobacter jejuni* and *Neisseria meningitidis* decorate specific surface proteins with N- or O-linked glycans. The mechanism by which this occurs has similarity with O-antigen synthesis, with specific oligosaccharides being assembled on the same lipid carrier prior to translocation and ligation to specific motifs on the target surface protein (Power et al. 2003; Feldman et al. 2005). The protein glycosylation machinery is fully functional when cloned into *E. coli,* but the relaxed specificity of the ligases, particularly the *Neisseria* enzyme PglL (Faridmoayer et al. 2008) means that a wide range of oligosaccharide structures can be attached. The density of oligosaccharides displayed on the surface may not be as high as that achieved by manipulation of the LPS core, but the modified glycoproteins may project further from the bacterial surface, providing steric advantages for blockade of receptor-ligand interactions.

Enteric infectious diseases remain a major cause of global human morbidity and mortality, and effective vaccines are not yet available for a number of important diarrhoeal diseases. Moreover, management with conventional antimicrobial therapy is being complicated by increasing rates of drug resistance. New approaches are needed to control infectious diseases in the twenty-first century, and receptor mimic probiotics such as those described here may play an important role, at least in some circumstances. Importantly, blockade of toxin-mediated host damage by the receptor mimic does not select for evolution of resistance by the targeted pathogen, because it would not have a significant effect on its capacity to survive and reproduce in the environment. Moreover, any spontaneous mutations in a toxin sequence that prevent binding to a receptor mimic would logically also attenuate virulence.

References

Angstrom J, Teneberg S, Karlsson KA (1994) Delineation and comparison of ganglioside-binding epitopes for the toxins of *Vibrio cholerae, Escherichia coli*, and *Clostridium tetani*: evidence for overlapping epitopes. Proc Natl Acad Sci U S A 91:11859–11863

Angstrom J, Bäckström M, Berntsson A, Karlsson N, Holmgren J, Karlsson KA, Lebens M, Teneberg S (2000) Novel carbohydrate binding site recognizing blood group A and B determinants in a hybrid of cholera toxin and *Escherichia coli* heat-labile enterotoxin B-subunits. J Biol Chem 275:3231–3238

Banerjee P, Merkel GJ, Bhunia AK (2009) *Lactobacillus delbrueckii* ssp. *bulgaricus* B-30892 can inhibit cytotoxic effects and adhesion of pathogenic *Clostridium difficile* to Caco-2 cells. Gut Pathog 1(1):8

Bovin NV, Tuzikov AB, Chinarev AA, Gambaryan AS (2004) Multimeric glycotherapeutics: new paradigm. Glycoconj J 21:471–478

Byres E, Paton AW, Paton JC, Löfling JC, Smith DF, Wilce MCJ, Talbot UM, Chong DC, Yu H, Huang S, Chen X, Varki NM, Varki A, Rossjohn J, Beddoe T (2008) Incorporation of a non-human glycan mediates human susceptibility to a bacterial toxin. Nature 456:648–652

Carey CM, Kostrzynska M, Ojha S, Thompson S (2008) The effect of probiotics and organic acids on Shiga-toxin 2 gene expression in enterohemorrhagic *Escherichia coli* O157:H7. J Microbiol Methods 73:125–132

DeGrandis S, Law H, Brunton J, Gyles C, Lingwood CA (1989) Globotetraosylceramide is recognized by the pig edema disease toxin. J Biol Chem 264:12520–12525

Faridmoayer A, Fentabil MA, Haurat MF, Yi W, Woodward R, Wang PG, Feldman MF (2008) Extreme substrate promiscuity of the *Neisseria* oligosaccharyl transferase involved in protein O-glycosylation. J Biol Chem 283:34596–34604

Feldman MF, Wacker M, Hernandez M, Hitchen PG, Marolda CL, Kowarik M, Morris HR, Dell A, Valvano MA, Aebi M (2005) Engineering N-linked protein glycosylation with diverse O antigen lipopolysaccharide structures in *Escherichia coli*. Proc Natl Acad Sci U S A 102:3016–3021

Focareta A, Paton JC, Morona R, Cook J, Paton AW (2006) A recombinant probiotic for treatment and prevention of cholera. Gastroenterology 130:1688–1695

Gilbert M, Karwaski MF, Bernatchez S, Young NM, Taboada E, Michniewicz J, Cunningham AM, Wakarchuk WW (2002) The genetic bases for the variation in the lipo-oligosaccharide of the mucosal pathogen, *Campylobacter jejuni*. Biosynthesis of sialylated ganglioside mimics in the core oligosaccharide. J Biol Chem 277:327–337

Godschalk PC, Kuijf ML, Li J, St Michael F, Ang CW, Jacobs BC, Karwaski MF, Brochu D, Moterassed A, Endtz HP, van Belkum A, Gilbert M (2007) Structural characterization of *Campylobacter jejuni* lipooligosaccharide outer cores associated with Guillain-Barre and Miller Fisher syndromes. Infect Immun 75:1245–1254

Guerry P, Szymanski CM (2008) *Camplylobacter* sugars sticking out. Trends Microbiol 16:428–435

Heinrichs DE, Yethon JA, Amor PA, Whitfield C (1998) The assembly system for the outer core portion of R1- and R4-type lipopolysaccharides of *Escherichia coli*. J Biol Chem 273:29497–29505

Hughes R (2004) *Campylobacter jejuni* in Guillain-Barré syndrome. Lancet Neurol 3:644

Karlsson K (1998) Meaning and therapeutic potential of microbial recognition of host glycoconjugates. Mol Microbiol 19:1–11

Kitov PI, Sadowska JM, Mulvey G, Armstrong GD, Ling H, Pannu NS, Read RJ, Bundle DR (2000) Shiga-like toxins are neutralized by tailored multivalent carbohydrate ligands. Nature 403:669–672

Linton D, Gilbert M, Hitchen PG, Dell A, Morris HR, Wakarchuk WW, Gregson NA, Wren BW (2000) Phase variation of a β-1,3 galactosyltransferase involved in generation of the ganglioside GM_1-like lipo-oligosaccharide of *Campylobacter jejuni*. Mol Microbiol 37:501–514

Lyras D, O'Connor JR, Howarth PM, Sambol SP, Carter GP, Phumoonna T, Poon R, Adams V, Vedantam G, Johnson S, Gerding DN, Rood JI (2009) Toxin B is essential for virulence of *Clostridium difficile*. Nature 458:1176–1179

Medellin-Peña MJ, Wang H, Johnson R, Anand S, Griffiths MW (2007) Probiotics affect virulence-related gene expression in *Escherichia coli* O157:H7. Appl Environ Microbiol 73:4259–4267

Merritt EA, Zhang Z, Pickens JC, Ahn M, Hol WG, Fan E (2002) Characterization and crystal structure of a high-affinity pentavalent receptor-binding inhibitor for cholera toxin and *E. coli* heat labile enterotoxin. J Am Chem Soc 124:8818–8824

Moran AP, Prendergast MM, Appelmelk BJ (1996) Molecular mimicry of host structures by bacterial lipopolysaccharides and its contribution to disease. FEMS Immunol Med Microbiol 16:105–115

Mulvey G, Kitov PI, Marcato P, Bundle DR, Armstrong GD (2001) Glycan mimicry as a basis for anti-infective drugs. Biochimie 83:841–847

Nachamkin I, Liu J, Li M, Ung H, Moran AP, Prendergast MM, Sheikh K (2002) *Campylobacter jejuni* from patients with Guillain-Barré syndrome preferentially expresses a GD(1a)-like epitope. Infect Immun 70:5299–5303

Nandi A, Mathew R, Visweswariah SS (1996) Expression of the extracellular domain of the human heat-stable enterotoxin receptor in *Escherichia coli* and generation of neutralizing antibodies. Protein Expr Purif 15:271–281

Nataro JP, Kaper JB (1998) Diarrhoeagenic *Escherichia coli*. Clin Microbiol Rev 11:142–201

Newburg DS (2009) Neonatal protection by an innate immune system of human milk consisting of oligosaccharides and glycans. J Anim Sci 87(Suppl 1):26–34

Parkes GC, Sanderson JD, Whelan K (2009) The mechanisms and efficacy of probiotics in the prevention of *Clostridium difficile*-associated diarrhoea. Lancet Infect Dis 9:237–244

Paton JC, Paton AW (1998) Pathogenesis and diagnosis of Shiga toxin-producing *Escherichia coli* infections. Clin Microbiol Rev 11:450–479

Paton AW, Ratcliff R, Doyle RM, Seymour-Murray J, Davos D, Lanser JA, Paton JC (1996) Molecular microbiological investigation of an outbreak of hemolytic uremic syndrome caused by dry fermented sausage contaminated with Shiga-like toxin-producing *Escherichia coli*. J Clin Microbiol 34:1622–1627

Paton AW, Morona R, Paton JC (2000) A new biological agent for treatment of Shiga toxigenic *Escherichia coli* infections and dysentery in humans. Nature Med 6:265–270

Paton AW, Morona R, Paton JC (2001a) Neutralization of Shiga toxins Stx1, Stx2c and Stx2e by recombinant bacteria expressing mimics of globotriose and globotetraose. Infect Immun 69:1967–1970

Paton JC, Rogers TJ, Morona R, Paton AW (2001b) Oral administration of formalin-killed recombinant bacteria expressing a mimic of the Shiga toxin receptor protects mice from fatal challenge with Shiga toxigenic *Escherichia coli*. Infect Immun 69:1389–1393

Paton AW, Srimanote P, Talbot UM, Wang H, Paton JC (2004) A new family of potent AB_5 cytotoxins produced by Shiga toxigenic *Escherichia coli*. J Exp Med 200:35–46

Paton AW, Jennings MP, Morona R, Focareta A, Roddam LF, Paton JC (2005) Recombinant probiotics for treatment and prevention of travellers' diarrhoea. Gastroenterology 128:1219–1228

Paton AW, Morona R, Paton JC (2006a) Designer probiotics for prevention of enteric infections. Nature Rev Microbiol 4:193–200

Paton AW, Beddoe T, Thorpe CM, Whisstock JC, Wilce MCJ, Rossjohn J, Talbot UM, Paton JC (2006b) AB_5 subtilase cytotoxin inactivates the endoplasmic reticulum chaperone BiP. Nature 443:548–552

Pinyon RA, Paton JC, Paton AW, Botten JA, Morona R (2004) Refinement of a therapeutic Shiga toxin-binding probiotic for human trials. J Infect Dis 189:1547–1555

Power PM, Roddam LF, Rutter K, Fitzpatrick SZ, Srikhanta YN, Jennings MP (2003) Genetic characterization of pilin glycosylation and phase variation in *Neisseria meningitidis*. Mol Microbiol 49:833–847

Prendergast MM, Moran AP (2000) Lipopolysaccharides in the development of the Guillain-Barré syndrome and Miller Fisher syndrome forms of acute inflammatory peripheral neuropathies. J Endotoxin Res 6:341–359

Raetz CR, Whitfield C (2002) Lipoploysaccharide endotoxins. Ann Rev Biochem 71:635–700

Reissbrodt R, Hammes WP, dal Bello F, Prager R, Fruth A, Hantke K, Rakin A, Starcic-Erjavec M, Williams PH (2009) Inhibition of growth of Shiga toxin-producing *Escherichia coli* by nonpathogenic *Escherichia coli*. FEMS Microbiol Lett 290:62–69

Ryan KA, O'Hara AM, van Pijkeren JP, Douillard FP, O'Toole PW (2009) *Lactobacillus salivarius* modulates cytokine induction and virulence factor gene expression in *Helicobacter pylori*. J Med Microbiol 58:996–1005

Sack DA, Sack RB, Nair GB, Siddique AK (2004) Cholera. Lancet 363:223–233

Schengrund C-L (2003) "Multivalent" saccharides: development of new approaches for inhibiting the effects of glycosphingolipid-binding pathogens. Biochem Pharmacol 65:699–707

Setia A, Bhandari SK, House JD, Nyachoti CM, Krause DO (2009) Development and in vitro evaluation of an *Escherichia coli* probiotic able to inhibit the growth of pathogenic *Escherichia coli* K88. J Anim Sci 87:2005–2012

Tien MT, Girardin SE, Regnault B, Le Bourhis L, Dillies MA, Coppée JY, Bourdet-Sicard R, Sansonetti PJ, Pédron T (2006) Anti-inflammatory effect of Lactobacillus *casei* on *Shigella*-infected human intestinal epithelial cells. J Immunol 176:1228–1237

Vaandrager AB (2002) Structure and function of the heat-stable enterotoxin receptor/guanylyl cyclase C. Mol Cell Biochem 230:73–83

Voth DE, Ballard JD (2005) *Clostridium difficile* toxins: mechanism of action and role in disease. Clin Microbiol Rev 18:247–263

Wacker M, Feldman MF, Callewaert N, Kowarik M, Clarke BR, Pohl NL, Hernandez M, Vines ED, Valvano MA, Whitfield C, Aebi M (2006) Substrate specificity of bacterial oligosaccharyltransferase suggests a common transfer mechanism for the bacterial and eukaryotic systems. Proc Natl Acad Sci U S A 103:7088–7093

Wang H, Paton JC, Paton AW (2007) Pathologic changes in mice induced by subtilase cytotoxin, a potent new *Escherichia coli* AB_5 toxin that targets the endoplasmic reticulum. J Infect Dis 196:1093–1101

Willison HJ, Yuki N (2002) Peripheral neuropathies and anti-glycolipid antibodies. Brain 125:2591–2625

Wolfson J, Thorpe CM, May KL, Paton JC, Jandhyala DM, Paton AW (2008) Subtilase cytotoxin activates PERK, ATF6 and IRE1 endoplasmic reticulum stress-signaling pathways. Cell Microbiol 10:1775–1786

Yang Q-L, Gotschlich EC (1996) Variation of gonococcal lipooligosaccharide structure is due to alterations in poly-G tracts in *lgt* genes encoding glycosyl transferases. J Exp Med 183:323–327

Yuki N, Susuki K, Koga M, Nishimoto Y, Odaka M, Hirata K, Taguchi K, Miyatake T, Furukawa K, Kobata T, Yamada M (2004) Carbohydrate mimicry between human ganglioside GM_1 and *Campylobacter jejuni* lipooligosaccharide causes Guillain-Barré syndrome. Proc Natl Acad Sci U S A 101:11404–11409

Zopf D, Roth S (1996) Oligosaccharide anti-infective agents. Lancet 347:1017–1021

Chapter 19
Future Prospects of Probiotics as Therapeutics Against Enteric Disorders

E. P. Culligan, C. Hill and R. D. Sleator

19.1 Introduction

The first suggestion of the health benefits of probiotics dates back to the early twentieth century, when Russian scientist Elias Metchnikoff attributed the long life of Bulgarian peasants to their consumption of fermented milk containing lactic acid bacteria (LAB) and believed that 'the dependence of the intestinal microbes on the food makes it possible to adopt measures to modify the flora in our bodies and to replace the harmful microbes with useful microbes' (Metchnikoff 1907). While early research yielded poor results and received little attention, in the last 20 years probiotic research has again come to the forefront of scientific research with significant progress being made in the development of and clinically effective probiotic strains.

Probiotics are defined as 'live microorganisms which when administered in adequate amounts confer a health benefit on the host' (FAO/WHO 2001). There is an increasing body of evidence to suggest that probiotics can be used in the treatment and prevention of infections and chronic inflammatory disorders of the gastrointestinal (GI) tract.

19.1.1 Diarrhoeal Disease

Each year enteric infections are responsible for significant morbidity and mortality worldwide. The World Health Organisation (WHO) estimates there to be more than four billion episodes of diarrhoeal disease annually, while there were 2.2 million deaths attributable to diarrhoeal disease in 2004, making it the fifth leading cause of death at all ages worldwide (WHO 2008) Probiotics have been used in the treatment and prevention of many forms of diarrhoeal disease.

R. D. Sleator (✉)
Department of Biological Sciences, Cork Institute of Technology,
Rossa Avenue, Bishopstown, Cork, Ireland
e-mail: roy.sleator@cit.ie

J. J. Malago et al. (eds.), *Probiotic Bacteria and Enteric Infections*,
DOI 10.1007/978-94-007-0386-5_19, © Springer Science+Business Media B.V. 2011

1. *Rotavirus diarrhoea:* A double-blind randomized placebo-controlled trial of 230 children by Dubey et al. (2008) using the probiotic formulation VSL#3 was found to significantly reduce stool frequency and requirement for oral rehydration salts (ORS) compared to the placebo group, resulting in reduced recovery time. In another study, administration of *Lactobacillus rhamnosus GG* to infants admitted to hospital with non-diarrhoeal complaints, reduced the risk of both nosocomial diarrhoea and symptomatic rotavirus gastroenteritis (Szajewska et al. 2001)
2. *Travellers' diarrhoea:* Travellers diarrhoea is a frequent problem among travellers to foreign countries, especially in South America, South East Asia and Africa. A meta-analysis of 940 studies by McFarland (2007) (12 of which met the inclusion criteria; various database searches from 1977 to 2005 to include randomization, controlled, blinded efficacy trials in humans from peer reviewed journals) concluded that probiotics are safe and effective for prevention of traveller's diarrhoea. However, contradictory results are often seen in studies of this kind due to differences in populations studied, type of probiotics used and the duration of treatment. Also, a number of factors may affect the efficacy of treatment such as travel destination, probiotic viability during the trip and traveller compliance with treatment.
3. *Antibiotic-associated diarrhoea (AAD):* AAD is often seen in patients receiving antibiotic therapy which results in the suppression of the normal host gastrointestinal microflora, thus facilitating the overgrowth of enteropathogens, which can cause diarrhoea and colonic inflammation (colitis). In extreme cases, *Clostridium difficile* can often cause pseudomembranous enterocolitis (Larson et al. 1978), which can be fatal. A clinical trial by Ruszczyński et al. (2008) assessed the efficacy of three *L. rhamnosus* strains in the prevention of AAD. In a double-blind, randomized, placebo-controlled trial of 240 children, 20 patients in the placebo group had diarrhoea compared to nine in the probiotic group. Furthermore, AAD diarrhoea was seen in nine placebo patients compared to three of those administered the probiotic preparation. Also, various meta-analyses have shown probiotics to be successful in the prevention of AAD (D'Souza et al. 2002; McFarland 2006).
4. *Clostridium difficile* is a major cause of nosocomial infection with symptoms ranging from mild diarrhoea to severe pseudmembranous enterocolitis, sepsis and death (Rupnik et al. 2009). Overall, adequate evidence is lacking to recommend the use of probiotics in the prevention or treatment of *C. difficile.* There have been some promising studies using the probiotic yeast, *Saccharomyces boulardii*, (Ooi et al. 2009; Surawicz et al. 2000). However, more research is needed encompassing large, standardised clinical trials with different probiotic strains.

19.1.1.1 Necrotizing Enterocolitits (NEC)

NEC is a serious gastrointestinal condition typically seen in premature infants. Symptoms include abdominal distension, bloody stool, and lethargy. Both Lin et al.

(2005) and Bin-nun et al. (2005) demonstrated that probiotic therapy reduced both the incidence and severity of NEC in a study of very low birth weight infants.

19.1.1.2 Inflammatory Bowel Disease (IBD)

IBD encompasses chronic inflammatory conditions of the gastrointestinal tract, characterized by unpredictable and spontaneous periods of remission and relapse. The most common types of IBD are ulcerative colitis (UC) and Crohn's disease (CD), with the prevalence of IBD estimated to be 1.4 million in the United States and 2.2 million in Europe (Loftus 2004). While the precise aetiology of IBD is unknown, genetic susceptibility, imbalances or disruption to the commensal host microflora (Mitsuyama and Sata 2008), especially a reduction in the Firmicutes phylum (Manichanh et al. 2006) and an abnormal intestinal immune response (Langhorst et al. 2009; Thompson-Chagoyan et al. 2005) are thought to play an important role in disease manifestation. Consequently, probiotics have been utilised in an attempt to re-establish the balance of the host microflora and attenuate an aberrant immune response. In a trial of UC patients investigating the effect of an oral capsule containing *Bifidobacteria* following treatment with sulfasalazine and glucocorticoids (a standard therapy for UC) it was found that 93.3% of patients in the placebo group suffered a relapse compared to only 20% in the probiotic group. A significant reduction in inflammation was also seen in those administered the probiotic compared to the control group (Cui et al. 2004). A recent study by (Sokol et al. 2008) has shown promising potential for the use of *Faecalibacterium prausnitzii* as a probiotic with anti-inflammatory properties in the treatment of CD. This bacterium was found in lower numbers in patients with recurrent CD. *F. prausnitzii* and its supernatant was found to have anti-inflammatory effects both *in vitro* and *in vivo*, inducing interleukin 10 (IL-10) production in peripheral blood mononuclear cells (PBMC's), reducing IL-8 and NF-κB (pro-inflammatory compounds) in Caco-2 cell lines and attenuating the severity of induced colitis in mice. There is, as yet, little evidence documenting the effectiveness of probiotics in the treatment of CD and as such further research is required. Shanahan (2006) suggests that differences in the composition of the host microflora and the locations of CD-associated lesions respectively along the GI tract may indicate one probiotic strain is not sufficient to exert a beneficial effect in different patients. Furthermore, the author poses the question of whether researchers are using the correct probiotic, at a high enough dose and for the correct indication. Overall, only a limited number of studies are available and often results are conflicting, but there is sufficient evidence to warrant further research. Future studies need to be randomized, double-blind placebo-controlled trials encompassing large subject bases and possibly using combination therapies with more than one probiotic strain. Probiotics have also been used in the treatment and prevention of other enteric disorders including; irritable bowel syndrome (IBS) (Whorwell et al. 2006) and *Helicobacter pylori* associated infection (for a review see; Lesbros-Pantoflickova 2007), as well as in non-gastrointestinal conditions such as urinary tract infections (UTI's) (Uehara et al. 2006) and atopic diseases (Kalliomaki et al. 2003).

19.2 Selection Criteria and Characteristics of Probiotics

Before bacteria can be considered for use as probiotics, it is recommended that they meet certain selection criteria and possess a number of intrinsic physico-chemical characteristics outlined in a joint report by the FAO and WHO in 2002 (FAO/WHO 2002). The report sets out a number of guidelines that should be followed so as to standardise procedures when assessing probiotics for use in foods and validating health claims thereof. The guidelines include: probiotic health effects are usually strain specific (Karimi and Pena 2003) so it is important to characterize probiotics to the strain level and with current nomenclature and subsequently deposit them in an international culture collection. It was recommended that *in vitro* tests be carried out before any subsequent *in vivo* animal or human trials were initiated. Furthermore, these tests require substantiation with *in vivo* performance. Common *in vitro* tests performed on probiotics include resistance to gastric acidity and bile, adherence to mucus and/or human epithelial cells, bile salt hydrolase activity, reduction of pathogen adherence to surfaces and antimicrobial activity against pathogenic bacteria such as *Heliocbacter pylori, Salmonella* sp., *Listeria monocytogenes* and *Clostridium difficile.* The report also outlines some safety considerations when dealing with probiotics. While probiotic bacteria, as a group, are generally regarded as safe (GRAS) organisms, safety tests should include the determination of antibiotic resistance profiles, evaluation of certain metabolic activities such as D-lactate production and bile salt deconjugation (an undesirable action in the small bowel), assessment of side effects in humans trials, post market epidemiological surveillance of adverse effects in consumers, toxin production and haemolytic activity. Furthermore, animal trials should be undertaken, where possible and appropriate, before commencing human studies. There have been very few reports of possible systemic infections attributed to probiotics (Rautio et al. 1999; Mackay et al. 1999; Oggioni et al. 1998) and in those rare instances all occurred in patients with serious underlying medical conditions. However, demonstration of lack of infectivity in immuno-compromised animals is also desirable to reinforce the safety profile of the probiotic strain. Large human clinical trials with probiotics are sorely lacking and the report recommends double-blind, randomized, placebo-controlled trials for humans with a sufficiently large participant base and preferably that the trial be repeated by an independent laboratory to confirm the outcome. The full report can be downloaded at: ftp://ftp.fao.org/es/esn/food/wgreport2.pdf.

19.2.1 In vivo Survival

Following oral administration, probiotics must survive transit through the gastrointestinal tract, facing host-associated stresses such as the low pH environment of the stomach (which can be as low as pH 1.5 when fasting) (Drasar et al. 1969) as well as bile and elevated osmolarity in the intestine. Once in the gut they must be able to colonise and proliferate and exert a beneficial effect on the host. In addition to host-associated stresses, probiotics encounter technological stresses during processing, formulation and packaging and must survive in sufficient numbers for an extended period of time during their shelf-life, often at refrigeration temperatures. They should also possess good organoleptic properties and be phage resistant (Mattila-Sandholm et al. 2002). Potential probiotic strains need to be both physiologically and technologically robust, but even the toughest strains are restricted in the variety of food applications to which they can be applied. In addition the most promising and clinically relevant probiotics are unfortunately often rendered unusable due to their physiological and technological fragility (Mattila-Sandholm et al. 2002). Pre-exposing probiotic strains to stresses such as sodium chloride (NaCl), elevated temperature, bile and low pH can increase their survival and viability (Gilliland and Rich 1990; Desmond et al. 2001). This pre-exposure to sublethal stresses can significantly increase their survival following subsequent exposure to lethal stress. Schmidt and Zink (2000) showed that exposing *Bifidobacterium adolescentis* to 47°C for 15 minutes prior to a lethal heat shock increased the strain's heat tolerance 128-fold. Collado and Sanz (2007) produced acid-resistant strains by subjecting acid-sensitive strains to prolonged exposure (16 hours) to pH2.0. The acid-resistant derivatives were seen to obtain a cross-protective effect from acid exposure, growing better in the presence of both bile and NaCl, whilst also having an increased fermentative ability and enzymatic activity. Such treatments however can result in a significant decrease in cell yield, as well as cellular activity and process volumetric productivity (Doleyres and Lacroix 2005). Ding and Shah (2007) compared the stress tolerance profiles of eight probiotic strains following microencapsulation challenged with acid, bile and heat stress. Microencapsulated strains survived three logs CFU/ml better than free probiotic control strains. Increased survival was also observed for acid and mild heat treatment in the microencapsulated strains. Other strategies such as immobilized-cell technology (Doleyres et al. 2004) have also been shown to enhance the stress tolerance profiles of certain probiotic strains. However further research is required before industry standards are reached (Doleyres and Lacroix 2005) and research on technologies such as microencapsulation and its benefits on the stability and release of bioactive compounds in the gastrointestinal tract (Champagne and Fustier 2007).

19.3 A Need for Alternative Strategies/Future Prospects of Bioengineered Probiotics

Whilst it has been clearly demonstrated that probiotics can be effective therapeutics in certain cases (Dubey et al. 2008; Ruszczyński et al. 2008; Uehara et al. 2006; Lin et al. 2005; Cui et al. 2004), results are often conflicting and can vary between and within individuals. This is due in part to different modes of action of probiotics as well as strain specific effects. Coupled with this, the increase in antibiotic resistance due to the indiscriminate use, overuse and misuse of antibiotics and the emergence of the so-called 'superbugs' (multi-antibiotic resistant strains), the need for novel and alternative therapies is paramount. The economic burden to the medical care sector in the United States for the treatment of patients with infections caused by antibiotic resistant organisms is estimated to be four billion $ US annually (Coast et al. 1998). In addition, the fact that no new antibiotic classes have been discovered (Projan 2003) and that pharmaceutical companies have severely reduced investment and in some cases, completely abandoned antimicrobial research and development (Alanis 2005), reinforces the point that radically new and innovative therapies are urgently needed. The design, creation and genetic modification of probiotic stains exclusively tailored to target a specific pathogen or toxin thereof, or as vaccine and drug delivery vehicles is a promising and rapidly expanding area of research.

In support of this, lactobacilli possessing desirable properties such as inherent immunogenicity, bile resistance and the ability to survive and proliferate in the gastrointestinal tract (Seegers 2002), have been successfully employed as oral, mucosal vaccine delivery vehicles. Such vaccine delivery vehicles offer an advantage over traditional live attenuated pathogenic strains in that there is no possibility of reversion to a virulent phenotype which always remains with the attenuated pathogenic strains (Seegers 2002). The majority of pathogens initiate their initial infection at a mucosal surface and vaccination against such pathogens is best accomplished through mucosal vaccination. Mucosal vaccines offer a number of functional advantages (De Magistris 2006) as well as practical benefits; they are non-invasive, easy to administer and do not require the presence of medically trained personnel, a significant advantage in the developing world (Levine 2003). Lactobacilli expressing Tetanus toxin fragment C (TTFC) have been shown to be able to elicit a positive immune response (Maassen et al. 1999). Such a strategy could be applied to the expression of other toxins from enteric pathogens, thus providing immune protection. In addition, a recombinant *Lactococcus lactis* strain producing human IL-10 has been investigated in a human clinical trial for treatment of Crohn's disease (Braat et al. 2006).

19.3.1 Designer Probiotics

As previously mentioned, diarrhoeal diseases are responsible for significant morbidity and mortality worldwide and treatment is becoming increasingly more difficult due to the rise in antibiotic resistance. Paton et al. (2006) described the development of designer probiotics for the prevention of enteric infections using a strategy involving the expression of host cell receptor-mimics on the surface of probiotic strains which can bind to the pathogen itself or neutralize and mop up secreted toxin (Fig. 19.1). This approach has a number of advantages; (1) the probiotic can be administered orally (2) numerous human receptors recognized by enteric pathogens or their toxins are well characterized (3) preventing pathogen adherence would inhibit development of infection (4) sequestration of a toxin would prevent clinical presentation of symptoms so the pathogen can be removed by the host immune system and possibly most importantly (5) this therapeutic strategy does not apply

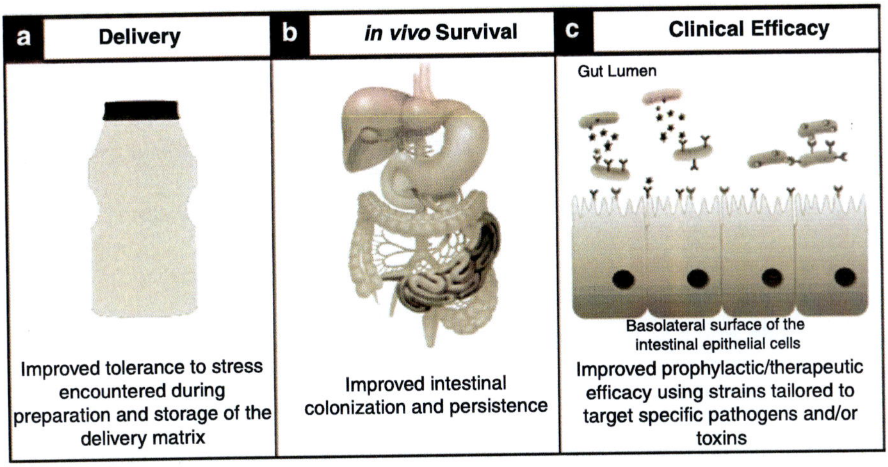

Fig. 19.1 Recent advances in the design of more effective probiotic cultures involve **a** improving probiotic tolerance to stresses encountered during food manufacture and storage. **b** Improving *in vivo* resistance to host specific stresses, thus facilitating improved gut colonization and persistence and **c** designer probiotics which specifically target pathogens and/or toxins; thus improving prophylactic and therapeutic effect. (Adapted, with permission from Sleator and Hill 2008b)

a selective pressure on the pathogen with the development of resistance extremely unlikely. Resistance usually results from a reduction in bacterial cells numbers due to the selective pressure and this does not occur with receptor mimic therapy.

Research in the same lab has resulted in the construction of a harmless *Escherichia coli* strain engineered to express two galactosyl-transferase genes from *Neisseria gonorrhoeae*. This resulted in a modified lipopolysaccharide (LPS) that mimics the Shiga-toxin (Stx) receptor and was found to effectively bind both Stx1 and Stx2. The engineered bacteria were shown to be 100% effective in treating mice infected with normally fatal shigatoxigenic *E. coli* (STEC) (Paton et al. 2000). In another study by the same research group, a probiotic was developed for the treatment of travellers' diarrhoea caused by enterotoxigenic *E. coli* (ETEC). Using a similar strategy to that mentioned above, an *E. coli* strain was engineered to produce a chimeric LPS receptor mimic capable of binding a heat-labile enterotoxin. *In vitro* tests showed more than 93% of the toxin could be neutralized by the recombinant strain, while *in vivo* studies showed the recombinant strain provided protection against fluid loss due to the toxin in rabbit ligated ileal loops (Paton et al. 2005). A molecular mimicry strategy was also used to create a recombinant probiotic for the treatment and prevention of cholera. Cholera, caused by the bacterium *Vibrio cholerae*, is a serious intestinal infection characterized by severe, watery diarrhoea resulting in rapid fluid loss and dehydration which can lead to death in just a few hours without treatment. Transmission is *via* the faecal-oral route, usually due to consumption of contaminated drinking water and the disease is epidemic in many developing countries. Current treatment is administration of a standard oral rehydration solution (ORS) to replace lost fluids, salts and electrolytes in combination with antibiotics in some cases. This is usually an effective treatment but must be administered promptly following infection. Focareta et al. (2006) designed a probiotic strain with an altered LPS which terminates in a structure that mimics the G_{M1} ganglioside terminus, which is the binding receptor for cholera toxin (CT). The bacteria were found to be able to bind CT, abolish >99% of its cytotoxicity and absorb more than 5% of its own weight of toxin. Murine studies demonstrated that administration of the recombinant probiotic immediately following infection with *V. cholerae* increased survival rates of the mice. Furthermore, when administration of the probiotic was delayed until after establishment of *V. cholerae* infection, all mice in the probiotic group survived compared to 1 out of 12 for the control group. Such powerful and significant results reinforce the potential of such alternative therapies. Further research to improve and fine-tune this approach may be necessary before such therapies can be applied to humans. Paton et al. (2006) suggests the introduction of genes to aid gastric transit or promote improved gut colonization would reduce dose regimes and thus costs, while the use of food grade bacteria such as lactococci and lactobacilli for receptor mimicry may also be possible. As mentioned earlier, evidence is lacking for the recommendation of probiotics for use in the treatment of *C. difficile.* However designer probiotics may offer a viable alternative (Sleator and Hill 2008a). The bacteriocin lacticin 3147 has been shown to have significant antibacterial activity against *C. difficile* (Rea et al. 2007) however it is acutely sensitive to gastric acidity *in vivo* (Gardiner et al. 2007). This sensitivity might be overcome by cloning bacteriocin production (and resistance) genes into a suitable host, such as a lactobacillus which could survive stomach passage and deliver the intact bacteriocin to the point of infection. Furthermore, designer probiotics meet the criteria laid out by McFarland (2005) for novel approaches to manage *C. difficile* (Sleator and Hill 2008a). With regard to inflammatory bowel disease, there have been some studies involving recombinant probiotics with some good preliminary results being seen in the treatment of induced-colitis in animal models (Han et al. 2006; Steidler et al. 2000; Foligne et al. 2007). However, more information is needed on the exact causes of such conditions before suitable treatments can be fully developed for use in human trials.

19.3.1.1 Patho-Biotechnology

The term patho-biotechnology was coined by Sleator and Hill (2006) to describe the concept of exploiting pathogenic bacteria or more precisely, exploitation of their stress adaptation, host evasion and virulence or virulence-associated characteristics for beneficial use in the biotechnology and food industries and in medicine. The patho-biotechnology concept encompasses a number of different areas. Firstly, it involves the use of pathogens such as *Listeria monocytogenes* as novel vaccine and drug delivery vehicles (Zhao et al. 2005). This may be approached either by the use of conditional auxotrophic mutants or the selective elimination of key virulence factors. Secondly, such a strategy may involve the isolation of certain immunogenic proteins from specific pathogens thus removing the necessity of using the pathogen itself as the carrier vehicle (Stier et al. 2005). The final area applicable to the

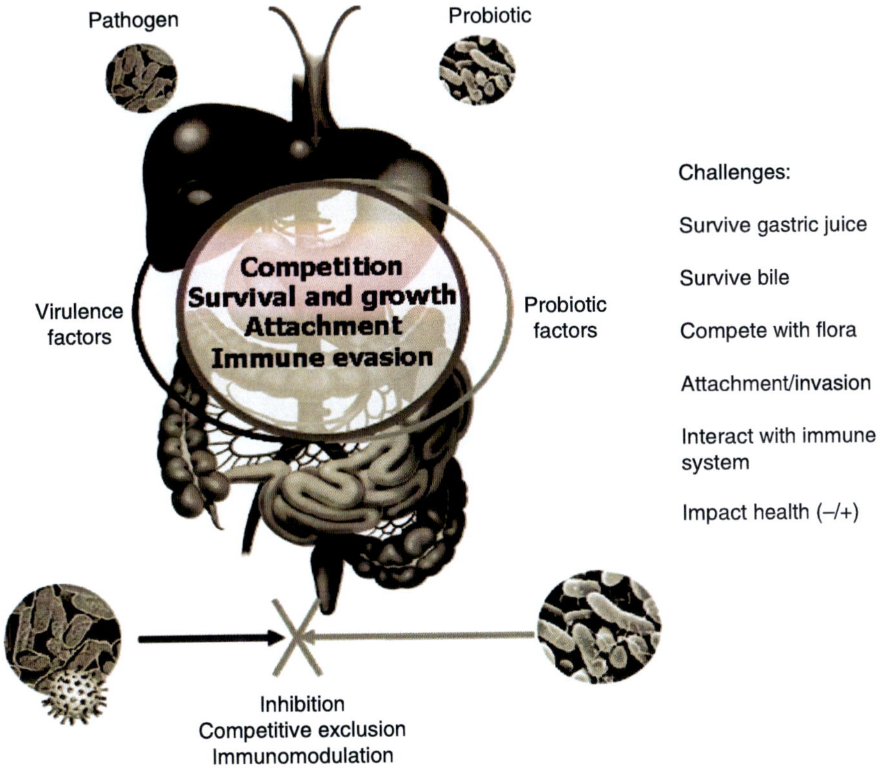

Fig. 19.2 Survival factors for beneficial applications. Pathogens and probiotics experience an almost identical set of challenges during gastrointestinal transit. A virulence associated factor in a pathogen may thus be exploited as a beneficial host adaptive system in a probiotic. (Adapted, with permission from Sleator and Hill 2007)

patho-biotechnology approach and the main focus of this chapter is the introduction of stress survival genes from pathogenic bacteria into non-pathogenic probiotic strains (Sleator and Hill 2006). Probiotic microorganisms encounter an identical set of stress conditions as pathogens upon encountering a host (Fig. 19.2). Thus from a human point of view, an undesirable element from a pathogen (e.g. genes that aid survival in stressful conditions such as gastric acid, bile, low iron, increased osmolarity) if introduced to a probiotic could prove to be beneficial by increasing its resistance to host-associated stresses as well as its technological robustness and clinical efficacy, a distinct advantage with many potentially promising probiotic strains (Mattila-Sandholm et al. 2002).

L. monocytogenes serves as an ideal candidate for the patho-biotechnology concept (Sleator and Hill 2007). Its genome has been fully sequenced (Glaser et al. 2001), *L. monocytogenes* is amenable to genetic manipulation (Hamon et al. 2006) while physiologically it is a robust pathogen capable of resisting numerous stresses (Gray et al. 2006) while also eliciting a strong host immune response (Lecuit 2005).

The patho-biotechnology approach has been successfully employed to increase the stress tolerance of the probiotic bacterium *Lactobacillus salivarius* UCC118 (Sheehan et al. 2006). *L. salivarius* has been shown to have desirable therapeutic properties (Dunne et al. 1999) and has recently been shown to protect mice from *L. monocytogenes* infection by the production of the bacteriocin, Abp118 (Corr et al. 2007). However the bacterium is technologically fragile. In an effort to improve the physiological robustness of the strain Sheehan et al. (2006) cloned the *betL* gene of *L. monocytogenes* (encoding the compatible solute betaine uptake system, BetL) into *L. salivarius* UCC118. Bacteria accumulate compatible solutes such as betaine, either from their environment or by *de novo* synthesis, to counter osmotic stress. BetL has been shown to increase *L. monocytogenes* tolerance to salt (Sleator et al. 1999), low temperature (Wemekamp-Kamphuis et al. 2004) and pressure stress (Smiddy et al. 2004), as well as increasing viability in certain foods (Sleator et al. 2003b). Thus, it was postulated that the introduction of *betL* into *L. salivarius* UCC118 might improve the strain's tolerance to a number of stresses. Indeed the strain harbouring the *betL* gene ($betL^+$) showed a significantly higher growth rate at 7% NaCl compared to the control strain lacking *bet* ($betL^-$) (Fig. 19.3). Also, when exposed to low temperature stress, $betL^+$ survival was 2 logs greater at −20°C and 0.5 logs greater at −70°C compared to wild type strain.

Furthermore, significantly higher survival rates for $betL^+$ were observed following both freeze and spray drying treatment compared to the control. Higher survival rates were also observed following high pressure processing (300 and 350 MPa). These results clearly demonstrate the potential of patho-biotechnology for improving the technological robustness of probiotic microorganisms. In addition to improving a strain's resistance to stresses encountered during food manufacture and storage, the patho-biotechnology concept has also been applied to tailor improved

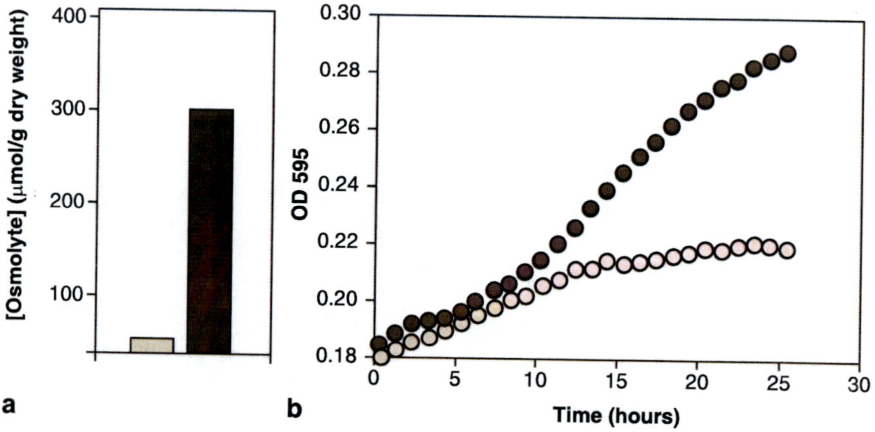

Fig. 19.3 **a** [14C]glycine betaine uptake in the *Lactobacillus salivarius* wild type (*light bar*) and the BetL complemented strain UCC118-BetL$^+$ (*dark bar*). **b** Growth of *L. salivarius* wild type (*light circles*) and UCC118-BetL$^+$ (*dark circles*) in MRS broth with 7% added NaCl. (Adapted, with permission, from Sheehan et al. 2006)

probiotic resistance to host specific stresses, significantly improving colonization, persistence and clinical efficacy. In support of this, Sheehan et al. (2007) cloned the *betL* gene into the probiotic bacterium *Bifidobacterium breve* UCC2003. *B. breve* UCC2003 strains expressing *betL* were shown to exhibit significantly increased tolerance to simulated gastric juice (pH 2.5) as well as osmotic stress. In addition, following successful colonization of the murine intestine, *B. breve* UCC2003 *betL*$^+$ strains were recovered at significantly higher levels in the faeces, large intestine and caecum of inoculated mice. Finally, the *B. breve* UCC2003 *betL*$^+$ strain was shown to protect against *L. monocytogenes* infection in a murine model. *L. monocytogenes* was recovered in significantly lower numbers from the spleens of mice fed with *B. breve* UCC2003 *betL*$^+$ compared the control strain (which lacked *betL*). To the best of our knowledge this study provides the first direct evidence for improved therapeutic efficacy using a patho-biotechnology based approach.

Following these initial experiments, subsequent research was carried out in relation to improving the bile tolerance of two common probiotic strains. Watson et al. (2008) cloned the bile exclusion system, BilE (genes *bilEA* and *bilEB*) from *L. monocytogenes* into both *L. lactis* NZ9000 and *B. breve* UCC2003. The BilE system functions as a cellular bile exclusion system and aids gastrointestinal transit and persistence in *L. monocytogenes* (Sleator et al. 2005). It was therefore postulated that heterologous expression of the BilE system would increase bile tolerance and subsequent gastrointestinal persistence of the probiotic strains. Indeed, the authors found that both *L. lactis* and *B. breve* expressing *bilE* exhibited a 2.5 logs greater survival rate compared to the wild type when grown in porcine bile at concentrations similar to that found in the intestine. Also, both *bilE*$^+$ strains persisted for longer in the gastrointestinal tract and murine faeces (Fig. 19.4a), while *B. breve bilE*$^+$ was recovered at significantly higher numbers (2 logs greater) directly from the murine intestine compared to the wild type (Fig. 19.4b). Furthermore *B. breve bilE*$^+$

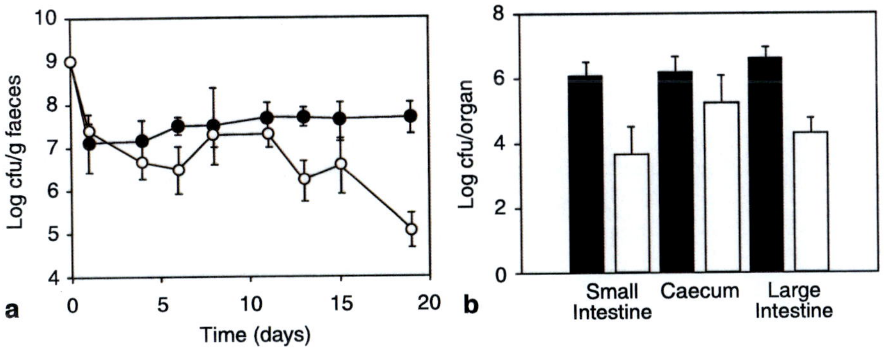

Fig. 19.4 **a** Effect of *bilE* on the gastrointestinal persistence of *Bifidobacterium breve bilE*$^+$ (*black*) and *Bifidobacterium breve bilE*$^-$ (*white*) were used for peroral inoculation of female BALB/c mice (n=5). *Bifidobacterium breve* counts were determined in stools at 48 hour intervals. **b** At day 19 mice were sacrificed and *Bifidobacterium breve* harbouring *bilE* (*white bars*) were recovered at significantly higher numbers in the intestines and caeca than the controls (*black*). (Adapted, with permission, from Watson et al. 2008)

significantly reduced *L. monocytogenes* numbers in the liver following oral infection with the pathogen. These results are significant in that increased bile tolerance also conferred increased gastrointestinal persistence which may enhance probiotic efficacy in therapeutic models (Watson et al. 2008).

An alternative approach to using pathogen-derived genetic elements to improve the physiological robustness of potential probiotics involves a relatively new and rapidly expanding area of scientific research; metagenomics.

19.4 Metagenomics

Metagenomics is the culture-independent analysis of the collective genomes of a population of microorganisms. The field of metagenomics is relatively new, with the first international conference on the topic being held in 2003. A metagenomic analysis involves the functional and sequence-based analysis of collective microbial genomes in an environmental sample (Handelsman et al. 1998). Metagenomics can be used to study as yet uncultured microbes, which represent more than 99% of the population in some environments. The total number of prokaryotic cells on earth has been estimated at $4-6\times10^{30}$ (Whitman et al. 1998), with the majority of these remaining unknown to science. This diversity represents a vast genetic pool that may be utilized for the discovery of novel genes, entire metabolic pathways and their products (Cowan et al. 2005). Functional metagenomic analysis is based on the construction of clones containing metagenomic DNA in a surrogate host and subsequent screening of the clones for the expression of a desired trait or phenotype. Functional metagenomics has identified novel antibiotics (Gillespie et al. 2002), Na^+/H^+ antiporter membrane proteins (Majernik et al. 2001), esterases (Heath et al. 2009), proteases (Waschkowitz et al. 2009) and lipases (Meilleur et al. 2009). Sequence based metagenomics can involve the complete sequencing of clones containing phylogenetic anchors that indicate the taxanomic group that is the probable source of the DNA fragment or it can involve the random sequencing of clones to identify a gene of interest and subsequent the search for phylogenetic anchors in the flanking DNA, which can provide a link of phylogeny with the functional gene (Handelsman 2004). Sequence based analysis has been used to identify a bacterial rhodopsin gene; the first evidence that rhodopsins are not exclusive to Archaea (Beja et al. 2000). Sequence based approaches have also been used to identify novel oxidative coupling enzymes (Banik and Brady 2008), chitinases (LeCleir et al. 2004) and a novel fibrinolytic metalloprotease (Lee et al. 2007) to name a few.

19.4.1 The Human Gut Metagenome and Meta-biotechnology

Research in our laboratory has focused on the human gut metagenome. The human distal gut is the highest density natural bacterial ecosystem known and the number of bacterial cells on or in our bodies is estimated to be 10 times greater than the number of human cells (Gordon et al. 2005). Furthermore, the number of genes in the representative species probably exceeds the number of human genes by 100-fold (Backhed et al. 2005). In this respect Lederberg (2000) suggests we as humans should be considered a superorganism in symbiosis with our vast microflora. This microbiome is largely untapped and contains a virtually limitless supply of novel genes to be exploited for use in medicine, science and industry. Exploiting or 'mining' the human gut metagenome for the development of novel therapeutics and designer probiotics, is essentially an extension of the patho-biotechnology concept. Considering these genes would be isolated from commensal species they are in essence 'self' genes and may alleviate some concerns regarding the use of genetic elements from pathogenic species. 'Mining' the intestinal flora to discover novel antimicrobial peptides, immunoregulatory molecules and stress tolerance genes for therapeutic purposes, is a promising and exciting area of research and a logical extension of the patho-biotechnology concept (Culligan et al. 2008). We thus coin the term 'meta-biotechnology' to describe the use of metagenomics as a robust tool to identify novel genes for the use in biotechnology, specifically building on the patho-biotechnology concept, by increasing the technological and physiological robustness and clinical efficacy of probiotic bacteria.

As mentioned above our research has focused on the human gut metagenome and more specifically the identification of novel osmotolerance loci used by bacteria in this environment. A metagenomic library of the human gut microbiome was screened to identify clones with increased osmotolerance compared to host strain. Representative clones were subsequently subjected to transposon mutagenesis in an attempt to disrupt osmotolerance loci. We have identified a number of such genes and current work is focused on testing their potential to increase osmotolerance in other bacterial strains. The ability to cope with fluctuations in environmental osmolarity is key to the survival and viability of all microorganisms and especially so to both bacteria transiently moving through the human gastrointestinal tract and to those who permanently colonize and proliferate in that environment. Mechanisms of osmotolerance have been well characterized for the pathogenic bacterium *L. monocytogenes* (for a review see Sleator et al. 2003a) and recently, osmolyte transporters have been shown to increase bile resistance (Watson et al. 2009), an important factor in gastrointestinal persistence. We hope that the identification of novel osmotolerance loci from our host microflora using a functional metagenomic approach will enable us to create probiotic strains with an increased physiological and technological robustness, as has been demonstrated previously with genes from the pathogen *L. monocytogenes* (Sheehan et al. 2006, 2007; Watson et al. 2008).

19.5 Biological Containment

The major disadvantage with designer probiotics is that they are genetically modified organisms (GMOs) and as such their use in the treatment of humans would essentially constitute the deliberate release of such a GMO into the environment.

Therefore the safety of such strains needs to be guaranteed and stringently monitored so that they do not; (1) possess antibiotic selection markers, (2) have the ability to accumulate in the environment and (3) transfer the genetic modification to other bacteria through lateral dissemination. Biological containment can be divided into active and passive forms. Active containment is conditionally controlled by the production of a compound that is toxic to the cells. Gene expression is tightly regulated and controlled by an environmentally responsive element. Passive containment meanwhile, is dependent on complementation of an auxotrophy by supplementation with either an intact gene or the essential metabolite (for a recent review see; Kimman et al. 2008). Perhaps the most elegant method developed to date, which combines the advantages of both active and passive containment, is that of the thymidylate synthase gene (*thyA*) (Steidler et al. 2003). The *thyA* gene form *L. lactis*, which is essential for growth, was replaced with the expression cassette for human interleukin-10 (*hIL-10*) by double crossover using a non-replicative plasmid (pORI19). An indigenous suicide system is induced by the activation of the SOS repair system and subsequent DNA fragmentation resulting from thymine and thymidine auxotrophy. Because thymine or thymidine is essential for growth, the *thyA*-deficient strain cannot accumulate in the environment in the absence of the essential growth factors. This approach deals completely with the biosafety issues mentioned above, in that no resistance marker is required to guarantee stable inheritance of the transgene, thus removing any concerns regarding the spread of antibiotic resistance. Also, the risk of the GMO accumulating in the environment is negligible due to rapid death in the absence of thymine or thymidine. Finally, in the unlikely event that a functional *thyA* gene is acquired by homologous recombination from a closely related bacterium, the transgene would be lost (Steidler et al. 2003). One of the strains developed using this method was approved in The Netherlands for the use in a human clinical trial for the treatment of IBD. This was the first clinical trial in which live genetically modified bacteria were used as bio-therapeutics in humans (Braat et al. 2006).

19.6 Conclusions and Future Outlook

Although described for over a century probiotics and research into their health benefits has only come to prominence in the past two decades. The health promoting benefits and efficacy of probiotics has been demonstrated in many models of gastrointestinal disease and indeed in diseases and conditions at other anatomically distinct locations. The use of probiotics in the treatment of many forms of Diarrhoeal disease appears especially promising. However, in some cases results have been conflicting and large randomized, double blind, placebo controlled human trials are disappointingly rare. The inherent physiological and technological fragility of what are often promising candidate probiotic strains can render them ineffective for clinical use. Coupled with this, the alarming rise in antibiotic resistance and the emergence of many multi-drug resistant strains emphasize the need for novel thinking and approaches for the development of alternative therapeutics for the treatment of enteric disorders. Such an alternative strategy, as outlined above, involves the creation of so called designer probiotics, exclusively tailored to target a specific condition, pathogen or toxin. We have also discussed the patho-biotechnology concept and the promising results seen in a number of proof of concept studies and introduced the idea of using metagenomics to identify novel genes for use in improving the robustness of probiotic strains for the treatment of enteric disorders ('meta-biotechnology'). The development of designer probiotics will also see a reduction in production, delivery and storage costs by circumventing the short half-life and fragility associated with conventional therapeutics, a distinct advantage in the developing world. However, consumer acceptance of genetically engineered designer probiotics remains a very significant hurdle. However, it is hoped that in addition to the utilization of rigorous biological containment protocols and the application of comprehensive risk-benefit analyses, the provision of balanced objective information and consumer education on the subject as well as clearly demonstrable medical benefits will ultimately allow such therapeutics to gain a broader acceptance in the general population. With advancements in technologies and further refinements and developments in new techniques, research in this area will continue to provide novel bio-therapeutics and therapeutic targets as well as novel probiotic strains for the treatment and prevention of enteric disorders.

Acknowledgements Dr. Roy Sleator is an Alimentary Pharmabiotic Centre and Health Research Board Principal Investigator. The authors also wish to acknowledge the continued financial assistance of the Alimentary Pharmabiotic Centre, funded by Science Foundation Ireland under the Centres for Science, Engineering and Technology Research Programme.

References

Alanis AJ (2005) Resistance to antibiotics: are we in the post-antibiotic era? Arch Med Res 36:697–705

Backhed F, Ley RE, Sonnenburg JL, Peterson DA, Gordon JI (2005) Host-bacterial mutualism in the human intestine. Science 307:1915–1920

Banik JJ, Brady SF (2008) Cloning and characterization of new glycopeptide gene clusters found in an environmental DNA megalibrary. Proc Natl Acad Sci U S A 105:17273–17277

Beja O, Aravind L, Koonin EV, Suzuki MT, Hadd A, Nguyen LP, Jovanovich SB, Gates CM, Feldman RA, Spudich JL, Spudich EN, Delong EF (2000) Bacterial rhodopsin: evidence for a new type of phototrophy in the sea. Science 289:1902–1906

Bin-nun A, Bromiker R, Wilschanski M, Kaplan M, Rudensky B, Caplan M, Hammerman C (2005) Oral probiotics prevent necrotizing enterocolitis in very low birth weight neonates. J Pediatr 147:192–196

Braat H, Rottiers P, Hommes DW, Huyghebaert N, Remaut E, Remon JP, Van Deventer SJ, Neirynck S, Peppelenbosch MP, Steidler L (2006) A phase I trial with transgenic bacteria expressing interleukin-10 in Crohn's disease. Clin Gastroenterol Hepatol 4:754–759

Champagne CP, Fustier P (2007) Microencapsulation for the improved delivery of bioactive compounds into foods. Curr Opin Biotechnol 18:184–190

Coast J, Smith RD, Millar MR (1998) An economic perspective on policy to reduce antimicrobial resistance. Soc Sci Med 46:29–38

Collado MC, Sanz Y (2007) Induction of acid resistance in Bifidobacterium: a mechanism for improving desirable traits of potentially probiotic strains. J Appl Microbiol 103:1147–1157

Corr SC, Li Y, Riedel CU, O'toole PW, Hill C, Gahan CG (2007) Bacteriocin production as a mechanism for the antiinfective activity of Lactobacillus salivarius UCC118. Proc Natl Acad Sci U S A 104:7617–7621

Cowan D, Meyer Q, Stafford W, Muyanga S, Cameron R, Wittwer P (2005) Metagenomic gene discovery: past, present and future. Trends Biotechnol 23:321–329

Cui HH, Chen CL, Wang JD, Yang YJ, Cun Y, Wu JB, Liu YH, Dan HL, Jian YT, Chen XQ (2004) Effects of probiotic on intestinal mucosa of patients with ulcerative colitis. World J Gastroenterol 10:1521–1525

Culligan EP, Hill C, Sleator RD (2008) Patho-biotechnology. In: Rai VR, Bhat R (eds) Biotechnology: concepts and applications. Narosa Publishing, New Delhi

De Magistris MT (2006) Mucosal delivery of vaccine antigens and its advantages in pediatrics. Adv Drug Deliv Rev 58:52–67

Desmond C, Stanton C, Fitzgerald GF, Collins K, Paul Ross R (2001) Environmental adaptation of probiotic lactobacilli towards improvement of performance during spray drying. Int Dairy J 11:801–808

Ding WK, Shah NP (2007) Acid, bile, and heat tolerance of free and microencapsulated probiotic bacteria. J Food Sci 72:M446-M450

Doleyres Y, Lacroix C (2005) Technologies with free and immobilised cells for probiotic bifidobacteria production and protection. Int Dairy J 15:973–988

Doleyres Y, Fliss I, Lacroix C (2004) Increased stress tolerance of Bifidobacterium longum and Lactococcus lactis produced during continuous mixed-strain immobilized-cell fermentation. J Appl Microbiol 97:527–539

Drasar BS, Shiner M, Mcleod GM (1969) Studies on the intestinal flora. I. The bacterial flora of the gastrointestinal tract in healthy and achlorhydric persons. Gastroenterology 56:71–79

D'souza AL, Rajkumar C, Cooke J, Bulpitt CJ (2002) Probiotics in prevention of antibiotic associated diarrhoea: meta-analysis. Br Med J 324:1361

Dubey AP, Rajeshwari K, Chakravarty A, Famularo G (2008) Use of VSL[sharp]3 in the treatment of rotavirus diarrhoea in children: preliminary results. J Clin Gastroenterol 42(Suppl 3, Pt 1):S126–S129

Dunne C, Murphy L, Flynn S, O'mahony L, O halloran S, Feeney M, Morrissey D, Thornton G, Fitzgerald G, Daly C, Kiely B, Quigley EM, O'sullivan GC, Shanahan F, Collins JK (1999) Probiotics: from myth to reality. Demonstration of functionality in animal models of disease and in human clinical trials. Antonie Van Leeuwenhoek 76:279–292

FAO/WHO (2001) Expert consultation on evaluation of health and nutritional properties of probiotics in food including milk powder with live lactic acid bacteria. FAO/WHO (Food and Agriculture Organization/World Health Organization) Cordoba, Argentina: WHO. http://www.who.int/foodsafety/publications/fs_management/en/probiotics.pdf

FAO/WHO (2002) Working group on drafting guidelines for the evaluation of probiotics in food. ftp://ftp fao.org/es/esn/food/wgreport2.pdf

Focareta A, Paton JC, Morona R, Cook J, Paton AW (2006) A recombinant probiotic for treatment and prevention of cholera. Gastroenterology 130:1688–1695

Foligne B, Dessein R, Marceau M, Poiret S, Chamaillard M, Pot B, Simonet M, Daniel C (2007) Prevention and treatment of colitis with Lactococcus lactis secreting the immunomodulatory Yersinia LcrV protein. Gastroenterology 133:862–874

Gardiner GE, Rea MC, O'riordan B, O'connor P, Morgan SM, Lawlor PG, Lynch PB, Cronin M, Ross RP, Hill C (2007) Fate of the two-component lantibiotic lacticin 3147 in the gastrointestinal tract. Appl Environ Microbiol 73:7103–7109

Gillespie DE, Brady SF, Bettermann AD, Cianciotto NP, Liles MR, Rondon MR, Clardy J, Goodman RM, Handelsman J (2002) Isolation of antibiotics turbomycin a and B from a metagenomic library of soil microbial DNA. Appl Environ Microbiol 68:4301–4306

Gilliland SE, Rich CN (1990) Stability during frozen and subsequent refrigerated storage of lactobacillus acidophilus grown at different pH. J Dairy Sci 73:1187–1192

Glaser P, Frangeul L, Buchrieser C, Rusniok C, Amend A, Baquero F, Berche P, Bloecker H, Brandt P, Chakraborty T, Charbit A, Chetouani F, Couve E, de Daruvar A, Dehoux P, Domann E, Dominguez-Bernal G, Duchaud E, Durant L, Dussurget O, Entian KD, Fsihi H, Garcia-del Portillo F, Garrido P, Gautier L, Goebel W, Gomez-Lopez N, Hain T, Hauf J, Jackson D, Jones LM, Kaerst U, Kreft J, Kuhn M, Kunst F, Kurapkat G, Madueno E, Maitournam A, Vicente JM, Ng E, Nedjari H, Nordsiek G, Novella S, de Pablos B, Perez-Diaz JC, Purcell R, Remmel B, Rose M, Schlueter T, Simoes N, Tierrez A, Vazquez-Boland JA, Voss H, Wehland J, Cossart P (2001) Comparative genomics of Listeria species. Science 294:849–852

Gordon J, Ley R, Wilson R, Mardis E, Xu J, Fraser C M, Relman DA (2005) Extending our view of self: the Human Gut Microbiome Initiative (HGMI). http://www.genome.gov/Pages/Research/Sequencing/SeqProposals/HGMISeq.pdf

Gray MJ, Freitag NE, Boor KJ (2006) How the bacterial pathogen Listeria monocytogenes mediates the switch from environmental Dr. Jekyll to pathogenic Mr. Hyde. Infect Immun 74:2505–2512

Hamon M, Bierne H, Cossart P (2006) Listeria monocytogenes: a multifaceted model. Nat Rev Microbiol 4:423–434

Han W, Mercenier A, Ait-Belgnaoui A, Pavan S, Lamine F, Van S II, Kleerebezem M, Salvador-Cartier C, Hisbergues M, Bueno L, Theodorou V, Fioramonti J (2006) Improvement of an experimental colitis in rats by lactic acid bacteria producing superoxide dismutase. Inflamm Bowel Dis 12:1044–1052

Handelsman J (2004) Metagenomics: application of genomics to uncultured microorganisms. Microbiol Mol Biol Rev 68:669–685

Handelsman J, Rondon MR, Brady SF, Clardy J, Goodman RM (1998) Molecular biological access to the chemistry of unknown soil microbes: a new frontier for natural products. Chem Biol 5:R245–R249

Heath C, Hu XP, Cary SC, Cowan D (2009) Identification of a novel alkaliphilic esterase active at low temperatures by screening a metagenomic library from antarctic desert soil. Appl Environ Microbiol 75:4657–4659

Kalliomaki M, Salminen S, Poussa T, Arvilommi H, Isolauri E (2003) Probiotics and prevention of atopic disease: 4-year follow-up of a randomised placebo-controlled trial. Lancet 361:1869–1871

Karimi O, Pena AS (2003) Probiotics: isolated bacteria strain or mixtures of different strains? Two different approaches in the use of probiotics as therapeutics. Drugs Today (Barc) 39:565–597

Kimman TG, Smit E, Klein MR (2008) Evidence-based biosafety: a review of the principles and effectiveness of microbiological containment measures. Clin Microbiol Rev 21:403–425

Langhorst J, Junge A, Rueffer A, Wehkamp J, Foell D, Michalsen A, Musial F, Dobos GJ (2009) Elevated human beta-defensin-2 levels indicate an activation of the innate immune system in patients with irritable bowel syndrome. Am J Gastroenterol 104:404–410

Larson HE, Price AB, Honour P, Borriello SP (1978) Clostridium difficile and the aetiology of pseudomembranous colitis. Lancet 1:1063–1066

LeCleir GR, Buchan A, Hollibaugh JT (2004) Chitinase gene sequences retrieved from diverse aquatic habitats reveal environment-specific distributions. Appl Environ Microbiol 70:6977–6983

Lecuit M (2005) Understanding how Listeria monocytogenes targets and crosses host barriers. Clin Microbiol Infect 11:430–436

Lederberg J (2000) Infectious history. Science 288:287–293

Lee DG, Jeon JH, Jang MK, Kim NY, Lee JH, Kim SJ, Kim GD, Lee SH (2007) Screening and characterization of a novel fibrinolytic metalloprotease from a metagenomic library. Biotechnol Lett 29:465–472

Lesbros-Pantoflickova D, Corthesy-Theulaz I, Blum AL (2007) Helicobacter pylori and probiotics. J Nutr 137:812S–818S

Levine MM (2003) Can needle-free administration of vaccines become the norm in global immunization? Nat Med 9:99–103

Lin HC, Su BH, Chen AC, Lin TW, Tsai CH, Yeh TF, Oh W (2005) Oral probiotics reduce the incidence and severity of necrotizing enterocolitis in very low birth weight infants. Pediatrics 115:1–4

Loftus EV (2004) Clinical epidemiology of inflammatory bowel disease: incidence, prevalence, and environmental influences. Gastroenterology 126:1504–1517

Maassen CB, Laman JD, Den Bak-Glashouwer MJ, Tielen FJ, van Holten-Neelen JC, Hoogteijling L, Antonissen C, Leer RJ, Pouwels PH, Boersma WJ, Shaw DM (1999) Instruments for oral disease-intervention strategies: recombinant Lactobacillus casei expressing tetanus toxin fragment C for vaccination or myelin proteins for oral tolerance induction in multiple sclerosis. Vaccine 17:2117–2128

Mackay AD, Taylor MB, Kibbler CC, Hamilton-Miller JM (1999) Lactobacillus endocarditis caused by a probiotic organism. Clin Microbiol Infect 5:290–292

Majernik A, Gottschalk G, Daniel R (2001) Screening of environmental DNA libraries for the presence of genes conferring Na(+)(Li(+))/H(+) antiporter activity on Escherichia coli: characterization of the recovered genes and the corresponding gene products. J Bacteriol 183:6645–6653

Manichanh C, Rigottier-Gois L, Bonnaud E, Gloux K, Pelletier E, Frangeul L, Nalin R, Jarrin C, Chardon P, Marteau P, Roca J, Dore J (2006) Reduced diversity of faecal microbiota in Crohn's disease revealed by a metagenomic approach. Gut 55:205–211

Mattila-Sandholm T, Myllärinen P, Crittenden R, Mogensen G, Fondén R, Saarela M (2002) Technological challenges for future probiotic foods. Int Dairy J 12:173–182

Mcfarland LV (2005) Alternative treatments for Clostridium difficile disease: what really works? J Med Microbiol 54:101–111

McFarland LV (2006) Meta-analysis of probiotics for the prevention of antibiotic associated diarrhoea and the treatment of Clostridium difficile disease. Am J Gastroenterol 101:812–822

McFarland LV (2007) Meta-analysis of probiotics for the prevention of traveler's diarrhoea. Travel Med Infect Dis 5:97–105

Meilleur C, Hupe JF, Juteau P, Shareck F (2009) Isolation and characterization of a new alkali-thermostable lipase cloned from a metagenomic library. J Ind Microbiol Biotechnol 36:853–861

Metchnikoff E (1907) Lactic acid as inhibiting intestinal putrefaction. In: Heinemann W (ed) The prolongation of life: optimistic studies. Kessinger Publishing, London, pp 161–183

Mitsuyama K, Sata M (2008) Gut microflora: a new target for therapeutic approaches in inflammatory bowel disease. Expert Opin Ther Targets 12:301–312

Oggioni MR, Pozzi G, Valensin PE, Galieni P, Bigazzi C (1998) Recurrent septicemia in an immunocompromised patient due to probiotic strains of Bacillus subtilis. J Clin Microbiol 36:325–326

Ooi CY, Dilley AV, Day AS (2009) Saccharomyces boulardii in a child with recurrent Clostridium difficile. Pediatr Int 51:156–158

Paton AW, Morona R, Paton JC (2000) A new biological agent for treatment of shigatoxigenic Escherichia coli infections and dysentery in humans. Nat Med 6:265–270

Paton AW, Jennings MP, Morona R, Wang H, Focareta A, Roddam LF, Paton JC (2005) Recombinant probiotics for treatment and prevention of enterotoxigenic Escherichia coli diarrhoea. Gastroenterology 128:1219–1228

Paton AW, Morona R, Paton JC (2006) Designer probiotics for prevention of enteric infections. Nat Rev Microbiol 4:193–200

Projan SJ (2003) Why is big Pharma getting out of antibacterial drug discovery? Curr Opin Microbiol 6:427–430

Rautio M, Jousimies-Somer H, Kauma H, Pietarinen I, Saxelin M, Tynkkynen S, Koskela M (1999) Liver abscess due to a Lactobacillus rhamnosus strain indistinguishable from L. rhamnosus strain GG. Clin Infect Dis 28:1159–1160

Rea MC, Clayton E, O'connor PM, Shanahan F, Kiely B, Ross RP, Hill C (2007) Antimicrobial activity of lacticin 3147 against clinical Clostridium difficile strains. J Med Microbiol 56:940–946

Rupnik M, Wilcox MH, Gerding DN (2009) Clostridium difficile infection: new developments in epidemiology and pathogenesis. Nat Rev Microbiol 7:526–536

Ruszczyński M, Radzikowski A, Szajewska H (2008) Clinical trial: effectiveness of Lactobacillus rhamnosus (strains E/N, Oxy and Pen) in the prevention of antibiotic-associated diarrhoea in children. Aliment Pharmacol Ther 28:154–161

Schmidt G, Zink R (2000) Basic features of the stress response in three species of bifidobacteria: B. longum, B. adolescentis, and B. breve. Int J Food Microbiol 55:41–45

Seegers JF (2002) Lactobacilli as live vaccine delivery vectors: progress and prospects. Trends Biotechnol 20:508–515

Shanahan F (2006) Probiotics: promise, problems and progress. Gastroenterol Hepatol Ann Rev 1:41–45

Sheehan VM, Sleator RD, Fitzgerald GF, Hill C (2006) Heterologous expression of BetL, a betaine uptake system, enhances the stress tolerance of Lactobacillus salivarius UCC118. Appl Environ Microbiol 72:2170–2177

Sheehan VM, Sleator RD, Hill C, Fitzgerald GF (2007) Improving gastric transit, gastrointestinal persistence and therapeutic efficacy of the probiotic strain Bifidobacterium breve UCC2003. Microbiology 153:3563–3571

Sleator RD, Hill C (2006) Patho-biotechnology: using bad bugs to do good things. Curr Opin Biotechnol 17:211–216

Sleator RD, Hill C (2007) Patho-biotechnology; using bad bugs to make good bugs better. Sci Prog 90:1–14

Sleator RD, Hill C (2008a) Designer probiotics: a potential therapeutic for Clostridium difficile? J Med Microbiol 57:793–794

Sleator RD, Hill C (2008b) New frontiers in probiotic research. Lett Appl Microbiol 46:143–147

Sleator RD, Gahan CG, Abee T, Hill C (1999) Identification and disruption of BetL, a secondary glycine betaine transport system linked to the salt tolerance of Listeria monocytogenes LO28. Appl Environ Microbiol 65:2078–2083

Sleator RD, Gahan CG, Hill C (2003a) A postgenomic appraisal of osmotolerance in Listeria monocytogenes. Appl Environ Microbiol 69:1–9

Sleator RD, Francis GA, O'beirne D, Gahan C G, Hill C (2003b) Betaine and carnitine uptake systems in Listeria monocytogenes affect growth and survival in foods and during infection. J Appl Microbiol 95:839–846

Sleator RD, Wemekamp-Kamphuis HH, Gahan CG, Abee T, Hill C (2005) A PrfA-regulated bile exclusion system (BilE) is a novel virulence factor in Listeria monocytogenes. Mol Microbiol 55:1183–1195

Smiddy M, Sleator RD, Patterson MF, Hill C, Kelly AL (2004) Role for compatible solutes glycine betaine and L-carnitine in listerial barotolerance. Appl Environ Microbiol 70:7555–7557

Sokol H, Pigneur B, Watterlot L, Lakhdari O, Bermudez-Humaran LG, Gratadoux JJ, Blugeon S, Bridonneau C, Furet JP, Corthier G, Grangette C, Vasquez N, Pochart P, Trugnan G, Thomas G, Blottiere HM, Dore J, Marteau P, Seksik P, Langella P (2008) Faecalibacterium prausnitzii is an anti-inflammatory commensal bacterium identified by gut microbiota analysis of Crohn's disease patients. Proc Natl Acad Sci U S A 105:16731–16736

Steidler L, Hans W, Schotte L, Neirynck S, Obermeier F, Falk W, Fiers W, Remaut E (2000) Treatment of murine colitis by Lactococcus lactis secreting interleukin-10. Science 289:1352–1355

Steidler L, Neirynck S, Huyghebaert N, Snoeck V, Vermeire A, Goddeeris B, Cox E, Remon JP, Remaut E (2003) Biological containment of genetically modified Lactococcus lactis for intestinal delivery of human interleukin 10. Nat Biotechnol 21:785–789

Stier EM, Mandal M, Lee KD (2005) Differential cytosolic delivery and presentation of antigen by listeriolysin O-liposomes to macrophages and dendritic cells. Mol Pharm 2:74–82

Surawicz CM, McFarland LV, Greenberg RN, Rubin M, Fekety R, Mulligan ME, Garcia RJ, Brandmarker S, Bowen K, Borjal D, Elmer GW (2000) The search for a better treatment for recurrent Clostridium difficile disease: use of high-dose vancomycin combined with Saccharomyces boulardii. Clin Infect Dis 31:1012–1017

Szajewska H, Kotowska M, Mrukowicz JZ, Armanska M, Mikolajczyk W (2001) Efficacy of Lactobacillus GG in prevention of nosocomial diarrhoea in infants. J Pediatr 138:361–365

Thompson-Chagoyan OC, Maldonado J, Gil A (2005) Aetiology of inflammatory bowel disease (IBD): role of intestinal microbiota and gut-associated lymphoid tissue immune response. Clin Nutr 24:339–352

Uehara S, Monden K, Nomoto K, Seno Y, Kariyama R, Kumon H (2006) A pilot study evaluating the safety and effectiveness of Lactobacillus vaginal suppositories in patients with recurrent urinary tract infection. Int J Antimicrob Agents 28(Suppl 1):S30–S34

Waschkowitz T, Rockstroh S, Daniel R (2009) Isolation and characterization of metalloproteases with a novel domain structure by construction and screening of metagenomic libraries. Appl Environ Microbiol 75:2506–2516

Watson D, Sleator RD, Hill C, Gahan CG (2008) Enhancing bile tolerance improves survival and persistence of Bifidobacterium and Lactococcus in the murine gastrointestinal tract. BMC Microbiol 8:176

Watson D, Sleator RD, Casey PG, Hill C, Gahan CG (2009) Specific osmolyte transporters mediate bile tolerance in Listeria monocytogenes. Infect Immun 77(11):4895–4904

Wemekamp-Kamphuis HH, Sleator RD, Wouters JA, Hill C, Abee T (2004) Molecular and physiological analysis of the role of osmolyte transporters BetL Gbu and OpuC in growth of Listeria monocytogenes at low temperatures. Appl Environ Microbiol 70:2912–2918

Whitman WB, Coleman DC, Wiebe WJ (1998) Prokaryotes: the unseen majority. Proc Natl Acad Sci U S A 95:6578–6583

WHO (2008) http://www.who.int/healthinfo/global_burden_disease/GBD_report_2004update_part2.pdf

Whorwell PJ, Altringer L, Morel J, Bond Y, Charbonneau D, O'mahony L, Kiely B, Shanahan F, Quigley EM (2006) Efficacy of an encapsulated probiotic Bifidobacterium infantis 35624 in women with irritable bowel syndrome. Am J Gastroenterol 101:1581–1590

Zhao X, Li Z, Gu B, Frankel FR (2005) Pathogenicity and immunogenicity of a vaccine strain of Listeria monocytogenes that relies on a suicide plasmid to supply an essential gene product. Infect Immun 73:5789–5798

Index

J. J. Malago et al. (eds.), *Probiotic Bacteria and Enteric Infections*,
DOI 10.1007/978-94-007-0386-5, © Springer Science+Business Media B.V. 2011

D

J

Q

R

S